Lecture Notes in Mathematics

Edited by A. Dold and B. Eckmann

834

Model Theory of Algebra and Arithmetic

Proceedings of the Conference on Applications of Logic to Algebra and Arithmetic Held at Karpacz, Poland, September 1 – 7, 1979

Edited by
L. Pacholski, J. Wierzejewski, and A. J. Wilkie

Springer-Verlag
Berlin Heidelberg New York 1980

Editors

Leszek Pacholski
Instytut Matematyczny PAN
Kopernika 18
51–617 Wrocław
Poland

Jedrzej Wierzejewski
Instytut Matematyki
Politechniki Wrocławskiej
Wybrzeże Wyspiańskiego 27
50-370 Wrocław
Poland

Alec J. Wilkie
Mathematical Institute
University of Oxford
24-29 St. Giles
Oxford OX1 3LB
England

AMS Subject Classifications (1980): 03Cxx

ISBN 3-540-10269-8 Springer-Verlag Berlin Heidelberg New York
ISBN 0-387-10269-8 Springer-Verlag New York Heidelberg Berlin

Printing and binding: Beltz Offsetdruck, Hemsbach/Bergstr.
2141/3140-543210

FOREWORD

The main part of this volume constitutes the Proceedings of the Conference on Applications of Logic to Algebra and Arithmetic held at the mountain resort of Bierutowice - Karpacz in Poland, September 1 - 7 , 1979. The volume contains papers contributed by the invited speakers and a few by other participants. Some papers by mathematicians who were invited but could not come have also been included.

The abstracts of all contributed papers will appear in the Journal of Symbolic Logic.

The conference was the fourth in the series Set Theory and Hierarchy Theory organized by the Institute of Mathematics of the Technical University of Wrocław. The conference was attended by 80 registrated participants from 12 countries.

The organizing committee consisted of A.Macintyre /Yale University/, L.Pacholski /Polish Academy of Sciences, Wrocław/, Z.Szczepaniak /Technical University, Wrocław/ and J.Wierzejewski /Technical University, Wrocław; chairman/.

On behalf of the organizing committee we wish to thank the Technical University of Wrocław and all the people who by their help contributed to the success of the conference and its good scientific and friendly atmosphere.

L.Pacholski
J.Wierzejewski
A.Wilkie

TABLE OF CONTENTS

FURTHER REMARKS ON THE ELEMENTARY THEORY OF FORMAL POWER SERIES RINGS

J. Becker[*], J. Denef[**] and L. Lipshitz

INTRODUCTION. §1 contains an elementary proof that the theory of the field $F_p((t))$ of formal power series over the p element field F_p with cross section is undecidable, and some extensions of this. This result is due to J. Ax (unpublished). The authors learned it from B. Jacob who had independently rediscovered it. The previous proofs were not elementary making use of a norm form and properties of the norm residue symbol. §2 contains some results on the existential theories of power series rings in $n \geq 2$ variables and §3 contains a result on definability in the rings $\mathbb{C}\{X\}$ of convergent power series.

§1. Let F be a field of characteristic p and let $K = F((t))$, the field of formal power series over F. Define

$$\mathrm{Sol}(\alpha) \leftrightarrow \exists x, y_1, \dots, y_{p-1} \in K \quad (\alpha = x^p - x + ty_1^p + t^2 y_2^p + \dots + t^{p-1} y_{p-1}^p).$$

1.1. LEMMA. Suppose that F is perfect and let $\alpha = \sum_{\substack{i \in \mathbb{Z} \\ i \geq n}} \alpha_i t^i$, $\alpha_i \in F$. Then

$$\mathrm{Sol}(\alpha) \leftrightarrow \exists x_0 \in F \quad (\alpha_0 = x_0^p - x_0).$$

PROOF. Since $(\sum a_i t^i)^p = \sum a_i^p t^{ip}$ we certainly have that if $\mathrm{Sol}(\alpha)$ then $\alpha_0 = x_0^p - x_0$, where $x = \sum x_i t^i$. For the converse notice that $\mathrm{Sol}(\alpha)$ is additive in α (i.e. $\mathrm{Sol}(\alpha), \mathrm{Sol}(\beta) \to \mathrm{Sol}(\alpha + \beta)$) so it suffices to prove (a) $\mathrm{Sol}(\sum_{i=1}^{\infty} \alpha_i t^i)$ and (b) $\mathrm{Sol}(at^{-k})$ for all $a \in F$, $k \in \mathbb{N}$, $k > 0$. For (a) if $\alpha = \sum_{i=1}^{\infty} \alpha_i t^i$ set $x = -\alpha + (-\alpha)^p + (-\alpha)^{p^2} + \dots$ and $y_1 = y_2 = \dots = y_{p-1} = 0$.

(b) Case 1. $p \nmid k$. Then $k = pq - j$ with $j \in \{1, 2, \dots, p-1\}$. Since F is perfect there is a $w \in F$ such that $w^p = a$. Hence $at^{-k} = (wt^{-q})^p t^j$ so we can set

[*] Supported in part by N.S.F. 8002789.

[**] Supported by the National Science Foundation of Belgium.

$x = 0$, $y_j = wt^{-q}$ and $y_i = 0$ for $i \neq j$.

(b) General case. The proof is by induction on k. $k = 1$ follows from case 1. Suppose $\mathrm{Sol}(wt^{-q})$ for all $0 < q < k$, $q \in \mathbb{N}$, and $w \in F$. We must prove $\mathrm{Sol}(at^{-k})$. If $p \nmid k$ we are in case 1, so suppose that $k = qp$. Let $w^p = a$. We have

$$at^{-k} = [(wt^{-q})^p - wt^{-q}] + wt^{-q}.$$

Now $\mathrm{Sol}[(wt^{-q})^p - wt^{-q}]$ (Set $x = wt^{-q}$ and $y_i = 0\ \forall i$) and by the induction hypothesis $\mathrm{Sol}(wt^{-q})$ so the result follows by the additivity of Sol. ∎

Define the following predicates (on $F((t))$)

$\mathrm{Cros}(u) \leftrightarrow u \in \{t, t^{-1}, t^2, t^{-2}, t^3, t^{-3}, \ldots\}$

$\mathrm{Con}(x) \leftrightarrow x \in F$

$\mathrm{Zer}(\alpha) \leftrightarrow$ the constant term of α is zero

$\mathrm{Int}(x) \leftrightarrow x = \sum_{i \geq 0} a_i t^i$, $a_i \in F$ (i.e. $\mathrm{ord}(x) \geq 0$).

We shall consider fields F which satisfy:

(*)
1) F is perfect of characteristic $p \neq 0$.
2) F is not closed under Artin-Schreier extensions (i.e. $\exists \alpha \in F$ such that $x^p - x = \alpha$ has no solution in F).

1.2. LEMMA. If F satisfies (*) then $\mathrm{Zer}(\alpha)$ is definable in $\langle F((t)); \mathrm{Con}, t \rangle$.

PROOF. We have by Lemma 1.1 that

$$\mathrm{Zer}(\alpha) \leftrightarrow \forall \beta (\mathrm{Con}(\beta) \rightarrow \mathrm{Sol}(\beta\alpha)).$$

1.3. LEMMA. Int is definable in $F((t))$.

PROOF. See [2].

1.4. LEMMA. $F[t]$ is definable in $\langle F((t)); \mathrm{Zer}, \mathrm{Cros} \rangle$.

PROOF. We have

$$x \in F[t] \leftrightarrow \mathrm{Int}(x) \wedge \exists s \forall w [(\mathrm{Cros}(w) \wedge \mathrm{Int}(\tfrac{s}{w})) \rightarrow \mathrm{Zer}(wx)].$$

1.5. LEMMA. If F satisfies (*) then $\langle F((t)); \mathrm{Cros}, \mathrm{Con} \rangle$ is undecidable.

PROOF. This follows immediately from Lemma 1.2 and 1.4 and the fact that $F[t]$ is undecidable for any field F. (Notice that t is definable from Cros.)

<u>1.6. THEOREM.</u> If F is a finite field then $\langle F((t));Cros\rangle$ is undecidable.

<u>PROOF.</u> Let q = the cardinality of F. Then $Con(x) \leftrightarrow x^q - x = 0$. ∎

Next we sharpen Lemma 1.5 to get the undecidability of $\langle F((t));Cros\rangle$ for any field F which satisfies (*).

Define

$$\overline{Zer}(x) \leftrightarrow \forall\beta[(\forall w(Cros(w) \to Sol(\beta w))) \to Sol(\beta x)].$$

<u>1.7. LEMMA.</u> If F satisfies (*) then

(a) $\overline{Zer}(x)$ is definable in $\langle F((t));Cros\rangle$,

(b) $\overline{Zer}(x) \to Zer(x)$,

(c) $[x \in F_p((t)) \wedge Zer(x)] \to \overline{Zer}(x)$.

<u>PROOF.</u> (b) $\overline{Zer}(x) \to \forall\beta(Con(\beta) \to Sol(\beta x))$.

(c) Let $x = \sum a_i t^i$, $a_i \in F_p$, $a_0 = 0$ and let $\beta \in F((t))$ be such that $Sol(\beta t^i)$ for all $i \neq 0$. We must show that $Sol(\beta x)$. Let $ord(\beta) > -n$. Then $ord(\beta \sum_{i \geq n} a_i t^i) > 0$ and hence $Sol(\beta \sum_{i \geq n} a_i t^i)$. By the additivity of Sol we also have $Sol(\sum_{i \leq n} a_i \beta t^i)$, since $a_i \in F_p$, and hence $Sol(\beta \sum a_i t^i)$. ∎

<u>1.8. LEMMA.</u> Suppose F satisfies (*). Then $F_p((t))$ is definable in $\langle F((t));Cros\rangle$.

<u>PROOF.</u> $x \in F_p((t)) \leftrightarrow \forall w \exists c \in F_p[(Cros(w) \vee w = 1) \to \overline{Zer}(wx - c)]$. ∎

From Theorem 1.6 and Lemma 1.8 we have:

<u>1.9. THEOREM.</u> If F satisfies (*) then $\langle F((t));Cros\rangle$ is undecidable.

<u>1.10. THEOREM.</u> For any prime p there are two perfect fields F_1 and F_2 of characteristic p such that $F_1 \equiv F_2$ but $\langle F_1((t));Cros,Con\rangle \not\equiv \langle F_2((t));Cros,Con\rangle$.

<u>PROOF.</u> If F satisfies (*) then $F_p[t]$ and $F[t]$ are definable in $\langle F((t));Cros,Con\rangle$ by a formula depending only on the characteristic p of F. The following formula ψ also depends only on p:

$$\forall x \exists P \exists Q(Con(x) \to P \in F_p[t] \wedge Q \in F[t] \wedge P = (t-x)Q).$$

$F((t))$ satisfies ψ iff F is algebraic over F_p. Let F_1 be an infinite algebraic extension of F_p which is not closed under Artin-Schreier extensions.

Let F_2 be any field which is elementarily equivalent with F_1 but not algebraic over F_p. Then $F_1((t)) \models \psi$ and $F_2((t)) \models \neg\psi$.

REMARK. It is well known [3] that $\langle F((t));\mathrm{Cros}\rangle$ is decidable when F is a decidable field of characteristic zero.

QUESTION. Is $F_p((t))$ (without crosssection) decidable?

§2. In this section we shall extend some of the results of [6] and [5] about the existential theories of power series rings. $\mathbb{C}$ will denote the complex numbers and F an arbitrary field. $F[\![X]\!]$ is the local ring of formal power series in the variables $X = (X_1,\dots,X_n)$, and (X) denotes the maximal ideal of $F[\![X]\!]$. $F\langle X\rangle$ is the subring of $F[\![X]\!]$ of all power series algebraic over $F[X]$. If F is a valued field (eg. $\mathbb{C}$) then $F\{X\}$ is the ring of all convergent power series (i.e. convergent on some neighbourhood of 0). In §2 and §3 we will always assume that there are symbols for $X_1,X_2,\dots$ and X_n in the first order language.

The following results follow immediately from the extension of the Artin Approximation Theorem given in §4 of [5]:

(1) $\langle F\langle X\rangle\,;\,F\langle X_1\rangle\rangle \prec_{\exists} \langle F[\![X]\!];F[\![F_1]\!]\rangle$ i.e. if $f_i(X,Y)$, $g(X,Y) \in F\langle X\rangle[Y]$, $Y = (Y_1,\dots,Y_m)$ and $\varphi = \exists Y_1,\dots,Y_m\{\bigwedge_i f_i(X,Y) = 0 \wedge g(X,Y) \neq 0 \wedge Y_1,\dots,Y_r$ depend only on $X_1\}$ then $F[\![X]\!] \models \varphi$ implies $F\langle X\rangle \models \varphi$.

(2) (1) is also true with $\langle\,\rangle$ replaced by $\{\,\}$ and F by $\mathbb{C}$. This requires the results of §5 of [8].

(3) Let $f_i(X,Y) \in F[X,Y]$, $Y = (Y_1,\dots,Y_m)$. If the system

"$\bigwedge_i f_i(X,Y) = 0 \wedge Y_1,\dots,Y_r$ depend only on X_1"

has a solution $Y \bmod (X)^k$ for all k, then it has a solution $Y \in F[\![X]\!]$.

(4) If the existential theory of F is decidable then so is the positive existential theory of $\langle F[\![X]\!];F[\![X_1]\!]\rangle$ and hence by (1) and (2) also of $\langle F\langle X\rangle;F\langle X_1\rangle\rangle$ and $\langle F\{X\};F\{X_1\}\rangle$ (if F is a valued field). See [5], §6, Remark (i).

On the other hand it was shown in [6] and [4] that:

(5) $\langle \mathbb{C}\langle X_1,X_2\rangle;\,\mathbb{C}\langle X_1\rangle,\,\mathbb{C}\langle X_2\rangle\rangle$ and $\langle \mathbb{C}[\![X_1,X_2]\!];\mathbb{C}[\![X_1]\!];\mathbb{C}[\![X_2]\!]\rangle$ have different positive existential theories, i.e. there is a system

$$\bigwedge_i f_i(X,Y) = 0 \wedge Y_1,\dots,Y_{r_1} \text{ depend only on } X_1$$
$$\wedge\, Y_{r_1+1},\dots,Y_{r_2} \text{ depend only on } X_2$$

(with $f_i(X,Y) \in F[X,Y]$, $Y = (Y_1,\dots,Y_m)$) which has a solution in $\mathbb{C}[\![X_1,X_2]\!]$ but not in $\mathbb{C}\langle X_1,X_2\rangle$. (The similar result is true with either $\langle\ \rangle$ or $[\![\ ]\!]$ replaced by $\{\ \}$ and can be easily established by the same method.)

(6) $\langle \mathbb{C}\{X,Y\};\mathbb{C}\{Y\}\rangle \not\prec_\exists \langle \mathbb{C}[\![X,Y]\!];\mathbb{C}[\![X]\!]\rangle$ where $X = (X_1,X_2)$ and $Y = (Y_1,\dots,Y_4)$. This uses a counter example of Gabrielov (see [6]).

2.1. PROPOSITION. Let F be an arbitrary field.

(i) The existential theory of $\langle F[\![X_1,X_2]\!];F[\![X_1]\!],F[\![X_2]\!]\rangle$ is undecidable, i.e. formulas of the form

$$\exists Y_1,\dots,Y_m \in F[\![X_1,X_2]\!]\,(\bigwedge_i f_i(X,Y) = 0 \wedge g(X,Y) \neq 0$$
$$\wedge\, Y_1,\dots,Y_{r_1} \in F[\![X_1]\!] \wedge Y_{r_1+1},\dots,Y_{r_2} \in F[\![X_2]\!])$$

are undecidable $(f_i(X,Y), g(X,Y) \in F[X,Y])$.

(ii) The existential theory of $\langle F[\![X_1,X_2,X_3]\!];F[\![X_1]\!],F[\![X_1,X_2]\!]\rangle$ is undecidable i.e. formulas of the form

$$(7)\quad \exists Y_1,\dots,Y_m \in F[\![X_1,X_2,X_3]\!]\,(\bigwedge_i f_i(X,Y) = 0$$
$$\wedge\, g(X,Y) \neq 0 \wedge Y_1,\dots,Y_{r_1} \in F[\![X_1]\!] \wedge Y_{r_1+1},\dots,Y_{r_2} \in F[\![X_1,X_2]\!])$$

are undecidable. (The same results hold with $[\![\ ]\!]$ replaced by $\langle\ \rangle$ or $\{\ \}$.)

PROOF. (i) We shall represent $\langle \mathbb{N};+,\cdot\rangle$ in $F[\![X_1]\!] - \{0\}$ by the correspondence $n \longmapsto [X_1^{\,n}] = \{\varepsilon X_1^{\,n} \mid \varepsilon$ is a unit in $F[\![X_1]\!]\}$. Addition in $\mathbb{N}$ corresponds to multiplication in $F[\![X_1]\!]$ and multiplication in $\mathbb{N}$ to exponentiation in $F[\![X_1]\!]$. Since multiplication is positive existentially definable from squaring it is sufficient to show that from $A = [X_1^{\,n}]$ we can define $B = [X_1^{\,n^2}]$ by a formula of the correct kind. The result will then follow from the undecidability of Hilbert's Tenth problem.

Notice: (a) If $A = [X_1^{\,n}]$ and $B = [X_2^{\,m}]$ and $A \equiv B \bmod (X_1 - X_2)$ then $m = n$ (i.e. if $A(X_2) = B$ then $m = n$).

(b) If $A \in [X_1^{\,n}]$ and $B \in [X_2^{\,n}]$ and $C \in [X_1^{\,m}]$ and $B \equiv C \bmod X_2 - A$ then $m = n^2$ (i.e. if $B(A) = C$ then $m = n^2$). The only inequalities which we needed were of the form $Y_i \neq 0$, because we represent $\mathbb{N}$ in $F[\![X_1]\!] - \{0\}$.

(ii) We shall do the same as (i) but using formulas of the required form.

Notice: (a) If $B \in F[\![X_1,X_2]\!]$ and $B \equiv C \in [X_1^n] \bmod X_2 - X_1$ then $\mathrm{ord}(B) \leq n$ and if $B \equiv C\, D(X_1,X_2,X_3) \bmod X_2 - X_1X_3$ and $B \equiv C^2 E(X_1,X_2,X_3) \bmod X_2 - X_1^2X_3$ then $B = bX_2^n$ + higher order terms.

(b) If $B = bX_2^m$ + higher order terms,

$C = cX_2^n$ + higher order terms,

$A \in [X_1^n]$, and

$A \equiv B \bmod X_1 - C$ (i.e. $A(C) = B(C,X_2)$)

then $m = n^2$.

This completes the proof of (ii). ∎

QUESTIONS

(1) Given a system of the form (7) does it follow that if it has a solution in $F[\![X]\!]$ then it has one in $F\langle X\rangle$? Also the obvious generalization to more than 3 variables. This problem was suggested by M. Artin in [1] and a positive answer would have applications in algebraic geometry (e.g. deformations of singularities). A positive answer to this question for F algebraically closed has been claimed in a number of places by T. Mostowski, but all the proofs to date have been completely confused (e.g. in: Die Approximationseigenschaft lokaler Ringe, Lecture Notes in Mathematics 634).

(2) Is the positive existential theory of $\langle \mathbb{C}[\![X_1,X_2,X_3]\!];\mathbb{C}[\![X_1]\!],\mathbb{C}[\![X_1,X_2]\!]\rangle$ decidable? This would follow from a positive answer to (1) (see [5, §4, Remark after Theorem 4.3 and §6, Remark (i)]).

(3) Is the existential theory of $\mathbb{C}[\![X]\!]$ decidable? (The positive existential theory is decidable, see e.g. §6 of [5].)

Consider the following condition on the field F.

(**) Let $f(X,Y) \in F[X,Y]$ and $X^{(i)} \subseteq \{X_1,\dots,X_n\}$. If the system $f(X,Y) = 0$, $Y_i \in F[\![X^{(i)}]\!]$ has a solution $\bmod (X)^k$ for all $k \in \mathbb{N}$ then it has a solution $Y \in F[\![X]\!]$.

(***) The same as (**) but with $f(X,Y) \in F[\![X,Y]\!]$.

It is clear that any $\aleph_0$-saturated field satisfies (**) and that any $\aleph_1$-saturated field satisfies (***) (e.g. $\mathbb{C}$, or the p element field F_p). In §5 of [5] it was shown that $\bar{F}_p$ (the algebraic closure of F_p) and $\mathbb{Q}$ (the rationals) do not satisfy (**).

2.2. PROPOSITION. $\mathbb{R}$ does not satisfy (**) (for $n = 3$).

PROOF. We shall express the conditions $Y \in \mathbb{R}[[X_1]]$ etc. by writing $Y(X_1)$ etc.. Let $Y_1(X_1) = \sum a_i X_1^i$, then if $Y_2(X_2) \equiv Y_1 \bmod X_2 - X_1$ we have $Y_2(X_2) = \sum a_i X_2^i$. Consider $Y_1 Y_2 = \sum_{i,j} a_i a_j X_1^i X_2^j$ and suppose that $Y_1 Y_2 \equiv Y_3(X_3) + X_1 Y_4(X_1, X_3) + X_2 Y_5(X_2, X_3) \bmod X_3 - X_1 X_2$. Then $Y_3 = \sum a_i^2 X_3^i$. Hence we can pick out elements of the form $\sum a_i^2 X_3^i$ by formulas of the correct form. If $Y_6(X_3) \equiv (1-X_1)^{-1} \bmod X_1 - 2X_3$ then $Y_6 = \sum 2^i X_3^i$. Consider the equation

$$a^2 \sum X_3^i - \sum 2^i X_3^i = \sum a_i^2 X_3^i \quad (= Y_3) \wedge a \in F.$$

This has a solution (for a and Y_3) $\bmod (X)^k$ for all k but has no solution in $\mathbb{R}[[X]]$. ∎

QUESTION. Which fields satisfy (**), (***)?

REMARKS

1) It is not necessary that F be $\aleph_0$-saturated in order for F to satisfy (**). Clearly if F realizes every type composed only of existential formulas, then F satisfies (**).

2) If F is an algebraically closed field satisfying (***), then F is $\aleph_1$-saturated, thus the transcendence degree of F is uncountable.

PROOF. Since F admits elimination of quantifiers we only have to consider types of the form $\{P_i(a) \neq 0\}_{i \in \mathbb{N}}$, with $P_i \in F[X_1]$. We can suppose that $\deg P_i = i$. Let $f(X_1, X_2) = \sum_i \bar{P}_i(X_1, X_2)$, where $\bar{P}_i(X_1, X_2)$ is the homogenization of $P_i(X_1)$, i.e. $\bar{P}_i(X_1, 1) = P_i(X_1)$. Then $f(aX_2, X_2) = \sum_i P_i(a) X_2^i$. Consider the equation below, with unknowns $a \in F$, $Y_1(X_1) = \sum_i f_i X_1^i$, $Y_4(X_1, X_3)$ and $Y_5(X_2, X_3)$:

$$Y_1(X_1) f(aX_2, X_2) \equiv \sum_i X_3^i + X_1 Y_4(X_1, X_3) + X_2 Y_5(X_2, X_3) \bmod X_3 - X_1 X_2.$$

This equation can be satisfied iff $b_i P_i(a_i) = 1$, thus iff $P_i(a) \neq 0$. ∎

3) If F is a real closed field satisfying (***), then F is $\aleph_1$-saturated. The proof follows easily by adapting the arguments in 2) and 2.2.

§3. In [7] several results are given on first order definability in the rings $\mathbb{C}[[X]]$, $\mathbb{C}\{X\}$ and $\mathbb{C}\langle X\rangle$ for $X = (X_1, \ldots, X_n)$, $n \geqslant 2$. (The elementary theories of these rings are undecidable.) Both $\mathbb{C}[[X]]$ and $\mathbb{C}\langle X\rangle$ have automorphisms γ which fix X, but move $\mathbb{C}$ (i.e. $\gamma(\mathbb{C}) \not\subseteq \mathbb{C}$). (Indeed, lift $\mathbb{C}$ to $\mathbb{C}\langle X\rangle$ by lifting first a transcendence basis $t_1, t_2, t_3, \ldots$ for $\mathbb{C}$ to $t_1 + X_1, t_2, t_3, \ldots$, and then using Hensel's Lemma, and extend this lifting to an automorphism of $\mathbb{C}[[X]]$ by mapping X to X) From this it follows that $\mathbb{C}$ is not first order definable (with symbols for X) in the rings $\mathbb{C}[\![X]\!]$ and $\mathbb{C}\langle X\rangle$. On the other hand $\mathbb{C}\{X\}$ has no such automorphisms $(n \geq 2)$. We show

3.1. PROPOSITION. $\mathbb{C}$ is first order definable in the ring $\mathbb{C}\{X\}$ $(X = (X_1,\dots,X_n), n \geq 2)$.

PROOF. In [7] it is shown how to define the second order theory of $\mathbb{N}$ in $\mathbb{C}\{X\}$. Consequently one can define the second order theory of $\mathbb{Q}$ in $\mathbb{C}\{X\}$. We shall use this to define $\mathbb{R}$ and hence $\mathbb{C}$. The Dedekind cuts in $\mathbb{Q}$ are definable. Hence it is sufficient to show that given a cut A (in $\mathbb{Q}$) we can define the $\alpha \in \mathbb{C}\{X\}$ (in fact $\alpha \in \mathbb{R}$) which corresponds to this cut. Given A one can define a sequence $a_i \in \mathbb{Q}$ such that $\lim_{i\to\infty}(a_i - A)i! = 0$. Let $f \in \mathbb{C}\{X\}$ be a unit and consider

$$g(X) = \sum_{i=0}^{\infty} (a_i - f)i!X_1^{\,i} \in \mathbb{C}[\![X]\!] .$$

Using the methods of [7] it is easy to see that $g(X)$ is definable in $\mathbb{C}\{X\}$ from f and A in the sense that the sequence of partial sums of $g(X)$ is definable. We shall show that $g(X) \in \mathbb{C}\{X\}$ iff $f = \alpha$. Clearly if $f = \alpha$ then $g(X) \in \mathbb{C}\{X\}$. Conversely suppose that $g(X) \in \mathbb{C}\{X\}$ and let $f = \alpha + \phi(X)$. Then

$$\begin{aligned} g(X) &= \sum (a_i - \alpha - \phi(X))i!X_1^{\,i} \\ &= \sum (a_i - \alpha)i!X_1^{\,i} - \phi(X) \sum i!X_1^{\,i} . \end{aligned}$$

Now since $(a_i - \alpha)i! \to 0$, $\sum (a_i - \alpha)i!X_1^{\,i} \in \mathbb{C}\{X\}$. Hence $\phi(X)\sum i!X_1^{\,i} \in \mathbb{C}\{X\}$. From this it follows easily that $\phi(X) = 0$ i.e. $f = \alpha$. ∎

Using the methods of [7] it is easy to establish:

3.2. COROLLARY. Let $X = (X_1,\dots,X_n)$, $n \geq 2$. In the ring $\mathbb{C}\{X\}$ the following are all first order definable (using symbols for X): (i) the subset $\mathbb{C}$, (ii) the subrings $\mathbb{C}\{Y\}$ for $Y \subset \{X_1,\dots,X_n\}$, (iii) $\frac{\partial}{\partial X_i}$, $i = 1,\dots,n$, (iv) composition.

In the rings $\mathbb{C}[\![X]\!]$ and $\mathbb{C}\langle X\rangle$ none of these are definable. (If we consider $\mathbb{C}[\![X]\!]$ and $\mathbb{C}\langle X\rangle$ as $\mathbb{C}$-algebras (i.e. with a predicate for the constants $\mathbb{C}$) then all of these are definable (see [7])).

REMARK. Proposition 3.1 is true even if $X_1,\dots,X_n$ are not in our language.

REFERENCES

1 ARTIN, M.: On the solutions of analytic equations, Inventiones Math. 5, 277-291 (1968).

2 AX, J.: On the undecidability of power series fields, Proc. Amer. Math. Soc. 16, p.846 (1965).

3 AX, J. & KOCHEN, S.: Diophantine problems over local fields III, Ann. Math. (2), 83, 437-456 (1966).

4 BECKER, J.: A counterexample to Artin approximation with respect to subrings, Math. Ann. 230, 195-196 (1977).

5 BECKER, J., DENEF, J., LIPSHITZ, L. & van den DRIES, L.: Ultraproducts and Approximation in local rings I, Inventiones Math. 51, 189-203 (1979).

6 BECKER, J. & LIPSHITZ, L.: Remarks on the elementary theories of formal and convergent power series, Fund. Math. (to appear).

7 DELON, F.: Résultats d'indécidabilité dans les Anneaux de Séries Formelles, Fund. Math. (to appear).

8 DENEF. J. & LIPSHITZ, L.: Ultraproducts and approximation in local rings II, Math. Ann. (to appear).

BRANDEIS UNIVERSITY, DEPT. OF MATH., WALTHAM, MASS. 02154, U.S.A.

KATHOLIEKE UNIVERSITEIT LEUVEN, DEPT. OF MATH., CELESTIJNENLAAN 200B, B-3030 LEUVEN (BELGIUM).

PURDUE UNIVERSITY, DEPT. OF MATH., WEST LAFAYETTE, INDIANA 47907, U.S.A.

ELIMINATION OF QUANTIFIERS FOR NON SEMI-SIMPLE RINGS OF CHARACTERISTIC p

Chantal BERLINE

CNRS - UNIVERSITE PARIS VII

Boffa, Macintyre and Point give in [B,M,P] the complete classification of semi-simple unitary rings of characteristic p which admit elimination of quantifiers (e.q.) in the language $(0,1,+,\times)$. Here we show that the classification problem for non semi-simple e.q. rings of char. p reduces to that of 3-nilpotent rings which have e.q. in $(0,+,\times)$.

As a corollary of the first step of the proof we get that there are no non semi-simple prime e.q. rings. This, together with [B,M,P], leads to the classification of prime e.q. rings, giving another proof of the result of [R_2].

Our result is :

Theorem.- (i) A non semi-simple unitary ring R of prime characteristic p which admits e.q. in the language $(0,+,\times,1)$, is local with residue field $\mathbb{F}_p$ or $\mathbb{F}_4$.

(ii) There are only two such rings which have residue field $\mathbb{F}_4$, namely the rings generated freely over $\mathbb{F}_2$ by two elements x and y subject to the relations $x^2 = 0$, $y^2 + y + 1 = 0$ and $xy - yx = 0$ (resp. $xy - yx = x$).

Here, as in [B,M,P], semi-simple means that J(R), the Jacobson radical of R, i.e. the intersection of all maximal left or right ideals, is trivial. We will assume familiarity with [B,M,P] and with the basic notions on e.q. (as presented for example in the first pages of [R]).

1.- INTRODUCTION

1.1.- Let R be a unitary ring. J(R) is definable in R without parameters, for instance by the formula $\forall y \exists z \ (1 + xy)z = 1$. If R has char. a prime p and if J(R) is algebraic over $\mathbb{F}_p$, then J(R) is nil (i.e. all its elements are nilpotent) : take $x \in J(R)$ and $P \in \mathbb{F}_p[X]$, $P \neq 0$, such that $P(x) = 0$; if m is the smallest integer such that X^m appears in P we have $x^m = 0$. Now assume that R is an e.q. ring of prime char. p. Then by lemma 1 (ii) of [B,M,P] every element of R transcendental over $\mathbb{F}_p$ is invertible so J(R) is algebraic, hence nil.

1.2.- A useful criterion for e.q. (in the general case) is the following (cf [S], p. 63). Let M be a realization of a language L. Then M admits e.q. iff, whenever M_1 and M_2 are two ultrapowers of M and f_1, f_2 are two monomorphisms from a realization P of L into M_1 and M_2 respectively, then there exist an ultrapower N of M and elementary monomorphisms g_1 and g_2 such that the following diagram is commutative.

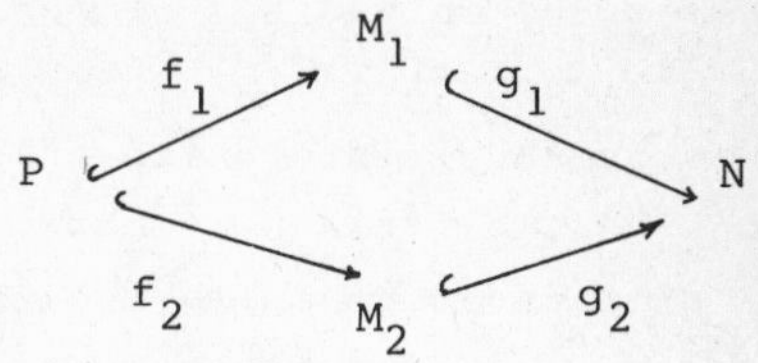

(This version is derived from the version in [S] by noticing that we are dealing with the complete theory T of M and that every model of T is, up to an isomorphism, an elementary substructure of some ultrapower of M).

1.3.- As a first consequence of 1.2 we note that every definable <u>substructure</u> of an e.q. structure M admits e.q. (in the same language as M does). So, if R is a unitary ring of prime char. p which admits e.q. in $L = (0,1,+,\times)$ the unitary subring $\mathbb{F}_p + J(R)$ of R admits e.q. in L.

With a little more subtle use of 1.2 one can show that J(R) itself admits e.q. in L' = (0,+,×) and that, conversely, if S is a nil ring of char. p which admits e.q. in L' the ring of char. p obtained by adding in the standard way a unit to S admits e.q. in L. (For these two facts a direct elimination of the quantifiers in the formulas is easy too).

From now on e.q. will mean e.q. in (0,+,×,1) if we are dealing with unitary rings and e.q. in (0,+,×) otherwise.

1.4.- Now, our theorem asserts that, exept for two finite rings with residual field $\mathbb{F}_4$, the non semi-simple e.q. unitary rings of char. p are exactly the rings obtained by adding a unit of char.p, in the standard way, to a nil e.q. ring of char. p. We shall see in lemma 2.1 that such nil rings are automatically 3-nilpotent (i.e. satisfy $\forall\, x,y,z \;\; xyz = 0$). There are recent results of Cherlin about the nil e.q. rings, providing in particular all the finite e.q. nil rings of char. p.

1.5.- A further consequence of 1.2 is that every ring R of char. $n = p_1^{n_1} \times \dots \times p_r^{n_r}$ where the p_i are distinct primes, admits e.q. iff all the $R_i = \{x \in R / p_i^{n_i} x = 0\}$ admit e.q. (see also lemma 2.1.1 of [R] or Prop. 1 of [B,M,P]). Also the only rings of char. 0 which admit e.q. are the algebraically closed fields of char. 0 [B]. Thus the classification problem for unitary rings which admit e.q. reduces to that of rings of char. p^n. Now, all rings of char. p^n, $n \geq 2$, are non semi-simple rings and we will never reach them with the results of this paper. However we can have a good control of these rings too (the corresponding results will appear elsewhere). In fact some phenomena are common to both cases as the existence of marginal cases when p = 2 and the fact that e.q. rings of char. p^n, $n \geq 2$, which have no idempotents are local with nilradical and residue field $\mathbb{F}_p$, except for marginal rings of char. 4.

2.- PROOF OF THE THEOREM

The following lemma is due to Boffa, Macyntire , Point and Cherlin

2.1.- Lemma.- Let S be a nil ring of char. p which admits e.q., then

(i) S is 3-nilpotent

(ii) The set of 2-nilpotents of S is the left (resp. right) annihilator of S - hence a two-sided ideal - .

Proof.- The point is that in an e.q. ring of prime characteristic all non trivial 2-nilpotents have the same type (as they clearly have the same quantifier-free type). Also all pairs of 2-nilpotents x,y which are linearly independent over $\mathbb{F}_p$ and such that $xy = yx = 0$ have the same type. This shows that every element x of S is 3-nilpotent (cf. lemma 2 of [B,M,P]) : if $x^n = 0$, $x^{n-1} \neq 0$ and $n > 3$ the pairs (x^{n-1}, x^{n-2}) and (x^{n-2}, x^{n-1}) have the same type by the preceding remark so there is an $u \in S$ such that $u^n = 0$, $x^{n-1} = u^{n-2}$ and $x^{n-2} = u^{n-1}$, then $x^{n-2} = ux^{n-1}$ and $x^{n-1} = ux^{n-1}.x = 0$, a contradiction. Let $x \neq 0$ be an element of the set $N_2 = \{x \in S / x^2 = 0\}$ and let a be any element of S. If $xax \neq 0$ then the pairs (x, xax) and (xax, x) have the same type so there is a $b \in S$ such that $x = (xax)\, b(xax)$, then $xax = (xaxbxax)\, a(xaxbxax) = xaxb(xa)^3\, xbxax = 0$ a contradiction. So $xax = 0$. If $xa \neq 0$ then (x, xa) and (xa, x) have the same type so there is a $b \in S$ such that $x = xab$. Then $x = x(ab)^3 = 0$ which is again a contradiction. As N_2 contains clearly the left annihilator of S we have just proved point (ii) of the lemma. Now, for all $x, y \in S$ we have $x^2, y^2, (x+y)^2 \in N_2$ so $xy + yx \in N_2$, $(xy + yx)\, xy = 0$, $xy \in N_2$ and $xyz = 0$ for all $z \in S$ thus S is 3-nilpotent.

2.2.- Remark.- There are no non semi-simple prime e.q. rings. Thus the only prime e.q. rings are the e.q. fields and the rings $M_2(\mathbb{F}_p)$ p prime.

Proof.- R is prime iff it satisfies :

$$\forall x \forall y \, [x \neq 0 \wedge y \neq 0 \Longrightarrow \exists z \quad xzy \neq 0].$$

Let R be a non semi-simple e.q. ring of char. a prime p, then J(R) admits e.q. (1.4) and has a non zero 2-nilpotent x. As in lemma 2.1 we show that $xax = 0$ for all a in R thus R cannot be prime. As the only prime e.q. rings of char. 0 are algebrically closed fields the classification of prime e.q. rings is reduced to that of semi-simple e.q. rings of char. p (cf. 1.6) so it suffices to read theorem 2 of [B,M,P] and to forget the $\mathbb{F}_{p^m} \times \mathbb{F}_{p^m}$ and the p^m-rings (which are not prime). Note that we need not here the full proof of theorem 2 as it is clear that a ring of continuous functions over a boolean space which is not a singleton cannot be prime.

2.3.- Let us return to our ring R. N_2 denotes now the set of 2-nilpotents of R and $N_2^* = N_2 - \{0\}$. We already know that N_2^* has no non-trivial definable subsets. In particular $J(R) \neq 0$ implies $N_2 \subset J(R)$ (if $J(R) \neq 0$ then $J(R)$ contains an element of N_2^*). More generally :

2.4.- For all $x \in N_2$ $\mathbb{F}_p x$ is a maximal proper subset of N_2 definable with the parameter x. Indeed let $y, y' \in N_2 - \mathbb{F}_p x$, then (x,y) and (x,y') have the same type so y and y' cannot be distinguished by a formula with parameter x.

A consequence of this is :

2.5 <u>Lemma</u>.-

(i) For all $x \in N_2^*$ $Rx = N_2$ or for all $x \in N_2$ $R_x = \mathbb{F}_p x$

(ii) Let x be in N_2^*. Then $\mathrm{Ann}_g\, x$, the left annihilator of x, is a maximal left ideal.

(iii) Let x be in N_2^* such that $Rx = \mathbb{F}_p x$ or $Rx = N_2 = xR$ then $\mathrm{Ann}_g\, x$ is two-sided, R is algebraic, and $R/\mathrm{Ann}_g\, x = \mathbb{F}_p$ or $\mathbb{F}_4$

(iv) The right versions of (i), (ii), (iii).

<u>Proof</u>.-

(i) We use 2.4 to claim that for all x in N_2 $Rx = N_2$ or $\mathbb{F}_p x$ and 2.3 to move the quantifiers.

(ii) Is trivial when we are in the second case of (i) so we assume that we are in the first case. Let $a \notin \mathrm{Ann}_g\, x$, then $ax \in N_2^*$ and there is a $\lambda \in R$ such that $x = \lambda ax$ thus $1 - \lambda a \in \mathrm{Ann}_g\, x$ and $\mathrm{Ann}_g\, x$ is a maximal left ideal.

(iii) Let $a \in \mathrm{Ann}_g\, x$ (i.e. $ax = 0$) and λ be any element of R, then $(a\lambda)x = a(\lambda x)$. If $Rx = \mathbb{F}_p x$ or if $Rx = xR = N_2$ then $a\lambda x \in axR = \{0\}$ so $\mathrm{Ann}_g\, x$ is a right ideal. Now, $\mathrm{Ann}_g\, x$ being two-sided we can speak of the ring $\bar{R} = R/\mathrm{Ann}_g\, x$. A consequence of e.q. (in R) is that all the elements of $\bar{R}$ which are not in the prime field $\mathbb{F}_p$ have the same type in $\bar{R}$: if $a,b \in R$ and $ax,bx \notin \mathbb{F}_p x$ then (x,ax) and (x,bx) have the same type ; let φ be a formula of $(0,1,+,\times)$ such that $\bar{R} \models \varphi(\bar{a})$ ($\bar{a},\bar{b}$ denote the classes of a and b modulo $\mathrm{Ann}_g\, x$). By e.q. there is a $c \in R$ such that $\bar{R} \models \varphi(\bar{c})$ and $bx = cx$. Then $\bar{b} = \bar{c}$ and $\bar{R} \models \varphi(\bar{b})$. This shows that $\bar{a}$ and $\bar{b}$ have the same type. Thus $\bar{R}$ has no nilpotents (otherwise $\bar{R}$ would have invertible elements not in $\mathbb{F}_p$ and a nilpotent cannot have the same type as an invertible element). By lemma 1 of [B,M,P] if R had a transcendental element t C(t) would be an algebraically closed field so its image in $\bar{R}$ would contain simultaneonsly transcendental and algebraic elements, a contradiction, so R is algebraic over $\mathbb{F}_p$. As $\bar{R}$ is an algebraic algebra over the finite field $\mathbb{F}_p$ and has no nilpotents it is commutative. As every non zero element is left invertible $\bar{R}$ is an algebraic field. Furthermore all elements in $\bar{R} - \mathbb{F}_p$ have the same minimal polynomial over $\mathbb{F}_p$ so $\bar{R}$ is isomorphic to $\mathbb{F}_{p^q}$ for a prime q and, if $q \neq 1$, we must have $p^q - p = q$ which implies $p = q = 2$ [use that a finite field $\mathbb{F}_{p^n}$, $n \in N^*$, is generated over $\mathbb{F}_p$ by an element α of degree n over $\mathbb{F}_p$ and is normal so has n elements satisfying the same minimal polynomial as α; furthermore if n is not prime $\mathbb{F}_{p^n}$ has a non trivial proper subfield thus an element of degree dividing n].

2.6. <u>Lemma</u>.- $R/J(R) = \mathbb{F}_p$ or $\mathbb{F}_4$. Thus $J(R) = \mathrm{Ann}_g x = \mathrm{Ann}_d x$ for all $x \in N_2^*$.

<u>Proof</u>.- Let $A = \mathrm{Ann}_g x$ if $Rx = \mathbb{F}_p x$ or $Rx = N_2 = xR$ and $A = \mathrm{Ann}_d x$ otherwise. By lemma 2.5 A is a two-sided ideal and $R/A = \mathbb{F}_p$ or $\mathbb{F}_4$. Also $A \supset J(R)$ and we have just to prove the reverse inclusion. From now on we suppose that we are in the first case so we know that A is a maximal left ideal. We first prove that R has no non trivial (i.e. $\neq 0,1$) idempotents. Let a be a non trivial idempotent. Then $ax = 0$ or $(1-a)x = 0$ (R/A has no non trivial idempotents). Suppose $ax = 0$. As two non trivial idempotents of an e.q. ring of char. p have the same

type we have an $y \in N_2^*$ such that $(1-a)y = 0$ and thus $a(x+y) = y$. If $y \in Rx$ ($= \mathbb{F}_p x$ or xR) $ay = 0$ and we already have a contradiction. If not then $Rx = \mathbb{F}_p x$ so $Rz = \mathbb{F}_p z$ for all $z \in N_2^*$, in particular for $z = x+y$. So there is an $m \in \mathbb{F}_p$ such that $a(x+y) = m(x+y) = y$ and we get a contradiction too. Now, by lemma 2.5 (iii) and 1.1 R is algebraic and J(R) nil.

It is a classical result that idempotents lift modulo nil ideals so R/J(R) has no non trivial idempotents. Since R/J(R) is algebraic over the finite field $\mathbb{F}_p$ every element x of R/J(R) has a power which is an idempotent (this is also classical ; hint : there are integers m,n such that $0 < m < n$ and $x^m = x^n$, then $x^{m(n-m)}$ is an indempotent). Thus each element of R/J(R) is invertible or nilpotent. It follows that R/J(R) is local. Therefore R is local and J(R) = A. Also J(R) is the set of all nilpotents of R.

To prove the theorem it remains to see what happens when $R/J(R) = \mathbb{F}_4$. It is the task of the two last lemmas.

2.7.- <u>Lemma</u>. If $R/J(R) = \mathbb{F}_4$ then R is isomorphic to the (associative) ring generated freely over $\mathbb{F}_2$ by two elements x, y subject to the relations $x^2 = 0$, $y^2 + y + 1 = 0$ and $xy - yx = 0$ (resp. $xy - yx = x$). Note that the first of these two rings is isomorphic to $\mathbb{F}_4[x]/x^2$.

<u>Proof</u>.- We first notice that R contains a copy of $\mathbb{F}_4$. Indeed, let $a \in R$ such that $w = a^2 + a + 1 \in J(R)$. Then w commutes with a, $w^3 = 0$ and $(a + w^2 + w)^2 + (a + w^2 + w) + 1 = 0$. Thus the subring generated by $\alpha = a + w^2 + w$ is isomorphic to $\mathbb{F}_4$ and $R = \mathbb{F}_2[\alpha] + J(R)$.

Another point is that under the hypothesis of the lemma J(R) has no 3-nilpotent v (this will occupy us for a while) : suppose $v^3 = 0$, $v^2 \neq 0$ and look at the pairs $(v, \alpha v)$ and $(v, \alpha v + v^2)$. One checks easily that they have the same quantifier -free type ; thus they have the same type and $\alpha v + v^2 = \beta v$ for a β with the same type as α. Then $\beta = \alpha + u$ or $\beta = (1+\alpha) + u$ with $u \in J(R)$. As v does not belong to N_2, the ideal of 2-nilpotents of R, the second case cannot occur. Showing that the first is impossible too will need more work. Suppose $\alpha v + v^2 = (\alpha+u)v$, then $uv = v^2 \neq 0$ so $u^2 \neq 0$. Furthermore if $\alpha + u$ has the same type as α we have $(\alpha+u)^2 + (\alpha+u) + 1 = 0$, i.e.

$u\alpha = \alpha u + u^2 + u$ then $u\alpha u = \alpha u^2 + u^2 = u^2\alpha + u^2$ and $u^2\alpha = \alpha u^2$. As $R = \mathbb{F}_2(\alpha) + J(R)$ and $u^2J(R) = J(R)u^2 = \{0\}$ (by lemma 2.1) u^2 commutes with every element of R. Also $(\alpha u)^2 = u^2$ and $(Ru)^2 = \{0,u^2\}$. Now with lemma 2.5 (i), $N_2 = Ru^2 = \{0,u^2,\alpha u^2,\alpha^2u^2\}$ so N_2 is a linear space of dimension 2 over $\mathbb{F}_2$. We deduce, in the same way as Cherlin for its finite e.q. nil rings, that $\dim_{\mathbb{F}_2} J(R)/N_2 \le 4$: let $a_1,\dots,a_n$ be elements of J(R) ; $x_1 = u^2$, $x_2 = \alpha u^2$ is a basis for N_2 and if $\varepsilon_1,\dots,\varepsilon_n$ are elements of $\mathbb{F}_2$ we have : $(\varepsilon_1a_1+\dots+\varepsilon_na_n)^2 = P(\varepsilon_i)x_1 + Q(\varepsilon_i)x_2$ where P and Q are quadratic polynomials in the ε_i. Then if $n \ge 5$ the Chevalley-Warnitz theorem yields a non trivial common zero of P and Q showing that the a_i are dependent over $\mathbb{F}_2$ modulo N_2.

We have already noticed that u, αu are square roots of u^2 and more generally that $(Ru)^2 = \{0,u^2\}$. By e.q. αu^2 must also have a square root u'. All true 3-nilpotents of R have the same type (by e.q.) so $(Ru')^2 = \{0,\alpha u^2\}$ and $u'^2 = (\alpha u')^2 = \alpha u^2$. It is now easy to check that $u,\alpha u,u',\alpha u'$ are independent over $\mathbb{F}_2$: if $\varepsilon_1u + \varepsilon_2\alpha u + \varepsilon_3u' + \varepsilon_4\alpha u' \in N_2 = Ru'^2$ for same $\varepsilon_i \in \mathbb{F}_2$ then $\varepsilon_1u + \varepsilon_2\alpha u \in Ru' \cap Ru$ so $(\varepsilon_1u + \varepsilon_2\alpha u)^2 \in (Ru)^2 \cap (Ru')^2 = \{0\}$ which is possible only if $\varepsilon_1 = \varepsilon_2 = 0$ then $(\varepsilon_3 + \varepsilon_4)u' = 0$ and $\varepsilon_3 = \varepsilon_4 = 0$. Thus $\dim_{\mathbb{F}_2} J(R)/N_2 = 4$ and $\dim_{\mathbb{F}_2} J(R) = 6$. Now, the ring $J(R)/N_2$ has $2^6/4 - 1 = 15$ non zero elements, these elements being the square roots of elements in $N_2^* = N_2 - \{0\}$. For $a \in N_2^*$ let S(a) be the set of square roots of a modulo N_2. By e.q. the three S(a) have the same cardinality so $|S(u^2)| = 5$ and there is a $w \in J(R)$ such that $w^2 = u^2$ and $w \ne u,\alpha u,\alpha^2u$ modulo N_2. Then, $u,\alpha u,\alpha^2u,w,\alpha w,\alpha^2w$ provide 6 distinct elements of $S(u^2)$, a contradiction. So we have proved that R has no "true" 3-nilpotents.

At this time we know that if R is a non semi-simple e.q. ring of char. 2 such that $R/J(R) = \mathbb{F}_4$ then R has an element α such that $\alpha^2 + \alpha + 1 = 0$ (equivalently such that $\mathbb{F}_2[\alpha] \simeq \mathbb{F}_4$ or $R = \mathbb{F}_2[\alpha] + J(R)$) and $J(R) = N_2$ is non trivial.

Let x be any element of N_2^*. Then, by lemma 2.5 (i), $N_2 = \mathbb{F}_2[\alpha]x = x\mathbb{F}_2[\alpha]$. Thus $x\alpha \in \mathbb{F}_2[\alpha]x = \{0,x,x\alpha,x\alpha^2\}$. As α and $\alpha^2 = \alpha + 1$ are invertible the two first cases cannot happen. So

$x\alpha = \alpha x$ or $x\alpha = \alpha x + x$. Now, every element of R is a linear combination of $1, \alpha, x, \alpha x$ with coefficients in $\mathbb{F}_2$. Also if we have $\lambda_1 + \lambda_2\alpha + \lambda_3 x + \lambda_4 \alpha x = 0$, $\lambda_i \in \mathbb{F}_2$, we have $\lambda_1 + \lambda_2\alpha \in N_2$; as $1, \alpha, \alpha+1$ are invertible this implies $\lambda_1 = \lambda_2 = 0$. But $(\lambda_3 + \lambda_4\alpha)x = 0$ implies, for the same reason, $\lambda_3 = \lambda_4 = 0$. Thus $1, \alpha, x, \alpha x$ are linearly independent over $\mathbb{F}_2$ and α, x verify only the relation generated by $x^2 = 0$, $\alpha^2 + \alpha + 1 = 0$ and $x\alpha = \alpha x$ (resp. $x\alpha = \alpha x + x$). Thus renaming $\alpha : y$, R is isomorphic to one of the two free rings generated over $\mathbb{F}_2$ by elements x and y subject to the above relations. In the first case R is isomorphic to $\mathbb{F}_4[x]/x^2$, in both cases R has 16 elements.

It remains to see that these two rings, say R_1 and R_2, admit e.q.. We isolate this part of the proof in a further lemma :

2.8. Lemma.- R_1 and R_2 admit e.q.

Proof.- Taking M finite in 1.2 we get : a finite realization M of a language L admits e.q. in L iff every isomorphism between two substructures of M can be lifted to an automorphism of M.

We shall use this to show that R_2 admit e.q. and leave the commutative case as an exercise.

So suppose R is generated over $\mathbb{F}_2$ by x and y satisfying $x^2 = 0$, $y^2 + y + 1 = 0$, $xy - yx = x$ and such that $1, y, x, yx$ are linearly independent over $\mathbb{F}_2$. Then R has dimension 4 as a $\mathbb{F}_2$ linear space and every proper non trivial subring (i.e. $\neq R, \mathbb{F}_2$) has dimension 2 or 3, so is generated , as a unitary subring of R, by 1 or 2 elements (which complete a base beginning by : 1). Furthermore we can take these elements of the form : $\lambda y + \mu x + \nu yx$ with $\lambda, \mu, \nu \in \mathbb{F}_2$. Now, if $t = \lambda y + \mu x + \nu yx$ it is easy to check that $t^2 = 0$ or $t^2 + t + 1 = 0$ according as $\lambda = 0$ or not. We take now a subring S of R generated by one element a, and write $S = \mathbb{F}_2[a]$. If $a^2 = 0$, $a \neq 0$ we have $a = (\mu + \nu y)x$ for some $\mu, \nu \in \mathbb{F}_2$, not both zero, and there is an automorphism of R bringing $(\mu + \nu y)x$ to x, y to y, thus $\mathbb{F}_2[a]$ to $\mathbb{F}_2[x]$ (easy to check). In the same way if $a^2 + a + 1 = 0$ then $a = y + \mu x + \nu yx$ for some $\mu, \nu \in \mathbb{F}_2$ and there is an automorphism of R bringing $y + \mu x + \nu yx$ to y, x to x, thus $\mathbb{F}_2[a]$ to $\mathbb{F}_2[y]$. Thus any two monogenerated subrings of R exchange by an automorphism of R. We turn now to the 2-generated subrings. It is easy to verify that if x',y' are two elements of R such that $x'^2 = 0$, $x' \neq 0$, $y'^2+y'+1 = 0$ then 1,x',y',y'x' are linearly independent over $\mathbb{F}_2$, thus there is no proper subring of R containing both x' and y' and every 2-generated subring S of R is generated by 2 distinct non trivial nilpotents. So $S = \mathbb{F}_2[a,b]$ with $a^2 = b^2 = 0$, $a \neq 0$, $b \neq 0$, $a \neq b$. Then S is isomorphic to $\mathbb{F}_2[x,yx]$ modulo an automorphism of R and we are finished.

Acknowledgement : Thanks are due to G. Cherlin who noticed that a previous proof (and statement) of lemma 2.7 was incomplete.

***

REFERENCES

[B] Ch. BERLINE : Rings which admit elimination of quantifiers, to appear in the Journal of Symbolic Logic.

[B,M,P] M. BOFFA, A. MACINTYRE, F. POINT : The quantifier elimination problem for rings without nilpotent elements and for semi-simple rings, in these Annals.

[M] A. MACINTYRE : On ω_1-categorical theories of fields, Fundamenta Mathematicae, vol. 71 (1971), pp. 1-25.

[R] B.I. ROSE : Rings which admit elimination of quantifiers, Journal of Symbolic Logic, vol. 43, N°1 (1978),p.92.112.

[R_2] B.I. ROSE : Prime quantifier eliminable rings, to appear in the Journal of the London Mathematical Society.

[S] G.E. SACKS : Saturated model theory, Benjamin, Reading, Massachussets, (1972).

THE QUANTIFIER ELIMINATION PROBLEM FOR RINGS WITHOUT NILPOTENT ELEMENTS AND FOR SEMI-SIMPLE RINGS*

M. Boffa, A. Macintyre, F. Point
University of Mons

0. Introduction.

Let R be a ring (=associative ring with identity). We will say that R has quantifier elimination (q.e.) when Th(R) has quantifier elimination in the natural language for rings, i.e. the language based on +, · , 0, 1. A complete classification of the rings having q.e. is still lacking, but this problem is solved for several particular classes of rings. The initial result in this direction was obtained by Macintyre [Fund. Math. 1971] who proved that the fields with q.e. are the algebraically closed fields and the finite fields. The proof was indirect, via $\aleph_1$-categoricity, but a straightforward proof is given in [Macintyre, McKenna, Van den Dries] . Subsequent results for wider classes of rings were obtained by Rose [J.S.L. 1978] and Berline [J.S.L., to appear] , namely :

(i) the division rings with q.e. are the algebraically closed and the finite fields (Rose);

(ii) the prime rings[1] with an infinite center and with q.e. are the algebraically closed fields (Rose);

(iii) the rings of characteristic 0 with q.e. are the algebraically closed fields of characteristic 0 (Berline).

---

* the main results of this paper were proved in the period march-may 1979 while the second author was visiting professor at the University of Mons.

[1] i.e. satisfying $(\forall x \neq 0)(\forall y \neq 0)(\exists z)(xzy \neq 0)$.

In this paper we describe completely the rings with q.e. in the case of rings with no nonzero nilpotent elements (Thm 1) and in the case of semi-simple rings (Thm 2). In the reduction of the first case we were inspired by [Macintyre-Rosenstein] for the use of an important generalization of the Stone representation due to Arens and Kaplansky [Trans. A.M.S. 1948]. In the second case we use a result of Levitzki on the existence of matrix rings of the form uRu (cf. [Jacobson], ch. X, § 11).

1. <u>The primary decomposition of a ring with q.e.</u>

Let us recall the notion of the primary decomposition of a ring of characteristic $\neq 0$. If R is any ring of characteristic $n \neq 0$ and if $n = n_1 n_2$ with n_1 prime to n_2, then it is easy to check that the ideals $R_1 = \{x \in R | n_1 x = 0\}$, $R_2 = \{x \in R | n_2 x = 0\}$ are rings of characteristic n_1, n_2 such that $R \cong R_1 \times R_2$. Thus, if $n = \prod_i p_i^{n_i}$ is the prime factorization of n, then the ideals $R_i = \{x \in R | p_i^{n_i} x = 0\}$ are rings of characteristic $p_i^{n_i}$ such that $R \cong \prod_i R_i$ (this is what we call the primary decomposition of R). When R has q.e. then it is easy to see that each R_i has q.e. (cf. Lemma 2.1.1 of [Rose]). The converse is also true, because of the following fact : if R_1, R_2 are rings with q.e. of characteristic n_1, n_2 different from zero and relatively prime, then $R_1 \times R_2$ has q.e. [proof : since n_1 is prime to n_2, we know that in $R_1 \times R_2$ the elements $e_1 = (1,0)$, $e_2 = (0,1)$ are multiples of the identity (1,1), sothat the result is a consequence of the following one : assuming only that n_1, n_2 are $\neq 0$, the q.e. for R_1, R_2 implies that $R_1 \times R_2$ has q.e. in the language of rings plus the constants e_1, e_2. This last result follows from the classical result of Feferman-Vaught : $R_1 \times R_2 \models \varphi(x,y,\ldots)$ is equivalent to a boolean combination of expressions of the form $R_i \models \varphi_{ij}(x_i,y_i,\ldots)$ where the φ_{ij} can be chosen quantifier-free and where $x_i = e_i x$, $y_i = e_i y,\ldots$, so that each expression $R_i \models \varphi_{ij}(x_i,y_i,\ldots)$ can be put in the form $R_1 \times R_2 \models \varphi'_{ij}(e_i,x,y,\ldots)$ with φ'_{ij} quantifier-free].

We conclude that a ring of characteristic $\neq 0$ has q.e. if and only if each factor of its primary decomposition has q.e..

In other words :

<u>PROP. 1</u>. Let n be any natural number $\neq 0$ with prime factorization $n = \prod_i p_i^{n_i}$; the rings of characteristic n with q.e. are the rings of the form $\prod_i R_i$ where R_i is a ring of characteristic $p_i^{n_i}$ with q.e..

2. <u>The case of rings with no nonzero nilpotent elements</u>.

In this case the characteristic is either 0 or a product of distinct prime numbers, so we have only to solve the characteristic p case. In order to do that we need the following result about rings of characteristic p with q.e. (where the terms "algebraic" and "transcendental" are relative to $\mathbb{F}_p$) :

<u>LEMMA 1</u>. Let R be a ring of characteristic p.

(i) if the center of R contains a transcendental element t and if R has q.e. in the language of rings plus the constant t, then R is an algebraically closed field;

(ii) if R has q.e., then the centralizer of any transcendental element of R is an algebraically closed field;

(iii) if R has q.e., then there is an algebraic ring which

is elementarily equivalent to R.

Proof :

(i)[1] let $D = \mathbb{F}_p[t]$; this is an infinite integral domain included in the center C of R. In R, each formula $\varphi(x)$ with just one free variable x is equivalent to a disjunction of conjunctions of equations and inequations of the form $p(x) = 0$, $q(x) \neq 0$ with $p(x), q(x) \in D[x]$.

This implies easily :

(★) let $E = \{x \in R \mid R \models \varphi(x)\}$; if $D \subset E$, then there is a nonzero polynomial $r(x) \in D[x]$ which vanishes on R-E.

For E = C this shows that there is a nonzero polynomial $r(x) \in D[x]$ of minimal degree which vanishes on R-C. But for each $d \in D$, $r(x+d)-r(x)$ also vanishes on R-C, thus $r(x+d)-r(x)$ is the zero polynomial, thus $r(d) = r(0)$. This implies that the polynomial $r(x)$ is a nonzero constant. Thus R-C is empty, i.e. R is commutative. Next we prove that every nonzero element d of D is invertible in R : in D, the ideal $I = (dR) \cap D$ is infinite(since it contains dD), thus cofinite by the q.e. hypothesis applied to $(\exists y)(x = dy)$, sothat $(I+1) \cap I \neq \phi$, thus $1 \in I$.

Using (★), we see now that there is a nonzero polynomial $r(x) \in D[x]$ of minimal degree n which vanishes on the set S of all non invertible elements of R. For each $d \in D$, $r(dx) - d^n r(x)$ also vanishes on S, thus $r(dx) - d^n r(x)$ is the zero polynomial, thus $r(d) = d^n r(1)$. The polynomial $r(x)$ is thus of the form cx^n where c is a nonzero element

[1] we adapt the ideas of [Berline] to the characteristic p case.

of D. This shows that the non invertible elements of R are nilpotent, so it remains to prove that R has no nonzero nilpotent elements. Let us first note that t has a square root in R, since the q.e. hypothesis implies that the set $A = \{x \in D | x \neq 0 \wedge (\exists\, y \in R)(x = y^2)\}$ is cofinite in D, so that $(tA) \cap A \neq \phi$, thus $t \in A$. So let $a \in R$ be a square root of t. Let us now suppose that R has a nonzero nilpotent element. Then R contains an element $\varepsilon \neq 0$ such that $\varepsilon^2 = 0$. Consider the subring $R_1 = D[\varepsilon, a\varepsilon] = \{d_o + d_1\varepsilon + d_2 a\varepsilon | d_i \in D\}$. It has an automorphism σ defined by $\sigma(d_o + d_1\varepsilon + d_2 a\varepsilon) = d_o + (d_1 + d_2)\varepsilon + d_2 a\varepsilon$. By q.e., we have $(R,c)_{c \in R_1} \equiv (R,\sigma c)_{c \in R_1}$, thus $(R,t,\varepsilon,a\varepsilon) \equiv (R,t,\varepsilon,(1+a)\varepsilon)$. This implies that there is an element $b \in R$ such that $b^2 = t$ and $(1+a)\varepsilon = b\varepsilon$. But then $(1-4t)\varepsilon = (1-2a)(1+2a)\varepsilon = (1-2a)(1+a+b)(1+a-b)\varepsilon = 0$, thus $\varepsilon = 0$, a contradiction.

(ii) It is an immediate consequence of (i) and the following fact (Lemma 2.1 of [Rose]) : if a ring R has q.e. and if S is a subring of R which is definable in R with parameters in S, then S has q.e. in the language of rings plus these parameters.

(iii) If R is algebraic, we have nothing to prove. If R contains a transcendental element, then by (ii) R contains algebraic elements of arbitrarily large degree, so that there is no quantifier-free formula $\varphi(x)$ such that $R \models (\exists x)\varphi(x)$ and $(\forall x \in R)(R \models \varphi(x) \rightarrow x$ is transcendental). Since R has q.e., this means that Th(R) locally omits the set of all formulas of the form $p(x) \neq 0$ where $p(x)$ runs over the nonzero polynomials of $\mathbb{F}_p[x]$. By the omitting types theorem ([Chang-Keisler] p. 79) there is a model of Th(R) which omits this set. ⊣

Let us apply this lemma when R is a ring or characteristic p without nilpotent elements and with q.e. : by (iii) and the fact that every algebraic algebra without nilpotent elements over a finite field is commutative ([Jacobson], p. 218), R is commutative; then, by (ii) and a compactness argument, R is either an algebraically closed field or an algebraic ring of bounded degree (in the sense that the degrees of its elements are bounded). In this last case, R satisfies a polynomial identity of the form $x^{p^n} = x$ (because each element of R generates a subring which is finite, commutative and without nilpotent elements, thus isomorphic to a finite product of finite fields). Let R' be a countable ring elementarily equivalent to R. By [Arens-Kaplansky] (Corollary of Thm 8.1 together with Thm 2.2), there exists a boolean space X, with a closed subset X_k for each proper divisor k of n, such that R' is the ring of all continuous functions from X to $\mathbb{F}_{p^n}$, restricted on X_k to values in $\mathbb{F}_{p^k}$. We take n minimal, so that each X_k is a proper subset of X. Moreover we can assume that $X_k = \{x \in X \mid (\forall f \in R')(f(x) \in \mathbb{F}_{p^k})\}$, and then (see [Macintyre-Rosenstein], Lemma 6) the set $E_k = \{f \in R' \mid f^2 = f$ and $(\forall x \in X_k)(f(x) = 0)\}$ is first-order definable in R'. Since the open set $X - X_k$ is non empty, there is a non empty clopen set $S \subset X - X_k$. The characteristic function f of S is a non trivial (i.e. $\neq 0,1$) idempotent of R' which belongs to E_k. By q.e. it is clear that all the non trivial idempotents of R' satisfy the same formulas, so that 1-f also belongs to E_k. Thus each X_k is empty, so that R' is the ring $C(X,\mathbb{F}_{p^n})$ of all continuous functions from X to $\mathbb{F}_{p^n}$. Thus the ring R' satisfies $x^{p^n} = x$ and there is an embedding $\mathbb{F}_{p^n} \to R'$. It is clear that the rings verifying these two conditions form an elementary class; we will call these rings the <u>p^n-rings</u>. R is thus a p^n-ring, so R is itself of the form $C(Y,\mathbb{F}_{p^n})$ for some boolean space Y, because of the following extension of the Stone representation ([Arens-Kaplansky], Corollary 2 of Thm 5.1 together with Thm 2.2) : if R is any p^n-ring then its structure space Y (the set of its maximal ideals with the Stone topology) is a boolean space; moreover if we fix an embedding $\mathbb{F}_{p^n} \xrightarrow{\varphi} R$ then each $M \in Y$ determines a unique morphism $R \xrightarrow{\varphi_M} \mathbb{F}_{p^n}$ such that $\ker \varphi_M = M$ and $\varphi_M \varphi$ is the identity on $\mathbb{F}_{p^n}$, and then the application $a \longmapsto f_a$ with $f_a(M) = \varphi_M(a)$ is an isomorphism from R to $C(Y,\mathbb{F}_{p^n})$. Now, since the atoms of the boolean algebra of idempotents of R are the characteristic functions of the isolated points of Y and since all the non trivial idempotents of R satisfy the same formulas, it is clear that Y is either a Cantor space (i.e. a boolean space without isolated points) or a discrete space with only 1 or 2 points. So we have proved that if R is a ring of characteristic p without nilpotent elements and with q.e., then R is either an algebraically closed field or a ring of the form $C(Y,\mathbb{F}_{p^n})$ where Y is either a Cantor space or a discrete space of cardinality 1 or 2.

Our analysis will be complete if we show that all the rings $C(Y,\mathbb{F}_{p^n})$ with Y as required above have q.e., but this is an immediate consequence of the following result due to Cherlin :

suppose Y is a boolean space and M is a model; if the boolean algebra of clopen sets of Y has q.e. and if in M each formula is equivalent to an atomic formula, then C(Y,M) has q.e.[1] (the proof uses an analogue of the Feferman-Vaught theorem for C(Y,M)). Remarks : (i) when Y is a discrete space of cardinality 1 or 2, $C(Y, \mathbb{F}_{p^n})$ coincides with $\mathbb{F}_{p^n}$ or $\mathbb{F}_{p^n} \times \mathbb{F}_{p^n}$; (ii) by the representation theorem for p^n-rings, the rings of the form $C(Y, \mathbb{F}_{p^n})$ where Y is a Cantor space are exactly the atomless p^n-rings (i.e. the p^n-rings in which the idempotents form an atomless boolean algebra); but all these rings are elementarily equivalent (by the analogue of Feferman-Vaught theorem), thus the theory of atomless p^n-rings has q.e.; this theory is thus the model completion of the theory of rings of characteristic p which satisfy $x^{p^n} = x$ (since each ring R of characteristic p satisfying $x^{p^n} = x$ extends to a p^n-ring $R_1 = R \otimes_{\mathbb{F}_p} \mathbb{F}_{p^n}$ which extends to an atomless p^n-ring $R_2 = C(2^\omega, R_1)$).

It is time to conclude our discussion :

<u>THM 1</u>. The rings without nilpotent elements with q.e. are :

(i) in characteristic 0 : the algebraically closed fields of characteristic 0;

(ii) in characteristic p : the algebraically closed fields of characteristic p, the finite fields of characteristic p, the products of two identical finite fields of characteristic p, and the atomless p^n-rings ($n \geqslant 1$);

(iii) in characteristic $p_1 \dots p_m$ (a product of distinct primes) : the products $R_1 \times \dots \times R_m$ where each R_i is a

---

[1] it is understood that M has the discrete topology.

ring of characteristic p_i in the list prescribed by (ii).

3. The case of semi-simple rings.

In this case again the characteristic is either 0 or a product of distinct prime numbers, so that the problem is reducible to the characteristic p case.

LEMMA 2. Let R be a ring of characteristic p with q.e.; if a is a nilpotent element of R, then $a^3 = 0$.

Proof :

if $a^3 \neq 0$, then $a^k = 0 \neq a^{k-1}$ for some natural number $k > 3$. Either $\frac{k}{2}$ or $\frac{k+1}{2}$ is a natural number n. Put $\varepsilon_1 = a^n$, $\varepsilon_2 = a^{n+1}$. It is easy to check that ε_1, ε_2 are linearly independent over $\mathbb{F}_p$ and that $\varepsilon_i\varepsilon_j = 0$ $(i, j = 1, 2)$. Thus the subring $\mathbb{F}_p[\varepsilon_1,\varepsilon_2]$ has an automorphism σ such that $\sigma(\varepsilon_1) = \varepsilon_1$, $\sigma(\varepsilon_2) = \varepsilon_1 + \varepsilon_2$. By q.e., $(R,\varepsilon_1,\varepsilon_2) \equiv (R,\varepsilon_1,\varepsilon_1+\varepsilon_2)$, i.e. $(R,a^n,a^{n+1}) \equiv (R,a^n,(1+a)a^n)$. Since 1+a is invertible in R, we conclude that $a^{n+1} = ba^n$ for some invertible element b of R, so that $a^{n+1} = ba^n$, $a^{n+2} = b^2a^n$, ... are all $\neq 0$, a contradiction. ⊣

It is proved in [Rose] that the matrix ring $M_2(\mathbb{F}_p)$ has q.e.. We will show that there is no other ring with q.e. in the class of semi-simple rings of characteristic p containing (nonzero) nilpotent elements. Indeed let R be such a ring. By Lemma 1 (iii) we can assume that R is algebraic over $\mathbb{F}_p$. Now (by [Jacobson], Thm 1 p. 237 together with Prop. 1 p. 210) any semi-simple ring R algebraic over $\mathbb{F}_p$ and with nilpotent elements contains a family of matrix units e_{ij} $(i,j = 1,2)$, i.e. nonzero elements satisfying $e_{ij}e_{k\ell} = \delta_{jk}e_{i\ell}$ where δ_{jk} is the Kronecker delta, and then $uRu \cong M_2(e_{11}Re_{11})$ where $u = e_{11} + e_{22}$. In our case, $u = 1$ (suppose $u \neq 1$; then u and e_{11} are non trivial idempotents of R, so they have the same first order properties; in particular, R contains a family of matrix units e'_{ij} such that $e_{11} = e'_{11} + e'_{22}$, and then $e_{11}Re_{11} \cong M_2(e'_{11}Re'_{11})$, so that $uRu \cong M_4(e'_{11}Re'_{11})$, in contradiction with Lemma 2). So R is of the form $M_2(B)$. B is semi-simple, algebraic over $\mathbb{F}_p$, and has no nilpotent elements (suppose B has nilpotent elements; then B contains a family of matrix units f_{ij}; put $a = \begin{pmatrix} f_{12} & f_{21} \\ 0 & f_{12} \end{pmatrix}$ and check that $a^4 = 0 \neq a^3$, in contradiction with Lemma 2). Thus B is commutative and so is the center of R. Moreover B has no non trivial idempotents (because R has a non central idempotent $\begin{pmatrix} 1 & 0 \\ 0 & 0 \end{pmatrix}$). Since B has q.e. (as a definable subring of R), Thm 1 (ii) shows that B is an algebraically closed or a finite field. By Lemma 1 (i) and a compactness argument, B is necessarily a finite field. We conclude by the argument used in the proof of Lemma 4.4 of [Rose] : the subring $\{\begin{pmatrix} x & 0 \\ 0 & x^p \end{pmatrix} \mid x \in B\}$ is an isomorphic image of B, thus (by q.e.) all its elements are in the center of R; this means that $x^p = x$ for $x \in B$, thus $B = \mathbb{F}_p$.

From this discussion we get our main result :

<u>THM 2</u>. The semi-simple rings with q.e. are :

(i) in characteristic 0 : the algebraically closed fields of characteristic 0;

(ii) in characteristic p : the algebraically closed fields of characteristic p, the finite fields of characteristic p, the products of two identical finite fields of characteristic p, the atomless p^n-rings ($n \geq 1$), and the matrix ring $M_2(\mathbb{F}_p)$;

(iii) in characteristic $p_1 \dots p_m$ (a product of distinct primes) : the products $R_1 \times \dots \times R_m$ where each R_i is a ring of characteristic p_i in the list prescribed by (ii).

REFERENCES

R.F. ARENS and I. KAPLANSKY, Topological representation of algebras, Trans. Amer. Math. Soc. 63 (1948), 457-481.

Ch. BERLINE, Rings which admit elimination of quantifiers, J. Symb. Logic, to appear.

C.C. CHANG and H.J. KEISLER, Model theory, North-Holland (1973).

N. JACOBSON, Structure of rings, Amer. Math. Soc. (1964).

A. MACINTYRE, On ω_1-categorical theories of fields, Fund. Math. 71 (1971), 1-25.

A. MACINTYRE, K. MCKENNA and L. VAN DEN DRIES, Elimination of quantifiers in algebraic structures, preprint.

A. MACINTYRE and J.G. ROSENSTEIN, $\aleph_0$-categoricity for rings without nilpotent elements and for boolean structures, J. Algebra 43 (1976), 129-154.

B.I. ROSE, Rings which admit elimination of quantifiers, J. Symb. Logic 43 (1978), 92-112.

EXISTENTIALLY CLOSED MODULES : TYPES AND PRIME MODELS

Elisabeth BOUSCAREN
UNIVERSITE PARIS VII

It is known that the theory of modules over a fixed unitary ring R has a model completion if and only if the ring R is coherent [E.S], that is if all finitely generated submodules of finitely presented modules are finitely presented. Let T_R^* be this model completion, i.e. the theory of existentially closed R-modules.

The question of the existence of prime models for T_R^* arises naturally. The only previous result was the following [E.S], [S] ; if R is noetherian, there is a prime model unique up to isomorphism over any set of parameters. This result fits in a general model theoretic framework (as in the case of differentially closed fields) since it is now known that the condition R noetherian is equivalent to T_R^* is totally transcendental (this can for example be deduced from more general results about modules of S. Garavaglia [G1], [G2]).

We have been looking at the case when R is not noetherian ; more precisely we have studied the case of a commutative Von Neumann regular ring, which gives an example when there is no prime model over the empty set.

The main result is the following :

Let R be a commutative Von Neumann regular ring, the following are equivalent :

i) the Boolean algebra of idempotents of R is atomic

ii) there is a unique prime model over any set of parameters

iii) the theory T_R^* is quasi totally transcendental (i.e. for all set A, the one types of $S_1(A)$ ranked by the Morley Rank are dense in $S_1(A)$).

This is obtained by characterizing, by algebraic conditions, in the general case of a coherent ring R, the types of T_R^* that are ranked. More precisely, a one-type p of T_R^* over a module A is determined by an ideal I_p of R and a homomorphism f_p from I_p into A ; most properties of p will be reduced to properties of the ideal I_p in the lattice of ideals of R.

To get these characterizations, we follow the presentation introduced by D. Lascar and B. Poizat in their paper "An introduction to forking" [L.P.].

0. PRELIMINARIES.

We assume throughout this paper that R is a unitary coherent ring, "module" means unitary left R-module and "ideal" means left ideal.
If $\lambda_1,\ldots,\lambda_n \in R$, we write $(\lambda_1\ldots\lambda_n)$ for the (left) ideal generated in R by $\{\lambda_1,\ldots,\lambda_n\}$.
We sometimes write $\bar{a} \in A$ instead of $(a_1,\ldots,a_n) \in A^n$.

0.1. *Satisfaction of a finite system of linear equations and inequations.*

Let S be a finite system of equations in one variable, with parameters in a module A,

$$S : \lambda_1 x = a_1 \wedge\ldots\wedge \lambda_n x = a_n \quad , \quad a_1,\ldots,a_n \in A \text{ and } \lambda_1,\ldots,\lambda_n \in R$$

We know (see eg. [E.S]) that S has a solution in an extension of A iff the following condition holds :

$$\text{for all } \alpha_1,\ldots,\alpha_n \in R \text{ , if } \sum_{i=1}^{n} \alpha_i \lambda_i = 0, \text{ then } \sum_{i=1}^{n} \alpha_i a_i = 0 .$$

This can be proved using the amalgamation property of modules. We write it $P_R(\lambda_1,\ldots,\lambda_n) = P(a_1,\ldots,a_n)$ (P for presentation).
The same proof shows that if S' is a system of equations and inequations,

$$S' : \lambda_1 x = a_1 \wedge\ldots\wedge \lambda_n x = a_n \wedge \mu_1 x \neq b_1 \wedge\ldots\wedge \mu_m x \neq b_m$$

where $\lambda_1,\ldots,\lambda_n, \mu_1,\ldots,\mu_m \in R$ and $a_1,\ldots,a_n, b_1,\ldots,b_m \in A$,
a sufficient condition for S' to have a solution in an extension of A is the following :

$$P_R(\lambda_1,\ldots,\lambda_n) = P(a_1,\ldots,a_n) \text{ and for all } j,\ 1 \leq j \leq m,\ \mu_j \notin (\lambda_1,\ldots,\lambda_n) .$$

0.2. *Axiomatization of the theory of existentially closed modules, T_R^* :*

T_R is the theory of modules over R in the usual language, $\mathcal{L}_R = \{+,0,\lambda\}_{\lambda\in R}$.
We give the idea of a short proof of the existence of a model completion of T_R, assuming R is coherent, which gives us directly a simple axiomatization.
As R is coherent, the sentence "$P_R(\lambda_1,\ldots,\lambda_n) = P(x_1,\ldots,x_n)$" is first-order.
Let T_R^* be the following theory :

$$T_R^* = T_R \cup \{\varphi_{\lambda_1,\ldots,\lambda_n,\mu_1,\ldots,\mu_m}\} \text{ for all } \lambda_1,\ldots,\lambda_n,\mu_1,\ldots,\mu_m \text{ in } R,$$

such that for all j, $1 \leq j \leq m$, $\mu_j \notin (\lambda_1,\ldots,\lambda_n)$,

where $\varphi_{\lambda_1,\ldots,\lambda_n,\mu_1,\ldots,\mu_m}$ is the following.

$$\forall z_1 \ldots \forall z_n \ \ \forall y_1 \ldots \forall y_m \ [(P_R(\lambda_1 \ldots \lambda_n) = P(z_1 \ldots z_n))$$

$$\Longrightarrow \exists x(\lambda_1 x = z_1 \wedge \ldots \wedge \lambda_n x = z_n \wedge \mu_1 x \neq y_1 \wedge \ldots \wedge \mu_m x \neq y_m)]$$

It can easily be shown by back and forth between two ω-saturated models of T_R^*, that T_R^* is complete and has elimination of quantifiers. All existentially closed modules are models of T_R^*, and T_R^* is model complete, so T_R^* is the model companion of T_R and in fact (T_R has the amalgamation property) the model completion of T_R.

I. THE SPACE OF ONE-TYPES OF T_R^*.

We use the definitions and theorems in [L.P.], to describe forking in T_R^* and characterize the types of T_R^* ranked by the Rank U and the Morley Rank.
We reduce most definitions to the case of a stable theory, T_R^* being stable from the general result that all theories of modules are stable [F.1].
We follow the usual convention that all the models of T_R^* we consider are elementary substructures of a saturated model S of inaccessible cardinal, and all sets of parameters are subsets of S. We can assume without any loss of generality that all sets of parameters are modules, by replacing a given set by the submodule it generates in S.

Types of T_R^* : All types will be complete one-types.
Let A be a module, $S_1(A)$ is the set of one-types of T_R^* over A, with the topology generated by the following basis of clopen sets :

$\langle\varphi(x,\bar{a})\rangle = \{p \in S_1(A) \ ; \ p \vdash \varphi(x,\bar{a})\}$ for all $\bar{a} \in A$ and all formulas $\varphi(x,\bar{y})$ of our language.
$S_1(A)$ is Hausdorff and compact.
$S_1(\phi) = S_1(\{0\})$ is the set of pure one-types.
As T_R^* has elimination of quantifiers, a basis for the topology is in fact given by the following formulas :

$$\langle\lambda_1 x = a_1 \wedge \ldots \wedge \lambda_n x = a_n \wedge \mu_1 x \neq b_1 \wedge \ldots \wedge \mu_m x \neq b_m\rangle$$

where $\lambda_1,\ldots,\lambda_n,\mu_1,\ldots,\mu_m \in R$, $a_1,\ldots,a_n$, $b_1,\ldots,b_m \in A$ are such that $P_R(\lambda_1,\ldots,\lambda_n) = P(a_1,\ldots,a_n)$ and for all j, $1 \leq j \leq m$, $\mu_j \notin (\lambda_1,\ldots,\lambda_n)$.

I.1. Description of types-Sons and heirs.

We associate with a type p, an ideal I_p of R and a homomorphism f_p in the following way :

PROPOSITION 1.

Let A be a module and p a consistent set of sentences of the language $\mathcal{L}_R \cup \{x\}$, with parameters in A, extending T_R^* ; p is a type of T_R^* if and only if there is an ideal I_p of R and a homomorphism f_p from I_p into A, necessarily unique, such that :

$$\text{if } \lambda \in I_p \quad p \vdash \lambda x = f_p(\lambda)$$

$$\text{if } \mu \notin I_p \quad p \vdash \mu x \neq a \quad \text{for all } a \in A.$$

Proof : Let p be a type of T_R^*, the set $\{\lambda \in R \; ; \exists a_\lambda \in R, p \vdash \lambda x = a_\lambda\}$ is an ideal of R and the map $f_p : \lambda \longmapsto a_\lambda$ is a homomorphism. Conversely, as T_R^* has elimination of quantifiers, if f is a homomorphism from J, ideal of R, into A, the following set of sentences p is complete :

$$p = T_R^* \cup \{\lambda x = f(\lambda)\}_{\lambda \in J} \cup \{\mu x \neq a\}_{\mu \notin J, a \in A} \quad .$$

If p is a pure type of $S_1(\phi)$, f_p is always the trivial homomorphism and there is a one-one correspondance between $S_1(\phi)$ and the set of ideals of R.
A type P of $S_1(A)$ is realized in A iff $I_p = R$.

PROPOSITION 2.

Let A,B be two modules, $A \subset B$, p a type of $S_1(A)$ and q a type of $S_1(B)$, q is a son of p (i.e. extension of p) if and only if :

(i) I_q contains I_p

(ii) f_q extends f_p

(iii) the inverse image of A, $f_q^{-1}(A)$ is equal to I_p.

The last condition implies that Ker f_q = Ker f_p.

Proof : q is a son of p iff $q\restriction_A = p$ iff for all $\lambda \in R$ and for all $a \in A$, $q \vdash \lambda x = a \leftrightarrow p \vdash \lambda x = a$.
If $\lambda \in I_p$, we must have $q \vdash \lambda x = f_p(\lambda)$, which is equivalent to (i) and (ii).
If $\lambda \notin I_p$, we must have $q \vdash \lambda x \neq a$ for all $a \in A$, which, assuming (i) and (ii), is equivalent to (iii).

Remark : Note that if p is a type over A, we can extend p to a type q over any $B \supset A$ such that $I_q = I_p$ and $f_q = f_p$.

PROPOSITION 3.

Let A be a module, and p a type of $S_1(A)$, for any ideal J of R, containing I_p, there is a module B containing A, and a type q of $S_1(B)$ such that q is a son of p and I_q is equal to J.

Proof : By amalgamation of J and A over I_p.
We can always extend q to a son of p,q' over an existentially closed module containing B, with $I_{q'} = I_q$ and $f_{q'} = f_q$, and we have Proposition 3 with the added condition that B is a model of T_R^*.

Heirs-definition : Let M,N be models of a complete theory T, N > M, p a type of $S_1(M)$, q a son of p in $S_1(N)$; q is a heir of p if and only if, for all $\bar{m}$ in M, all $\bar{n}$ in M and for all formula $\varphi(x,\bar{y},\bar{z})$, if $q \vdash \varphi(x,\bar{m},\bar{n})$, there is $\bar{m}'$ in M such that $p \vdash \varphi(x,\bar{m},\bar{m}')$.
T is a stable theory iff a type over a model M has exactly one heir over any elementary extension of M.

PROPOSITION 4.

Let M,N be models of T_R^*, M < N, and p a type of $S_1(M)$; the heir of p in $S_1(N)$ is the unique son q of p, such that I_q is equal to I_p.

Proof : From the characterization of a son of p (Prop. 2), p has only one son over any N > M, q such that $I_q = I_p$.
We know p has a heir over N, and if $I_q \supsetneq I_p$, there is $\lambda \in I_q$, $\lambda \notin I_p$ such that $q \vdash \lambda x = n$, $n \in N$, while for all $m \in M$, $p \vdash \lambda x \neq m$; from the definition, q cannot be a heir of p.

I.2. The fundamental order of T_R^* and the Rank U.

We describe now the fundamental order of T_R^* and show it is the reverse order of the lattice of ideals of R. This will then enable us to characterize the types of T_R^* ranked by the Rank U.
Let p be a type over a model M of a theory T, and $\varphi(x,\bar{y})$ a formula, $\varphi(x,\bar{y})$ is said to be represented in p if there is $\bar{m}$ in M such that $p \vdash \varphi(x,\bar{m})$.
The following preorder is defined on types over models of T : $p \in S_1(M)$, $q \in S_1(N)$, $p \geq q$ if every formula represented in p is represented in q. The quotient of this preorder by the equivalence relation $p \sim q$ = "$p \geq q$ and $q \geq p$" is the fundamental order of T. If T is a stable theory and p is a type over a set A, all the sons of p over models of T, which are maximal for the fundamental order among the sons of p over models of T, are in the same equivalence class, which is called the bound of p (p has always a maximal son over any model containing A).

Definition of forking : Let p be a type over a set A, and q a son of p over B, $B \supset A$, q is a non-forking extension of p if and only if the bound of p is equal to the bound of q.
This presentation of forking [L.P.] is equivalent to the presentation introduced by S. Shelah [Sh1].

LEMMA 1.

Let M,N be models of T_R^*, p a type of $S_1(M)$ and q a type of $S_1(N)$,

i) if Ker f_q = Ker f_p and $I_p = I_q$, then $p \sim q$

ii) if Ker f_q = Ker f_p and $I_p \subset I_q$, then $p \geq q$.

Proof : i) From our hypothesis, $f_p(I_p)$ is isomorphic with $f_q(I_q)$. Let P be a saturated model of T_R^* in which M and N are embedded, and let p_1, q_1 be the heirs of p and q over P. There is an automorphism of P, extending the isomorphism between $f_p(I_p)$ and $f_q(I_q)$, which exchanges p_1 and q_1. Two isomorphic types are equivalent and a type is equivalent to its heir.

ii) Let r be a son of p over a model of T_R^* such that I_r is equal to I_q. There is such a type from Prop. 3. From i), $r \sim q$ and as r is a son of p, $p \geq r$.

PROPOSITION 5.

Let M,N be models of T_R^*, p a type of $S_1(M)$, q a type of $S_1(N)$

i) $p \geq q$ iff Ker f_p = Ker f_q and $I_p \subset I_q$

ii) $p \sim q$ iff Ker f_p = Ker f_q and $I_p = I_q$

iii) $p > q$ iff Ker f_p = Ker f_q and $I_p \subsetneq I_q$.

Proof : If two types p and q are related in the fundamental order, they have the same restriction to the empty set, which means that Ker f_p = Ker f_q. The proposition follows from the lemma and the definition of the fundamental order.

The following theorem [L.P.] will give us the types ranked by the Rank U : Let T be a stable theory, p a type over a set A, p is ranked by the Rank U ($U(p) < \infty$) if and only if the set of predecessors of the bound of p in the fundamental order of T is well founded. The Rank U, which was introduced by D. Lascar is the smallest notion of Rank and a theory is superstable iff all types are ranked by the Rank U.

LEMMA 2.

Let A be a module, p a type of $S_1(A)$, q a son of p over a model of T_R^* containing A, q is in the bound of p if and only if I_q is equal to I_p.

Proof : From Prop. 5, q will be maximal for the fundamental order if and only if its associated ideal I_q is minimal among the possible ideals corresponding to sons of p, that is iff $I_q = I_p$.

A type p over a module A has only one non-forking extension to any B containing A, the unique extension q of p with associated ideal equal to I_p ; if p is ranked this will also be its unique extension with the same rank.

PROPOSITION 6.

Let A be a module, p a type of $S_1(A)$; p is ranked by the Rank U if and only if the module R/I_p is noetherian.

Proof : The sets of predecessors of the bound of p is well founded if and only if there is no infinite ascending chain of ideals in R, containing I_p.

The Rank U of a type p will only depend on the place of the ideal I_p in the lattice of ideals of R.
If p is ranked by Rank U and q is such that $I_q \supsetneq I_p$, then $U(q) < U(p)$.

I.3. The types of T_R^* ranked by the Morley Rank - Isolated types.

The existence of prime models is linked with the Morley Rank in the following way : if all the one-types of a theory T are ranked by the Morley Rank (T is totally transcendental), there is a prime model unique up to isomorphism over any set of parameters [M] [Sh2] . But a weaker condition gives the same result : it is enough for T to be quasi totally transcendental i.e. for all set A, the types of $S_1(A)$ ranked by the Morley Rank are dense in $S_1(A)$ [Sa].

We give the definition by induction of the Morley Rank which we are going to use :
Let T be a complete theory, p a type of $S_1(T)$ over a set of parameters A,

- $RM(p) \geq 0$

- if α is a limit ordinal $RM(p) \geq \alpha$ if $RM(p) \geq \beta$ for all $\beta < \alpha$

- $RM(p) \geq \alpha+1$ if p has a son q over a set B, which is accumulation point in $S_1(B)$ of types of $RM \geq \alpha$.

If $RM(p) \geq \alpha$ for all α, we write $RM(p) = \infty$; if not, $RM(p) = \alpha$, where α is the first ordinal such that $RM(p) \not\geq \alpha+1$.
If $RM(p) = \alpha$, p is isolated in $S_1(A)$ among the types of $RM \geq \alpha$.

PROPOSITION 7.

Let A be a module, p a type of $S_1(A)$, p is ranked by the Morley Rank if and only if

(*) the ideal I_p is finitely generated and the module R/I_p is noetherian.

The Morley Rank of p is then equal to the Rank U of p.

(*) is equivalent to : every ideal containing I_p is finitely generated.

Proof : If p is ranked by the Morley Rank, p is ranked by the Rank U and (Prop. 6) R/I_p is noetherian.
We show that if I_p is not finitely generated, p is accumulation point in $S_1(A)$ of types of infinite RU (and therefore of infinite RM) and cannot be ranked by RM.
Let $<\varphi>$ be a basic open set of $S_1(A)$, $p \in <\varphi>$

$$<\varphi> = <\lambda_1 x = a_1 \wedge \ldots \wedge \lambda_n x = a_n \wedge \mu_1 x \neq b_1 \wedge \ldots \wedge \mu_m x \neq b_m>,$$

$a_1,\ldots,a_n, b_1,\ldots,b_m \in A$, $\lambda_1,\ldots,\lambda_n, \mu_1,\ldots,\mu_m \in R$ and for all j, $1 \leq j \leq m$, $\mu_j \notin (\lambda_1\ldots\lambda_n)$.
As p is in $<\varphi>$, I_p contains $(\lambda_1\ldots\lambda_n)$ but it is not itself finitely generated. Take q in $S_1(A)$, such that I_q is $(\lambda_1\ldots\lambda_n)$ and f_q is the restriction of f_p to $(\lambda_1\ldots\lambda_n)$; q is in $<\varphi>$ and $U(q) = \infty$ as R/I_q is not noetherian (I_p/I_q is not finitely generated).

Conversely, assume that I_p satisfies condition (*), p is then ranked by RU. We show by induction on the rank of p that RM(p) = U(p).
If U(p) = 0, p is realized in A and RM(p) = 0.
If $U(p) = \alpha$, we show that p and all the sons of p are isolated among the types of $RM \geq \alpha$. Let q be a son of p over B containing A : If $I_q \supsetneq I_p$, U(q) < U(p) and by induction hypothesis $RM(q) = U(q) < \alpha$, and q cannot be accumulation point of types of $RM \geq \alpha$. If $I_p = I_q$ (in particular for p itself), let $\lambda_1,\ldots,\lambda_n$ be the generators of I_q in R, and $a_1,\ldots,a_n$ in A such that $q \vdash \lambda_1 x = a_1 \wedge \ldots \wedge \lambda_n x = a_n$.
Let $<\varphi>$ be the following open set of $S_1(B)$:

$$<\varphi> = <\lambda_1 x = a_1 \wedge \ldots \wedge \lambda_n x = a_n> .$$

q is in $<\varphi>$ and $<\varphi>$ isolates q among the types of $RM \geq \alpha$: let r be any type of $S_1(B)$, also in $<\varphi>$. If $I_r = (\lambda_1\ldots\lambda_n)$, $f_r = f_q$ and r = q ; we can therefore assume that $I_r \supsetneq (\lambda_1\ldots\lambda_n) = I_p$, but then U(r) < U(p) and by induction hypothesis $RM(r) = U(r) < \alpha$.
From the definition of the Morley Rank, $RM(p) \ngeq \alpha+1$ and as $RU(p) \leq RM(p)$, $RM(p) = \alpha$.

We get as a corollary this result we have already mentioned : the theory T_R^* is totally transcendental iff it is superstable iff the ring R is noetherian.

Remark : T_R^* has α_T finite iff R is artinian.
We will also need the characterization of the isolated (or principal) types of $S_1(\phi)$ which we give here without proof.

PROPOSITION 8.

Let p be a type of $S_1(\phi)$, p is isolated in $S_1(\phi)$ if and only if the ideal I_p

satisfies the following condition :

I_p is finitely generated and there is a finite sequence of ideals, $J_1,\ldots,J_n$, necessarily unique, pairwise incomparable for the inclusion relation, containing I_p and such that any ideal containing I_p contains one at least of these ideals. We call these ideals the successors of I_p.

II. EXISTENCE OF PRIME MODELS WHEN R IS COMMUTATIVE VON NEUMANN REGULAR.

A ring R is Von Neumann regular if for all $a \in R$, there is $x \in R$ such that $axa = a$. From now on we assume that R is commutative Von Neumann regular. We are going to use the following properties [B1] :

- R is coherent
- R is without nilpotent elements
- All prime ideals are maximal and any ideal is the intersection of the maximal ideals which contain it.
- Every element is equal to the product of an idempotent and a unit, every finitely generated ideal is principal and generated by an idempotent.

Stone space of R : Spec R is the set of maximal ideals of R, carrying the topology generated by the following basis of clopen sets :

$$O_e = \{M \in \text{Spec } R \ ; \ e \notin M\} \text{ for all } e \text{ idempotent of } R.$$

Spec R is Hausdorff compact. If B(R) is the Boolean algebra of idempotents of R, the following properties of Spec R are consequence of properties of Boolean algebras :

- an isolated point of Spec R is a principal maximal ideal, that is an ideal generated by the complement of an atom of B(R)
- the isolated points are dense in Spec R if and only if B(R) is atomic.

Notations : Let e,f be idempotents of R, we write $e \leq f$ if $ef = e$ (or $e \in (f)$) and $1-e$ for the complement of e in B(R).

Since every finitely generated ideal is principal, we need only consider formulas of a certain type : let A be a module and φ a consistent formula with parameters in A

$$\varphi : \lambda_1 x = a_1 \wedge \ldots \wedge \lambda_n x = a_n \wedge \mu_1 x \neq b_1 \wedge \ldots \wedge \mu_m x \neq b_m$$

where $\lambda_1,\ldots,\lambda_n,\mu_1,\ldots,\mu_m \in R$, $a_1,\ldots,a_n,b_1,\ldots,b_m \in A$, $P_R(\lambda_1\ldots\lambda_n) = P(a_1\ldots a_n)$ and

for all j, $1 \le j \le m$, $\mu_j \notin (\lambda_1 \dots \lambda_n)$.
The formula φ is equivalent to a formula φ' of the following type :

$$\varphi' : ex = a' \wedge f_1x \neq b_1' \wedge \dots \wedge f_mx \neq b_m'$$

where $e, f_1, \dots, f_m \in B(R)$, $a', b_1', \dots, b_m' \in A$, $ea' = a'$ and for all j, $1 \le j \le m$, $f_j \notin (e)$.

We now apply to this case the characterizations we have given of types ranked by the Morley Rank and of isolated types.

PROPOSITION 9.

Let A be a module and p a type of $S_1(A)$, p is ranked by the Morley Rank if and only if the ideal I_p is equal to R or is an intersection of a finite number of principal maximal ideals.
If I_p is the intersection of n distinct principal maximal ideals, RM(p) = n.

Proof : If I_p is equal to R, p is realized in A and RM(p) = 0.
From Prop. 7, if p is ranked by the Morley Rank, all ideals containing I_p are finitely generated. As an ideal of R is the intersection of the maximal ideals that contain it, I_p must be an intersection of principal maximal ideals, that is of ideals generated by the complement of an atom of B(R). This intersection has to be finite as an ideal contained in infinitely many maximal ideals, is included in at least one not finitely generated maximal ideal : given an ideal J, let us consider the following subset of Spec R $F = \{M \in \text{Spec } R ; J \subset M\}$. F is closed in Spec R, therefore it is compact and, if it is infinite, must contain at least one non isolated point, that is one non principal maximal ideal.
Conversely, we show by induction that if $I_p = \bigcap_{i=1}^{n} (1-a_i)$ with $a_1, \dots, a_n$ distinct atoms of B(R), then RM(p) = n.
If $I_p = (1-a)$, I_p is maximal, p is ranked by RM and any forking extension of p is realized. Therefore U(p) = 1 = RM(p).
If $I_p = \bigcap_{i=1}^{n} (1-a_i)$, I_p is the principal ideal generated by $(\prod_{i=1}^{n} 1-a_i)$. It is easy to check that the ideals containing I_p are exactly the principal ideals $\bigcap_{i \in I} (1 - a_i)$ where $I \in 2^n$, $I \neq n$. By induction hypothesis, any forking extension of p is of Rank U < n, but p has a forking extension of Rank U equal to n-1, therefore U(p) = n and as p satisfies condition (*) of Prop. 7, RM(p) = U(p) = n.

If the Boolean Algebra of idempotents of R is atomless, the only types ranked by RM are the types of Rank 0.

PROPOSITION 10.

It the Boolean algebra of idempotents of R is atomic, the theory T_R^* is quasi totally transcendental.

Proof : Let A be a module, $\langle\varphi\rangle$ a basic open set of $S_1(A)$, we show that there is in $\langle\varphi\rangle$ a type ranked by the Morley Rank.
We know we can assume φ to be of the following type :

$$\langle\varphi\rangle = \langle ex = m_o \wedge f_1x \neq m_1 \wedge \ldots \wedge f_n(x) \neq m_n\rangle$$

$e, f_1, \ldots, f_n \in B(R)$, $em_o = m_o$ and for all j, $1 \leq j \leq m$, $f_j \notin (e)$.
B(R) is atomic, therefore the ideals generated by the complement of an atom are dense in Spec R and for all i, $1 \leq i \leq m$, there is an atom a_i such that $e \in (1-a_i)$ and $f_i \notin (1-a_i)$.
We consider the following type p of $S_1(A)$:

$$\begin{cases} I_p = \bigcap_{i=1}^{n} (1-a_i) \\ f_p : I_p \longrightarrow A \quad \text{such that} \quad f_p(\lambda) = \lambda m_o \,. \end{cases}$$

This definition is consistent as $(\prod_{i=1}^{n} 1-a_i)\, m_o = m_o$ and p is in $\langle\varphi\rangle$ as $p \vdash ex = e(\prod_{i=1}^{n} 1-a_i)x = em_o = m_o$ and by construction, for all i, $1 \leq i \leq n$, f_i is not in I_p, therefore $p \vdash f_ix \neq m_i$.
By Prop. 9 the type p is ranked by the Morley Rank.

COROLLARY 1.

If B(R) is atomic, T_R^* has a prime model unique up to isomorphism over any set of parameters.

Proof : There is a Morley prime model over any set [Sa] and the proof of the uniqueness of prime models for totally transcendental theories [Sh2] can be extended to this case.

We now look at the case when B(R) is not atomic that is when there is an idempotent of R with no atom beneath it.

PROPOSITION 11.

If the Boolean algebra of idempotents of R is not atomic, the isolated types are not dense in $S_1(\phi)$.

Proof : Let e be an idempotent of R such that 1-e has no atom beneath it and let

$<\varphi>$ be the following open set of $S_1(\phi)$:

$$<\varphi> = <ex = 0 \wedge x \neq 0> .$$

There is no isolated type in $<\varphi>$:
p is in $<\varphi>$ iff $I_p \neq R$ and $I_p \supset (e)$; p will be isolated (Prop. 8) iff I_p is finitely generated and, in the lattice of ideals of R, I_p has a finite number of successors. If p is in $<\varphi>$ and p is isolated, there is f in B(R), $f \geq e$ and $I_p = (f)$. A successor of I_p would be an ideal (g) such that $g \in B(R)$, $g > f$ and there is no h in B(R), $g > h > f$; g-f would then be an atom of B(R) such that $g\text{-}f \leq 1\text{-}e$.

COROLLARY 2.

If B(R) is not atomic, T_R^* has no atomic models and if R is countable, T_R^* has no prime model over the empty set.

Proof : A model M is atomic iff for all $\bar{m}$ in M the type of $\bar{m}$ is isolated. This implies that the isolated types are dense in $S_1(\phi)$.
If a theory T is countable a model of T is prime over the empty set iff it is atomic.

Further results : An explicit structure for prime models and extension to non commutative rings.

Let N be a semi-simple module, i.e. a direct sum of copies of R/M, where M is a maximal ideal of R [B2]. N is a model of T_R^* if and only if the maximal ideals appearing in the decomposition of N are dense in Spec R.

From this result we get the fact that T_R^* has no prime model when B(R) is not atomic, with no restriction on the cardinality of R, as well as the following :

Assume B(R) is atomic, let A be a module, the prime model of T_R^* over A, M(A), has the following structure :

$$M(A) \cong A \oplus \bigoplus_{J \in F_A} R/_J^{(\varepsilon_J)} \quad \text{with} \quad \varepsilon_J = \begin{cases} \omega \text{ if } R/J \text{ is finite} \\ 1 \text{ otherwise} \end{cases}$$

where F_A is the set of principal maximal ideals of R annihilating only a finite number of elements of A.

There is also a straightforward generalization of §II to the case of non commutative strongly regular rings (or non commutative regular Von Neumann rings without nilpotents).

REFERENCES.

[B1] N. BOURBAKI - Algèbre commutative, Chap. 1, Hermann, Paris 1961.

[B2] N. BOURBAKI - Algèbre, Chap. 8, Hermann, Paris 1958.

[E-S] P. EKLOF - G. SABBAGH - Model completions and modules, Annals of Math. Logic Vol. 2 N°3 (1971), 251-295.

[F.1] E. FISHER - Powers of saturated modules, abstract, J. Symb. Logic Vol. 37 (1972) 777.

[G.1] S. GARAVAGLIA - Direct product decomposition of theories of modules, J. Symb. Logic Vol. 44 N°1 (1979) 77-88.

[G.2] S. GARAVAGLIA - Decomposition of totally transcendental modules, Preprint.

[L-P] D. LASCAR and B. POIZAT - An introduction to forking, J. Symb. Logic Vol. 44 N°3 (1979).

[M] M.D. MORLEY - Categoricity in power, Trans. Am. Math. Soc., Vol. 114 (1965) 514-538.

[S] G. SABBAGH - Sous-modules purs, existentiellement clos et élémentaires, C.R. Acad. Sc. Paris 272 (1971) Ser. A 1289-1292.

[Sa] G.E. SACKS - Saturated Model Theory, Benjamin, Reading, Mass., 1972.

[Sh.1] S. SHELAH - Classification Theory and the number of non-isomorphic models, North-Holland, Amsterdam, 1978.

[Sh.2] S. SHELAH - Uniqueness and characterization of prime models over sets for totally transcendental first-order theories, J. Symb. Logic 37 N°1 (1972).

Rings of Continuous Functions: Decision Problems[1]

Gregory Cherlin
Rutgers University
New Brunswick, N.J. 08903

Abstract. $R = C(X;\mathbb{R})$ is the ring of continuous functions from the topological space X to the real field

Theorem I. If X is a nondiscrete metric space then second order arithmetic is interpretable in R.

Theorem II. If X is the Stone-Cech compactification of a discrete set then the theory of R is decidable.

Introduction.

The problem studied in the present paper is the determination of the class of topological spaces X such that the ring $C(X;\mathbb{R})$ has a decidable theory. Our methods appear adequate in all naturally occurring cases, but the definitive topological criterion remains elusive.

Our proof of Theorem I uses a simple coding trick applicable to a broad class of rings, including the rings of smooth functions $C^{(n)}(X;\mathbb{R})$ for $n \leq \infty$ with X a differential manifold. This coding also provides a good deal of extra information about definability in such rings, yielding as a byproduct:

Corollary 1. For $0 \leq m < n \leq \infty$ $C^{(m)}(\mathbb{R};\mathbb{R}) \not\equiv C^{(n)}(\mathbb{R};\mathbb{R})$.

Furthermore the proof of Theorem I yields the undecidability of $C(X;\mathbb{R})$ whenever X has a nonisolated point with a metrizable neighborhood. On the other hand if X is discrete, or more generally if $C(X;\mathbb{R})$ contains only locally constant functions, then $C(X;\mathbb{R})$ is decidable by the Feferman-Vaught Theorem or one of its extensions. At this point we are naturally led to ask about the situation in the case of the ring $C(\beta X;\mathbb{R})$ with X a discrete space.

Now the general philosophy behind the Feferman-Vaught-Mostowski idea is that in favorable cases it is possible to reduce "global" assertions to "local" assertions, or in more precise terms: statements about sections of sheaves can sometimes be reduced to statements about elements of the stalks and certain open subsets of the base space. Even our proof of Theorem I is entirely compatible with this point of view, as we take care to describe our coding in a way which d rives the undecidability

---

1. This research was supported by the NSF Grant MCA 76-06484.

of $C(X;\mathbb{R})$ back to the undecidability of the ring of germs at a suitable point of X. In the case of $C(\beta X;\mathbb{R})$ the situation is somewhat different. It is necessary to consider essentially global statements, prototypically: "$\lim_{x\to\infty} f(x) = 0$", which can be expressed ring-theoretically in $C(\beta X;R)$. However, it still turns out to be possible to make an exhaustive list of all the "irreducible" global statements, and thus to prove Theorem II.

It seems that the decidability of the real field has aspects of interest to some computer scientists, and perhaps the decidability of $C(\beta\mathbb{N};\mathbb{R})$, the ring of bounded sequences of reals, is also of some relatively concrete interest. In any case we have produced some reasonably effective machinery for determining whether a ring $C(X;\mathbb{R})$ is decidable.

My thanks go to Macintyre and Winkler for discussions in connection with Theorem I, and to Weispfenning for discussions of the Feferman-Vaught Theorem. My proof of Theorem II uses an extension of Weispfenning's formalism, which lends itself nicely to the sort of explicit quantifier elimination required here.

Theorem I and Corollary 1 are proved in the first part, which is agreeably brief. Theorem II and some closely related analogs are proved in the second part, which is unpleasantly long, but which resists intelligible compression. The two parts of the paper are entirely independent of one another.

Part I. Undecidable Rings of Continuous Functions

1. Preliminaries

We let X be a metric space, and R is the ring of continuous functions $C(X;\mathbb{R})$. If X is discrete then R is decidable [1]. Suppose therefore that $p \varepsilon X$ is a nonisolated point of X, and that G_p is the ring of germs of continuous functions at p, that is $G_p = R/I_p$ where I_p is the ideal of functions which vanish near p. If J_p is the ideal of germs corresponding to functions which vanish at p then we have surjections:

$$\text{germ} : R \to G_p$$

$$e: \ G_p \to G_p/J_p \simeq \mathbb{R} \ \text{ canonically.}$$

Using e we may speak of the value of one or more germs at p, even though we cannot easily compare the values of a single function at two different points p and q (since the canonical isomorphism $G_p/J_p \simeq G_q/J_q$ is not in general definable over R).

For $f \varepsilon R$ let $Z(f)$ be the zero-set of f in X.

Lemma A. The following notions are definable in R:

1. $Z(f) = \phi$
2. $Z(f) \cap Z(g) = \phi$
3. $Z(f) \subseteq Z(g)$
4. $Z(f) = Z(g)$
5. $f \varepsilon I_p$.

Parts 1-4 come from [4], and are easily verified for completely regular spaces X. The point p may be coded by any element of R whose zero-set is $\{p\}$.

Corollary. The ring G_p is interpretable over R.

Since we are going to prove that ${}^2\mathbb{N}$ is interpretable over G_p, we need an analog of Lemma A for G_p. Consequently we will speak of the germ of a set (such as a zero-set) at p. Furthermore for $g \varepsilon G_p$ define

$$O(g) = \text{germ [interior } (Z(f))] \text{ where } g = \text{germ } f.$$

Lemma B. The following notions are definable in G_p:

1. $g \varepsilon J_p$
2. $O(g) = \phi$
3. $O(g) \cap O(h) = \phi$

4. $O(g) \subseteq O(h)$
5. $O(g) = O(h)$.

<u>Proof</u>:

1. J_p consists of the noninvertible elements.
2. $O(g) = \phi$ iff g is not a zero-divisor.
3. $O(g) \cap O(h) = \phi$ iff $O(g^2+h^2) = \phi$.
4. $O(g) \subseteq O(h)$ iff $\forall k\ [O(h) \quad O(k) = \phi \Longrightarrow O(h) \quad O(g) = \phi]$.

As is explained in detail in [4] this lemma allows us to interpret the complemented lattice generated by the set germs $O(g)$ in the structure G_p.

2. Interpretation of ${}^2\mathbb{N}$ over G_p

Set $\mathbb{R}(p) = G_p/J_p \simeq \mathbb{R}$. We intend to code arbitrary closed subsets of $\mathbb{R}(p)$ by elements of G_p. We will then be able to define the set $N \subseteq \mathbb{R}(p)$ via the induction axiom and then automatically provide an interpretation of ${}^2\mathbb{N}$ over G_p (since all subsets of $\mathbb{N}$ are closed in $\mathbb{R}(p)$).

It is useful to extend G_p to the ring of quotients G_p^* obtained by inverting all non-zero-divisors. Then G_p^* is interpretable over G_p (and G_p may be retained as a distinguished subset).

Now fix an element $a \in R$ such that $Z(a)$ is a union of disjoint closed balls clustering at p. Let $b = \text{germ}\ (a)$. For $c \in G_p^*$ let $c[p]$ be the set of real numbers of the form $g(p)$ where g satisfies:

$$(*) \qquad g \in G_p \text{ and } O(g-c) \cap O(b) \neq \phi.$$

Then $c[p]$ is a closed set in $\mathbb{R}(p)$ and we claim that any closed set $C \subseteq \mathbb{R}(p)$ is coded in this way by some $c \in G_p$; to find c start with a countable dense subset $D \subseteq C$ and choose functions f, g so that:

$$D = \{r : p \in \text{closure } [(f/g)^{-1}(r) \cap O(b)]\}.$$

Then let $c = \text{germ } f/\text{germ } g$.

In particular $\mathbb{N} \subseteq \mathbb{R}(p)$ is coded by some $c \in G_p$, and all subsets of $\mathbb{N}$ are coded by elements of G_p. Thus:

<u>Theorem</u>. ${}^2\mathbb{N}$ is interpretable over G_p.

The interpretation of ${}^2\mathbb{N}$ over G_p depends on the parameter b as well as the parameter c actually coding the set $\mathbb{N}$. These parameters are eliminable. First consider all closed sets coded as above relative to any parameter b. Then we have coded all closed sets. Next look at a specific closed set which is a subgroup of $\mathbb{R}(p)$ containing 1 as its smallest element. Then this is $\mathbb{N}$.

Thus what we have is a way of associating to each nonisolated point p of X a canonical coding of ${}^2\mathbb{N}$ in G_p, the whole affair being definable over R without parameters. In particular $Th({}^2\mathbb{N})$ is reducible to Th(R) as well as to each $Th(G_p)$ for p not isolater. This establishes Theorem I.

3. Subrings of $C(X;\mathbb{R})$

The preceding arguments evidently apply to many subrings $R' \subseteq C(X;\mathbb{R})$. For example if R' is a subring defined by very general growth conditions then $C(X;\mathbb{R})$ will in general be interpretable in a ring of quotients of R'. Similarly if X is a differential manifold and $R' = C^{(n)}(X;\mathbb{R})$ with $0 \leq n \leq \infty$ then the arguments of section 2 still apply.

Indeed in this case R' is still well-supplied with zero-sets so that only the actual coding of closed subsets of $\mathbb{R}(p)$ requires attention. Letting G_p = germs of R' at p and G_p^* = the localization of G_p at non-zero-divisors, the essential point is that for any countable set D there are functions $f,g \in R'$ and a germ b so that

$$D = \{r:p \in \text{closure } [(f/g)^{-1}(r) \cap O(b)]\}.$$

One may first construct f and g, both going to zero at p, so that f/g oscillates wildly and takes on all the values in D on open sets clustering at r, and then pick a suitable germ b afterwards. (Compare the remarks after the theorem of §2).

4. A definability theorem

Theorem. Let p,q lie in the same connected component of X and $R = C(X;\mathbb{R})$. Then the canonical isomorphism:

$$\mathbb{R}(p) \simeq \mathbb{R}(q)$$

is definable over R.

Proof: We may assume that X is connected, since restriction to a connected component is interpretable over R. Then the following notions are definable over R:

1. f is integer valued (hence constant)
2. f is rational valued (hence constant)
3. f is constant: namely, for any rational constant r we have $r \leq f$ everywhere or $f \leq r$ everywhere.

Now observe that the constant functions define the canonical isomorphism from $\mathbb{R}(p)$ to $\mathbb{R}(q)$.

5. Corollaries

Corollary A. The set of degrees of theories of rings $C(X;\mathbb{R})$ with X nondiscrete metric is the cone above $\deg(\mathrm{Th}({}^2\mathbb{N}))$.

Corollary B. For X a compact 2-manifold

$$\deg \mathrm{Th}(C(X;\mathbb{R})) = \deg \mathrm{Th}({}^2\mathbb{N}).$$

Corollary C. For X a differential manifold and for $0 \leq m < n \leq \infty$ $\mathrm{Th}(C^{(m)}(X;\mathbb{R})) \neq \mathrm{Th}(C^{(n)}(X;\mathbb{R}))$.

Corollary D. The models of $\mathrm{Th}(C(\mathbb{R};\mathbb{R}))$ are classified up to isomorphism by the isomorphism types of models of $\mathrm{Th}({}^2\mathbb{N})$.

Proof sketches:

A. Any degree can be the degree of unsolvability of a nondiscrete metric space in the sense of [3], as any of various simple codings demonstrate, and in general $\deg \mathrm{Th}(C(X;\mathbb{R}))$ will be the join of $\deg(\mathrm{Th}(\mathrm{Top}\ X))$ and $\deg({}^2\mathbb{N})$, for X a nondiscrete metric space.

B. Compact 2-manifolds are all coded in ${}^2\mathbb{N}$.

C. According to the theorem of §4 we may identify the various fields $\mathbb{R}(p)$ for p varying over any connected component of X. Then $C^{(m)}(X;\mathbb{R})$ satisfies:

"There is a basis for the topology of X consisting of sets U such that there are [coordinate] functions $u_1,\dots,u_n$ satisfying: the map $p \to (u_1(p),\dots,u_n(p))$ is a bijection onto $\mathbb{R}^n$ and the map $f \to f^*$ defined by $f(p) = f^*(u_1(p),\dots,u_n(p))$ induces an isomorphism between $R|_U$ and $C^{(m)}(\mathbb{R}^n;\mathbb{R})$."

This is a statement in the language of R.

D. Let $X = \mathbb{R}$ as a topological space. Then $C(X;\mathbb{R})$ satisfies:
"For some f: f is a homeomorphism from X to $\mathbb{R}$.
After identifying X with $\mathbb{R}$ via f I turn out to be $C(X;\mathbb{R})$ as coded by ${}^2\mathbb{N}$."

Thus any model of the theory of $X(X;\mathbb{R})$ has the same property relative to the model of ${}^2\mathbb{N}$ which it encodes.

6. Other target fields

There are results analogous to Theorem I in which $\mathbb{R}$ is replaced by other locally compact fields. The use of C rather than $\mathbb{R}$ introduces trivial variations. If a totally disconnected nondiscrete locally compact field K is used in place of $\mathbb{R}$ then in order to ensure a rich supply of functions in $C(X;K)$ it is appropriate to consider base spaces X which are also nondiscrete locally compact totally disconnected spaces. Then one may again obtain without difficulty the results of §§1,2 (interpretability of ${}^2\mathbb{N}$ in nonisolated stalks) and the first remark in §3 (passage to subrings determined by growth conditions). Most of the remaining material becomes meaningless in this context, while Corollary D fails for the ring $C(K;K)$ if K is totally disconnected, because in this case the theory is closed under direct products.

Part II. The Decidability of $C(\beta Y;\mathbb{R})$

We will prove the decidability of the ring $C(\beta Y;\mathbb{R})$ for Y infinite discrete via a primitive recursive elimination of quantifiers in a two-sorted language. It turns out that for Y_1, Y_2 uncountable discrete $C(\beta Y_1;\mathbb{R}) \equiv C(\beta Y_2;\mathbb{R})$ [3]. We first treat the simpler case in which Y is countable, and then indicate the very similar treatment of uncountable Y. We note that the topology of βY will not interest us, and we view $C(\beta Y;\mathbb{R})$ consistently as the ring of bounded functions on Y; equivalently we think of the ring $C(Y;\mathbb{R})$ of all real-valued functions on Y equipped with the predicate "bounded".

Section A. The countable case

We will give the proof of Theorem II at length for the case of countable discrete Y. Subsequently we will discuss the modifications necessary in the case of uncountable discrete Y, and other generalizations.

1. A formal system

It is convenient to use an extended first order language with two sorts of variables and enough functions and predicates for an elimination of quantifiers. Our language is specified schematically as follows (we include information regarding intended interpretations):

Language L

1. Variables:

   1.1 Boolean variables: $e_1, e_2, \ldots$ denoting [characteristic functions of] subsets of Y

   1.2 Function variables: $f_1, f_2, \ldots$ denoting functions from Y to $\mathbb{R}$.

2. Nonlogical Constants:

   2.1 Boolean notions:

   2.1.1 Boolean constants: 0, 1

   2.1.2 Boolean functions: $\cup$, $\cap$, -

   2.1.3 Boolean predicates: =, $\subseteq$; Small, Cardn - meaning "finite" or "finite of size n"

2.2 Local function notions:

2.2.1 Function operations Θ: for each definable function t in the language of real closed fields we have a symbol Θ for the functional defined on $C(Y;\mathbb{R})$ by:

$$[\Theta(f_1,\ldots,f_k)](y) = t(f_1(y),\ldots,f_k(y)).$$

2.2.2 Truth value operators $e(F;f_1,\ldots,f_n)$: for each formula $F(x_1,\ldots,x_n)$ in the language of real closed fields we have an operation $e(F;f)$ to be interpreted as follows:

$$e(F;f_1,\ldots,f_n) = \{y:F(f_1(y),\ldots,f_n(y)\}.$$

2.3 Global function notions:

2.3.1 Function predicates a, ℓ, m, c, m^k to be interpreted momentarily (they arise in the process of eliminating function quantifiers).

In addition we make use of two special abbreviations, which are also construed as part of the formal apparatus of L:

3. Additional Notation

3.1 Relativization $[\phi(f_1,\ldots,f_k)]e$: if ϕ is a formula of L and e is a Boolean variable or truth value operator then

$$[\phi(f_1,\ldots,f_k)]e$$

denotes the relativization of ϕ to e, that is the assertion that $C(\beta e;\mathbb{R})$ satisfies $\phi(f_1|e,\ldots,f_k|e)$ where $f|e$ is the restriction of f to e.

3.2 The connective $+$: for any formulas ϕ_1,ϕ_2 of L we define:

$$\phi_1+\phi_2 = \text{"}\exists e([\phi_1]e \ \& \ [\phi_2](-e))\text{."}$$

We conclude our description of L with the intended interpretations of the predicates a, ℓ, m, c. The predicates m^k form the subject matter of the next paragraph.

The predicates a, ℓ, m, c are defined and interpreted as follows:

1. $af = \text{"}\exists g\ e(\text{"}x_1x_2 = 1\text{"}; f,g) = 1\text{"}$ means:
   f is bounded away from zero
2. $\ell f = \text{"}\forall e\ \ \text{Small}(e) \vee [\neg af]e\text{"}$ means:
   $\lim_{y\to\infty} f(y) = 0.$

3. mf = "$\neg(af+\ell f)$" means:

f has mixed behavior.

4. $c(f,g)$ = "$f \leq g$ & mf & $\forall e\,([\ell f]e \Rightarrow [\ell g]e)$" means:

$f \leq g$ and the behavior of f, though mixed, controls the behavior of g.

2. The predicates m^k

Definition. Define by simultaneous induction on $k \geq 0$:

1. the notion: basic formula of rank k
2. the notion: reducible formula of rank k
3. the predicate $m^{k+1}(f_1,\ldots,f_{k+1})$

This is done as follows:

1. The basic formulas $B(f_1,\ldots,f_n)$ of rank k are conjunctions of formulas of the form:

i. $f_j = f_{j+1}$ or $f_j < f_{j+1}$

ii. af_j or ℓf_j

iii. $c(f_j,f_{j+1})$

and at most one formula of the form:

iv. $m^k(f_{j_1},\ldots,f_{j_k})$ with $j_1 < \ldots < j_k$ (for $k \geq 1$).

We require that each variable f_j which occurs in B should occur exactly once in a formula of type (i) (if $j < n$) and in addition should occur either:

a. just once more, in a formula of type (ii)

or b. once in the formula $c(f_{j-1},f_j)$ and possibly once more in the formula $c(f_j,f_{j+1})$

or c. once in a formula of type (iv) and possibly once more within the formula $c(f_j,f_{j+1})$.

As there is no real loss in generality in assuming that $f_1<\ldots<f_n$, we will usually do so in the sequel.

2. A reducible formula of rank k is a formula $R(f_1,\ldots,f_n)$ of the form:

$$B_1^{\sigma_1}(f_1,\ldots,f_n)+\ldots+B_r^{\sigma_r}(f_1,\ldots,f_n)$$

where the B_i are basic of rank at most k and the σ_i represent permutations acting on $\{f_1,\ldots,f_n\}$.

It will be proved in §4 that there are only finitely many reducible formulas of rank k, up to logical equivalence in $C(\beta Y;\mathbb{R})$. This observation is necessary for the final clause of our definition:

$$3.\quad m^{k+1}(f_1,\dots,f_{k+1}) = \text{"}f_1<\dots<f_{k+1}\,\&\,(\mathop{\&}_{\substack{R \text{ reducible}\\ \text{of rank } k}} \neg R(f_1,\dots,f_{k+1}))\text{"}$$

Notice that $m^1 f \equiv mf$.

3. The inductive interpretation of m^k

We have noted that the definition of §2 depends on the corollary of §4. Since the theorem of this paragraph depends on the theorem of §4 (and conversely) it is clear that the above definition together with the results of this and the next paragraph are to be treated within a single simultaneous induction on k.

Theorem. For $k \geq 0$ $m^{k+1}(f_1,\dots,f_{k+1})$ is equivalent to:

$$\forall\varepsilon > 0\, \exists\delta < \varepsilon\ (\delta > 0\&[m^k(f_1,\dots,f_k)]e(\text{"}\delta < x \leq \varepsilon\text{"};f_{k+1})).$$

(For $k = 0$ take $[m^k]e$ to mean: Small(e).)

Proof: Taking negations and recalling the definition of m^{k+1} we must prove the equivalence of:

$$1.\quad \bigvee_{\substack{R \text{ reducible}\\ \text{of rank } k}} R(f_1,\dots,f_{k+1})$$

and

$$2.\quad \exists\varepsilon > 0\ \forall\delta < \varepsilon\ (\delta > 0 \Longrightarrow [\neg m^k(f_1,\dots,f_k)]e(\text{"}\delta < x \leq \varepsilon\text{"};f_{k+1})).$$

$1 \Longrightarrow 2.$ Suppose $Rf_1\dots f_{k+1}$ where $R = B_1+\dots+B_r$ and the B_i are basic of rank at most k. Partition $1 = e_1+\dots+e_r$ so that

$$[B_i f_1\dots f_{k+1}]e_i.$$

It will suffice if we prove (2) on each e_i, so assume $R = B = B_i$, $e_i = 1$.

Since any $k-1$ functions satisfy a reducible formula of rank at most $k-1$ (at worst: m^{k-1}), it follows that if B contains a conjunct ℓf_j, af_j, or $cf_{j-1}f_j$, with $j \leq k$, then ε may be taken arbitrarily. If none of these occur then we have one of the following conjuncts in B:

a. ℓf_{k+1}: then ℓf_j for all j and ε is arbitrary

b. af_{k+1}: take $\varepsilon < \inf_1 f_{k+1}(y)$

c. $c(f_k,f_{k+1})$: with ε arbitrary and $0 < \delta < \varepsilon$ we have $[af_k]e("\delta < x \leq \varepsilon";f_{k+1})$, which yields the desired conclusion.

<u>2 $\Longrightarrow$ 1</u>. Fix $\varepsilon > 0$ so that:

$$\forall\delta < \varepsilon\ \delta > 0 \Longrightarrow [\neg m^k f_1 \dots f_k]e("\delta < x \leq \varepsilon";f_{k+1}).$$

On the set $e = f_{k+1}^{-1}(\varepsilon,\infty)$ we of course have af_{k+1}, and then $f_1 \dots f_{k+1}$ satisfy a reducible formula. Hence we may assume $-e = 1$. Choose a sequence ε_n decreasing monotonically to zero with $\varepsilon_0 = \varepsilon$ so that the sets:

$$e_n = \{y:\varepsilon_{n+1} < f_{k+1}(y) \leq \varepsilon_n\}$$

are infinite; if this is impossible then already

$$af_{k+1}$$

and the result is clear.

Let the basic formulas of rank $k-1$ be:

$$\{B_i: i \leq i \leq K\}.$$

By the corollary in §4 K is finite. We also need the theorem of §4:

(*) $\qquad B_i + B_i \equiv B_i \quad$ for $\quad i \leq K.$

It then follows that there are partitions:

$$e_n = e_n^1 + \dots + e_n^K$$

with each e_n^i empty or infinite, such that for $e_n^i \neq 0$:

$$[B_i(f_1,\dots,f_k)]\ e_n^i \quad \text{for all} \quad n.$$

Set $e^i = \bigcup_n e_n^i$, so that $e^1,\dots,e^K$ partition 1.

If $e^i \neq 0$ then $f_1,\dots,f_{k+1}$ satisfy a reducible formula of rank k on e^i, since B_i contains one of the following conjuncts:

a. ℓf_j: hence ℓf_j holds on e^i

or b. $cf_{j-1}f_j (j \leq k)$: hence $cf_{j-1}f_j$ holds on e^i

or c. af_j: hence af_k, and hence $cf_k f_{k+1}$ holds on e^i.

This completes the argument.

<u>4. Basic formulas are idempotent</u>

<u>Theorem</u>. If $Bf_1 \dots f_n$ is basic then $(B+B) \equiv B$.

Proof: As stated in §3 we are proceeding by induction on the rank k of B. We will now proceed within rank k by induction on n.

Case 1. Suppose $Bf_1 \ldots f_n = B'(f_1,\ldots,f_j,\ldots,f_n)$ & H where B' is basic and H is the conjunction of $f_{j-1} < f_j < f_{j+1}$ with one of the formulas ℓf_j, af_j, or $cf_{j-1}f_j$. (Here $1 \leq j \leq n$.)

In this case the following formulas are equivalent:

1. $B + B$
2. B' & $H + B'$ & H
3. $(B' + B')$ & H
4. B' & H

The equivalence of 3 and 4 is given by induction hypothesis, and the equivalence of 2 and 3 is clear in view of the meaning of H.

Case 2. If Case 1 does not apply then $Bf_1 \ldots f_n = m^n(f_1,\ldots,f_n)$ and $n = k$.

Now $m^k + m^k \Longrightarrow m^k$ by the theorem of §3 combined with induction on k. For the converse assume $m^k(f_1,\ldots,f_k)$ and fix ε_j decreasing monotonically to zero so that for the sets

$$e_j = \{y: \varepsilon_{j+1} < f_k(y) \leq \varepsilon_j\}$$

we have:

$$[m^{k-1}(f_1,\ldots,f_{k-1})]e_j \quad \text{for all } j.$$

Since by induction $m^{k-1} \Longrightarrow m^{k-1} + m^{k-1}$ we may partition

$$e_j = e_j^1 + e_j^2$$

so that $[m^{k-1}(f_1,\ldots,f_{k-1})]e_j^i$ $(i = 1,2;\ j<\infty)$. Then take $e^i = \bigcup_j e_j^i$ for $i = 1,2$ and note that

$$[m^k(f_1,\ldots,f_k)]e^i,$$

so that $(m^k + m^k)(f_1,\ldots,f_k)$, as desired.

Corollary. There are up to logical equivalence only finitely many reducible formulas in the variables $f_1,\ldots,f_n$.

This corollary would of course be false if we considered reducible formulas in n terms rather than in n variables. Starting in the next paragraph we will apply the term reducible in this other, broader sense, but we will do so with caution.

5. Statement of the main theorem

Roughly speaking our claim is that each formula in our language is equivalent to a disjunction of reducible formulas. To be more accurate we must take into account the expressive power of function terms and Boolean terms.

Notation. Given Boolean terms $S_1,\dots,S_j$ and $\varepsilon:\{1,\dots,j\} \to \{+1,-1\}$ we let $\varepsilon(\overline{S}) = \bigwedge\{s_i:\varepsilon(i) = +1\} \wedge \bigcap\{-s_1:\varepsilon(i) = -1\}$. Such a term is called a bit of $\overline{S}$.

Given Boolean variables $e_1,\dots,e_m$ and function variables $f_1,\dots,f_n$, we define a generalized reducible formula in the variables $\overline{e};\overline{f}$ to be any conjunction $\underset{b}{\&} F_b$ where:

1. F_b is a formula of one of the forms:
   a. $[Rt_1\dots t_k]b \ \& \ \neg \mathrm{Small}(b)$
   b. $\mathrm{Cardn}(b)$
   c. $\mathrm{Small}(b) \ \& \ \underset{i<n}{\&} \ \neg \mathrm{Cardi}(b)$
2. The terms $t_1,\dots,t_k$ vary over a fixed set of function terms in $f_1,\dots,f_m$
3. b varies over all bits of a fixed finite set of Boolean terms and truth value operators involving at most $\overline{e};\overline{f}$
4. R is a reducible formula.

Finally, we call a formula composite iff it is a disjunction of generalized reducible formulas.

At this point we modify our previous terminology and call a formula $R(t_1,\dots,t_k)$ reducible if the formula $R(f_1,\dots,f_k)$ is reducible in the strict sense.

The Main Theorem. Every formula $F(\overline{e};\overline{f})$ is equivalent to a composite formula in $\overline{e};\overline{f}$.

The main step in the proof of this theorem will be the elimination of function quantifiers. We proceed by induction on the complexity of formulas, with quantifier-free formulas and the passage to disjunctions causing no difficulties. We deal accordingly with negations and the existential Boolean and function quantifiers.

As far as the existential Boolean quantifier is concerned we restrict our attention to a formula:

$\exists e \mathop{\&}_{b} F_b$ (b varies over bits of $e, e_1, \ldots, e_j$ for some $e_1, \ldots, e_j$) with the F_b as above. This formula is equivalent to:

$$(*) \qquad \mathop{\&}_{b'} (F_{b'.e} + F_{b'.-e})$$

where b' varies over the bits of $e_1, \ldots, e_j$. It is clear that (*) is a generalized reducible formula since the sum of reducible formulas is reducible, and because one can split off a finite set of any desired size from an infinite set. This disposes of the elimination of the Boolean quantifier.

We will be occupied with the proof of the main theorem through §14. The reader will notice that our reductions are consistently primitive recursive, and that a complete axiomatization for $\mathrm{Th}(C(\beta Y;\mathbb{R}))$ can be obtained from our proof.

6. Equivalence of reducible formulas

We will show in this paragraph that a conjunction of reducible formulas (in a fixed set of variables) is again reducible. It will then follow easily that the negation of a composite formula is [equivalent to] a composite formula.

Lemma. Let $Bf_1 \ldots f_n$ be basic and $Rf_1 \ldots f_n = \Sigma B_i f_1 \ldots f_n$ reducible. If $B \,\&\, R$ is consistent then $B \equiv R$.

Proof. Suppose first that for some j $B = B'(f_1 \ldots \hat{f_j} \ldots f_n) \,\&\, H$ where H is the conjunction of $f_{j-1} < f_j < f_{j+1}$ and a formula ℓf_j, af_j, or $cf_{j-1}f_j$ $(j \leq n)$. Assuming $B \,\&\, R$ is consistent we can write

$$B_i = B_i'(f_1, \ldots, \hat{f_j}, \ldots, f_n) \,\&\, H$$

for each i. Then $B' \,\&\, \Sigma B_i'$ is consistent, so we may assume inductively that $B' \equiv \Sigma B_i'$. Then:

$$B = B' \,\&\, H \equiv (\Sigma B_i') \,\&\, H \equiv \Sigma(B_i' \,\&\, H) = R.$$

The only remaining case is that in which $B = m^n f_1 \ldots f_n$. It will suffice therefore to prove that for any basic B_1 we have:

$$m^n + B_1 \equiv m^n.$$

The proof that $m^n + B_1 \Longrightarrow m^n$ is immediate on the basis of the inductive interpretation of m^n in §3, using induction on n. Assume conversely that $m^n f_1 \ldots f_n$. The proof of $(m^n + B_1(f_1, \ldots, f_n))$ will break up into a number of cases, depending upon what B_1 says about f_n.

As a preliminary step use the inductive interpretation of m^n to choose a sequence ε_k decreasing monotonically to zero with $\varepsilon_1 = \infty$ so that after setting

$$e_k = \{y: \varepsilon_{k+1} < f_n(y) \leq \varepsilon_k\}$$

we have:

$$[m^{n-1} f_1 \ldots f_{n-1}] e_k.$$

We may ignore the trivial case in which the conjunct ℓf_n occurs in B_1.

<u>Case 1</u>. $B_1 = B_1'(f_1,\ldots,f_{n-1})$ & $f_{n-1} < f_n$ & af_n.

Then partition $e_1 = e_1^1 + e_1^2$ so that:

$$[m^{n-1} f_1 \ldots f_{n-1}] e_1^1 \;\&\; [B_1' f_1 \ldots f_{n-1}] e_1^2.$$

This is possible since $m^{n-1} \Longrightarrow m^{n-1} + B_1'$ by induction hypothesis. Then we have:

$$[m^n f_1 \ldots f_n](e_1^1 \cup \bigcup_{k>1} e_k) \;\&\; [B_1 f_1 \ldots f_n] e_1^2,$$

so that $(m^n + B_1)(f_1,\ldots,f_n)$, as desired.

<u>Case 2</u>. $B_1 = B_1'(f_1,\ldots,f_{n-1})$ & $cf_{n-1}f_n$ & $f_{n-1} < f_n$. Then $B_1' = B_1''$ & $m^i(f_{j_1},\ldots,f_{j_i})$ & $\underset{j_i \leq j < n-1}{\&} c(f_j f_{j+1})$ & $f_{n-1} < f_n$. Let $B_0 = B_1''$ & $m^i(f_{j_1},\ldots,f_{j_i})$ & $\underset{j_i < j \leq n-1}{\&} af_j$. By induction hypothesis we have:

$$[(m^{n-1} + B_0)(f_1,\ldots,f_{n-1})] e_k,$$

so partition $e_k = e_k^1 + e_k^2$ in such a way that:

$$[m^{n-1} f_1 \ldots f_{n-1}] e_k^1 \;\&\; [B_0 f_1 \ldots f_{n-1}] e_k^2.$$

Let $e^i = \bigcup_k e_k^i$ $(i = 1,2)$. Then we have

$$[m^n f_1 \ldots f_n] e^1 \;\&\; [B_1 f_1 \ldots f_n] e^2$$

as desired; the first conjunct is obvious while the second will be equally obvious after verification of the clauses:

$$[cf_j f_{j+1}] e^2 \qquad (j_i \leq j < n).$$

These are quite clear since $f_{j_1} < \ldots < f_n$, $[af_{j_i}] e_k^2$ for each k, and

$\lim_{k\to\infty} (\sup_{e_k} f_n(y)) = 0.$

Case 3. $B_1 = B_1' \ \& \ m^{i+1}(f_{j_1},\ldots,f_{j_i},f_n) \ \& \ f_{n-1} < f_n$. Then let $B_0 f_1 \ldots f_{n-1} = B_1' \ \& \ m^i f_{j_1} \ldots f_{j_i}$. Partition $e_k = e_k^1 + e_k^2$ so that:

$$[m^{n-1} f_1 \ldots f_{n-1}] e_k' \quad \text{and} \quad [B_0 f_1 \ldots f_{n-1}] e_k^2.$$

and set $e^i = \bigcup_k e_k^i$ $(i = 1,2)$. Then we have:

$$[m^n \overline{f}] e^1 \ \& \ [B_1 \overline{f}] e^2.$$

This completes the argument.

Example. If $f < g$ then

$$cfg \equiv cfg + (af \ \& \ ag) + (\ell f \ \& \ \ell g).$$

Theorem. Any two reducible formulas $Rf_1 \ldots f_n$, $Sf_1 \ldots f_n$ are either equivalent or mutually inconsistent.

Proof. We will say that R is maximal iff for every basic formula B such that $R \equiv R + B$, B actually occurs explicitly as a summand of R, and that R is irredundant iff no basic formula occurs as a summand of R with multiplicity exceeding 1. Using the formula $B + B \equiv B$ of §4 we see that any reducible formula is equivalent to a maximal irredundant formula. We may assume therefore that R, S are maximal irredundant, and it will suffice to prove that every summand of R occurs as a summand of S.

Assume then that B is a summand of R and $S = \Sigma B_i$. Fix $f_1,\ldots,f_n$ such that $(R \ \& \ S)(\overline{f})$. Then in particular we have a set e and a partition $1 = \Sigma e_i$ so that:

$$[B\overline{f}]e \quad \text{and for each } i \quad [B_i \overline{f}] e_i.$$

There are reducible formulas R_i such that

$$[R_i \overline{f}]\ (e \cap e_i).$$

Then $B \ \& \ \Sigma R_i$ is consistent, so $B \equiv \Sigma R_i$ by our lemma. On the other hand $S \equiv \Sigma R_i + S'$ for some reducible S', so;

$$B + S \equiv B + \Sigma R_i + S' \equiv B + B + S' \equiv B + S' \equiv \Sigma R_i + S' \equiv S,$$

and thus B occurs in S since S is maximal. This completes the argument.

Corollary 1. A consistent conjunction of reducible formulas in a fixed set of variables is reducible.

Corollary 2. A consistent conjunction of generalized reducible formulas involving a fixed set of terms is a generalized reducible formula.

Corollary 3. The negation of a composite formula is composite.

Proof. It is sufficient, in view of the preceding corollary, to treat the negation of a single generalized reducible formula, and this reduces to the case of a formula of the form:

$$F_b$$

with F_b as in §5. With the help of the above Theorem such formulas are easily seen to be composite (one also uses the finiteness of the set of reducible formulas in a fixed set of variables).

7. The formula $\ell(\Theta(f,\overline{f}))$.

We know that the set of composite formulas is closed under the connectives and Boolean quantification, and we may accordingly devote all our attention to the elimination of the function quantifier. It suffices to remove the function quantifier from an expression of the form:

$$\text{(E)} \qquad \exists f\ B(e_1,\ldots,e_m;\ f,f_1,\ldots,f_n),$$

where B is in general composite. It will be convenient to specify that f ranges over all real-valued functions. Boundedness can be explicitly required when desired by including the clause $a(1/f)$.

Now since the existential quantifier commutes with disjunctions as well as with conjunctions over bits we may assume that B is simply a reducible formula (not necessarily in the variables f, $\overline{f}$ but in some finite set of terms $\Theta_1,\ldots,\Theta_p$ in these variables). Before eliminating the function quantifier we will look more closely at function terms $\Theta(f,\overline{f})$ in connection with familiar special features of the theory of $\mathbb{R}$.

Hypothesis and Notation. We suppose that Θ is an operation symbol corresponding to the algebraic function defined as the $k^{\underline{th}}$ real solution of:

$$(*) \qquad r_d(f,\overline{f})y^d + r_{d-1}(f,\overline{f})y^{d-1}+\ldots+r_1(f,\overline{f})y + r_0(f,\overline{f}) = 0$$

where the r_i are rational functions and for some fixed j we have $r_j(f,\overline{f}) \equiv 1 \geq r_i(f,\overline{f})$ for $0 \leq i \leq d$.

For any ℓ let $\Theta_\ell(f,\overline{f})$ be the following formula in the language of real closed fields:

"The ℓ complex solutions of (*) which lie nearest to zero - counted with multiplicites - include the $k^{\underline{th}}$ real solution of (*)."

We may take Θ_ℓ to be quantifier-free.

Lemma. With the above hypotheses and notation:

1. If $\ell \leq j$ is such that: $(\underset{i<\ell}{\&}\ \ell r_i(f,\overline{f}) \ \&\ ar_\ell(f,\overline{f}))$

then

$$\ell\Theta(f,\overline{f}) \equiv \text{"} \exists e(\mathrm{Small}(e)\ \&\ -e \subseteq e(\Theta_\ell;f,\overline{f}))\text{"}$$

2. $\ell\Theta(f,\overline{f}) \equiv$ "$\forall e \ \underset{\ell \leq j}{\&}\ [(\underset{i<\ell}{\&}\ \ell r_i(f,\overline{f}) \ \&\ ar_\ell(f,\overline{f})) \Longrightarrow \ell\Theta(f,\overline{f})]e$.

Proof.

1. Under the given conditions ℓ roots of (*) tend to zero and the rest are bounded away from zero.

2. The necessity of the given condition follows from 1. If we now assume that $\neg\ell(\Theta(f,\overline{f}))$ and more specifically that $[a\Theta(f,\overline{f})]e$ for some fixed e, then by shrinking e we may arrange that for some $\ell \leq j$:

$$[\underset{i<\ell}{\&}\ \ell(r_i(f,\overline{f}))\ \&\ a(r_\ell(f,\overline{f}))]e,$$

which contradicts the right side of (2).

We may apply these observations to simplify our problem:

Reduction. We may assume in expression (E) above that:

$$B = B(r_1,\ldots,r_d)$$

is a basic formula in the rational functions $r_i(f,\overline{f})$, each of which is globally defined.

Proof. Notice that a, c and the predicates m^k are definable in terms of ℓ using Boolean quantification. Imagine (E) transformed accordingly so that a, c, m^k are eliminated in favor of ℓ. Now every definable function t in the language of real closed fields has a definition of the form:

(T) $$t(\overline{x}) = t_i(\overline{x}) \quad \text{if} \quad h_i(\overline{x})$$

where $h_i(\overline{x})$ are quantifier-free conditions on $\overline{x}$, exactly one of which holds for each $\overline{x}$, and the t_i are functions defined as in (*) above. Hence for any $f, \overline{f}$ 1 may be partitioned so that the preceding lemma applies; in other words we may eliminate the function operations Θ in favor of the various rational functions involved in defining the operations Θ_i corresponding to the functions t_i in (T), at the cost of introducing some additional Boolean quantifiers. We then turn around and remove these Boolean quantifiers, arriving at a composite formula involving only terms corresponding to rational functions, each everywhere defined on an appropriate bit. Finally, as indicated at the beginning of this paragraph, we may consider reducible formulas separately on the different bits. We are then reduced to the problem of eliminating the function quantifier from an expression:

(E) $$\exists f \; B(e_1,\ldots,e_m; f, f_1,\ldots,f_n)$$

where B is reducible and all terms occurring in B are everywhere defined rational functions of $f, \overline{f}$. Since the function quantifier commutes with the connective $+$, we may assume B is basic, as claimed.

8. More on $\ell\Theta(f,\overline{f})$

In this paragraph we intend to show that the rational functions $r_i(f,\overline{f})$ occurring in $B(r_1,\ldots,r_d)$ in the expression (E) can in fact be taken to be of the form:

$$f_i, \quad f_i f, \quad \text{or} \quad f_i/f.$$

This may require minor changes in the notation, but involves no loss of generality.

Lemma. If g, $h > 0$ and $n \geq 0$ then:

1. $\ell(f(g+h)^n) \equiv \ell(fg^n) \;\&\; \ell(fh^n)$
2. $\ell(f/(g+h)^n) \equiv (\ell(f/g^n)) + (\ell(f/h^n))$.

Proof.

1. $\sup(|f|g^n, |f|h^n) \leq |f|(g+h)^n \leq 2^n \sup(|f|g^n, |f|h^n)$
2. $2^{-n} \inf(|f|/g^n, |f|/h^n) \leq |f|/(g+h)^n \leq \inf(|f|/g^n, |f|/h^n)$.

<u>A second reduction</u>. In expression (E) above we may assume that:

$$Bf, f_1, \dots, f_n$$

is a basic formula involving at most the terms f_j, $f_j f$, f_j/f.

<u>Proof</u>. Our analysis proceeds in five steps.

1. Consider first a formula $\ell r(f,\bar{f})$ with r rational. Using definable functions we can write:

$$r(f,\bar{f}) = r_d(\bar{f}) \prod_i (f-\Theta_i)^{k_i} \prod_j ((f-\Theta_j)^2 + \Theta_j')^{\ell_j}$$

with k_i, $\ell_j \in \mathbb{Z}$ and Θ_i, Θ_j, Θ_j' definable functions of $\bar{f}$ (but not of f).

Then applying the above lemma and using Boolean quantifiers we may express $\ell r(f,\bar{f})$ in terms of a number of formulas of the form:

$$\ell(\Theta(\bar{f}) \prod_i (f-\Theta_i(\bar{f}))^{k_i}).$$

2. Consider accordingly a formula of the form:

$$\text{(F)} \qquad \ell(g \prod_i (f-g_i)^{k_i}).$$

Using Boolean quantifiers we may assume that:

$$g_1 < \dots < g_j < f < g_{j+1} < \dots < g_p \qquad \text{(possibly } j=0 \text{ or } p).$$

Now observe:

$$\text{for } i < j\text{:}\quad f-g_i = (f-g_j) + (g_j-g_i)$$
$$\text{for } i > j+1\text{:}\quad g_i-f = (g_i-g_{j+1}) + (g_{j+1}-f)$$

and thus the previous lemma may be applied to express formula (F) in terms of formulas of the form:

$$\text{(F')} \qquad \ell(\Theta(g,\bar{g})(f-g_j)^k(g_{j+1}-f)^\ell) \qquad (k,\ell \in \mathbb{Z}).$$

This is a good moment to recall that Θ, f, g_j, g_{j+1} represent possibly unbounded functions; this is unavoidable even if $f_1, \dots, f_n$ were bounded.

3. Consider a formula

$$\text{(F")} \qquad \ell(g(f-g_1)^k(g_2-f)^\ell) \qquad (k,\ell \in \mathbb{Z}).$$

Assuming $g_1 < f < g_2$ we may replace f by:

$$f' = (f-g_1)/(g_2-g_1).$$

Then our formula becomes:

$$\ell(g(g_2-g_1)^{k+\ell}(f')^k(1-f')^\ell)$$

which is an instance of the formula:

$$\ell(g\ f^k(1-f)^\ell).$$

4. Consider a formula $\ell(g\ f^k(1-f)^\ell)$.

Using Boolean quantifiers we may treat separately the cases:

a. $|f| > 2$: replace $f^k(1-f)^\ell$ by $f^{k+\ell}$

b. $\frac{1}{2} < f \leq 2$: drop the term f^k

c. $-2 \leq f \leq \frac{1}{2}$: drop the term $(1-f)^\ell$

Thus replacing f by $1-f$ if necessary the formula $\ell(g\ f^k(1-f)^\ell)$ is expressible using Boolean quantifiers and specializations of formulas:

(F''') $\qquad \ell g\ f^k \qquad (k \in \mathbb{Z}).$

If $k \neq 0$ then writing f for $g^{1/|k|}$ this formula is equivalent to:

$$\ell gf \ (\text{if } k > 0), \quad \ell g \ (\text{if } k = 0), \quad \ell g/f \ (\text{if } k < 0).$$

5. These steps may be assembled as follows. To begin with the original formula:

(A) $\qquad B(f, f_1, \ldots, f_n)$

is expressed, using Boolean quantification, in terms of a number of formulas of the form

(B) $\qquad \ell r_i(f, \overline{f})$

with the r_i rational. There will also be side conditions $f > \Theta(\overline{f})$ or $f < \Theta(\overline{f})$ for various definable functions occurring at each step of the reduction.

All of the formulas of type (B) may be analyzed simultaneously as in steps 1 and 2; we must take the sequence

$$g_1 < g_2 < \ldots < g_j < f < g_{j+1} < \ldots < g_p$$

to include all parameters arising from the different formulas ℓr_i. It is then possible to make the substitution of step 3 unless $j = 0$ or p, in which case it is unnecessary. The point here is that a single transformation of variables suffices to simplify all of the formulas arising from the various ℓr_i. Then step 4 finishes the reduction, in the sense that the variables of B have the desired form. Of course at this point B may no longer be composite, much less basic, so the simple reduction indicated at the beginning of §7 may need to be repeated (together with the reductions of §§5,6, now largely taken for granted).

This completes our discussion of the second reduction.

At this point our problem is the elimination of the function quantifier from the expression:

(E') $$\exists f(h_1 < f < h_2 \;\&\; B(t_1,\ldots,t_p))$$

where t_i is a function term of the form g_i, $g_i f$, or g_i/f and B is essentially a basic formula, except that for notational convenience we have removed the inequalities from B, since we may simply assume that:

$$0 < t_1 < \ldots < t_j = h_1 < h_2 = t_{j+1} < \ldots < t_p.$$

Possibly $j = 0$ or p (and h_1 or h_2 may be absent). We assume $0 < t_1$ in order to avoid a phalanx of absolute value symbols.

Some further simplifications are possible. We may replace any single term $t_i = g_i f$ or g_i/f by the variable f and alter the other terms accordingly. Other reductions may be made for the sake of notational simplicity. Since $t_1 < \ldots < t_p$ we may assume that no conjunct involving ℓ or a occurs other than ℓt_1 or at_p. Similarly any conjunction $c(t_i,t_{i+1}) \;\&\; c(t_{i+1},t_{i+2}) \& \ldots \& c(t_{i+k-1},t_{i+k})$ may be condensed to $c(t_i,t_{i+k})$. After omitting superfluous terms, our basic B of rank k may be taken to be a conjunction of formulas of the form:

1. ℓt_1
2. $m^k(t_2,t_4,\ldots,t_{2k})$: if this term is present and $t_{2k} \neq g_{2k}$ take $t_{2k} = f$.
3. $c(t_{2i},t_{2i+1})$ $\quad (1 \leq i \leq k)$
4. at_{2k+2}.

Any of these conjuncts may be missing. Notice that $p \leq 2k+2$. It will be convenient to set

$$Q(f,\overline{g}) = \text{"}h_1 < f < h_2 \;\&\; B(f,\overline{g})\text{"}.$$

Thus our problem is to eliminate the quantifier from:

(E') $$\exists f\; Q(f,\overline{g})$$

where Q is normalized as we have described above.

9. The case of rank 0.

The elimination of quantifiers from expressions of type (E') proceeds by induction on the rank of B. If B has rank 0 then there are four cases, all of them easy.

Lemma.

1. If $B(f,\bar{g}) = "\ell f"$ then $\exists f\, Q(f,\bar{g}) \equiv \ell h_1$
2. If $B(f,\bar{g}) = "af"$ then $\exists f\, Q(f,\bar{g}) \equiv ah_2$
3. If $B(f,\bar{g}) = "\ell g_1 f \ \&\ af"$ then $\exists f\, Q(f,\bar{g}) \equiv "\ell g_1 \ \&\ \ell g_1 h_1 \ \&\ ah_2"$
4. If $B(f,\bar{g}) = "\ell g_1/f \ \&\ af"$ then $\exists f\, Q(f,\bar{g}) \equiv "\ell g_1/h_2 \ \&\ ah_2"$

Proof.

1, 2, 4. Clear.

3. The given condition is evidently necessary. Conversely if $\ell(g_1)$ & $\ell(g_1 g)$ & $a(h_2)$ let $\varepsilon = \inf_y h_2(y)$ and choose f satisfying:

$$f < h_2 \ \&\ \max(h_1,\varepsilon/2) < f < \max(2h_1,\varepsilon).$$

Then $h_1 < f < h_2$ & $\ell(g_1 f)$ & $a(f)$.

The quantifier elimination is more complicated when $k = \text{rank } B$ is positive and m^k occurs in B. We will need to develop a way to describe certain infinite partitions within our formalism.

10. Partitionable formulas

Definition. The formula F is said to be:

1. idempotent iff $F + F \equiv F$
2. convex iff $\forall e' \subseteq e \subseteq e''$ $([F]e' \ \&\ [F]e'' \Longrightarrow [F]e)$.
3. partitionable iff for every covering $\{e_n\}$ of Y such that $[F]e_n$ for all n there is a partition $\{e_n'\}$ of Y refining $\{e_n\}$ so that $[F]e_n'$ for all n.

Remark. Basic formulas are idempotent by §4 and are clearly convex.

Definition. For any function f a sequence $\{e(i):i < \omega\}$ of subsets of Y is an f-sequence iff $\lim_e \sup_{e(i)} f = 0$. If the f-sequence is a covering or a partition of Y it is called an f-covering or f-partition.

To prove that basic formulas are partitionable requires an explicit interpretation of m^k.

<u>Theorem</u>. For any functions $f_1,\ldots,f_k$ the following are equivalent:

1. $m^k(f_1,\ldots,f_k)$

2. There is a function: $e:(\bigcup_{\ell\leq k}\omega^\ell) \to$ infinite subsets of Y such that for $\ell \leq k$ and $i_1,\ldots,i_{\ell-1} < \omega$ the sequence $\{e(i_1,\ldots,i_{\ell-1},i):i < \omega\}$ is an $f_{k-\ell+1}$-partition of $e(i_1,\ldots,i_{\ell-1})$ $(\ell > 0)$ or Y $(\ell = 0)$.

<u>Proof</u>.

$1 \Longrightarrow 2$: Proceed by induction on ℓ, starting with $e = Y$ for $\ell = 0$. At stage ℓ construct a function

$$e:\omega^\ell \to \text{ subsets of } Y$$

so that for all $i_1,\ldots,i_{\ell<\omega}$:

$$[m^{k-\ell}(f_1,\ldots,f_{k-\ell})]e(\bar{i})$$

and the sequence $\{e(i_1,\ldots,i_{\ell-1,i}):i < \omega\}$ is an $f_{k-\ell+1}$-partition of $e(i_1,\ldots,i_{\ell-1})$. In view of the inductive interpretation of m^k in §3 the induction step is immediate. For $\ell = k$ we may interpret

$$[m^0]e(\bar{i})$$

as signifying $\text{Small}(e(\bar{i}))$.

$2 \Longrightarrow 1$: With e as specified, prove by downward induction on $\ell \leq k$ that for all ℓ and all $i_1,\ldots,i_{\ell<\omega}$:

$$[m^{k-\ell}(f_1,\ldots,f_{k-\ell})]e(\bar{i}).$$

For $\ell = 0$ this is the desired conclusion.

<u>Theorem</u>: Basic formulas are partitionable.

<u>Proof</u>. For obvious reasons it may be assumed without loss of generality that our basic formula B is simply $m^k(f_1,\ldots,f^k)$ for some $k > 0$.

Let $\{e_n\}$ be a covering of Y such that $[B]e_n$ for all n and let:

$$E_n: \bigcup_{\ell\leq k}\omega^\ell) \to \text{ subsets of } Y$$

be the functions provided by the previous theorem. (In particular $E_n[\omega^k]$ is a partition of e_n.)

Let $\mathcal{C} = \bigcup_n E_n[\omega^k]$. Then $\mathcal{C}$ is a covering of Y by $\aleph_0$ sets.

It is trivial that $\mathcal{C}$ can be refined to a partition $\mathcal{C}'$ of Y by replacing every element e of $\mathcal{C}$ by an infinite subset e' of e. Let the subset associated to $E_n[\bar{1}]$ be denoted $E'_n[\bar{1}]$ and let

$$e'_n = \bigcup_{\omega^k} E'_n[\bar{1}].$$

Then $\{e'_n\}$ refines $\{e_n\}$ and $[B]e'_n$ for all n.

11. The construction of partitions.

We now show how to exploit the expressive capacity of the Boolean quantifier in order to assert the existence of certain f-partitions. This will be the basis for the elimination of the function quantifier.

Definition. For t a term and F a formula set

Part (t:F) = "$\forall e, e'$ if $e \cap e' = 0$ & $[a(t)](e \cup e')$ then $\exists e'' \supseteq e'$ so that $e \cap e'' = 0$ & $[F]e''$ "

Theorem. Let F be an idempotent convex partitionable formula. Let $h > 0$. Then the following are equivalent:

1. Part (h:F)
2. $Y = \bigcup_n e_n$ with $\limsup_n h_e = 0$ and $[F]e_n$.
3. There is a partition $Y = \Sigma e_n$ with $\limsup_n h_{e_n} = 0$ and $[F]e_n$.

Proof. We will show that $3 \Longrightarrow 1 \Longrightarrow 2$.

$1 \Longrightarrow 2$. For $n \geq 1$ let $e^n = \{y : 1/n < h(y) \leq 1/n-1\}$, $\bar{e}^n = \{y : 1/n < h(y)\}$. Then for all n we have:

$$\bar{e}^n \cap e^{n+1} = 0 \quad \text{and} \quad [ah](\bar{e}^n \cup e^{n+1}).$$

By assumption there exist $e_n \supseteq e^{n+1}$ with:

$$e_n \cap \bar{e}^n = 0 \quad \text{and} \quad [F]e_n.$$

Then $Y = \bigcup e_n$, $\limsup_n h_{e_n} = 0$ and $[F]e_n$.

$3 \Longrightarrow 1$. Let $Y = \sum e_n$ be a partition satisfying $\limsup_n h_{e_n} = 0$ and $[F]e_n$ for each n. Assume also that e, e' are given with $e \cap e' = 0$ and $[a\,h](e \cup e')$. Then for some n $e \cup e' \subseteq e_1 + e_2 + \ldots + e_n$. Let

$$e'' = (e_1 + \ldots + e_{n+1}) - e.$$

Then $e'' \wedge e = 0$, $e' \subseteq e''$, and $e_{n+1} \subseteq e'' \subseteq e_1 + \ldots + e_{n+1}$. Since F is idempotent we have $[F](e_1 + \ldots + e_{n+1})$. Then since F is convex we get $[F]e''$, as desired.

12. Quantifier elimination: $t_{2k}=f$, $t_{2k+1} \neq g_{2k+1}f$

Our problem is the elimination of the function quantifier from an expression

$$(E') \qquad \exists f(h_1 < f < h_2 \ \& \ B(t_1,\ldots,t_p))$$

where B is basic of positive rank k and normalized as in §8.

In particular B contains a conjunct $m^k(t_2,t_4,\ldots,t_{2k})$ and $t_{2k} = f$ or $t_{2k} = g_{2k}$. We may write:

$$B(f,\overline{f}) = B_1(f,\overline{f}) \ \& \ m^k(t_2,\ldots,t_{2k}) \ \& \ B_2(f,\overline{f})$$

where B_1 involves at most $t_1,t_2,\ldots,t_{2k-1}$ and B_2 is either empty, $c(t_{2k},t_{2k+1})$, at_{2k+2}, or the conjunction $c(t_{2k},t_{2k+1}) \ \& \ at_{2k+2}$.

Definition.

1. $Q(f,\overline{f})$ = "$h_1 < f < h_2 \ \& \ B(f,\overline{f})$"
2. $Q'(f,\overline{f})$ = "$h_1 < f < h_2 \ \& \ B_1(f,\overline{f}) \ \& \ m^{k-1}(t_1,\ldots,t_{2k-2}) \ \& \ at_{2k}$"
3. $h_1' = \inf\ (2h_1, \frac{1}{2}(h_1+h_2)$
4. $h_2' = \sup(\frac{1}{2}\ h_2,\ \frac{1}{2}(h_1+h_2))$.

Remark. $h_1 < h_1' \leq h_2' < h_2$, $h_1' \leq 2h_1$, $h_2 \leq 2h_2'$.

In this paragraph we will assume that $t_{2k} = f$ and that t_{2k+1}, if present, is not $g_{2k+1}f$. The quantifier elimination in this case conforms to a single pattern, although a multitude of subcases require attention.

As a preliminary observation notice that

$$" \exists fQ(f,\overline{f})" \equiv " \exists fQ'(f,\overline{f}) + (\exists fQ(f,\overline{f}) \ \& \ f \leq 1)".$$

Since rank $Q' <$ rank Q we need only eliminate the quantifier from $\exists f(Q(f,\overline{f}) \ \& \ f \leq 1)$, or in other words in terms of our current notation we assume $h_2 \leq 1$.

Theorem. With the above notation and hypotheses, and in particular assuming $t_{2k} = f$, t_{2k+1} is absent or not of the form $g_{2k+1}f$, and $h_2 \leq 1$:

$$\exists f\ Q(f,\overline{f}) \equiv \text{"}a(u_1) \ \& \ \text{Part}\ (u_2\text{: } \exists f\ Q'\ (f,\overline{f}))\text{"}$$

where u_1, u_2 are defined as follows:

1. If B_2 is empty then $u_1 = 1$, $u_2 = h_1$
2. If $B_2 = c(f,t_{2k+1})$ then $u_1 = 1$ and
   a. if $t_{2k+1} = g_{2k+1}$ then $u_2 = \max(h_1, g_{2k+1})$
   b. if $t_{2k+1} = g_{2k+1}/f$ then $u_2 = g_{2k+1}/h_2$
3. If $B_2 = at_{2k+2}$ and
   a. $t_{2k+2} = g_{2k+2}f$ then $u_1 = g_{2k+2}h_2$, $u_2 = \max(h_1, 1/g_{2k+2})$
   b. $t_{2k+2} = g_{2k+2}/f$ then $u_1 = g_{2k+2}/h_1$, $u_2 = h_1$.
4. If $B_2 = c(f,t_{2k+1})$ & at_{2k+2} and
   a. $t_{2k+2} = g_{2k+2}f$, $t_{2k+1} = g_{2k+1}$ then
      $u_1 = g_{2k+2}h_2$, $u_2 = \max(h_1, g_{2k+1}, 1/g_{2k+2})$
   b. $t_{2k+2} = g_{2k+2}f$, $t_{2k+1} = g_{2k+1}/f$ then
      $u_1 = g_{2k+2}h$, $u_2 = g_{2k+1}/h_2$
   c. $t_{2k+2} = g_{2k+2}/f$, $t_{2k+1} = g_{2k+1}$ then
      $u_1 = g_{2k+2}/h_1$, $u_2 = \max(h_1, g_{2k+1})$
   d. $t_{2k+2} = g_{2k+2}/f$, $t_{2k+1} = g_{2k+1}/f$ then
      $u_1 = g_{2k+2}/h_1$, $u_2 = \max(h_1, g_{2k+1}/h_2, g_{2k+1}/g_{2k+2})$.

<u>Proof</u>.

$\Longrightarrow$: Assume $Q(f,\overline{f})$. Then $a(u_1)$ is clear in all cases. Now partition $Y = \Sigma e_n$ with $\lim_{n} \sup_{e_n} f = 0$ and $[Q'(f,\overline{f})](e_n)$ for each n. Our claim is then that in each case $\lim_{n} \sup_{e_n} u_2 = 0$ as well, so that

$$\mathrm{Part}\,(u_2 : \exists f Q'(f,\overline{f})).$$

For this it suffices to prove:

$$\forall e[au_2 \Longrightarrow a(f)](e).$$

This depends in the various cases on diverse combinations of the following observations:

A. $h_1 < f$

B. When $t_{2k+1} = g_{2k+1}$ use $c(f, g_{2k+1})$.

C. When $t_{2k+1} = g_{2k+1}/f$ use $c(f, g_{2k+1}/f)$ and
   $g_{2k+1}/h_2 < g_{2k+1}/f$

D. When $t_{2k+2} = g_{2k+2}f$ use $f = (g_{2k+2}f) \cdot 1/g_{2k+2}$

E. When $t_{2k+1} = g_{2k+1}/f$ and $t_{2k+2} = g_{2k+2}/f$ then argue as follows:

$$[\ell f]e \Longrightarrow [\ell t_{2k+1} \,\&\, a(t_{2k+2})]e \Longrightarrow [\ell t_{2k+1}/t_{2k+2}]e \text{ so that}$$
$$[\ell g_{2k+1}/g_{2k+2}]e \quad \text{(case 4d above).}$$

$\Longleftarrow$: Assume that a partition $Y = \Sigma e_n$ is given so that $\limsup_n \sup_{e_n} u_2 = 0$ and that for each n we have:

$$[\exists f\, Q'(f,\bar{f})]e_n.$$

It then follows that any function f satisfying:

$$[af \,\&\, h_1 < f < h_2]e_n$$

also satisfies $[Q'(f,\bar{f})]e_n$ (recall the bound $h_2 \leq 1$).

We will define such a function f on a case-by-case basis, with the additional properties specified by B_2. The numbering of the cases in the following definition corresponds to the numbering of cases in the statement of the theorem.

1,2a. $f = \min(h_2', \sup_{e_n} h_1')$ on e_n.

2b,4b. $f = h_2'$

3a. $f = \min(h_2', \max(h_1', 1/g_{2k+2}, 1/n)$ on e_n.

3b. $f = \max(h_1', \min(h_2', g_{2k+2}, \sup_{e_n} h_1'))$ on e_n.

4a. $f = \min(h_2', \max(1/g_{2k+2}, \sup_{e_n} h_1'))$ on e_n.

4c. $f = \min(h_2', \sup_{e_n} h_1', \max(h_1', g_{2k+2}))$ on e_n.

4d. $f = \max(h_1', \min(h_2', ng_{2k+1}, g_{2k+2}))$.

It is then necessary to verify the following facts.

A. $h_1 < f < h_2$, in fact $h_1' \leq f \leq h_2'$.

B. $[a(f)]e_n$ for each n:

Since $[\exists fQ'(f,\bar{f})]e_n$ we have $[a(h_2)]e_n$. $[a(f)]e_n$ follows in cases 1, 2, 3a, and 4a-b. Now when $t_{2k+2} = g_{2k+2}/f$ then

$$\forall f(a(f) \,\&\, f < g_{2k+2}/f \Longrightarrow a(g_{2k+2}))$$

and hence we deduce $[a(g_{2k+2})]e_n$. $[a(f)]e_n$ follows in cases 3b and 4c. A similar argument applies when $t_{2k+1} = g_{2k+1}/f$, disposing of case 4d.

C. $[Q'(f,\bar{f})]e_n$. This follows from A, B.

D. $\limsup_n f_{e_n} = 0$

We know this is true of u_2, and this yields the desired result for f directly in cases 1, 2a, 3, 4a, and 4c. In cases 2b, 4b, and 4d, since $t_{2k+1} = g_{2k+1}/f$ and $f < t_{2k+1}$ we have $f^2 < g_{2k+1}$, so $h_2 \leq \sqrt{g_{2k+1}}$ (without loss of generality) and thus:

$$g_{2k+1}/h_2 \geq \sqrt{g_{2k+1}} \geq h_2.$$

Thus in these cases $u_2 \geq h_2$, and the desired result follows also here.

E. In cases 2 and 4 we have $c(f,t_{2k+1})$:

In other words we claim that $\limsup_n t_{2k+1\,e_n} = 0$. We consider the different subcases:

2a,4a,4c $\qquad u_2 \geq g_{2k+1} = t_{2k+1}$

2b,4b $\qquad t_{2k+1} = g_{2k+1}/f = g_{2k+1}/g_2' \leq 2u_2$

4d $\qquad t_{2k+1} = g_{2k+1}/f$, $f \geq \min(h_2', ng_{2k+1}, g_{2k+2})$,

and $g_{2k+1}/g_2 \leq u_2$, $g_{2k+1}/g_{2k+2} \leq u_2$.

F. In cases 3 and 4 we have at_{2k+2}:

3a,4a,4b $\qquad$ Since $f \geq \min(h_2', 1/g_{2k+2})$ we have

$t_{2k+2} = g_{2k+2}f \geq \min(\frac{1}{2}\, g_{2k+2}h_2, 1)$. Now

$u_1 = g_{2k+2}h_2$ so $au_1 \Longrightarrow at_{2k+2}$.

3b,4c,4d $\qquad$ Since $f \leq \max(h_1', g_{2k+2})$ we have

$t_{2k+2} = g_{2k+2}/f \geq \min(\frac{1}{2}\, g_{2k+2}/h_1, 1)$.

Since $u_1 = g_{2k+2}/h_1$ therefore $au_1 \Longrightarrow at_{2k+2}$.

This completes the proof of the theorem.

13. Quantifier elimination: $t_{2k+1} = fg_{2k+1}$ or $t_{2k} = g_{2k}$

In this paragraph we will treat the remaining cases of quantifier elimination by the methods of the previous paragraph. We begin by assuming that $t_{2k} = f$ and $t_{2k+1} = fg_{2k+1}$, and proceed much as before. The numbering of subcases is a continuation of that used in §12.

Theorem. With the notation and general hypotheses of 12, but assuming that $t_{2k} = f$ and $t_{2k+1} = fg_{2k+1}$, the following are equivalent:

L. $\exists f Q(f,\overline{f})$

R. $(\exists f Q'(f,\overline{f})\ \&\ a(f)) + (a(u_1)\ \&\ \mathrm{Part}\ (u_2$:"$a(1/g_{2k+1})\ \&\ \exists f Q'(f,\overline{f})$") where u_1, u_2 are defined as follows:

2c. If $B_2 = c(f,t_{2k+1})$ then $u_1 = 1$, $u_2 = g_{2k+1}h$

4. If $B_2 = c(f,t_{2k+1})$ & at_{2k+2} and

4e. $t_{2k+2} = g_{2k+2}f$ then $u_1 = g_{2k+2}h_2$ and

$u_2 = \max(g_{2k+1}h_1, 1/f_{2k+2}, g_{2k+1}/g_{2k+2})$.

4f. $t_{2k+2} = g_{2k+2}/f$ then $u_1 = g_{2k+2}/h_1$ and $u_2 = g_{2k+1}h_1$.

Proof.

L $\Longrightarrow$ R: The proof is exactly as before except that when we partition $Y = \Sigma e_n$ so that $\limsup_n {}_{e_n} f = 0$ the clause $c(f,fg_{2k+1})$ ensures that

$$[a\ 1/g_{2k+1}]e_n$$

for all sufficiently large n.

L $\Longleftarrow$ R: Partition $Y = \sum_{n=0}^{\infty} e_n$ so that $\limsup_n {}_{e_n} u_2 = 0$ and:

$$[a(1/g_{2k+1})\ \&\ \exists f Q'(f,\overline{f})]e_n \quad \text{for } n \geq 1$$

$$[\exists f(Q'(f,\overline{f})\ \&\ a(f))]e_0.$$

There is no problem defining f on e_0; for $n \geq 1$ define f on e_n by:

2c. $f = \max(h_1', \min(h_2', \sup_{e_n} h_1', 1/(ng_{2k+1})))$

4e. $f = \min(h_2', \max(h_1', 1/g_{2k+2}, 1/(ng_{2k+1})))$

4f. $f = \max(h_1', \min(h_2', g_{2k+2}, 1/(ng_{2k+1})))$.

We then verify:

A. $h_1 < f < h_2$

B. $[af]e_n$ for $n \geq 1$: this is argued as in §12, using

$[a(1/g_{2k+1})]e_n$.

C. $[Q'(f,\overline{f})]e_n$ $(n \geq 1)$

D. $\limsup_n {}_{e_n} f = 0$.

Since $f < t_{2k+1} = g_{2k+1}f$, $1 < g_{2k+1}$. Since $\limsup_n {}_{e_n} u_2 = 0$ the claim follows by comparing the definition of f with the definition of u_2.

E. $c(f, g_{2k+1}f)$

2c. $f \leq \max(h_1', 1/(ng_{2k+1}))$ on e_n and $g_{2k+1}h_1' \leq 2u_2$.

4e. $f \leq \max(h_1', 1/g_{2k+2}, 1/(ng_{2k+1}))$ and $u_2 \geq g_{2k+1}h_1$,

g_{2k+1}/g_{2k+2}

4f. $f \leq \max(h_1', 1/(ng_{2k+1}))$ and $g_{2k+1}h_1 \leq u_2$.

F. In case 4 we have $a(t_{2k+2})$;

4e. $t_{2k+2} = g_{2k+2}f$ and $f \geq \min(h_2', 1/g_{2k+2})$ while

$u_1 = g_{2k+2}h_2$; thus $a(u_1) \Rightarrow a(t_{2k+2})$.

4f. $t_{2k+2} = g_{2k+2}/f$ and $f \leq \max(h_1', g_{2k+2})$ while

$u_1 = g_{2k+2}/h_1$; thus $a(u_1) \Rightarrow a(t_{2k+2})$.

This disposes of the elimination of function quantifiers apart from the case $t_{2k} = g_{2k}$, which is treated in the following theorem. When $t_{2k} = g_{2k}$ some other term may be taken to be f. There are relatively few cases involved, but they fall into a less coherent pattern.

<u>Theorem</u>. Under the general assumptions of §12, but with $t_{2k} = g_{2k}$, the formula $fQ(f,\overline{f})$ is equivalent to one of the following:

5. If B_2 is empty: $\mathrm{Part}(g_{2k}: ag_{2k} \ \& \ \exists fQ'(f,\overline{f}))$

6. If $B_2 = c(g_{2k},f):(ag_{2k} \ \& \ \exists fQ'(f,\overline{f})) + \mathrm{Part}(h_1: \exists fQ'(f,\overline{f}))$

7. If $B_2 = af{:}ah_2 \ \& \ \mathrm{Part}(g_{2k}: ag_{2k} \ \& \ \exists fQ'(f,\overline{f}))$

8. If $B_2 = c(g_{2k},f) \ \& \ at_{2k+2}$ and:

a. $t_{2k+2} = g_{2k+2}f$:

$(ag_{2k} \& \exists fQ'(f,\overline{f}))+(ag_{2k+2}h_2 \& c(g_{2k},\max(h_1,2g_{2k}))\&\mathrm{Part}(g_{2k}+1/g_{2k+2}:ag_{2k}\&\exists fQ'(f,\overline{f})))$

b. $t_{2k+2} = g_{2k+2}/f$:

$(ag_{2k} \& \exists fQ'(f,\overline{f}))+(a(g_{2k+2}/h_1)\&c(g_{2k},\max(h_1,2g_{2k}))\&\mathrm{Part}(g_{2k}:ag_{2k}\&\exists fQ'(f,\overline{f})))$.

<u>Proof</u>.

$\Rightarrow$: Given f so that $Q'(f,\overline{f})$, partition $Y = \Sigma e_n$ so that $\limsup_n \sup_{e_n} g_{2k} = 0$ and $[ag_{2k} \ \& \ Q'(f,\overline{f})]e_n$. Then the implication is evident in case 5. In case 6 since $c(g_{2k},f)$ we have $a(g_{2k}) + a(1/f)$. Thus we may as well assume $a(1/f)$, hence $a(1/h_1)$. Then we need only show that $\limsup_n \sup_{e_n} h_1 = 0$; but $h_1 < f$ and $c(g_{2k},f)$, so this is clear.

Case 7 is trivial. In case 8 as in case 6 we may as well assume that f is bounded. Then case 8b is essentially trivial and in case

8a we claim that $\limsup_n \limsup_{e_n} 1/g_{2k+2} = 0$. The argument for this rests on the fact that for any e:

$$[\ell(g_{2k})]e \Longrightarrow [\ell(f) \ \& \ a(g_{2k+2}f)]e \Longrightarrow [\ell(1/g_{2k+2})]e.$$

$\Longleftarrow$: In cases 6, 8 we may ignore the clause $ag_{2k} \ \& \ \exists fQ'(f,\overline{f})$, which is neither help nor hindrance. Let us assume at the outset that $Y = \Sigma e_n$ has been partitioned according to the relevant clause of case 5, 6, 7, or 8. Then:

5. For each n choose f_n so that $[Q'(f_n,\overline{f})]e_n$ and let $f = f_n$ on e_n. Then $Q(f,\overline{f})$.

6. Let $f = h_1'$. Then $[af]e_n$ and $Q(f,\overline{f})$.

7. For each n choose f_n so that:

$$[Q'(f_n,\overline{f})]e_n,$$

and let $c = \inf_1 \frac{1}{2} h_2$. Let $f = \max(f_n,c)$ on e_n. Then $h_1 < f < h_2$ and easily $[Q'(f,\overline{f})]e_n$. Thus $Q'(f,\overline{f})$.

8b. Let $f = h_1'$. Then $g_{2k} < f < h_2$, so $[Q'(f,\overline{f})]e_n$. Also $c(g_{2k}, \max(h_1,2g_k))$ yields $c(g_{2k},f)$. Finally $t_{2k+2} = g_{2k+2}/h_1'$, so at_{2k+2}.

8a. Let $f = \min(h_2', \max(h_1',1/g_{2k+2}))$. Then $h_1 < f < h_2$. Since $g_{2k} < f$ we have $[af]e_n$, so $[Q'(f,\overline{f})]e_n$. Now $f \leq \max(h_1',1/g_{2k+2})$ and $c(g_{2k}, \max(h_1,2g_{2k}))$, while $\limsup_n \limsup_{e_n} (g_{2k} + 1/g_{2k+1}g_{2k+2}) = 0$, so we have $c(g_{2k},f)$. Finally since

$$t_{2k+2} = g_{2k+2}f \geq \min(g_{2k+2}h_1',1)$$

it follows that at_{2k+2}.

14. Decidability of $C(\beta Y;\mathbb{R})$

At this point the proof of the main theorem is complete. A summary seems appropriate here.

Main Theorem. Every formula $F(\overline{e};\overline{f})$ is equivalent to a composite formula in the variables $\overline{e};\overline{f}$.

Proof. Let $\mathcal{C}$ be the closure of the class of composite formulas under logical equivalence. In §§5,6 we saw that $\mathcal{C}$ contains the quantifier-free formulas and is closed under the connectives and Boolean quantification. Hence it suffices to prove that $\mathcal{C}$ is closed under existential function quantification.

In §§9-12 we saw that for any formula of the form

$$(E') \qquad \exists f Q(f,t_1,\ldots,t_p)$$

such that Q is basic and the t_i are terms in f and other variables subject to certain reductions and normalizations we have:

1. If rank $Q = 0$ then E' is equivalent to a formula without function quantifiers
2. If rank $Q = k > 0$ then E' is equivalent to a formula obtainable from similar formulas of lower rank using only the connectives and Boolean quantification.

We conclude immediately by induction: every formula of type (E') is in $\mathcal{C}$. In §§7-8 we showed how this result implies that $\mathcal{C}$ is closed under the existential function quantifier.

Thus we have proved, as desired, that all formulas are in $\mathcal{C}$.

Corollary. If Y is countable and discrete then $Th(C(\beta Y;\mathbb{R})$ is primitive recursively decidable.

Proof. Let S be a sentence in the language of $C(\beta Y;\mathbb{R})$. By the fact that the foregoing elimination of quantifiers is primitive recursive we may assume that S is composite. Furthermore S, having no free variables, involves as function terms only constants. Hence the formulas a, ℓ, c, and m^k are vacuous. The Boolean terms involve truth value symbols whose values (0 or 1) may be obtained using the primitive recursive elimination of quantifiers for real closed fields. The rest is propositional calculus.

Section B. The uncountable case

We now take Y to be any uncountable discrete set and we prove that $C(\beta Y;\mathbb{R})$ has a decidable theory. Our analysis requires fairly minor alternations. Of course the formal system will require some slight extension, but the discussion in §§12-13 will be useful without alteration.

15. A formal system

We incorporate the formal system of §1 into an extended system whose new features are as follows (our numbering is compatible with that of §1):

2.1.3 There are additional Boolean predicates: Medium, Large meaning countably and uncountably infinite respectively.

2.1.2 We use as global function predicates the symbols: a, ℓ, c, m^k, A, C, M^k. At the same time we incorporate relativization (3.1) into the notation for these predicates, which are accordingly to be construed as relations between Boolean terms and function terms.

The present lower-case global function predicates may be defined in terms of the old ones by:

$$a(e;f) = \text{Medium}(e) \ \& \ [af]e$$
$$\ell(e;f) = \text{Medium}(e) \ \& \ [\ell f]e$$
$$c(e;f,g) = \text{Medium}(e) \ \& \ [c(f,g)]e$$
$$m^k(e,f_1,\ldots,f_k) = \text{Medium}(e) \ \& \ [m^k(f_1,\ldots,f_k)]e.$$

We also define:

$$A(e;f) = \text{"Large}(e) \ \& \ \forall e' \subsetneq e \ \ \ell(e';f)$$
$$C(e;f,g) = \text{"}M^1(e,f) \ \& \ f \leq g \ \& \ \forall e' \subsetneq e \ \ell(e';f) \Longrightarrow \ell(e';g)\text{"}$$

The definition of the predicates M^k - and hence of C - is involved, being in the style of §2 (and §§3-4).

<u>Definition</u>.

1. A basic formula $E(e;\overline{f})$ of rank 0 is either:
   a. Cardn(e), Small(e), or Medium(e)
   b. "$f_1 < \ldots < f_n$ & Af_1"
2. If the notion of basic formula of rank $\leq k$ is defined, then a reducible formula of rank $\leq k$ is an arbitrary finite sum of such.
3. $M^0(e) = \text{Large}(e)$

$$M^{k+1}(e;f_1,\ldots,f_{k+1}) = \text{"}f_1 < \ldots < f_{k+1} \ \& \ \& \ \{\neg R(e;\overline{f}): \ R \text{ reducible of rank } k\}\text{"}$$

4. A basic formula of rank $k > 0$ is the conjunction of a single formula $M^k(f_{j_1},\ldots,f_{j_k})$ with "$f_1 < \ldots < f_n$" and possible additional conjuncts:
   a. $C(f_j, f_{j+1})$
   b. Af_j

subject to the conditions laid down in §2.

<u>16. The inductive interpretation of M^k; reducible formulas</u>

<u>Theorem</u>. $M^{k+1}(e;f_1,\ldots,f_{k+1}) \equiv \forall \varepsilon > 0 \ \ \exists 0 < \delta < \varepsilon[M^k(e',f_1,\ldots,f_k)]$ where $e' = e(\text{"}\delta < x \leq \varepsilon\text{"}; \ f_{k+1}) \ \ e$.

The proof is much as in §3, although when Y is partitioned into finitely many pieces the countable and uncountable pieces are considered separately. In particular the fact that Medium (e) is basic is used. As before one needs to know that there are only finitely many reducible formulas of given rank k (as is indeed necessary for the very definition of M^{k+1}). We prove in fact as in §§4,6:

<u>Theorem</u>.

1. If R, S are two reducible formulas in the same variables then they are either equivalent or mutually inconsistent.
2. For B basic $B + B \equiv B$.

We dwell on this momentarily. Statement 2 is proved as previously. Statement 1 also follows similar lines: define <u>maximal</u> or <u>irredundant</u> as in §6, and reduce the proof (easily) to:

<u>Lemma</u>. If B is basic, $R = \Sigma B_i$ is reducible, and B & R is consistent then $B \equiv R$.

<u>Proof</u>. When $B \neq M^n(e,f_1,\ldots,f_n)$ proceed as previously. When $B = M^n$ claim that for any basic B_1:

$$M^n + B_1 \equiv M^n,$$

from which the result will follow easily. That $M^n + B_1 \Longrightarrow M^n$ is immediate by induction on n using the inductive interpretation of M^n. We therefore need only assume $M^n(e;\overline{f})$, and prove $(M^n + B_1(e;\overline{f})$. To simplify notation take $e = 1$ and write $M^n(\overline{f})$ for $M^n(e;\overline{f})$.

Choose a sequence ε_k decreasing monotonically to zero, with $\varepsilon_1 = \infty$, and so that with

$$e_k = \{y: {}_{k+1} < f_n(y) \leq \varepsilon_k\}$$

we have:

$$[M^{n-1} f_1 \ldots f_{n-1}] e_k.$$

We have to deal with two cases.

1. $B_1(e';\overline{f}) \Longrightarrow \text{Medium}(e')$.

In this case it is enough to prove $M^n \Longrightarrow M^n + m^n$, since $m^n \Longrightarrow m^n + B_1$ is known. Proceed by induction on n, partitioning

$$e_k = e_k' + e_k''$$

with

$$m^{n-1}(e_k';\overline{f}) \;\&\; M^{n-1}(e_k'';\overline{f}).$$

Letting $e' = \Sigma e_k'$ and $e'' = \Sigma e_k''$ we have

$$m^n(e';f) \;\&\; M^n(e'',f)$$

as desired.

2. $B_1 \not\Rightarrow$ Medium e'

Then there are three cases entirely analogous to those treated in §6.

17. The formula Part; elimination of $\exists f$

The main theorem is unchanged:

Main Theorem. Every formula is equivalent to a composite formula.

The definition of "composite" may stand unaltered, and the treatment of connectives and the elimination of Boolean quantifiers is valid as it stands. We can also retain the reductions of the problem of elimination of the function quantifier as described in §§7-8, since $\ell(e;f)$ may be treated as the primary global function predicate in terms of which the others are defined.

We will again use the formula:

$$\mathrm{Part}(f;F) = \text{"}\forall e,e'[e\wedge e'=0\ \&\ A(e\vee e';f) \Rightarrow\ e''\supseteq e'(e\wedge e''=0\ \&\ [F]e'')]\text{"}$$

The notions of idempotent, convex, and partitionable formula are defined as in §10. Basic formulas have these properties for much the same reasons as before. Thus the formula Part (h;F) introduced in §11 continues to have its intended meaning for basic formulas, and the elimination of quantifiers proceeds as in §§12-13, using the theorems proved in §§12-13 as well as their analogues for uncountable sets.

Corollary. The theory of all rings of the form:

$$C(\beta Y;\mathbb{R})$$

with Y uncountable discrete is complete and primitive recursively decidable.

Section C. Valued Fields

In this part we study Th(C(βY;K)) with Y discrete and K a field with nonarchimedean valuation. To fix notation, let ord: K → Z be the valuation. The image Z of ord is an ordered abelian group.

Let C(βY;K) be the ring of functions f:Y → K such that ord(f) is bounded below. Similarly let C(βY;Z) be the group of functions α:Y → Z which are bounded below. We take as a standing assumption:

Hypothesis.

1. Z is a $\mathbb{Z}$-group or divisible
2. Z has countable cofinality.

Then we will see that $Th(C(\beta Y;K))$ is primitively recursively reducible to $Th(K)$. Hence in particular:

<u>Theorem II'</u>. If X is the Stone-Cech compactification of a discrete set and K is a locally compact field of characteristic zero then $C(X;K)$ is decidable.

18. $C(\beta Y;Z)$

We use a formal system based on that of Section B, but with two sorts of function variables (we continue the numbering of §§1,15):

1.2 f_1, f_2 K-function variables representing $f: Y \to K$

1.3 α_1, α_2 $\mathbb{Z}$-function variables representing $\alpha: Y \to Z$

Note that α varies over functions bounded below in Z (i.e. over $C(\beta Y;Z)$) and f varies over functions bounded in K (i.e. $\text{ord } f \in C(\beta Y;Z)$).

Use local function notions appropriate to the similarity types of K and Z (clause 2.2). Retain the global function predicates $a, \ell, c, m^k, A, C, M^k$ but apply them only to Z-valued functions. For instance:

$\ell(e;\alpha)$ means that $\alpha \to +\infty$ in Z (hence Medium (e)).

$a(e;\alpha) =$ "Medium (e) & $\forall e' \subseteq e \quad \ell(e';\alpha)$" meaning that α is bounded above

$A(e;\alpha) =$ " Medium(e) & $\forall e' \subseteq e \quad \ell(e';\alpha)$".

The usual complicated definition by induction of §§2-4, 15-16 is used to define c, m^k, C, M^k.

Call the resulting formal system L, or for emphasis L_K, and let L_Z be the fragment of L appropriate to a discussion of $C(\beta Y;Z)$ alone (including of course the Boolean notions). Then we have:

<u>Theorem</u>. Each formula of L_Z is equivalent to a composite formula of L_Z.

The arguments of Sections A, B are for the most part adequate for the proof of the present theorem as they stand. The heart of the problem is of course the elimination of the function quantifier from an expression of the form:

(E) $$\exists\alpha \quad B(e;\alpha,\alpha_1,\ldots,\alpha_n)$$

where B is basic, but involves terms t_i in $\alpha,\alpha_1,\ldots,\alpha_n$. As in §§7-8 we now want to show that these terms can be taken to have the particularly simple forms:

$$\alpha_i, \quad \alpha_i + \alpha, \quad \alpha_i - \alpha.$$

The analogy with the result of §8 is apparent.

We first have to discuss the definable functions in the language of $\mathbb{Z}$-groups. These are generated by the following primitive functions, adequate for the quantifier-elimination in $\mathbb{Z}$-groups:

1. Constants 0, 1
2. +,-
3. For each integer $n \geq 1$ there is an "integral part" function $[z/n]$ defined by: $0 \leq z-n[z/n] < n.1$.

It will be simpler to deal with the order of magnitude of a function. For any term t define an associated term $\hat{t}$ (outside the formal system L_K) by induction:

1. $\hat{0} = \hat{1} = \hat{0}$
2. $\hat{\alpha} = \alpha$
3. $(t_1 \pm t_2)\hat{} = \hat{t}_1 \pm \hat{t}_2$
4. $[t/n]\hat{} = \hat{t}/n$

We think of $\hat{t}$ as the order of magnitude of t, and we view it more concretely as denoting the function from Y to:

$$\hat{Z} = Z/\mathbb{Z} \cdot 1$$

induced by t. The point of this notion is that the order of magnitude of a term $t(\alpha_1,\ldots,\alpha_n)$ can be put in the form:

$$\frac{1}{M} \Sigma m_i \alpha_i$$

for suitable $M, m_1, \ldots, m_n$, and that as far as the global function predicates are concerned we may replace t by $\hat{t}$. Unfortunately we must still deal with the expression "$t_1 > \ldots > t_p$" as part of B.

To facilitate the discussion rewrite (E) as:

(E') $\qquad \exists \alpha\, (t_1 > \ldots > t_p \;\&\; B(t_1,\ldots,t_p))$

by removing the inequalities from B. We continue to refer to the new formula B as basic.

Now $\ell(e;t)$ is equivalent to $\ell(e;\hat{t})$. Since the other global predicates are defined in terms of ℓ (E') is equivalent to:

(E") $\qquad \alpha\, (t_1 > \ldots > t_p \;\&\; B(\hat{t}_1,\ldots,\hat{t}_p))$.

At this stage we are working well outside our formal system, and certain ambiguities occur. $B(\hat{t}_1,\ldots,\hat{t}_p)$ is to be interpreted in a way compatible with the possibility that $\hat{t}_i = \hat{t}_{i+1}$, so for example the clause $C(e;\hat{t}_i,\hat{t}_{i+1})$ is understood to mean:

$$\text{"Large(e) \& } \forall e' \subseteq e(\ell(e';t_1) \Longrightarrow \ell(e';\hat{t}_{i+1}))\text{"}$$

without implying that $\hat{t}_i > \hat{t}_{i+1}$ on e.

Now $t_i = A_i(\alpha,\overline{\alpha})$ and the clause "$t_1 > \ldots > t_p$" may be replaced by a clause:

$$h_1 > \alpha > h_2$$

with h_1, h_2 terms in $\alpha_1,\ldots,\alpha_n$. (There is in general another clause not involving α which we can ignore.)

Hence (E") is equivalent to:

(E‴) $$(h_1 > \alpha > h_2 \ \& \ B(\hat{t}_1,\ldots,\hat{t}_p))$$

Write the terms $\hat{t}_i$ using rational coefficients as:

$$\hat{A}_i = q_i\,\alpha + \sum_{j=1}^{n} q_{ij}\alpha_j.$$

Define $r_i =$

$$|q_i|^{-1} \quad \text{if} \quad q_i \neq 0$$

$$1 \quad \text{if} \quad q_i = 0$$

and set $\beta_i = r_i \sum_j q_{ij}\alpha_j$.

Then $\ell(e;\hat{t}_i)$ is equivalent to:

$\ell(e;\alpha+\beta_i)$ if $q_i > 0$

$\ell(e;\beta_i)$ if $q_i = 0$

$\ell(e;\beta_i-\alpha)$ if $q_i < 0$.

Make a change of notation $\alpha_i = \beta_i$ and:

$$t_i = \alpha_i, \quad \alpha + \alpha_i, \quad \text{or} \quad \alpha_i - \alpha$$

so that (E‴) retains the same meaning. Then (E‴) is equivalent to:

(F) $$\alpha\ (h_1 > \alpha > h_2 \ \& \ B(t_1,\ldots,t_p))$$

with $t_i = \alpha_i$, $\alpha_i + \alpha$, or $\alpha_i - \alpha$. (If Z is divisible to begin with this discussion is largely irrelevant.)

Now that we have carried out the desired initial reduction, we introduce the formula Part from §§11, 17, and apply it as in §§12-13. We illustrate this program by proving the analog of the final theorem of §13. It will be seen that the changes involved are entirely superficial.

As we noted at the end of §8, B may be taken to be a conjunction of formulas:

1. Lt_1
2. $M^k(t_2,t_4,\ldots,t_{2k})$: if $t_{2k} \neq \alpha_{2k}$ take $t_{2k} = \alpha$
3. $C(t_{2i},t_{2i+1})$ $(1 \leq i \leq k)$

4. At_{2k+2}

some of which may be absent.

We recall the notation from §12:

<u>Definition</u>.

1. $Q(\alpha,\bar{\alpha})$ = "$h_1 < \alpha < h_2$ & $B(\alpha,\bar{\alpha})$"
2. $Q'(\alpha,\bar{\alpha})$ = "$h_1 < \alpha < h_2$ & $B_1(\alpha,\bar{\alpha})$ & $M^{k-1}(t_2,\ldots,t_{2k-2})$ & At_{2k}"

where $B(\alpha,\bar{\alpha}) = B_1(\alpha,\bar{\alpha})$ & $M^k(t_2,\ldots,t_{2k})$ & $B_2(\alpha,\bar{\alpha})$, B_1 involving at most $t_1,\ldots,t_{2k-1}$, B_2 involving at most t_{2k},t_{2k+1},t_{2k+2}.

3. $h_1' = \inf(2h_1,[\frac{1}{2}(h_1+h_2)])$
4. $h_2' = \sup([\frac{1}{2}\, h_2],[\frac{1}{2}(h_1+h_2)])$.

In 3, 4 we may assume $h_2 - h_1 \geq 2$.

<u>Theorem</u>. With the above hypotheses and notations, and assuming $t_{2k} = \alpha_{2k}$, the formula $\exists\alpha\, Q(\alpha,\bar{\alpha})$ is equivalent to one of the following:

5. If B_2 is empty: Part$(\alpha_{2k}$:$A\alpha_{2k}$ & $\exists\alpha\, Q'(\alpha,\bar{\alpha}))$
6. If $B_2 = C(\alpha_{2k},\alpha)$: $(A\alpha_{2k}$ & $\exists\alpha\, Q'(\alpha,\bar{\alpha}))$+Part $(h_1$: $\exists\alpha\, Q'(\alpha,\bar{\alpha}))$.
7. If $B_2 = A\alpha$: Ah_2 & Part$(\alpha_{2k}$:$A\alpha_{2k}$ & $\exists\alpha\, Q'(\alpha,\bar{\alpha}))$
8. If $B_2 = C(\alpha_{2k},\alpha)$ & At_{2k+2} and
   a. $t_{2k+2} = \alpha_{2k+2} + \alpha$:
      $(A\alpha_{2k}$ & $\exists\alpha\, Q'(\alpha,\bar{\alpha}) + (A(\alpha_{2k+2}+\alpha)$ & $C(\alpha_{2k},\inf(h_1,\alpha_{2k}))$
      & Part$(\sup(\alpha_{2k},-\alpha_{2k+2})$:$A\alpha_{2k}$ & $\exists\alpha\, Q'(\alpha,\bar{\alpha})))$
   b. $t_{2k+2} = \alpha_{2k+2} - \alpha$:
      $(A\alpha_{2k}$ & $\exists\alpha\, Q'(\alpha,\bar{\alpha}))+(A(\alpha_{2k+2}-h_1)$ & $C(\alpha_{2k},\inf(h_1,\alpha_{2k}))$
      & Part$(\alpha_{2k}$:$A\alpha_{2k}$ & $\exists\alpha\, Q'(\alpha,\bar{\alpha})))$.

<u>Proof</u>.

$\Longrightarrow$. Fix α so that $Q(\alpha,\bar{\alpha})$ and partition $Y = \Sigma e_n$ so that $\liminf_{n}{}_{e_n} \alpha_{2k} = \infty$ and $[A\alpha_{2k}$ & $Q'(\alpha,\bar{\alpha})]e_n$. Then the claim is evident in cases 5, 7. In case 6 $C(\alpha_{2k},\alpha)$, so $(A\alpha_{2k} + A(-\alpha))$. Thus we may as well assume $A(-\alpha)$, and $A(-h_1)$. Then we need only: $\liminf_{n}{}_{e_n} h_1 = \infty$; but $h_1 > \alpha$ and $C(\alpha_{2k},\alpha)$, so this is clear.

In case 8 as in case 6 we may assume $A(-\alpha)$. Then case 8b is essentially trivial and in case 8a we claim that $\lim_n \inf_{e_n} - \alpha_{2k+2} = \infty$

This follows by observing that for any e:

$$[L\alpha_{2k}]e \Longrightarrow [L\alpha \ \& \ A(\alpha_{2k+2}+\alpha)]e \Longrightarrow [L(-\alpha_{2k+2})]e.$$

$\Leftarrow$. In cases 6, 8 we may ignore the clause $A\alpha_{2k} \ \& \ \exists\alpha\, Q'(\alpha,\bar{\alpha})$, which is neither aid nor obstacle. Assuming that $Y = \Sigma e_n$ has been partitioned according to the relevant clause of case 5, 6, 7, or 8, and choosing for each n a function α_n so that

$$[Q'(\alpha_n,\bar{\alpha})]e_n,$$

a suitable α is defined as follows:

5. $\alpha = \alpha_n$ on e_n.

6. $\alpha = h_1'$

7. $\alpha = \inf(\alpha_n, e)$ on e_n where $c = \sup_1 [h_2/2]$

8b. $\alpha = h_1'$

8a. $\alpha = \sup(h_2', \inf(h_1', -\alpha_{2k+2}))$.

The verification given in §13 will not be repeated here.

We remark that in 1, 2a and elsewhere the parallel to our earlier treatment suggests that in the proof of sufficiency of the quantifier-free condition one should make use of the quantity:

(i) $$\inf_{e_n} h_1.$$

This is not feasible, since $\inf_{e_n} h_1$ belongs not to Z but to its Dedekind completion. We must therefore adopt some artifice. We can take a sequence $a_n \in Z$ increasing monotonically to ∞ and replace (i) by:

$$\sup \{a_n : a_n < \inf_{e_n} h_1\}$$

One additional point is worth mentioning in case 3a and elsewhere. Where in §§12-13 a function f is defined in terms of $1/n$, ng_i, or the like, the corresponding term in our case is $-a_n$, $a_n + \alpha_i$, etc. where $a_n \in Z$, $\lim_{n\to\infty} a_n = \infty$.

It should also be noted that the nature of our earlier reductions entails that the previous theorem also has to be proved for basic formulas $B(e;\alpha,\bar{\alpha})$ which imply Medium(e) (lower-case basic formulas). What has been done here adequately illustrates the parallel to the previous work.

19. Elimination of quantifiers in L_K

Every formula of L_K is equivalent to a formula without function quantifiers.

Proof. Our formalism is such that a function symbol can occur in a formula only within a term of the form

$$\text{ord } t \quad \text{or} \quad e(F;t_1,\ldots,t_n).$$

Hence any formula of the form:

$$(Q) \qquad \exists f\ Q(\overline{e},\overline{\alpha},f,\overline{f})$$

is equivalent to:

$$(Q') \qquad \exists f,\overline{\alpha}',\overline{e}'(Q(\overline{e},\overline{e}',\overline{\alpha},\overline{\alpha}',\overline{f})\ \&\ \underset{i}{\&}\ \varepsilon_i(f,\overline{f},\overline{e},\overline{\alpha}) = e_i'$$
$$\&\ \underset{j}{\&}\ a_j(f,f,\alpha) = \alpha_j')$$

where the ε_i are Boolean terms and the a_i are function terms denoting elements of Z. The term a_j may involve expressions ord $t_k(f,\overline{f})$ where t_k is any term corresponding to a definable function in the language of valued fields.

Since f no longer occurs in Q we need only omit the function quantifier from:

$$(Q'') \qquad \exists f(\underset{i}{\&}\ \varepsilon_i(f,\overline{f},\overline{e},\overline{\alpha}) = e_i'\ \&\ \underset{j}{\&}\ a_j(f,\overline{f},\overline{\alpha}) = \alpha_j')$$

Since the existential function quantifier commutes with conjunction over bits (cf. the beginning of §7) we can reduce (Q") to:

$$(Q''') \qquad \exists f(\underset{i}{\&}\ \varepsilon_i(f,\overline{f},\overline{\alpha}) = 1\ \&\ \underset{j}{\&}\ a_j(f,f,\alpha) = \alpha_j').$$

With a slight change in notation we may assume that ε_i has the form:

1. $e(F_i,f,\overline{f})$ with F_i a formula in the language of valued fields

or 2. $e(F_i$, ord $t_1(f,\overline{f}),\ldots,$ord $t_k(f,\overline{f}),\overline{\alpha})$ with F_i a formula in the language of $\mathbb{Z}$-groups (or divisible groups, as the case may be).

By abuse of notation we write in both cases: $F_i = f_i(f,\overline{f},\overline{\alpha})$. Then Q''' is equivalent to:

$$(Q^*) \qquad e(\text{"}\ \exists f(\underset{i}{\&}\ F_i(f,\overline{f},\overline{\alpha})\ \&\ \underset{j}{\&}\ a_j(f,\overline{f},\overline{\alpha}) = \alpha_j')\text{"};\overline{f},\overline{\alpha}) = 1$$

This is a quantifier free formula of L_K (we are using a language of valued fields having variables ranging over the value group).

Theorem. If Z is a $\mathbb{Z}$-group (or divisible) of countable cofinality then $Th(C(\beta Y;K))$ is Turing equivalent to $Th(K)$.

Proof. By the preceding primitive recursive elimination of quantifiers, all formulas can be taken to be quantifier-free. In particular the truth value of a sentence of L_K can be determined from the value of certain truth value functions:

$$e(F;)$$

with F a sentence in the language of valued fields, which proves the theorem.

Theorem II' follows. A curious situation arises if we now consider $C(\beta Y;K)$ with K a countably generated nonarchimedean extension of an archimedean ordered field. Let K_0 be the ring of "finite" or "bounded" elements of K. Then as is well known K is a valued field with valuation ring K_0 and maximal ideal the set of infinitesimals. Since K is countably generated over an archimedean field the value group Z of K is countable.

If K is real closed as ordered field then as a valued field K is Henselian with divisible value group. Hence by the preceding:

Theorem II'. With K as above, $C(\beta Y;K)$ is decidable for Y discrete. In fact we already know more than this. An examination of the proof of Theorem II (coupled with a modified treatment of cases like 1 and 2a as suggested at the end of §18) establishes:

Theorem II'''. For K as above $C(\beta Y;K) \equiv C(\beta Y;\mathbb{R})$.

The proof of Theorem II' is somewhat simpler than the proof of Theorem II, and proves the decidability of the same theory; unfortunately we need the proof of Theorem II to see that the theories are indeed the same.

Problem. Give a simple proof of Theorem II''' .

Section D. Removing countability assumptions

To summarize our decidability results:

Theorem IIA. If K is an archimedean real closed field then $C(\beta Y;K)$ is decidable for Y discrete.

Theorem IIB. If K is a valued field with value group Z divisible or a $\mathbb{Z}$-group and Z is of countable cofinality then $Th(C(\beta Y;K))$ is reducible to $Th(K)$ for Y discrete.

The proof of Theorem II proves Theorem IIA. We will lift the countability restriction in Theorem IIB, so that in particular we need not assume that K is archimedean in Theorem IIA.

20. The formal system

We study $C(\beta Y;K)$ with Y discrete and K a valued field whose value group Z is a $\mathbb{Z}$-group or divisible. The cofinality of Z is called γ.

20.1 L_K

Our formal system incorporates the features described in sections A-C.

1. Variables:
   - 1.1 Boolean variables e.
   - 1.2 K-function variables f.
   - 1.3 Z-function variables α_1 representing functions $\alpha: Y \to Z$ which are bounded below.
2. Nonlogical Constants:
   - 2.1 Boolean notions:
     - 2.1.1 Constants 0,1
     - 2.1.2 Functions $\cup, \cap, -$
     - 2.1.3 Predicates $=, \subseteq$, Card n, Small, Medium, Large (the last referring to cardinality $<, =,$ or $> \gamma$).
   - 2.2 Local function notions:
     - 2.2.1 K-function operations θ corresponding to definable functions in the language of valued fields.
     - 2.2.2 Z-function operations τ corresponding to definable functions in the language of Z-groups.
     - 2.2.3 Truth value operators $e(F;\overline{f},\overline{\alpha})$.
   - 2.3 Global function notions:
     - 2.3.1 Z-function predicates $a, \ell, c, m^k, A, C, M^k$.
3. Additional notation
   - 3.1 Relativization: unnecessary. This will be incorporated into 2.3.1. We will, however, continue to write $[F]e$ on occasion for brevity.
   - 3.2 The connective + (used to eliminate certain kinds of Boolean quantifiers).

20.2 Semantics

The interpretation of this system is largely self-evident. As always the interpretation of the global function notions $M^k(e;\bar{\alpha})$ and $M^k(e;\bar{\alpha})$ depends on certain preliminary results. Lower case global function predicates refer to medium sized e and upper case predicates refer to large e. Both a and A mean "bounded above in Z" while $\ell(e;\alpha)$ means:

$$\forall a \in Z \mid \{y \in e : \alpha(y) < a\} \mid < \gamma$$

and this of course implies Medium(e). All global function predicates as well as Small, Medium, and Large can be defined in terms of ℓ.

20.3 M^k and m^k

We may take for granted the definition of basic or reducible formula and of M^k, m^k along the lines of §2. The inductive interpretation of M^k and m^k is essential. We state it for M^k:

Theorem. For $k \geq 0$ the following are equivalent:

1. $M^{k+1}(\alpha_1,\ldots,\alpha_{k+1})$
2. $\forall a \in Z \ \exists b > a \ (M^k(e(\text{"}a \leq x < b\text{"};\alpha_{k+1});\alpha_1,\ldots,\alpha_k))$.

   If $k = 0$, note that by fiat $M^0(e;\ ) = \text{"Large (e)"}$.

Proof. There is some value in carrying through the verification that $2 \Longrightarrow 1$ in the present context.

$2 \Longrightarrow 1$: Fix $a \in Z$ so that for all $b > a$;

$$M^k(\alpha_{k+1}^{-1}[a,b];\alpha_1,\ldots,\alpha_k).$$

We may assume $\alpha_{k+1} \geq a$ on 1 (cf. §3). Choose a sequence $a_i \to \infty$ $(i < \gamma)$ with $a_0 = a$ so that the sets

$$e_i = \alpha_{k+1}^{-1}[a_i,a_{i+1}]$$

are large; if this is impossible our claim is clear, because $A\alpha_{k+1}$ iff a set is of at most medium size.

Now we need the two results that the set of basic formulas of rank $k-1$:

$$\{B_i : 1 \leq i \leq K\}$$

is finite and that

$$B_i + B_i \equiv B_i \ .$$

Then partition:

$$e_i = e_i^1 + \ldots + e_i^k$$

so that

$$[B_j(e_1^j; f_1, \ldots, f_k)] \quad \text{for all} \quad i.$$

Set $e^j = \bigcup_i e_i^j$ so that $1 = e^1 + \ldots + e^k$.

If e^j is large then $f_1, \ldots, f_{k+1}$ satisfy a reducible formula of rank k on e^j (cf. §3). This suffices to complete the argument.

As part of the same induction on k, one proves the results of §4 in our setting. This is straightforward.

21. Elimination of quantifiers

We define composite formulas in the style of §5 and let $\mathcal{C}$ be the class of formulas equivalent to composite formulas. We must show primarily that $\mathcal{C}$ is closed under:

1. negation
2. Boolean quantification
3. Z-function quantification
4. K-function quantification.

Of course 4 is an easy consequence of 3 and 2 is easy. In treating negations we repeat the material of §6, the main point being that for B basic and k arbitrary:

Lemma. $M^k(e;\overline{f}) \Longrightarrow (M^k + B)(e;\overline{f})$.

If the proof given in §6 for this lemma is read carefully it will prove adequate in the present context.

Thus only the elimination of the Z-function quantifier requires attention, and here too everything runs smoothly. First note that the argument of §18 is independent of cardinality considerations. Secondly the discussion of partitionability in §10 generalizes to the present context, the combinatorial point being that if one covers κ with $\leq\kappa$ sets of size κ, the cover can be refined to a partition by shrinking each set of the cover to a subset which still has size κ. (There are two applications of this remark: $\kappa = \gamma$ and $\kappa = \gamma^+$, i.e. Medium and Large; in both cases one is concerned with a covering by γ sets.)

Thus the formula Part has its intended meaning, and the quantifier elimination may be completed.

Concluding remarks

On the technical side, it would be interesting to lift the restriction on Z in the case of valued fields, presumable using the detailed information in [2]. It would also be interesting to find topological spaces X such that the decidability of $C(X;\mathbb{R})$ is not settled by our methods.

A more significant question is whether these methods have any relevance to decidability questions for rings of analytic functions.

References

1. S. Feferman and R. Vaught, "The first order properties of products of algebraic systems," Fund. Math. 47 (1959), 57-103.

2. Y. Gurevich, "Elementary properties of ordered abelian groups," AMS Translations, 46, 165-192.

3. W. Henson, C. Jockusch, C. Rubel, G. Takeuti, "First order topology," Dissertationes Math. 143(1977) 40pp.

4. A. Macintyre, "On the elementary theory of Banach algebras," Ann. Math. Logic 3 (1971), 239-269.

5. A. Macintyre, "Model-completeness for sheaves of structures," Fund. Math. 81 (1973), 73-89.

6. V. Weispfenning, "Elimination of quantifiers for subdirect products of structures," J. Alg. 36 (1975), 252-277.

Weak partition relations, finite games, and independence results in Peano arithmetic

Peter Clote[1]

U.E.R. de Mathématiques
Université Paris VII
2 Place Jussieu
75005 Paris, France

Abstract A basis and an anti-basis result is given for the complexity of infinite weakly homogeneous sets and for winning strategies of finite games. Then using Kirby-Paris indicator theory, we give some independence results for Peano arithmetic.

Introduction

We use standard recursion theoretic notation as found in Rogers [6]. For convenience, we briefly outline what we will use. Let $\sigma \in \{0,1\}^{<\omega} = \bigcup_{n\in N} {}^n\{0,1\}$ be a string of 0's and 1's. $\{e\}^\sigma$ is the oracle Turing machine with index e. $\{e\}^\sigma_s(x) = y$ iff the oracle machine with index e, when given x calculates y in $<$ s steps in asking questions of the form "$\sigma(m)=0$" for $m \in$ domain σ. If a question "$\sigma(m) = 0$" is asked where $m \notin$ domain σ then $\{e\}^\sigma_s(x)$ is not defined. If L is a total function, then $[L(0,\ldots,L(s)] = \sigma$ where length $lh(\sigma) = \max\{L(i): i \leq s\} + 1$ and $\sigma(m)=0 \Leftrightarrow \exists i \leq s\ L(i)=m$. For F and G partial functions, $F \cong G$ means that both have the same domain and $\forall x \in$ domain F $F(x)=G(x)$. For $X \subseteq N$ $\{e\}^X(x)=y$ iff $\exists \sigma((\forall x \in \text{domain}\,\sigma\ \ \sigma(x)=0 \leftrightarrow x \in X)\ \&\ \{e\}^\sigma_s(x)=y)$
For $X,Y \subseteq N$ $X \leq_T Y$ iff $\exists e \forall x$ $\{e\}^X(x) = \begin{cases} 0 & \text{if } x \in Y \\ 1 & \text{if } x \notin Y \end{cases}$. We systematically confuse sets with their characteristic functions.

1 Except for theorem 1.5 this material appears in my thesis from Duke University 1979. I want to thank most warmly Prof. K. McAloon.

Let P denote the first order theory of Peano arithmetic with language $\{0,S,+,\cdot\}$. I is an <u>initial</u> <u>segment</u> or <u>cut</u> in $M \models P$, denoted $I \subseteq_e M$, if $\emptyset \neq I \subsetneq M$ and $\forall x,y \in M(\ x \in I \ \&\ y \leq x \rightarrow y \in I)$ and $\forall x \in M(\ x \in I \longrightarrow S(x) \in I)$. If $A \subseteq I \subsetneq_e M$ then A is <u>coded</u> in M if $A = I \cap B$ where B is definable with parameters in M. $I \subseteq_e M$ is <u>semi-regular</u> in M if $\forall a \in I\ \forall$coded functions $f:\{0,\ldots,a\} \rightarrow M$ $\exists b \in I(\ \ (f''\ \{0,\ldots,a\}) \cap I \subseteq \{0,\ldots,b\})$. $I \subseteq_e M$ is <u>regular</u> in M if $\forall a \in I\ \forall$coded partitions $f\colon I \rightarrow a\ \exists b < a(\ f^{-1}(\{b\})$ is unbounded in I). $I \subseteq_e M$ is <u>strong</u> in M if I is semi-regular and $\forall$ coded partitions $f\colon [I]^3 \rightarrow 2$ $\exists$coded I-unbounded set X such that $f''[X]^3$ has cardinality 1. Here for $A \subseteq I, [A]^n$ is the set of increasing n-tuples drawn from A. The above relation is written $I \longrightarrow (I)^3_2$. X is called <u>homogeneous</u> if card($f''[X]^3$) = 1. An <u>indicator</u> Y for a class of initial segments C of M is a function $Y\colon M^2 \longrightarrow M$ definable in M such that $\forall a,b \in M$ ($Y(a,b) > \mathbb{N} \longleftrightarrow \exists I \in C(a < I < b)$).

<u>Section</u> 1 Square-bracket partition relations

First, as a tool for the recursion theoretic analysis of recursively presented problems, we generalize the well-known limit lemma [7]p.23.

1.1 <u>Definition</u> If F is an m+1-ary partial function then

$$\lim_s F(x_1,\ldots,x_n,s) = y \quad \text{iff} \quad \exists t\, \forall s \geq t\ \ F(x_1,\ldots,x_n,s) = y.$$

If F is an m+n-ary partial function then

$$\lim_{s_1} \ldots \lim_{s_n} F(x_1,\ldots,x_n,s_1,\ldots,s_n) \cong \lim_{s_1} (\lim_{s_2} \ldots \lim_{s_n} F(x_1,\ldots,x_n,s_1,\ldots,s_n))$$

1.2 <u>Lemma</u> For all $n \geq 1$ there is a recursive partial function F_n such that

$$\forall e,x \quad \lim_{s_1} \ldots \lim_{s_n} F_n(e,x,s_1,\ldots,s_n) \cong \{e\}^{0^n}(x).$$

Moreover, regardless of whether $F_n(e,x,s_1,\ldots,s_n)$ is defined or not, the computation terminates in a finite number of steps.

<u>Proof</u> By induction on $n \geq 1$. There is a fixed c_0 such that $\forall X \quad \{c_0\}^{X}{''}N = X'$.

n=1: This is just the limit lemma. $\{c_0\}''N$ is a recursive enumeration of $0'$. Let

$F_1(e,x,s_1) \simeq \{e\}_{s_1}^{[\{c_0\}(0),\ldots,\{c_0\}(s_1)]}(x)$. If $\{e\}^{0'}(x) = y$ then, since only finitely many questions are asked of the oracle $0'$,
$\exists t \forall s_1 \geq t \;\; F_1(e,x,s_1) = F_1(e,x,t) = y$ so that $\lim_{s_1} F_1(e,x,s_1) \simeq \{e\}^{0'}(x)$.

$n=k+1$: $\{c_0\}^{0^k}$ "$N = 0^{k+1}$.

$$\{e\}^{0^{k+1}}(x) \simeq \lim_{s_1} \{e\}_{s_1}^{[\{c_0\}^{0^k}(0),\ldots\ \{c_0\}^{0^k}(s_1)]}(x)$$

$$\simeq \lim_{s_1} \{e\}_{s_1}^{[\lim_{s_2}\ldots\lim_{s_{k+1}} F_k(c_0,x,0,s_2,\ldots,s_{k+1}),\ldots,\lim_{s_2}\ldots\lim_{s_{k+1}} F_k(c_0,x,s_1,\ldots s_{k+1})]}(x)$$

$$\simeq \lim_{s_1}\ldots\lim_{s_{k+1}} \{e\}_{s_1}^{[F_k(c_0,x,0,s_2,\ldots,s_{k+1}),\ldots,F_k(c_0,x,s_1,\ldots,s_{k+1})]}(x)$$

Define $F_{k+1}(e,x,s_1,\ldots,s_{k+1})$ to be the expression to the right of $\lim_{s_{k+1}}$.

Then F_{k+1} satisfies the properties. (Note: in my thesis the limit lemma has been extended into the hyperarithmetic hierarchy: in the place of n-tuples for the approximation of 0^n, one uses a set of tuples of order type ω^α for the approximation of 0^α.)

In [1] Jockusch considered the recursion theoretic version of Ramsey's theorem: given a recursive partition $G:[N]^n \to m$ for $n,m \geq 2$ what can be said about the complexity of the infinite homogeneous sets? In the positive direction he showed that there is always a Π^0_n infinite homogeneous set. In the negative direction he constructed a recursive partition $G_n:[N]^n \to 2$ for each $n \geq 2$ without any Σ^0_n infinite homogeneous set. Schematically

$$N \longrightarrow (\Pi^0_n)^n_m \quad \text{for } n,m \geq 2 \qquad N \not\longrightarrow (\Sigma^0_n)^n_2 \quad \text{for } n \geq 2$$

Here we consider square-bracket partition relations.

1.3 <u>Definition</u> For $I \subseteq_e M \models P$, $I \longrightarrow [I]^n_m$ means for all coded partitions $G: [I]^n \longrightarrow m$ there is a coded I-unbounded set $X \subseteq I$ such that $G''[X]^n$ has cardinality $< m$. X is called <u>weakly-homogeneous</u> if $\text{card}(G''[X]^n) < m$ and $I \longrightarrow [I]^n_m$ is called a <u>square-bracket partition relation</u>.

It now immediately follows that for any recursive partition $G: [N]^n \longrightarrow m$ there is a Π^0_n infinite weakly-homogeneous set (since any homogeneous set is weakly-homogeneous). It is natural to ask whether a weakly homogeneous set could be found with lower complexity. Surprisingly, however, a minor modification of Jockusch's original argument yields the answer no. So schematically we have

$$N \longrightarrow [\Pi^0_n]^n_m \quad \text{and} \quad N \not\longrightarrow [\Sigma^0_n]^n_m \quad \text{for all } n,m \geq 2.$$

1.4 <u>Theorem</u> For each $n,m \geq 2$ there is a recursive partition $G: [N]^n \longrightarrow m$ without any infinite Σ^0_n weakly-homogeneous sets.

<u>Proof</u> Fix $n,m \geq 2$. By 1.2 let F be a recursive partial function such that $\lim_{s_1} \ldots \lim_{s_{n-1}} F(e,x,s_1,\ldots,s_{n-1}) \simeq \{e\}^{0^{n-1}}(x)$. G will be constructed in stages. For all $x < s_1$ $G(x,s_1,\ldots,s_{n-1})$ will be defined by the end of "stage" $(s_1,\ldots,s_{n-1})$.

$$\text{Let } A_e = \begin{cases} \text{the set with characteristic function } \{e\}^{0^{n-1}} \text{ if the latter is a characteristic function} \\ \text{undefined otherwise} \end{cases}$$

$$A_{e,s_1,\ldots,s_{n-1}} = \begin{cases} \text{the set with characteristic function } F(e,x,s_1,\ldots,s_{n-1}) \text{ provided that} \\ \quad \forall x < s_1 (F(e,x,s_1,\ldots,s_{n-1}) \text{ defined} \rightarrow \\ \qquad F(e,x,s_1,\ldots,s_{n-1}) = 0 \text{ or } 1) \\ \text{undefined otherwise} \end{cases}$$

$$D_e = \begin{cases} \text{the least } me+m \text{ elements of } A_e \text{ if such exist} \\ \text{undefined otherwise} \end{cases}$$

$$D_{e,s_1,\ldots,s_{n-1}} = \begin{cases} \text{the least } me+m \text{ elements of } A_{e,s_1,\ldots,s_{n-1}} \\ \text{which are less than } s_1 \\ \text{undefined otherwise} \end{cases}$$

For $(s_1,\dots,s_{n-1}) \in [N]^{n-1}$ we have the "stage" $(s_1,\dots,s_{n-1})$ and the substages $0,\dots,e,\dots,s_1$.

Substage $e < s_1$: If $D_{e,s_1,\dots,s_{n-1}} \neq \emptyset$ then pick the least n elements $a_1,\dots,a_m$ in $D_{e,s_1,\dots,s_{n-1}} - \{x\colon G(x,s_1,\dots,s_{n-1})$ is already defined at some substage $i < e\}$ and define $G(a_i,s_1,\dots,s_{n-1}) = i-1$.

At substage s_1 define $G(x,s_1,\dots,s_{n-1}) = 0$ for all $x < s_1$ for which G is not yet defined. This ends the construction.

Let X be an infinite Σ_n^0 set. X is r.e. in 0^{n-1} so there is an infinite set $Y \subseteq X$, where $Y \leq_T 0^{n-1}$ with characteristic function $\{e\}^{0^{n-1}}$ for some e. By 1.2 let $s_1 \in Y - \{0,\dots,\max D_e\}$ such that

$$\forall x \leq \max D_e \lim_{s_2} \dots \lim_{s_{n-1}} F(e,x,s_1,\dots,s_{n-1}) = \begin{cases} 0 & \text{if } x \in D_e \\ 1 & \text{if not} \end{cases}$$

Inductively, given $s_1 < s_2 < \dots < s_i$ where $i < n-1$ let $s_i < s_{i+1} \in Y$ such that $\forall x \leq \max D_e \lim_{s_{i+2}} \dots \lim_{s_{n-1}} F(e,x,s_1,\dots,s_{n-1}) = \begin{cases} 0 & \text{if } x \in D_e \\ 1 & \text{if not} \end{cases}$

Hence we have $(s_1,\dots,s_{n-1}) \in [Y]^{n-1}$, $\max D_e < s_1$ and $\forall x \leq \max D_e$ $F(e,x,s_1,\dots,s_{n-1}) = 0 \iff x \in D_e$. So by the construction there are $a_1,\dots,a_m < s_1$ such that $F(a_i,s_1,\dots,s_{n-1}) = i-1$ and so Y is not weakly homogeneous. Hence X is not weakly homogeneous.

In order to obtain an independence result in P from an infinite combinatorial problem, it is essential to show not simply that there is no recursive solution of the recursively presented problem, but also that all solutions reconstruct 0'. In [1] Jockusch showed that for each $n \geq 1$ there is a recursive partition $G\colon [N]^{n+2} \to 2$ such that $\forall X$(X infinite & homogeneous $\to X \geq_T 0^n$). Here we extend this result to weak partition relations.

1.5 Theorem $\forall n \geq 1$ $\forall m \geq 2$ there is a recursive partition $G\colon [N]^{m+n} \longrightarrow m$ such that $\forall X$(X infinite & weakly homogeneous $\longrightarrow X \geq_T 0^n$).

Proof Fix $n \geq 1$ $m \geq 2$. As in lemma 5.9 of [1] let $h \equiv_T 0^n$ where

$h = \{e_n\}^{0^n}$ such that $\forall g(\forall x(g(x) > h(x)) \to g \geq_T h)$. For $a_{m+1} < \ldots < a_{m+n}$ define the "recursive approximation"

$h_{a_{m+1},\ldots,a_{m+n}}(x) =_{df} F_n(e_n, x, a_{m+1}, \ldots, a_{m+n})$, where F_n is as in 1.2.

For $(a_1,\ldots,a_{m+n}) \in [N]^{m+n}$ define

$$G(a_1,\ldots,a_{m+n}) = \begin{cases} 0 & \text{if } h_{a_{m+1},\ldots,a_{m+n}}(a_1) < a_2 \\ \vdots & \\ i & \text{if } h_{a_{m+1},\ldots,a_{m+n}}(a_1) \not< a_{i+1} \text{ and} \\ & \quad h_{a_{m+1},\ldots,a_{m+n}}(a_1) < a_{i+2} \\ \vdots & \\ m-1 & \text{if } h_{a_{m+1},\ldots,a_{m+n}}(a_1) \not< a_m \end{cases}$$

Let $X \subseteq N$ be infinite and weakly homogeneous, i.e. $\text{card}(G''[X]^{m+n}) < m$).

Case 1: $\exists a_1 < \ldots < a_{m-1}$ in X such that $h(a_1) \not< a_{m-1}$. Then let $b_1 < \ldots < b_{m-1}$ in X such that $a_{m-1} < b_1$ and $h(a_1) < b_1$. By 1.2 let $a_{m+1} < \ldots < a_{m+n}$ be in X such that $h_{a_{m+1},\ldots,a_{m+n}}(a_1) = h(a_1)$ and $b_{m-1} < a_{m+1}$. Then

$G(a_1,\ldots,a_{m-1},b_1,a_{m+1},\ldots,a_{m+n}) = m-2$

$G(a_1,a_3,a_4,\ldots,a_{m-1},b_1,b_2,a_{m+1},\ldots,a_{m+n}) = m-3$

$\vdots$

$G(a_1,a_{m-1},b_1,\ldots,b_{m-2},a_{m+1},\ldots,a_{m+n}) = 1$

$G(a_1,b_1,\ldots,b_{m-1},a_{m+1},\ldots,a_{m+n}) = 0$

Then $m-1 \notin G''[X]^{m+n}$ since X is weakly homogeneous. It follows that $\forall a_1 < \ldots < a_m$ in X $h(a_1) < a_m$. Let p_X be the function which enumerates X in increasing order. Then $h(x) \leq h(p_X(x)) < p_X(x+m-1)$. Define $g(x) = p_X(x+n-1)$. Then $\forall x(g(x) > h(x))$ so $g \geq_T h \equiv_T 0^n$. Since $X \equiv_T p_X \equiv_T g$ we have $X \geq_T 0^n$.

Case 2: $\forall a_1 < \ldots < a_{m-1}$ in X $h(a_1) < a_{m-1}$. Then $h(x) \leq h(p_X(x)) < p_X(x+m-2)$. Define $g(x) = p_X(x+m-2)$. Then we have $\forall x(g(x) > h(x))$ so $g \geq_T h \equiv_T 0^n$. Since $X \equiv_T p_X \equiv_T g$ we have $X \geq_T 0^n$.

Taking m=2 in the above theorem yields Jockusch's original result. It is still an open question whether $\forall n \geq 1\ \forall m \geq 2$ there is a recursive partition $G:[N]^{m+n} \to m$ such that $\forall X$(X infinite & weakly homogeneous $\to X \geq_T 0^n$). Jockusch has proved this when m=2.

Now we turn to models of arithmetic. Weak-bracket partition relations have been considered before by Kleinberg and Shore [4] in the context of large cardinals in set theory. Shore showed that $ZF + V=L \vdash \kappa$ regular & $\kappa \to [\kappa]^n_m \Rightarrow \kappa \to (\kappa)^2_2$ for $n,m \in N$. Kleinberg and Shore showed among other things that $ZFC \vdash \kappa \to [\kappa]^n_{2^{n-1}} \Rightarrow \kappa \to (\kappa)^2_2$. We would like to show that if I is semi-regular and satisfies some weak partition property then I is strong. However, Kleinberg and Shore's proof does not immediately extend to the case of models of arithmetic, because they use the existence of a set which cannot be coded in M. Hence we have to resort to another combinatorial argument which can be formalized in M.

1.6 <u>Proposition</u> If I is semi-regular and $I \to [I]^4_3$ then $I \to (I)^3_2$.

<u>Proof</u> Let $F:[I]^3 \to 2$ be coded. This induces a coded partition $G:[I]^4 \to 4$ by $G(a,b,c,d) = \langle F(a,b,c), F(b,c,d)\rangle$. Get an unbounded coded set $X \subseteq I$ such that $G"[X]^4$ has cardinality ≤ 2 by first regrouping to form $G_1:[X]^4 \to 3$ where

$$G_1(a,b,c,d) = \begin{cases} 0 & \text{if } G(a,b,c,d)=0 \\ 1 & \text{if } G(a,b,c,d)=1 \\ 2 & \text{otherwise} \end{cases}$$

then getting X_1 weakly homogeneous for G_1 and then if $2 \in G_1"[X_1]^4$ defining $G_2:[X_1]^4 \to 3$ by $G_2(a,b,c,d) = G(a,b,c,d)$. By semi-regularity X_1 is order isomorphic to I so we can get X coded unbounded in I and weakly homogeneous for G_2, hence $G"[X]^4$ has cardinality ≤ 2. There are $\binom{4}{2}$ possibilities. These are

00	00	00	11	11	10
11	10	01	10	01	01

where eg. $\begin{matrix}00\\11\end{matrix}$ means given $a,b,c,d \in X$ with $a<b<c<d$, one either has $F(a,b,c)=0=F(b,c,d)$ or $F(a,b,c)=1=F(b,c,d)$.

The last possibility $\begin{matrix}10\\01\end{matrix}$ is the only one which does not immediately give rise to a coded unbounded set Y (even cofinite in X) which is homogeneous for F.

Claim $\overset{10}{01}$ is not obtained.

Proof of claim If not, then there are $a < b < c < d < e < f$ in X with $F(a,b,c)=0=F(d,e,f)$. Then $F(a,b,c)=0$ $F(b,c,d)=1$ $F(c,d,e)=0$ and $F(d,e,f)=1$. For notational shortcut we write

This yields a contradiction thus proving the claim.

1.7 <u>Proposition</u> Let I be semi-regular and $m,n \in \mathbb{N}$. Then $I \to [I]^{n+1}_{m+1} \Rightarrow I \to [I]^n_m$.

<u>Proof</u> Given any coded $F:[I]^n \to m$ let the induced partition $G:[I]^{n+1} \to m^2$ be defined by

$$G(a_1,\dots,a_{n+1}) = \langle F(a_1,\dots,a_n), F(a_2,\dots,a_{n+1})\rangle$$

Let $X \subseteq I$ be coded unbounded and $G''[X]^{n+1}$ have cardinality $\leq m$ (obtained by regrouping and iterating the partition relation $I \to [I]^{n+1}_{m+1}$, using semi-regularity of I). Suppose that $F''[X]^n$ has cardinality m (otherwise X is weakly homogeneous for F) and that $\{\langle s_1,t_1\rangle,\dots,\langle s_m,t_m\rangle\}$ is obtained.

<u>Claim</u> $\{0,\dots,m-1\} \subseteq \{s_1,\dots,s_m\}$

<u>Proof of claim</u> Suppose $i \notin \{s_1,\dots,s_m\}$. Then $i \notin \{t_1,\dots,t_m\}$ for otherwise there is $(a_1,\dots,a_{n+1}) \in [X]^{n+1}$ such that $F(a_1,\dots,a_{n+1}) = i$. Then $\langle F(a_2,\dots,a_{n+1}), F(a_3,\dots,a_{n+2})\rangle = \langle i,t_j\rangle$ contradicting the hypothesis that $i \notin \{s_1,\dots,s_m\}$. So if $i \notin \{s_1,\dots,s_m\}$ then $F''[X]^n$ has cardinality $< m$ which contradicts the supposition that X is not weakly homogeneous.

<u>Claim</u> Either $X - \{\min X\}$ is weakly homogeneous for F or $\{0,\dots,m-1\} \subseteq \{t_1,\dots,t_m\}$.

<u>Proof of claim</u> obvious.

Now if $\{0,\dots,m-1\} \subseteq \{t_1,\dots,t_m\}$ then this yields a cyclic permutation $\sigma : \{0,\dots,m-1\} \to \{0,\dots,m-1\}$ of length m.

(If not, then σ is a product of cycles and a variant of this argument works.)

<u>Case 1</u>: m divides n. Then as X is not weakly homogeneous for F, choose $(a_1,\dots,a_{2n+1}) \in [X]^{2n+1}$ such that $F(a_1,\dots,a_n) = 0$ and $F(a_{n+2},\dots,a_{2n+1}) = 0$. Then we also have $F(a_{n+2},\dots,a_{2n+1}) = \sigma(0) \neq 0$. Contradiction.

<u>Case 2</u>: $n = mq+r$ where $0 < r < m$. Then choose $(a_1,\dots,a_{2n}) \in [X]^{2n}$ such that $F(a_1,\dots,a_n) = 0 = F(a_{n+1},\dots,a_{2n})$. Then we also have $F(a_{n+1},\dots,a_{2n}) = \sigma^r(0) \neq 0$. Contradiction.

Hence $X - \{\min X\}$ is coded unbounded and weakly homogeneous for F.

1.8 <u>Corollary</u> If $n \in \mathbb{N}$ and I is semi-regular and $I \to [I]^{n+1}_n$ then I is strong.

1.9 Corollary If $m < n \in \mathbb{N}$ and I is semi-regular and $I \to [I]^n_m$ then I is strong.

Corollary 1.9 gives another indicator for models of P. Fix $n, m \in \mathbb{N}$ $m < n$ and $M \models P$. We define by induction the notion of k-dense $[n,m]$ for sets X finite in the sense of M: X is 1-dense $[n,m]$ if $|X| > \min X + 2$; X is k+1-dense$[n,m]$ if (i) for all coded $F:X \to d$ either $\{0,\ldots,d-1\} \cap X$ is k-dense$[n,m]$ or there is a subset $Y \subseteq X$ such that Y is k-dense$[n,m]$and card(F"Y)=1 (ii) for all coded $G:[X]^n \to m$ there is a subset $Y \subseteq X$ such that Y is k-dense$[n,m]$and weakly homogeneous for G. Set $Y(a,b)=c=\max\{k: [a,b] \text{ is k-dense}[n,m]\}$ Y(a,b) is well-behaved in the sense of [2]p.22. The usual proof shows that Y(a,b) is an indicator for initial segments I such that I is regular and $I \to [I]^n_m$. So by indicator theory, we know that $P \nvdash \forall a \forall c \exists b \ Y(a,b) \geqslant c$.

We end this section by stating some open problems.

1.10 Open problems

(1) If I is semi-regular, $m < n$ (both possibly infinite) and $I \to [I]^n_m$ then is I strong?

(2) If I is semi-regular then does $I \to [I]^n_m$ imply $I \to [I]^{n+1}_{m+1}$?

(3) If I is semi-regular and $I \to [I]^3_3$ then is I strong? This is related to the open question of [2]p. 56: if I is semi-regular and $I \to (I)^2_2$ then is I strong?

(4) Let $n \underset{*}{\to} [2m]^m_m$ mean for all partitions $F: [\{0,\ldots,n\}]^m \to m$ there is a subset $X \subseteq \{0,\ldots,n\}$ such that $|X| > \min X$ and $|X| > 2m$ and X is weakly homogeneous. Is the sentence $\forall m \exists n \ n \underset{*}{\to} [2m]^m_m$ unprovable in P?

Section 2 Finite Games

Let $n \geqslant 1$ and $N^n = A \cup \bar{A}$. The game G_A associated with A is played as follows: players I and II alternately take turns writing an integer

$$\begin{array}{lcccccl} \text{I} & a_1 & & a_3 & & & \\ \text{II} & & a_2 & & a_4 & \cdots & \text{until the } n^{th} \text{ move.} \end{array}$$

I wins if $(a_1,\ldots,a_n) \in A$; otherwise II wins. The game G_A is <u>recursive</u> if $A \subseteq N^n$ is recursive. A game G_A lasting n moves is called an n-game. A <u>strategy</u> is a function $\sigma : \omega^{<\omega} \to \omega$. $(a_1, a_2, \ldots)$ is a <u>play according to the strategy for I</u> if $a_1 = \sigma(\emptyset)$ & $\forall m \geqslant 1 \; a_{2m+1} = \sigma(a_1,\ldots,a_{2m})$. σ is a winning strategy for I (w.s.) if $(a_1,\ldots,a_n) \in A$ for any play according to σ. The analogous notions for player II are similarly defined.

In [5] using the existence of a simple set, Rabin displayed a recursive 3-game without any recursive winning strategy for either player. Here we give a recursive 3-game such that any w.s. reconstructs 0' and in fact we give for each $n \geqslant 1$ a recursive (n+2)-game such that $\forall \sigma\,(\sigma \text{ is w.s.} \to \sigma \geqslant_T 0^n)$.

2.1 <u>Proposition</u> There is a recursive set $A \subseteq N^3$ such that in the associated game G_A II can always win and $\forall \sigma\,(\sigma \text{ is w.s.} \to \sigma \geqslant_T 0')$.

<u>Proof</u> Let $(a,b,c) \in \bar{A}$ iff

$\forall e \leqslant a\,(T_1(e,e,b) \leftrightarrow T_1(e,e,b+c))$ where T_1 is Kleene's primitive recursive predicate (see [8] p. 157).

G_A is clearly a recursive game where II can always win.

$e \in 0' \leftrightarrow \exists z\, T_1(e,e,z) \leftrightarrow T_1(e,e,\sigma(e))$ hence $0' \leqslant_T \sigma$.

2.2 <u>Theorem</u> $\forall n \geqslant 1$ there is a recursive set $A \subseteq N^{n+2}$ such that II can always win the associated game G_A and $\forall \sigma\,(\sigma \text{ w.s.} \to \sigma \geqslant_T 0^n)$.

<u>Proof</u> Let $n \geqslant 1$. An (n+2)-tuple $(a_1,\ldots,a_{n+2})$ is called <u>allowable</u> if $a_1 < (a_2)_0 < (a_2)_1 < (a_3)_0 < (a_3)_1 < \ldots < (a_{n+2})_0 < (a_{n+2})_1$

where we use the recursive projection functions described in [8]p. 117

$\forall n \geqslant 1$ let c_n be an index where $\{c_n\}^{0^n}$ is the characteristic function for 0^n.

Now define $b_{i+1} = (a_1)_i$ for $i=0,\dots,n-2$. F_n is as in 1.2.

Let $(a_1,\dots,a_{n+2}) \in \bar{A}$ iff <u>either</u> $(a_1,\dots,a_{n+2})$ is allowable and

(1) $\forall e \le a_1 (F_n(c_n,e,(a_2)_0,(a_4)_0,(a_5)_0,\dots,(a_{n+2})_0)$

$= F_n(c_n,e,(a_3)_0,(a_4)_0,(a_5)_0,\dots,(a_{n+2})_0))$ and

(2) $\forall e \le a_1 (F_n(c_n,e,b_1,(a_2)_0,(a_4)_0,(a_5)_0,\dots,(a_{n+2})_0)$

$= F_n(c_n,e,b_1,(a_3)_0,(a_4)_0,(a_5)_0,\dots,(a_{n+2})_0))$ and

(3) $\forall e \le a_1 (F_n(c_n,e,b_1,b_2,(a_2)_0,(a_4)_0,(a_5)_0,\dots,(a_{n+2})_0)$

$= F_n(c_n,e,b_1,b_2,(a_3)_0,(a_4)_0,(a_5)_0,\dots,(a_{n+2})_0))$ and

⋮

(n) $\forall e \le a_1 (F_n(c_n,e,b_1,b_2,\dots,b_{n-1},(a_2)_0)$

$= F_n(c_n,e,b_1,b_2,\dots,b_{n-1},(a_3)_0))$

<u>Or</u> $(a_1,\dots,a_{n+2})$ is not allowable and if $i_0 =$ least $i((a_1,\dots,a_i)$ is not allowable) then i_0 is odd.

By induction on $n \ge 1$ we can show that II can always win. Let σ be a w.s. for II. Given e define $d_1(e)=(\sigma(e))_0$, $d_2(e)=(\sigma(\langle d_1(e),e\rangle))_0,\dots,d_n(e)=(\sigma(\langle d_1(e),d_2(e),\dots,d_{n-1}(e),e\rangle))_0$

Then $e \in 0^n \longleftrightarrow \lim_{s_1} \dots \lim_{s_n} F_n(c_n,e,s_1,\dots,s_n)=0$

(by 1) $\longleftrightarrow \lim_{s_2} \dots \lim_{s_n} F_n(c_n,e,d_1(e),s_2,\dots,s_n)=0$

(by 2) $\longleftrightarrow \lim_{s_3} \dots \lim_{s_n} F_n(c_n,e,d_1(e),d_2(e),s_3,\dots,s_n)=0$

⋮

(by n) $\longleftrightarrow F_n(c_n,e,d_1(e),\dots,d_n(e))=0$.

Since the d_i's can be calculated recursively in σ, $0^n \le_T \sigma$.

The above proof using 1.2 seems to be notationally the most clear and succinct. The intuition behind it for the case n=2 is as follows: a first attempt to define a recursive partition $N^4 = A \cup \bar{A}$ where $\forall\sigma(\sigma \text{ w.s.} \rightarrow \sigma \geq_T 0'')$ is this -- let $(a,b,c,d) \in \bar{A}$ iff $\forall\varphi\,(\ulcorner\varphi\urcorner \leq a \;\&\; \varphi = \exists x \forall y \psi \rightarrow (N \models \exists x < b \forall y < d \psi \leftrightarrow N \models \exists x < b{+}c \forall y < d \psi))$ where φ is a formula in the language of arithmetic and ψ is bounded. Then II can always win the game. If σ is a w.s. for II then $N \models \exists x \forall y \varphi \leftrightarrow N \models \exists x < \sigma(a) \forall y \psi$ where $a = \ulcorner \exists x \forall y \psi \urcorner$. $\exists x < \sigma(a) \forall y \psi$ is equivalent to a Π^0_1 formula in P, so we incorporate a subgame into the above partition in order for II to bound the socond quantifier. This is the purpose of (1),...,(n).

On the other hand, we have the positive results:

2.3 <u>Proposition</u> For all $n \geq 2$ if $A \subseteq N^n$ is recursive then in the associated game G_A either I has a w.s. $\sigma \leq_T 0^{n-3}$ or II has a w.s. $\sigma \leq_T 0^{n-2}$.

<u>Proof</u> Either (i) $\exists x_1 \forall x_2 \ldots Qx_n \; (x_1,\ldots,x_n) \in A$

or (ii) $\forall x_1 \exists x_2 \ldots Q'x_n \; (x_1,\ldots,x_n) \notin A$

where Q is $\exists$ or $\forall$ and Q' is a quantifier of the opposite kind as Q. Using the minimum operator μ it is easy to see that in case (i) I has a w.s. $\sigma \leq_T 0^{n-3}$ and in case (ii) II has a w.s. $\sigma \leq_T 0^{n-2}$.

For $n \geq 0$ 2.2 and 2.3 completely describe the situation for recursive games G_A lasting n+2 moves:

(i) $\forall n \geq 0$ there is a recursive (n+2)-game G_A such that II can always win and $\forall\sigma(\sigma \text{ w.s.} \rightarrow \sigma \geq_T 0^n)$

(ii) $\forall n \geq 0$ for all recursive (n+2)-games G_A there is a w.s. $\sigma \leq_T 0^n$.

This is a striking parallel with Jockusch's analysis of Ramsey's theorem.

2.4 <u>Definition</u> Let n-AD be the Π^1_2 sentence "for all n-games either I or II has a w.s.". $I \subset_e M$ is <u>n-determined</u> if $(I, \text{coded}_M I) \models$ n-AD.

2.5 <u>Proposition</u> If $I \subseteq_e M$ is closed under multiplication then I is 2-determined.

<u>Proof</u> By [2] if I is closed under multiplication then

$(I, \mathrm{coded}_M I) \models \Delta_1^0$ comprehension. If $I^2 = A \cup \bar{A}$ is coded then there is a w.s. σ which is Δ_1^0 in A hence coded.

2.6 <u>Proposition</u> For $I \subseteq_e M$, I is strong iff I is 3-determined and semi-regular.

<u>Proof</u> $\Rightarrow$ By definition of strong, I is semi-regular. Given a coded 3-game $I^3 = A \cup \bar{A}$ either (i) $\exists x \forall y \exists z \ (x,y,z) \in A$ or (ii) $\forall x \exists y \forall z \ (x,y,z) \notin A$. Hence there is a w.s. which is Δ_2^0 in A. By [2] $(I, \mathrm{coded}_M I) \models \Sigma_\infty^0$ comprehension, so $\sigma \in \mathrm{coded}_M I$. Hence I is 3-determined.

$\Leftarrow$ We show that $(I, \mathrm{coded}_M I) \models \Sigma_1^0$ comprehension. Since I is semi-regular, by [2] this implies that I is strong. Let $A \in \mathrm{coded}_M I$. We must show that $B = \{x \in I: (I,A) \models \exists y\, \varphi(A,x,y)\}$ is coded where φ is a bounded formula in the language of arithmetic. Define $I^3 = A \cup \bar{A}$ by $(a,b,c) \in \bar{A}$ iff

$$(I,A) \models \forall x \leq a(\exists y < b \ \varphi(A,x,y) \longleftrightarrow \exists y < b+c \ \varphi(A,x,y)).$$

This game is coded and II wins. Let σ be a coded w.s. for II. Then $x \in B$ iff $(I,A) \models \exists y < \sigma(x) \ \varphi(A,x,y)$ hence B is coded.

Note that with the notion of n-determined initial segments for n non-standard, there is a possibility of developing non-standard syntax. The interpretation for $I \models \exists x_1 \ldots Qx_n \varphi(x_1,\ldots,x_n)$ where $n \in I$ would be

$\exists \sigma$ (σ is a coded w.s. for I in the coded n-game given by φ).

If I is n-determined then we have

$I \models \neg \exists x_1 \ldots Qx_n \varphi(x_1,\ldots,x_n) \iff I \models \forall x_1 \ldots Q'x_n \neg \varphi(x_1,\ldots,x_n)$.

ie. $\exists \sigma$ (σ is coded w.s. for II in the coded n-game given by $\neg \varphi$).

We do not yet know if there is a density associated with n-determined cuts, but using Kirby-Paris game indicators, we can give an indicator for the set of semi-regular 3-determined cuts. Define a Kirby-Paris game $G_c(a,b)$ which lasts c moves as follows: at each move I asks either "is $e \in I$?" (for $e \in [a,b]$);"if so, then for which $i < e$ in the coded partition $\{B_i : i < e\}$ of $\{0,\ldots,b\}$ is B_i unbounded in I?"

or "who wins the coded game $\{0,\dots,b\}^3 = A \cup \bar{A}$ and what is a code for a w.s. σ of the named player?". Player 2 answers. Player 1 wins iff by the cth move, player 2 has produced a contradiction or if $\{\sigma(\emptyset), \sigma(x), \sigma(x,y): x,y \in I\} \nsubseteq I$. As in [3] this gives the indicator $Y(a,b) = \max\{c: 2 \text{ has a w.s. for } G_c(a,b)\}$ such that $Y(a,b)=c > N$ iff there is a semi-regular 3-determined cut I with $a \in I < b$. Since semi-regular 3-determined cuts are strong (and hence models of P), we have

2.7 <u>Proposition</u> $P \nvdash \forall a\, \forall c\, \exists b\;\; Y(a,b) \geq c$.

2.8 <u>Open Problems</u>

(1) For $I \subseteq_e M$ and $n \in I$ what is the relation between $I \to (I)^n_2$ and I is n-determined?

(2) Is there a finer result concerning a game-theoretical property $R(a,b,c,n)$ such that

$\forall c \in N \quad P^- + \Sigma^0_n$ induction $\vdash \forall a\, \exists b\; R(a,b,\bar{c},\bar{n})$

but $\quad P^- + \Sigma^0_n$ induction $\nvdash \forall a\, \forall c\, \exists b\; R(a,b,c,\bar{n})$?

References

1 C.G. Jockusch, Jr. "Ramsey's Theorem and Recursion Theory" Journal of Symbolic Logic 37-2 June 1972

2 L.A. Kirby Ph.D. Thesis Manchester University 1977

3 L.A. Kirby and J. Paris "Initial Segments of Models of Peano's Axioms" Lecture Notes in Mathematics 619 Springer Verlag

4 E. Kleinberg and R. Shore "Square Bracket Partition Relations" Journal of Symbolic Logic 37-4 1972

5 M. Rabin "Effective Computability of Winning Strategies" Contributions to the Theory of Games 3, Annals of Math. Study 39 Princeton 1957

6 H. Rogers, Jr. Theory of Recursive Functions and Effective Computability McGraw-Hill 1967

7 J.R. Shoenfield Degrees of Unsolvability North Holland 1971

8 J.R. Shoenfield Mathematical Logic Addison-Wesley 1967

Added in proof

After submitting this paper, in a conversation P.Aczel explained to the author that Theorem 2.2 was an apparently long known result with a trivial proof: suppose f is a total function such that $f \equiv_T 0^n$. Then

$$\forall x \exists y\ f(x) = y \longleftrightarrow \forall x \exists y \exists y_1 \forall y_2 \ldots Q y_{n+1}\ R(x,y,y_1,\ldots,y_{n+1})$$
$$\longleftrightarrow \forall x_1 \exists x_2 \forall x_3 \ldots Q x_{n+2}\ S(x_1,x_2,\ldots,x_{n+2})$$

Then $\overline{S} = \mathbb{N}^{n+2} - S$ is a recursive set and in the $(n+2)$-game $G_{\overline{S}}$, II wins and every winning strategy of II reconstructs 0^n.

HENSEL FIELDS IN EQUAL CHARACTERISTIC $p > 0$

Françoise DELON

This talk is an attempt to be a survey of the problem of Hensel fields in equal characteristic. No paper has appeared since the works of Ershov [E] and Ax - Kochen ([AK],[KO]). The only positive result in char. $p > 0$ is due to Ershov, based on an important algebraic work of Kaplansky [Ka] . Our talk doesn't countain new results but gives many examples and counter-examples which show the limits and difficulties of a possible generalization.

1.- We consider in this talk <u>valued fields</u>, that is fields with a surjection "the valuation" $K^* \to G$, where G is an ordered group, extended in 0 by putting $\mathrm{val}(0) = \infty$, an infinite element adjoined to G, and with the usual properties

$$\mathrm{val}(xy) = \mathrm{val}(x) + \mathrm{val}(y)$$

and the stronger form of triangular inequality

$$\mathrm{val}(x + y) \geq \mathrm{Min}\,[\mathrm{val}(x), \mathrm{val}(y)] .$$

In this field we define the valuation ring

$$A = \{x \in K \; ; \; \mathrm{val}(x) \geq 0\}$$

which is a local ring, with maximal ideal

$$I = \{x \in K \; ; \; \mathrm{val}(x) > 0 \}$$

and the residue field $\overline{K} = A/I$. In our cas K and $\overline{K}$ have the same characteristic.

The Hensel property for the valued field K is the following : If $f(x) \in A[x]$ has a simple residual root, then it has a root in A, whose residue agrees with the root of f. It is a first-order property.

The important result of Ax-Kochen and Ershow is the following :

Proposition.- If K is a henselian field of equal characteristic 0, then $\mathcal{H} \cup Th(\overline{K}) \cup Th(\text{val } K) \vdash Th(K)$; where $\mathcal{H}$ are the axioms saying that K is henselian.

It is well known that this result doesn't generalize to the case of characteristic p. The following counter-example is often given :

Definition.- If k is a field and G an ordered group, we define the field of generalized power series with coefficients in k and exponents in G :

$$k((T^g \; ; \; g \in G)) = \{\Sigma \, a_i T^i \; ; \; a_i \in k, (i) \text{ is a well-ordered subset of } G\}$$

The operations are the usual over the series, multiplication being possible by the condition of well-ordered support :

$$(\sum_{k_o} a_k)(\sum_{\ell_o} b_\ell) = (\sum_{k+\ell=i} a_k . b_\ell)$$

then when k increases from k_o to $i - \ell_o$, ℓ decreases from ℓ_o to $i - k_o$ ans so takes only a finite number of values.

Example.- Let k be an algebraically closed field with characteristic $p > 0$. It is known that $k((T^g \; ; \; g \in \mathbb{Q}))$ is an algebraically closed field. Let us now look at the subfield :

$$K = \bigcup_{n \in \mathbb{N}} k((T^{\frac{1}{n!}}))$$

(K is generalization of the Puisieux series over $\mathbb{C}$) ; K is not algebraically closed as we see it by looking at the Artin Schreier equation $x^p - x + \frac{1}{T} = 0$ whose solutions are

$$x_i = \frac{1}{T} + \frac{1}{T^p} + \ldots + \frac{1}{T^{p^n}} + \ldots + i, \quad i = 0,1,\ldots,p-1.$$

The fields $k((T^g\ ;\ g \in \mathbb{Q}))$ and K are then two Hensel fields with same residue field k and valuation group $\mathbb{Q}$, but they are not elementary equivalent.

We draw out of Ax and Kochen's proof two facts which are true for characteristic 0 but false for characteristic p :

1°.- We give first some definitions :

If $K \subset L$ are two valued fields, we can look at the residue extension and the group extension. We say that the extension is immediate when $\overline{K} = \overline{L}$ and val K = val L. It is the case for k(T) and k((T)), and more generally for a field and its completion. A field is said to be maximal when it has no immediate proper extension ; an example is k((T)) or all generalized power series fields.

The valuation is an homomorphism $(K^*,.) \to (\text{val } K,+)$; a cross-section is a section of this mapping. It allows us to see the valuation group as included in K.

For the characteristic 0, we have an isomorphism theorem : two maximal fields with cross-section and having same residue field and same valuation group are isomorphic.

For the characteristic p, Kaplansky in [Ka] has studied the uniqueness of the immediate maximal extension of a valued field K ; he has given conditions on K, called Kaplansky conditions or conditions A guaranteeing the uniqueness :

- val K is p-divisible
- for all $a_{n-1},\ldots,a_o,\ b \in \overline{K}$ the equation $x^{p} + a_{n-1}x^{p^{n-1}} + \ldots + a_1x^p + a_ox + b = 0$ has a solution in $\overline{K}$.

With the same conditions the isomorphism theorem is true.

2°.- An henselian field of equal characteristic 0 doesn't admit any immediate algebraic extension. In the example of the Puisieux serie, at the opposite extreme, $K[x_o]$ where $x_o = \sum_{i \in \mathbb{N}} \frac{1}{T^{\frac{1}{p^i}}}$ is an immediate extension of K.

This difference is easy to be taken up : we remark that the

property for a valued field of having no immediate algebraic proper extension - K is then called <u>algebraically maximal</u> - is first order ; K has only to satisfy for all integer n the sentence : "For all polynomial $P(X) \in K[X]$ of degree n

$$\{\forall v\ [\ \exists x(\text{val } P(x) = v)] \rightarrow [\ \exists x'(\text{val } P(x') > v)]\}$$

$$\rightarrow \{\exists y(P(y) = 0)\} \quad "$$

Here we use again the work of Kaplansky ; if there is in K a pseudo-convergent sequence $(u_\alpha)_{\alpha<\alpha_0}$ and a polynomial P(X) (of minimal degree) such that val $(P(u_\alpha))$ is not eventually constant, then K has an immediate extension K[u] where P(u) = 0. With the terminology of Kaplansky, K is algebraically maximal iff all pseudo-convergent sequence of algebraic type have a pseudo-limit in K.

A direct proof allows us to avoid the reference to Kaplansky : algebraic maximality is equivalent to the sentences which say :

- "K is henselian" ; we then have uniqueness of the extension of the valuation in all algebraic extension of K ; if c is algebraic over K with minimal polynomial $x^m + c_{m-1}x^{m-1} + \ldots + c_1x + c_o$, we must have $\text{val}(c) = \frac{1}{m}\text{val}(c_o)$.

- For all n : "for each polynomial $P(X) \in K[X]$ of degree n, there is $a \in K(x) = K[X]/P(X)$ such that $\text{val}(a) \notin \text{val } K$ or $[\text{val}(a) = 0$ and $\bar{a} \in \bar{K}]$. "

Now by elimination of x between P(x) and the decomposition of a in K(x), we know how to characterize a by its minimal polynomial over K. Hence the expression inverted commas is equivalent to :

"There exists $A(X) = X^r + a_{r-1}X^{r-1} + \ldots + a_1X + a_o$ minimal polynomial of an $a \in K(x)$ such that

$$[\bigwedge_{i=0}^{r-1} \text{val}(a_i) \geq 0 \wedge \forall y\ \bar{A}(\bar{y}) \neq \bar{0}] \vee [r \nmid \text{val}\ (a_o)]"$$

As far as we know, this notion of algebraic maximality is only studied in [Z].

With these two precisions, the same proof as for char. 0 works and gives the following result :

Proposition.- (Ershov) : When K_1 and K_2 are two algebraically maximal Kaplansky fields, we have $K_1 \equiv K_2$ iff $\overline{K}_1 \equiv \overline{K}_2$ and val $K_1 \equiv$ val K_2.

2.- On the contrary, if K is not Kaplansky, the system $Th(\overline{K}) \cup Th(\text{val } K) \cup$ ("K is algebraically maximal") is in general not complete ; we shall give a counterexample which shows other interesting facts.

Proposition.- Let k be a field, char. $k = p > 0$, and $a = a_o^p + T.a_1^p \in k((T))$ be such that $a_o, a_1 \in k((T))$ are algebraically independant over $k(T)$; K is the relative algebraic closure for $k(T,a)$ in $k((T))$.

Then K is algebraically maximal and $K \not\equiv k((T))$.

Proof.- The field K is valued in $\mathbb{Z}$ and hence its completion is its unique immediate maximal extension. It is easy to see that a valued field is algebraically maximal iff it is relatively algebraically closed in each immediate maximal extension ; therefore K is algebraically maximal by construction. The first-order property which distinguishes the theories of $k((T))$ and K is the algebraic completeness, notion introduced by Ershov : a valued field K is algebraically complete iff

1) it is henselian

2) each finite algebraic extension $L \supset K$ satisfies $[L : K] = [\overline{L} : \overline{K}]$ (val L : val K).

(To be sure this property is first order the reader can use the same kind of proof that we gave directly for the algebraic maximality). The field $k((T))$ is algebraically complete as it is a maximal field (see for example [R]). On the other hand the extension $L = K[T^{\frac{1}{p}} ; a^{\frac{1}{p}}]$ of K satisfies $[L : K] = p^2$, $\overline{L} = \overline{K}$, (val L : val K) = p.

Remarks : 1°) We draw as a lesson from this example the fact that an algebraic extension, even a finite one, of an algebraically maximal field is not necessarily algebraically maximal. So $a^{1/p}$ is immediate over $K[T^{\frac{1}{p}}]$ but is not in this field.

2°) We see the limits of the algebraic maximality : we have definitely the implication

$$\left\{\begin{array}{l} K \text{ alg. maximal} \subset L \\ \text{val } K \text{ pure subgroup of val } L \\ \bar{K} \text{ relat. alg. closed in } \bar{L} \end{array}\right.$$

$$\Rightarrow K \text{ relativ alg. closed in } L,$$

but nothing works for transcendental extensions ; for example $L = K[a_o, a_1]$ is an immediate but inseparable extension of K (it is known that an elementary extension is separable).

3°) The previous remark gives another first-order property distinguishing K and k((x)). In this particular case (where the valuation group is $\mathbb{Z}$) , the following sentences express the fact that there is no immediate inseparable extension :

$$\text{(for all n)} \quad \text{"}\forall k_1,\ldots,k_n \; [\; \forall v \; \exists \; x_1,\ldots,x_n \; (\text{val}(\Sigma k_i x_i^p) > v)]$$

$$\rightarrow [\; \exists \; y_1,\ldots,y_n \; (\Sigma k_i y_i^p = 0)]\text{"}.$$

3.- We have defined different properties of maximal fields. If we refine the algebraic maximality into separable or inseparable alg. max., we have the implications :

alg. complete $\Rightarrow$ alg. max. $\Rightarrow$ sep. alg. max. $\Rightarrow$ henselian

with coïncidence of all there notions in char. 0, of the two first for Kaplansky fields and of the two last when the valuation group is $\mathbb{Z}$:

Proposition.- Let K be a field valued in $\mathbb{Z}$, then K is henselian iff it is separably algebraically maximal.

Proof.- One of the implications is obvious. Conservely let K be a field valued in $\mathbb{Z}$ with an immediate algebraic extension K(a), where the minimal polynomial A of a over K is separable ; a is then a limit of a Cauchy sequence $(a_\alpha)_{\alpha<\alpha_o}$ in K, such that $\text{val}(P(a_\alpha))$ increases with

α with no limit. Now the only initial segment ($\neq \emptyset$) of $\mathbb{Z}$ without supremum is $\mathbb{Z}$; hence A admits a root approached to each order. In particular, since $A'(a) \neq 0$ we have eventually

$$\mathrm{val}(A(a_\alpha)) > 2\ \mathrm{val}\ A'(a) = 2\ \mathrm{val}\ A'(a_\alpha)$$

and then, by the strong form of Hensel lemma, A has a root in K.

Q.E.D.

This equivalence doesn't generalize when the valuation group is finitely generated or has the same theory as $\mathbb{Z}$.

<u>Example of a Hensel field valued in a $\mathbb{Z}$ -group, with an immediate Artin Schreier extension</u>

$$K = \bigcup_{i \in \mathbb{N}} k((T^g\ ;\ g \in \mathbb{Z}[\tfrac{\alpha}{i!}])) \subset k((T^g\ ;\ g \in \mathbb{Z}^*))$$

where k is a field with characteristic p, $\mathbb{Z}^*$ is a non-standard model of $\mathbb{Z}$ and $\alpha \in \mathbb{Z}^* - \mathbb{Z}$ is positive and divisible by all standard integers ; K is an henselian field as it is an increasing union of henselian fields ; val K is the divisible envelope of $\mathbb{Z}[\alpha]$ in $\mathbb{Z}^*$ and is hence a $\mathbb{Z}$ -group. Now the root $\sum_{i \in \mathbb{N}} T^{-\frac{\alpha}{p^i}}$ of the equation $x^p - x + T^{-1} = 0$ is not in K.

<u>Example of a Hensel field valued in $\mathbb{Z}[\beta]$, with an immediate Artin-Schreier extension</u>

Let K be the henselisation of $k(T,T^\beta,a^p)$, where k is a field of characteristic p, β a non-standard positive integer, p doesn't divide β and $a = a'.T^{-\beta}$, with $a' \in k((T))$ transcendental series over k(T). The solutions of the equation $x^p - x + a^p = 0$ are

$$x_i = a + a^{\frac{1}{p}} + \ldots + a^{\frac{1}{p^n}} + \ldots + i,\ i = 0,1,\ldots,p-1\ ;$$

this notation is valid because in the decomposition of each term $a^{\frac{1}{p^n}} = T^{-\frac{\beta}{p^n}} . a'^{\frac{1}{p^n}}$ as a series in T with exponents in $\mathbb{Z}[\frac{\beta}{p^n}\ ;\ n \in \mathbb{N}]$

all the monomials have valuation at standard distance from $-\beta p^{-n}$ (the reader will note that the support of the series x_i has order-type ω^2). Because of the inclusion $K \subset k((T^g, g \in \mathbb{Z}[\beta]))$ x_i is not in K ; to have that x_i is immediate over K, it is enough to note that a is not in K, as it is not in $k(T)(T^\beta)(a^p)$ and as it is radical over this field.

4.- We can ask ourselves whether algebraic completeness is the good first-order characterization of completeness or not. The fact that we do not know of two algebraically complete fields K and L satisfying $\overline{K} \equiv \overline{L}$, val K $\equiv$ val L and $K \not\equiv L$ may tempt us to give a positive answer. On the other hand we have the following result (unpublished result of Kochen and Jacob ; see for example a proof in [BDL]) :

Proposition.- In the language of valued fields with cross-section, $\mathbb{F}_p((T))$ is undecidable.

So in this enriched language, not only is the system "$\overline{K} \equiv \mathbb{F}_p$" $\cup$ "val $K \equiv \mathbb{Z}$" $\cup$ "K is algebraically complete" incomplete but but so also are all systems obtained by replacing algebraic maximality by a recursively enumerable system of axioms.

Along the same lines, we may have two maximal fields with the same residual and valuational theories but not elementary equivalent in this language :

Proposition.- In the language of valued fields with cross-section, if char. $k = p > 0$, $k((T)) \not\equiv k((T^g ; g \in \mathbb{Z}^*))$

Proof.- Let π be the cross-section, let us consider the sentence

$$\exists a [\exists b\ a = \pi(b)] \wedge [\mathrm{val}(a) < 0] \wedge [\exists x (x^p - x + a = 0)]$$

which is false in $k((T))$ but true in $k((T^g ; g \in \mathbb{Z}^*))$. We have only to take $a = T^{-\alpha}$ where α is a positive non-standard integer, infinitely divisible by p.

The question is now to determine the importance of cross-section in the language. But it remains true that in characteristic 0, even if we adjoin it in the language, $k((X))$ is decidable iff k is.

BIBLIOGRAPHY

[AK] J. AX and S. KOCHEN : Diophantine problems over local fields I Am. Journal of Math., vol 187 (1965), pp. 605-630, pp. 631-648.

Diophantine problems over local fields III, decidable fields, Annals of Math. 83 (1966), pp.437-456.

[BDL] J. BECKER, J. DENEF and L. LIPSCHITZ : Further remarks on the elementary theory of formal power series rings, this volume.

[E] J.L. ERSHOV : On the elementary theory of maximal normed fields Doklady 1965, Tome 165 N°1, pp.1390-1393.

[Ka] I. KAPLANSKY : Maximal fields with valuation, Duke Math. Journal 9 (1942), pp. 303-321.

[Ko] S. KOCHEN : The model theory of local fields, Logic Conference, Kiel 1974, Lecture notes in Mathematics, 499 Berlin, Springer Verlag 1975.

[Z] M. ZIEGLER : Die elementare Theorie der henselschen Körper, Inaugural Dissertation Köln 1972.

A basic book about valued fields is :

[R] P. RIBENBOÏM : Théorie des valuations, Les Presses de l'Université de Montréal, (1964).

ON POLYNOMIALS OVER REAL CLOSED RINGS

M.A. DICKMANN

CNRS - Université de Paris VII

INTRODUCTION.

The first order theory RCR of real closed rings (= ordered commutative unitary rings with the intermediate value property, which are not fields) was introduced in Cherlin-Dickmann [3], and some of its basic metamathematical properties studied there; see Part I, §1.B below for further details.

In this paper we deal with polynomial rings $A[X_1,\dots,X_n]$ ($=A[\overline{X}]$) where A is a model of RCR. In Part I we prove a "nullstellensatz" for ideals of a certain type in rings of this form. To be precise, we prove the following result:

THEOREM 1. Let A be a real closed ring, and let M_A denote its maximal ideal. Let I be an ideal of $A[\overline{X}]$ satisfying the following conditions:

(1) I is finitely generated.

(2) For every $k \geq 1$, $P_1,\dots,P_k,G_1,\dots,G_k \in A[\overline{X}]$ and $a_1,\dots,a_k \in M_A$,

$$\sum_{i=1}^{k}(1-a_iP_i)\cdot G_i^2 \in I \implies G_1,\dots,G_k \in I.$$

(3) For every $Q \in A[\overline{X}]$ and $b \in M_A$, $b\neq 0$:

$$bQ \in I \implies Q \in I.$$

Then, for any $P \in A[\overline{X}]$, the following are equivalent:

(a) $V_A(I) \subseteq V_A(P)$,

(b) $P \in I$.

Here $V_A(S) = \{\bar{a} \in A^n \mid Q(\bar{a}) = 0 \text{ for all } Q \in S\}$ denotes the variety over A of the set S of n-variable polynomials with coefficients in A.

0.1. Remarks.

(i) Since every positive element of a real closed ring is a square (cf. §1.B below), conditions (2) and (3) can be recast into the single condition:

(4) $$\sum_{i=1}^{k} b_i \cdot (1-a_i P_i) \cdot G_i^2 \in I \wedge P_i, G_i \in A[\overline{X}] \wedge a_i \in M_A \wedge b_i \in A \wedge b_i > 0 \implies G_1, \ldots, G_k \in I.$$

(ii) Condition (2) implies that the ideal I is radical ($G^n \in I \implies G \in I$, for $n \geq 1$), and that no polynomial of the form $1-aP$, with a infinitesimal (i.e., $a \in M_A$), belongs to I, unless I is improper. In other words, this last condition means that no non-invertible element of A becomes invertible in $A[\overline{X}]/_I$; i.e., infinitesimal elements of A stay infinitesimal in $A[\overline{X}]/_I$.

(iii) Likewise, condition (3) says that no non-zero element of A becomes a zero-divisor in $A[\overline{X}]/_I$. This condition replaces and strengthens the more common requirement $I \cap A = (0)$. If I is prime both conditions are obviously equivalent.

Theorem 1 presents many analogies with the Dubois-Risler "reel-nullstellensatz" for polynomial rings over ordered fields (cf. [5], [6], [9], [10]). For example, our assumptions correspond to their requirement that the ideal I be "real"; in the case of real closed fields, [10], this means:

$$\sum_{i=1}^{n} G_i^2 \in I \implies G_1, \ldots, G_n \in I,$$

which corresponds to our condition (2); in the case of an arbitrary ordered field k, this requirement takes the form:

$$\sum_i b_i G_i^2 \in I \wedge b_i \in k \wedge b_i > 0 \implies G_i \in I,$$

(cf. [5] and [9,p.17-12]), which clearly relates to our condition (4).

The presence in our assumptions of additional factors of the form $1-aP$ ($a \in M_A$) is due, firstly, to the reasons given in Remark 0.1(ii) above and, secondly, to the fact that polynomials of this type are positive definite in A.

Following these analogies we shall call real a polynomial ideal verifying condition (2) of Theorem 1, and strongly real one which satisfies conditions (2) and (3).

In Part II of this paper we give an algebraic characterization of polynomials in $A[\overline{X}]$ which are positive definite (i.e., always take on non-negative values) in A.

In the context of real closed rings the formulation of this problem differs from that of Hilbert's classical 17th problem for fields: the presence of infinitesimal (i.e., non-invertible) elements makes positive definite certain polynomials other than sums of squares of rational functions; for example, the polynomial $aX+1$, where $a \in M_A$, $a \neq 0$, is positive definite (cf. property I.e, §1.B below). Likewise, all polynomials $Q \in A[\overline{X}]$ representable in the form:

$$Q \cdot \sum_j (1-b_j Q_j) H_j^2 = \sum_i (1-a_i P_i) G_i^2 ,$$

(with all $a_i, b_j \in M_A$ and $Q_j, H_j, P_i, G_i \in A[\overline{X}]$), are positive definite.

Our main result in this section, THEOREM 2, proves, conversely, that polynomials which are positive definite in A are necessarily representable in this form. This can be interpreted as saying that the polynomials $1-aP$ with $a \in M_A$ form, together with the squares, a "basis" for the class of all positive definite polynomials over a real closed ring A.

The methods used in this paper are inspired from the model-theoretic proofs of the corresponding classical results due to Robinson and some of his disciples (cf. Cherlin [1,pp.541-543] and [2,pp.22-30]). Application of these methods to the present case is possible thanks to a simple characterization of elementary inclusion between models of RCR (Fact 1.2, §1.B below), although —contrary to the cases of algebraically closed and of real closed fields— this theory is not model-complete.

It is an open question whether the results of this paper admit purely algebraic proofs. Notice, however, that some of the tools employed in known algebraic proofs of the Artin-Schreier theorem and of the Dubois-Risler "reelnullstellensatz" do not seem to have an appropriate analog in the theory of real closed rings.

The main results of this paper were announced in [4]. The author wishes to acknowledge useful discussions with Ch. Berline, F. Delon and A. Macintyre.

PART I. THE NULLSTELLENSATZ.

§1. PRELIMINARIES.

A. Some known forms of the nullstellensatz for polynomial rings.

In [2,pp.103-104] Cherlin proves the following generalized version of a nullstellensatz:

THEOREM A. Let T be an inductive theory of rings, A an existentially complete model of T, I a finitely generated ideal of $A[\overline{X}]$ such that $I \cap A = (0)$, and $P \in A[\overline{X}]$. Then the following are equivalent:

(1) $V_A(I) \subseteq V_A(P)$,

(2) P is in the T-radical of I.

As before, $V_A(S)$ denotes the variety of S in A; the T-radical of I is defined as follows:

$$\text{T-rad}(I) = \bigcap\{J \mid J \text{ is an ideal of } A[\overline{X}],\ I \subseteq J,\ J \cap A = (0) \text{ and } A[\overline{X}]/_J \text{ is embeddable in a model of } T\}.$$

The assumption that T be inductive is used only to secure the existence of existentially complete models.

Hilbert's nullstellensatz for algebraically closed fields follows immediately from Theorem A; Risler's version of the "reelnullstellensatz" for real closed fields is also a relatively simple consequence of it. It suffices to put, respectively:

$$T = F = \text{the theory of fields},$$
$$T = OF = \text{the theory of ordered fields},$$

and prove that:

$$\text{F-rad}(I) = \{P \in A[\overline{X}] \mid P^m \in I \text{ for some } m \geq 1\},$$

$$\text{OF-rad}(I) = \{P \in A[\overline{X}] \mid \text{there are } r \geq 1 \text{ and } Q_1,\ldots,Q_k \in A[\overline{X}] \text{ such that } P^{2r} + \sum_j Q_j^2 \in I\}.$$

We remark in passing that more general versions of Hilbert's nullstellensatz (e.g., that of Lang [7,p.256]), as well as Dubois' version of the "reelnullstellensatz" for ordered fields (cf. Dubois-Efroymson [6,pp.114-115]) can be similarly derived from the following generalization of Theorem A:

THEOREM B. Let T be an inductive theory of rings, B an existentially complete model of T, $A \subseteq B$ (A is not necessarily a model of T!), I a finitely generated ideal of $A[\overline{X}]$ such that $I \cap A = (0)$, and $P \in A[\overline{X}]$. Then the following are equivalent:

(1) $V_B(I) \subseteq V_B(P)$,

(2) $P \in \bigcap\{J \mid J$ is an ideal of $A[\overline{X}]$, $J \cap A = (0)$, $I \subseteq J$ and $A[\overline{X}]/_J$ is embeddable in a model C of T so that the diagram

$$\begin{array}{ccccc} & & B & & \\ & \nearrow & & \searrow & \\ A & & & & C \\ & \searrow & & \nearrow & \\ & & A[\overline{X}]/_J & & \end{array}$$

commutes$\}$.

However, Theorem 1 cannot be derived directly from either of the preceding results; in fact, none of the above definitions of the radical is the appropriate one for models of RCR. Secondly, RCR is not an inductive theory; furthermore, it does not have existentially complete models. We shall see, however, that using the correct model-theoretic notion of radical the argument proving Theorems A and B can still be used to prove Theorem 1.

B. Some properties of real closed rings.

Here we summarize the elementary algebraic properties of real closed rings that will be used in the sequel. For more information see Cherlin-Dickmann [3] and [3a].

DEFINITION 1.1. A real closed ring is an ordered, commutative, unitary ring A which is not a field, having the intermediate value property:

(IVP) if $Q \in A[X]$ is a polynomial in one variable which changes sign in A, i.e., $Q(a)Q(b) < 0$ for some $a,b \in A$, $a < b$, then Q has a root $c \in A$ such that $a < c < b$.

Numerous examples of such rings occur in practice; e.g.:

(i) rings of the form $C(Y)/_P$ for large classes of prime, non-maximal ideals P in rings $C(Y)$ of real-valued continuous functions on completely regular topological spaces Y;
(ii) convex subrings of real closed fields.
The reader is referred to [3] for a comprehensive analysis of examples.

The following conditions, added to the axioms for commutative, ordered, unitary rings having a non-invertible element, provide an alternative axiomatization for RCR's:

(I) $0 \leq a \leq b \implies b|a$ (b divides a);

(II) every positive element has a square root;

(III) every monic polynomial of odd degree has a root.

The crucial property here is (I); it is equivalent to each of the following:

(I.a) $|a| \leq |b| \implies b|a$ (where $|\cdot|$ denotes absolute value);

(I.b) the intermediate value property (IVP) for linear polynomials Q.

Obviously (I.a) implies:

(I.c) for all a,b, $a|b$ or $b|a$;

i.e., any RCR is a valuation ring, and hence a local ring.

(I.d) a invertible and $|a| \leq |b|$ imply b invertible;

(I.e) a non-invertible and b invertible imply $|ax| < |b|$ for all $x \in A$.

This says that the (unique) maximal ideal M_A of A is a convex subset of A; in particular, if $a \in M_A$, the polynomials $1-aP$ take on only positive values in A.

Example 10 of [3a] shows that the first order theory RCR is not model-complete. However, we have the following simple but important characterization of elementary inclusion for models of RCR:

<u>FACT 1.2</u>. Let $A \subseteq B$ be models of RCR. The following are equivalent:

(a) $A \prec B$;
(b) if $a,b \in A$ and $B \models a|b$, then $A \models a|b$;
(c) $M_A = M_B \cap A$;
(d) if $b \in B$ and $b > a$ for all $a \in A$, then $b^{-1} \notin A$.

§2. MODEL-THEORETIC ARGUMENTS.

Throughout this paper A will stand for an arbitrary model of the theory RCR. All embeddings of rings extending A will be A-embeddings, i.e., leave A pointwise fixed.

The RCR-radical of an ideal $I \subseteq A[\bar{X}]$ is defined as follows:

$$\text{RCR-rad}(I) = \bigcap\{J \mid J \text{ is an ideal of } A[\bar{X}],\ I \subseteq J,\ J \cap A = (0) \text{ and } A[\bar{X}]/_J \text{ is A-embeddable in a model } B \text{ of RCR such that } A \prec B\}.$$

2.1. Remarks.

(i) An ideal J satisfying the requirements of the preceding definition is necessarily prime, since $A[\bar{X}]/_J \subseteq B$ and B is an integral domain. Thus, the adjective "prime" can be added to the definition above without altering the result.

(ii) If J is prime, $J \cap A = (0)$ is equivalent to condition (3) of Theorem 1. Hence RCR-rad(I) has this property too.

Next we prove:

PROPOSITION 2.2. For a finitely generated ideal $I \subseteq A[\bar{X}]$ and $P \in A[\bar{X}]$, the following are equivalent:

(a) $V_A(I) \subseteq V_A(P)$,

(b) $P \in$ RCR-rad(I).

Proof. Let I be generated by $P_1,\dots,P_k \in A[\bar{X}]$.
(a)$\Longrightarrow$(b). Assume that $V_A(I) \subseteq V_A(P)$, $I \subseteq J$, $J \cap A = (0)$ and $C = A[\bar{X}]/_J \subseteq B$, with $B \models$RCR and $A \prec B$. We must show that $P \in J$. If not, the point $\langle X_1/_J,\dots,X_n/_J\rangle$ viewed as a point of C^n lies on $V_C(I)$ (since $I \subseteq J$) but not in $V_C(P)$ (since $P \notin J$). Consider the following sentence σ:

"there is a point in $V(P_1,\dots,P_k)$ which is not in $V(P)$".

This is an existential sentence with parameters in A, true in C; hence $B \models \sigma$. Since $A \prec B$, we have $A \models \sigma$, contradicting (a).

(b)$\Longrightarrow$(a). This is exactly as the proof of Theorem A. Assume that $P \in$ RCR-rad(I) and $\bar{a} \in A^n$ is a point in $V_A(I)$. We have to show that $\bar{a} \in V_A(P)$. Let $e: A[\bar{X}] \to A$ be the evaluation map at $\bar{a}$: $e(Q) = Q(\bar{a})$ for $Q \in A[\bar{X}]$. The map e obviously is a surjective ring homomorphism, and $e|A$ is the identity. Let $J = \text{Ker}(e)$. Evidently

$J \supseteq I$ (since $\bar{a} \in V_A(I)$), $J \cap A = (0)$ and $A[\bar{X}]/_J$ is A-isomorphic to A, which is a model of RCR and an elementary extension of itself. It follows that RCR-rad$(I) \subseteq J$, and hence $P \in J$, i.e., $e(P) = 0$; therefore $\bar{a} \in V_A(P)$.

Now we study more closely the condition:

(*) $A[\bar{X}]/_J$ is A-embeddable in a model B of RCR such that $A \prec B$, where $J \supseteq I$, $J \cap A = (0)$.

The equivalence (a)$\Longleftrightarrow$(c) of 1.2 yields:

PROPOSITION 2.3. Condition (*) is equivalent to:

(**) $A[\bar{X}]/_J$ has an order $<$ such that $aP/_J < 1$ holds for all $a \in M_A$ and $P \in A[\bar{X}]$.

Proof. (*)$\Longrightarrow$(**). Let $C = A[\bar{X}]/_J$. If $B \models$ RCR, $C \subseteq B$, $A \prec B$, in the order that B induces on C we have $aP/_J < 1$, since this holds for all $x \in M_B$ and $a \in M_A \subseteq M_B$ implies $aP/_J \in M_B$.

(**)$\Longrightarrow$(*). Endow C with an order $<$ as in (**). Let K be the field of fractions of C provided with the order induced by that of C. Let $\tilde{K}$ be a real closure of K and

$$B = \{x \in \tilde{K} \mid \text{there is } c \in C \text{ such that } |x| \le |c|\}.$$

B is a convex subring of the real closed field $\tilde{K}$ and therefore it satisfies the intermediate value property.

Notice that any order of C has to extend that of A, because every positive element of A is a square in A (Property (II), §1.B) and hence in C (since $J \cap A = (0)$). This implies that the inclusion of C into B leaves A pointwise fixed, and that $A \subseteq B$ as ordered rings; in particular $M_A \supseteq M_B \cap A$.

To prove the reverse inclusion, suppose there is $x \in M_A$ such that $x \notin M_B$; we can also assume that $x > 0$. Then x is invertible in B: $xy = 1$ for some $y \in B$, whence $y \le c$ for some $c \in C$; this implies that $1 = xy \le xc$, contradicting the assumption (**). This proves that $M_A = M_B \cap A$, and by 1.2 that $A \prec B$.

Condition (**) can be rephrased as follows:

(***) there is an order for $A[\bar{X}]/_J$ which makes the ideal $M = M_A \cdot A[\bar{X}]/_J$ infinitesimal.

[Property (I.c) of §1.B implies that the set M is, indeed, an ideal of $A[\bar{X}]/_J$.]

§3. SOME ALGEBRAIC RESULTS.

We pause now to review some algebraic notions and results which we shall need to complete the proof of Theorem 1 and to prove Theorem 2. Most of these are slight generalizations or simply variants of known arguments, conveniently adapted to our purposes.

A. Multiplicative sets modulo squares.

DEFINITION 3.1. Let R be an integral domain and $F \subseteq R$. F is called *multiplicative modulo squares* iff for all $a,b \in F$ there are $c \in F$ and $y \in R$ such that $ab = cy^2$.

Obviously, every multiplicative subset of R is multiplicative modulo squares (put $y = 1$). The converse is not true, as shown by the example $F = \{x,-x,-1,1\}$, where $x \neq 0,1,-1$. Another example is provided by the sets

$$M_x = \{x^{\varepsilon}y \mid y \in M, \varepsilon = 0,1\},$$

where M is a multiplicative set, $x \in R$ is a fixed non-zero element, and $x^{\varepsilon} = 1$ if $\varepsilon = 0$, $x^{\varepsilon} = -x$ if $\varepsilon = 1$. Notice that M_x has the property:

$$a,b \in M_x \implies ab \in M_x \text{ or there is } c \in M_x \text{ such that } ab = cx^2$$

(i.e., the y of Definition 3.1 is x or 1). Likewise, the product of finitely many members of M_x is in M_x or of the form cx^{2k} for some $c \in M_x$, $k \geq 1$.

The next result generalizes Theorem 1 of Ribenboim [8,pp.145-147]; we state it in abbreviated form.

PROPOSITION 3.2. Let R be an integral domain of characteristic $\neq 2$ and $F \subseteq R$ a set multiplicative modulo squares such that $0,1 \in F$. The following are equivalent:

(1) there is an order $\leq$ of R such that $a \geq 0$ for all $a \in F$;

(2) if $\sum_{i=0}^{n} a_i x_i^2 = 0$, where $a_i \in F$, $a_i \neq 0$, $x_i \in R$ $(0 \leq i \leq n)$, then $x_0 = \ldots = x_n = 0$.

Proof. A slight modification of the proof of Theorem 1 of [8] yields the result when R is a field. The general case is, then, easily derived by applying the result to the field of fractions of R.

PROPOSITION 3.3. Let R be an integral domain, $J \subseteq R$ a proper prime ideal, and $M \subseteq R$ a multiplicative set modulo squares such that $1 \in M$.

The following are equivalent:

(1) There is an order $<$ of $R/_J$ such that $a/_J > 0$ for all $a \in M$.

(2) No element of the form $\sum_i a_i/_J \, x_i^2/_J$ with $a_i \in M$, $x_i \notin J$, is the negative of a sum of squares in $R/_J$.

(3) Same as (2) with $R/_J$ replaced by its field of fractions.

(4) For $a_i \in M$, $x_i \in R$,

$$\sum_i a_i x_i^2 \in J \implies x_i \in J \quad (0 \le i \le n,\ n \in \omega).$$

<u>Proof</u>. (2)$\iff$(3) is immediate using that J is prime.

(1)$\implies$(2). Since a square is non-negative in any order, $x_i \notin J$ implies $x_i^2/_J > 0$. Then (1) implies that for all $y_j \in R$ $(0 \le j \le k;\ k \in \omega)$:

$$\sum_i a_i/_J \, x_i^2/_J > 0 \ge -\sum_j y_j^2/_J.$$

(2)$\implies$(4). If $\sum_i a_i/_J \, x_i^2/_J = 0 = -0^2$, then (2) gives at once $x_i \in J$.

(4)$\implies$(1). Note that (4) implies that $a/_J \ne 0$ for all $a \in M$ (otherwise $0 = a/_J = a/_J \cdot 1^2/_J$ and (4) would give $1 \in J$, contradicting that J is proper) and that $\mathrm{char}(R/_J) \ne 2$ (otherwise $2 = 1\cdot 1^2 + 1\cdot 1^2$ together with (4) would contradict again the assumption that J is proper). Since (4) is just condition (2) of Proposition 3.2 with $R/_J$ instead of R and $F = \{a/_J \mid a \in M\} \cup \{0\}$, we conclude the existence of an order $\le$ of $R/_J$ such that $a/_J \ge 0$ for all $a \in M$; by the above, $a/_J > 0$ for $a \in M$.

<u>Remark</u>. Note that (4) obviously implies that J is a radical ideal.

B. <u>A generalized radical</u>.

Generalizing the "real radical" of an ideal (cf. [10,p.114]), we introduce the following:

<u>DEFINITION 3.4</u>. Let R be a commutative ring with unit, I an ideal of R and $M \subseteq R$ a multiplicative set such that $1 \in M$. The <u>M-radical of I</u> is the set:

$$\sqrt[M]{I} = \{r \in R \mid \text{for some } k > 0,\ n \ge 0,\ a, b_1, \dots, b_n \in M, \text{ and } x_1, \dots, x_n \in R,\ r^{2k}a + \sum_i b_i x_i^2 \in I\}.$$

<u>PROPOSITION 3.5</u>. With R, I and M as in Definition 3.4, we have:

(1) $\sqrt[M]{I}$ is an ideal and $I \subseteq \sqrt[M]{I}$.

(2) $\sum_i b_i x_i^2 \in \sqrt[M]{I}$ and $b_i \in M \implies x_i \in \sqrt[M]{I}$.

(3) If $J \supseteq I$ is an ideal such that

$$\sum_i b_i x_i^2 \in J \text{ and } b_i \in M \implies x_i \in J,$$

then $J \supseteq \sqrt[M]{I}$.

(4) $\sqrt[M]{I} = \sqrt[M]{\sqrt[M]{I}}$.

Proof. The only possible difficulty may arise in the verification that $\sqrt[M]{I}$ is closed under addition. This is done as in [9,p.17-22], by estimating the term $(r + s)^{2\ell} + (r - s)^{2\ell}$, where $r,s \in \sqrt[M]{I}$, for an appropriate $\ell > 0$.

We shall call M-radical any ideal I such that $I = \sqrt[M]{I}$. By (1) and (3) of the preceding proposition this is equivalent to:

$$\sum_i b_i x_i^2 \in I \text{ and } b_i \in M \implies x_i \in I.$$

Note that if I is M-radical, then I is proper iff $I \cap M = \emptyset$. Indeed, if $a \in I \cap M$, then $a = a \cdot 1^2 \in I$, whence $1 \in I$; the converse is obvious because $1 \in M$. In particular, if I is M-radical and proper, then $0 \notin M$.

With this notation we have:

3.7. EXAMPLES. The following holds for polynomial ideals over a real closed ring A:

(i) M_0-radical = real,

for the multiplicative set $M_0 = \{1-aP \mid a \in M_A, P \in A[\overline{X}]\}$.

(ii) M_1-radical = strongly real,

for the multiplicative set

$$M_1 = M_0 \cdot A^+ = \{b(1-aP) \mid b \in A, b > 0, a \in M_A, P \in A[\overline{X}]\}.$$

LEMMA 3.8. Let R be an integral domain, M a multiplicative subset of R containing 1, and I a proper M-radical ideal.

(i) If P is maximal among proper M-radical ideals containing I, then P is prime.

(ii) If, in addition, A is a subring of R, $I \cap A = (0)$, and P is maximal among M-radical ideals J containing I and such that $J \cap A = (0)$, then P is prime.

Proof. We prove (ii); (i) will get proved along the way.
Assume $a_1, a_2 \notin P$ but $a_1 a_2 \in P$. Let $J_i = \sqrt[M]{\langle P, a_i \rangle}$, where $\langle P, a_i \rangle$ denotes the ideal generated by P and a_i $(i = 1,2)$. Since J_i is M-radical and contains P properly, it follows that either $J_i = R$ or $J_i \cap A \neq (0)$. Analyzing the possible combinations of these two situa-

tions, we get:

Case 1) $J_1 = J_2 = R$. Then $1 \in J_i$, i.e.,

$$c_i + \sum_j c_{ji}x_{ji}^2 = a_ih_i+p_i$$

for some $c_i, c_{ji} \in M$, $p_i \in P$, $h_i, x_{ji} \in R$ $(i = 1,2)$. Since $a_1a_2 \in P$ and M is multiplicative, the product of the preceding expressions gives:

$$c_1c_2 + \sum_\ell d_\ell y_\ell^2 \in P$$

for some $d_\ell \in M$, $y_\ell \in R$. Since $c_1c_2 \in M$ and P is M-radical, we obtain $1 \in P$, contradiction. In particular, this proves (i).

Case 2) $J_1 = R$, $J_2 \cap A \neq (0)$. Since $1 \in J$ as before we have $c + \sum_j c_jx_j^2 = a_1h_1+p_1$, with $c, c_j \in M$, $h_1, x_j \in R$, $p_1 \in P$. Let $d \in J_2 \cap A$, $d \neq 0$; then $d^2 \in J_2 \cap A$, whence $d^2 = a_2h_2+p_2$ for some $h_2 \in R$, $p_2 \in P$. Multiplying these expressions and recalling that $a_1a_2 \in P$, we obtain:

$$d^2c + \sum_j c_j(dx_j)^2 \in P.$$

Since P is M-radical and $c \in M$, this implies $d \in P$; hence $P \cap A \neq (0)$, contradiction.

Case 3) $J_i \cap A \neq (0)$ for $i = 1,2$. We leave this as an exercise.

C. Results for certain multiplicative sets modulo squares.

The results of paragraph B above generalize, with minor modifications, to the multiplicative sets modulo squares of type M_x introduced following Definition 3.1. We state these results without proofs. It is assumed throughout that the multiplicative set M underlying M_x contains 1.

The M_x-radical of an ideal I is defined by:

$$\sqrt[M_x]{I} = \{r \in R \mid \text{for some } k_1,k_2 > 0,\ n \geq 0,\ a,b_1,\dots,b_n \in M_x \text{ and } y_1,\dots,y_n \in R,\ r^{2k_1}x^{2k_2}a + \sum_i b_iy_i^2 \in I\}.$$

PROPOSITION 3.9. With R, I, M and x as above, we have:

(1) $\sqrt[M_x]{I}$ is an ideal and $I \subseteq \sqrt[M_x]{I}$.

(2) i) $\sum_i b_iy_i^2 \in \sqrt[M_x]{I}$ and $b_i \in M_x \implies y_i \in \sqrt[M_x]{I}$.

ii) $rx \in \sqrt[M_x]{I} \implies r \in \sqrt[M_x]{I}$.

(3) If $J \supseteq I$ is an ideal such that

$\sum_i b_iy_i^2 \in J$ and $b_i \in M_x \implies y_i \in J$,

$$rx \in J \implies r \in J,$$

then $J \supseteq {}^{M_x}\!\sqrt{I}$.

(4) $${}^{M_x}\!\sqrt{I} = {}^{M_x}\!\sqrt{{}^{M_x}\!\sqrt{I}}\ .$$

The notion of an M_x-radical ideal is defined as in paragraph B. By (1) and (3) of Proposition 3.9, I is M_x-radical if and only if:

$$\sum_i b_i y_i^2 \in I \text{ and } b_i \in M_x \implies y_i \in I,$$

$$rx \in I \implies r \in I.$$

Notice that an M_x-radical ideal is necessarily radical. Hence we have

$$rx^n \in I \implies r \in I \quad \text{for all } n \geq 1,$$

i.e., no power of $x/_I$ is a zero-divisor in $R/_I$.

Notice too that if I is M_x-radical, then I is proper iff $I \cap M = \emptyset$ (rather than $I \cap M_x = \emptyset$). Indeed, if $a \in I \cap M$, then $a \in M_x$ and $ax^2 \in I$; from 3.9(2.i), $x \in I$, and then by 3.9(2.ii), $1 \in I$.

The following analog of Lemma 3.8(i) is valid.

LEMMA 3.10. Let R, M be as in Lemma 3.8, $x \in R$, $x \neq 0$, and I a proper M_x-radical ideal. Let P be maximal among proper M_x-radical ideals containing I. Then P is prime.

As an exercise the reader can formulate and prove the corresponding analog of 3.8(ii).

§4. ALGEBRAIC CHARACTERIZATION OF THE RCR-RADICAL.

Now we combine the results of §2 with those of §3 applied to the integral domain $R = A[\overline{X}]$ and the multiplicative sets M_0 and M_1 considered in 3.7, to complete the proof of Theorem 1. This is an immediate consequence of Proposition 2.2 and

PROPOSITION 4.1. If $I \subseteq A[\overline{X}]$ is a strongly real ideal, then $I = \text{RCR-rad}(I)$.

Proof. Since I is strongly real, it is also real, and we have $I = {}^{M_1}\!\sqrt{I} = {}^{M_0}\!\sqrt{I}$ with M_0, M_1 as defined in 3.7

On the other hand, Proposition 2.3 proves that RCR-rad(I) is the intersection of all prime ideals J containing I such that $J \cap A = (0)$ and $A[\overline{X}]/_J$ has an order making $1-aP/_J$ positive for all $a \in M_A$ and $P \in A[\overline{X}]$. The equivalence between (1) and (4) of

Proposition 3.3 applied to $R = A[\overline{X}]$ and the set M_0 entails, then, that RCR-rad(I) is the intersection of all real prime ideals J containing I and such that $J \cap A = (0)$. It follows that RCR-rad(I) is a real ideal too. By remark 2.1(ii) it is strongly real. By Proposition 3.5(3) we conclude that $\sqrt[M_1]{I} \subseteq$ RCR-rad(I).

To show the reverse inclusion, assume that $P \notin \sqrt[M_1]{I}$; then $P^{2k}b(1-aQ) \notin I$ for all $k > 0$, $b \in A$, $b > 0$, $a \in M_A$, $Q \in A[\overline{X}]$. Since I is strongly real this relation is also valid for $k = 0$; in other words, $I \cap N = \emptyset$, where N is the multiplicative set

$$\{P^{2k}b(1-aQ) \mid k \geq 0,\ b \in A^+,\ Q \in A[\overline{X}]\}.$$

By 3.5 and the remark following it we have that $\sqrt[N]{I}$ is a proper N-radical ideal. Taking J to be a maximal, proper, N-radical ideal containing $\sqrt[N]{I}$, Lemma 3.8(i) ensures that J is prime. Since $M_1 \subseteq N$, J is M_1-radical and therefore strongly real. Since $P^2 \in N$ and $J \cap N = \emptyset$, we have $P^2 \notin J$ and hence $P \notin J$. It follows that $P \notin$ RCR-rad(I), as we wanted to prove.

§5. EXAMPLES AND OPEN PROBLEMS.

It is natural to ask whether the assumptions (1)-(3) of Theorem 1 are mutually independent.

5.1. EXAMPLES.

(a) The obvious example $I = \langle 1-aX \rangle$, where $a \in M_A$, $a \neq 0$, gives a proper, finitely generated, non-real ideal satisfying condition (3). To check this last assertion, equalize coefficients on the polynomial identity $bF = (1-aX)Q$, where $b \in M_A$, $b \neq 0$, and $Q(X) = c_nX^n+\ldots+c_0$. One obtains the relations $b|c_0$, $b|c_i-ac_{i-1}$ $(1 \leq i \leq n)$, $b|-ac_n$, which inductively yield $b|c_i$ for all i. Hence $F \in I$.

(b) The ideal $I = \sqrt[M_0]{\langle aX \rangle}$, where $a \in M_A$, $a > 0$, is an example of a proper, non-finitely generated, real but not strongly real ideal. Indeed, I is generated by the monomials b_nX, where b_n is a 2^n-th root of a $(b_0 = a)$, and it is easy to see that no finite subset of these generates I. Moreover, $X \notin I$.

To check the preceding assertion, since I is radical and contains aX, it also contains b_nX for all $n \in \omega$. To prove the converse it suffices to verify that $J = \langle b_nX \mid n \in \omega \rangle$ is a real ideal. If $\sum_i (1-a_iP_i)G_i^2 \in J$, then $\sum_i (1-a_iP_i)G_i^2 = b_nXQ$ for some $n \in \omega$ and $Q \in A[\overline{X}]$. For $X = 0$ this polynomial identity gives $G_i(0) = 0$ and then $X|G_i$ for all i. Let now c be largest in absolute value among the coefficients of all the G_i's; by property (I.a) of §1.B, c divides all coefficients of all the G_i's and we have:

$$\sum_i (1-a_iP_i)G_i^2 = c^2\cdot\sum_i (1-a_iP_i)H_i^2,$$

where some H_i has a term with coefficient 1. Choosing X equal to some invertible element d so that $Q(d) \neq 0$ (such an element exists because Q has a finite number of roots), and using property (I.d) of §1.B, the two polynomial identities above give $b_n|c^2\cdot e$, where $e = \sum_i (1-a_iP_i(d))(H_i(d))^2$ is invertible. It follows that $b_n|c^2$, and then $b_{n+1}|c$. Thus $b_{n+1}X|G_i$, and $G_i \in J$ for all i.

Note that Theorem 1 fails in both these examples.

Apparently, the situation becomes much more involved when one considers the mutual independence of assumptions (1) and (3) of Theorem 1 in the presence of condition (2). I have only been able to decide this question in the simple case of principal ideals of polynomials in one variable; in this case (3) follows from (1) and (2); see Corollary II.2 below.

5.2. OPEN PROBLEMS.

(a) Does finitely generated and real imply strongly real?

(b) Is a strongly real ideal necessarily finitely generated?

(c) Are conditions (1) and (3) equivalent for a real ideal?

The following question seems to be related to those above:

(d) Is there a real ideal such that $\sqrt[M_1]{I} \subsetneq \text{RCR-rad}(I)$?

An answer to any of these problems may give valuable insight concerning the behaviour of algebraic curves over real closed rings; this behaviour may be specially interesting in the case of a positive answer.

PART II. REPRESENTATION OF POSITIVE DEFINITE POLYNOMIALS.

Let A be a real closed ring. A polynomial $Q \in A[\overline{X}]$ is positive definite on A iff $Q(\bar{a}) \geq 0$ for all $\bar{a} \in A^n$; otherwise, we shall say that Q changes sign in A^n. For $Q \in A[\overline{X}]$ and any ordered ring B extending A let us write:

$$V_B(Q < 0) = \{\bar{b} \in B^n \mid Q(\bar{b}) < 0\}.$$

Now we shall prove:

THEOREM 2. Let $A \models RCR$ and $Q \in A[\overline{X}]$ be a non-zero polynomial. Then Q is positive definite on A if and only if there are $a_i, b_j \in M_A$ and $P_i, G_i, Q_j, H_j \in A[\overline{X}]$, $G_i, H_j \neq 0$ such that

$$(*) \qquad Q \cdot \sum_j (1 - b_j Q_j) H_j^2 = \sum_i (1 - a_i P_i) G_i^2 .$$

Proof. As remarked in the Introduction all polynomials Q representable in this form are positive definite.

For the converse, suppose that Q is not representable in the form (*); we shall prove that Q is not positive definite on A. To achieve this we shall construct an ideal $J \subseteq A[\overline{X}]$ such that:

(i) J is prime, $J \cap A = (0)$ and J is real.

Thus, by the Example 3.7(i), the equivalence between (1) and (4) of Proposition 3.3 and Proposition 2.3, $C = A[\overline{X}]/_J$ has an order which makes it A-embeddable in a model B of RCR such that $A \prec B$; in addition we shall require this order to satisfy:

(ii) $Q/_J < 0$.

Assuming that such an ideal J exists, the proof is completed as follows; the value of the polynomial $Q(\overline{X})$ at a point $\langle F_1/_J, \ldots, F_n/_J \rangle$ of C^n is given by $Q(F_1/_J, \ldots, F_n/_J) = Q(F_1, \ldots, F_n)/_J$. Thus, we have: $Q(X_1/_J, \ldots, X_n/_J) = Q/_J < 0$, which shows that $V_C(Q < 0) \neq \emptyset$. Since C is A-embeddable in B, we obtain $V_B(Q < 0) \neq \emptyset$. As this is expressed by a sentence of the language of ordered rings with parameters in A, and $A \prec B$, it follows that $V_A(Q < 0) \neq \emptyset$, i.e., Q is not positive definite on A.

In order to construct an ideal J satisfying (i) and (ii), consider the multiplicative set modulo squares N_Q defined by Q and the multiplicative set $N = M_1$ of Example 3.7(ii). We claim that any maximal element J of the family $\mathcal{J}$ of proper N_Q-radical ideals of $A[\overline{X}]$ has the required properties. Indeed, Lemma 3.10 ensures that J is prime, and the remark preceding it that $J \cap N = \emptyset$; in particular, $J \cap A = (0)$. Since J is N_Q-radical and $M_0 \subseteq N_Q$, it is M_0-radical and, by 3.7(i), real. Finally, by Proposition 3.3 ((1) and (4)), $A[\overline{X}]/_J$ has an order such that $P/_J > 0$ for all $P \in N_Q$; in particular, $-Q/_J > 0$, i.e., $Q/_J < 0$.

Zorn's lemma ensures that the family $\mathcal{J}$ has a maximal element, provided it is non-empty. To prove this we check that $(0) \in \mathcal{J}$, i.e., that (0) is N_Q-radical.

It is clear that $QF = 0 \implies F = 0$, because $Q \neq 0$. On the other hand, if $\sum_i Q^{\varepsilon_i} b_i (1 - a_i P_i) G_i^2 = 0$ with $a_i \in M_A$, $b_i > 0$, $\varepsilon_i \in \{0,1\}$ and

$G_i \neq 0$, by considering c_iG_i instead of G_i, where $b_i = c_i^2$, we are reduced to the case where all $b_i = 1$. Separating the terms with $\varepsilon_i = 0$ from those with $\varepsilon_j = 1$ and noting that the ε_i's do actually take on both of these values (since $Q, G_i \neq 0$), we obtain:

$$\sum_i (1-a_iP_i)G_i^2 - Q \cdot \sum_j (1-a_jP_j)G_j^2 = 0,$$

which contradicts the assumption that Q is not representable in the form (*). Hence

$$\sum_i Q^{\varepsilon_i} b_i(1-a_iP_i)G_i^2 = 0 \implies G_i = 0.$$

This completes the proof of Theorem 2.

We use now Theorem 2 to obtain information about principal real ideals.

<u>COROLLARY II.1</u>. Let $F \in A[\overline{X}]$ be a polynomial which generates a real ideal. Then in any decomposition $F = F_1 \cdot \ldots \cdot F_k$ where the polynomials F_i have total degree ≥ 1, each F_i changes sign in A^n. In particular, any such decomposition is square-free. Furthermore, no F_i is a multiple of a different F_j by a constant factor.

<u>Proof</u>. Assume false and let $F = F_1 \cdot \ldots \cdot F_k$ be a decomposition of F violating the conclusion. Grouping together all factors which do not change sign in A^n we obtain a decomposition $F = G \cdot H$ where G is positive definite on A and has total degree $\deg(G) \geq 1$. By Theorem 2 G has a representation

$$(*) \qquad G \cdot \sum_j (1-b_jQ_j)H_j^2 = \sum_i (1-a_iP_i)G_i^2$$

($a_i, b_j \in M_A$), and therefore one such representation with minimal index $m = \sum_i \deg(G_i) + \sum_j \deg(H_j)$. Multiplying by H^2 we get

$$\sum_i (1-a_iP_i)(HG_i)^2 \in \langle F\rangle,$$

and since $\langle F\rangle$ is real, $HG_i \in \langle F\rangle$. Then $HG_i = FD_i = HGD_i$, and $G_i = GD_i$. Substituting back into (*) gives

$$\sum_j (1-b_jQ_j)H_j^2 = G \cdot \sum_i (1-a_iP_i)D_i^2.$$

Now $\deg(G) \geq 1$ implies $\deg(D_i) < \deg(G_i)$ and we have a representation of G of type (*) with smaller index $m' = \sum_i \deg(D_i) + \sum_j \deg(H_j)$, contradiction.

Using this corollary it is easy to determine the principal real ideals in $A[X]$. These are generated by polynomials $F = \alpha \cdot \prod_i (X-\beta_i)$

where α is invertible and the β_i's pairwise distinct. Indeed, if $\langle F\rangle$ is real, by the corollary F changes sign and (since $A \models RCR$) it has a root β_1; by [7,Theorem 4,p.121], $F = (X-\beta_1)G$. As $\deg(G) < \deg(F)$ either G is a constant ($= \alpha$) or, by the corollary, it changes sign and hence it has a root β_2. By induction $F = \alpha\cdot\prod_i(X-\beta_i)$, and by the corollary again the β_i's are pairwise different. In addition, since $\langle F\rangle$ is radical, we have $\alpha_n F' \in \langle F\rangle$ for all $n \in \omega$, where α_n is a 2^n-th root of α and $F' = \prod_i(X-\beta_i)$. This gives $\alpha|\alpha_n$, which implies that α is invertible. Conversely, it is clear that polynomials F of the form considered above generate real ideals.

<u>COROLLARY II.2</u>. Every principal real ideal of $A[X]$ is strongly real.

<u>Proof</u>. Assume $F \in A[X]$ generates a real ideal. By the preceding argument $F = \alpha\cdot\prod_i(X-\beta_i)$ with the β_i's pairwise different and α invertible. Obviously we can assume $\alpha = 1$. Then $F = X^n - s_1X^{n-} + \ldots + (-1)^n s_n$ where $s_i = s_i(\beta_1,\ldots,\beta_n)$ are the symmetric polynomials of the β_i's.

Let $bP = FQ$ with $Q = c_0X^m + \ldots + c_m$ and $b \neq 0$. Equalizing coefficients gives the relations $b|c_0$ and $b|c_\ell + \sum_{j=1}^{\ell}(-1)^j s_j c_{\ell-j}$ for $1 \leq \ell \leq n$; inductively this yields $b|c_\ell$ for all ℓ. Hence $P \in \langle F\rangle$.

<u>REFERENCES</u>.

[1] G. L. CHERLIN, Model theoretic algebra, J. Symb. Logic 41 (1976), 537-545.

[2] G. L. CHERLIN, <u>Model Theoretic Algebra; Selected Topics</u>, Lecture Notes in Math. 521, Springer, Berlin, 1976.

[3] G. L. CHERLIN and M. A. DICKMANN, Real closed rings and rings of continuous functions, to appear. Cf. also:

[3a] G. L. CHERLIN and M. A. DICKMANN, Anneaux réels clos et anneaux des fonctions continues, C. R. Acad. Sc. Paris 290 (1980), 1-4.

[4] M. A. DICKMANN, Sur les anneaux de polynômes à coefficients dans un anneau réel clos, C. R. Acad. Sc. Paris 290 (1980), 57-60.

[5] D. W. DUBOIS, A nullstellensatz for ordered fields, Arkiv für Math. 8 (1969-1970), 111-114.

[6] D. W. DUBOIS and G. EFROYMSON, Algebraic theory of real varieties, I., Studies and essays presented to Yu Why Chen on his 60th birthday, Acad. Sinica, Taipei, 1970, 107-135.

[7] S. LANG, Algebra, Addison-Wesley, Reading, Mass., 1971.

[8] P. RIBENBOIM, L'arithmétique des corps, Hermann, Paris, 1972.

[9] P. RIBENBOIM, Le théorème des zéros pour les corps ordonnés, Séminaire Dubreil-Pisot (Algèbre et Théorie des nombres), 24 (1970-71), Exposé No. 17.

[10] J.-J. RISLER, Le théorème des zéros en géométries algébrique et analytique réelles, Bull. Soc. math. France 104 (1976), 113-127.

UER de Mathématiques
Université de Paris VII
Tour 45-55, 5e. étage
2, Place Jussieu
75005 Paris - FRANCE

LES CORPS FAIBLEMENT ALGEBRIQUEMENT CLOS NON SEPARABLEMENT CLOS ONT LA PROPRIETE D'INDEPENDANCE

Jean-Louis DURET

Après que Saharon Shelah eut fondé la théorie de la stabilité, les modèles-théoriciens ont naturellement été amenés à chercher parmi les structures naturelles (groupes, anneaux, corps,...) celles dont les théories complètes sont stables.

Le premier théorème important concernant la stabilité des corps * est dû à Angus Macintyre [Ma]. Ce résultat a été généralisé par Saharon Shelah et Gregory Cherlin (à paraître) et on sait maintenant qu'un corps infini est ω-stable si et seulement s'il est superstable si et seulement s'il est algébriquement clos. Macintyre et Shelah ont en outre remarqué qu'un corps séparablement clos et non algébriquement clos est stable et non superstable [Sh, Th 8-8], [Wo]. Ce sont là les seuls exemples de corps stables actuellement connus.

Désespérant de démontrer la conjecture (à laquelle il semble bien que beaucoup de logiciens croient) que ce sont là les seuls corps stables, on peut chercher à attaquer le problème "par l'autre bout", c'est-à-dire en cherchant des corps instables. On remarque d'abord, à ce sujet, que les démonstrations de l'indécidabilité de corps montrent l'instabilité de ces corps et même qu'ils ont la propriété d'indépendance ; en effet, ces résultats ont été obtenus (quoique théoriquement ce ne soit pas nécessaire) en définissant (ou en traduisant) l'arithmétique élémentaire dans ces corps. On sait donc, que les corps de

------

* Tous les corps considérés ici sont commutatifs ; nous dirons qu'un corps est stable pour dire que sa théorie complète est stable.

dimension finie sur $\mathbb{Q}$ [RJ],les extensions transcendantes pures d'un corps formellement réel [RR], les extensions transcendantes simples d'un corps fini [Er] [Pe] ont la propriété d'indépendance. De plus, Ax ayant remarqué que l'anneau des séries formelles à une indéterminée sur un corps quelconque est définissable dans le corps des séries formelles à une indéterminée sur le même corps [Al], on sait (d'après [Fe]), que les corps de séries formelles à une indéterminée sont instables ; il est faux que ces corps aient en général, la propriété d'indépendance [De]. Enfin, les corps formellement réels sont instables [Du] ; il est faux qu'ils aient en général la propriété d'indépendance, car les corps réels clos ne l'ont pas.

Dans une discussion avec Bruno Poizat, nous amorçâmes la démonstration du fait que les corps pseudo-finis non séparablement clos ont la propriété d'indépendance ; Daniel Lascar termina la démonstration le lendemain. Le but de ce travail était de généraliser ce résultat aux corps "faiblement algébriquement clos" non séparablement clos ; cette notion de corps "faiblement algébriquement clos" coïncide avec celle des corps "faiblement pseudo-algébriquement clos" ("weakly pseudo-algebrically closed") introduite par William Wheeler [Wh,C-O]. En fait, les résultats les plus importants sont le théorème 6-4 et ses deux corollaires. Les résultats algébriques utilisés sont (outre les plus élémentaires) l'existence et la propriété fondamentale de la restriction de Weil d'une variété affine et les théorèmes sur les extensions cycliques. Ces résultats sont énoncés au début.

On pourrait espérer étendre la méthode exposée ici aux corps quasi-algébriquement clos ; c'est malheureusement impossible, car $\mathbb{C}((X))$ (où X est une indéterminée) est quasi-algébriquement clos et n'a pas la propriété d'indépendance [De].

Signalons que les exemples de corps instables cités plus haut sont différents de ceux résultant du théorème 6-4 (à l'exception peut-être des séries formelles). Il nous suffit de voir que ces corps n'ont pas de sous-corps faiblement algébriquement clos. Soient K une extension de degré fini de $\mathbb{Q}$, k un sous-corps de K, v le prolongement à K d'une valuation p-adique de $\mathbb{Q}$; la restriction de v à k est une valuation discrète à corps résiduel fini, non triviale (car non triviale sur $\mathbb{Q}$) ; or un corps avec une telle valuation n'est pas faiblement algébriquement clos [JM, Lemme 2-9] . Soient k un corps fini, t un élément transcendant sur k, K un sous-corps de k(t) ; ou bien K est fini et n'est donc pas faiblement algébriquement clos [Ax, p. 245] ; ou bien K contient un élément x de k(t) - k ; il y a une valuation P-adique (où P est un élément premier de k[t]) tel que la valuation de x n'est pas nulle ; la restriction de cette valuation à K est non triviale, discrète, à corps résiduel fini, donc K n'est pas faiblement algébriquement clos [JM, Lemme 2-9]. Pour les corps formellement réels (et notamment les extensions transcendantes pures d'un corps formellement réel), comme un sous-corps d'un corps formellement réel est formellement réel, il suffit de remarquer que l'idéal engendré par $X^2 + Y^2 + 1$ est absolument premier et que donc, un corps faiblement algébriquement clos n'est pas formellement réel. Un corps de séries formelles k((T)) peut évidemment contenir un corps faiblement algébriquement clos (par exemple si k est faiblement algébriquement clos) ; signalons quand même que k((T)) n'est pas faiblement algébriquement clos, car l'idéal engendré par $Y^3(T + 2X^2) + T^2(T + X^2)$ est absolument premier, mais n'a pas de zéro dans k((T)) (Delon).

Remarquons enfin qu'il résulte de la démonstration de 6-2 que si k est un corps faiblement algébriquement clos, et p un nombre premier différent de la caractéristique de k tels qu'il existe un élément de k sans racine $p^{\text{ième}}$ dans k, alors il y a 2^λ types au-dessus de tout ensemble de paramètres de k de cardinal λ.

Notation.- Si K est un corps commutatif, Th(K) désignera la théorie complète de K.

0.1.- Théorème. [Ja, p.96, th. 4 et p. 97, ex. 1] . Soient p un nombre premier ; k un corps commutatif de caractéristique différente de p, contenant les racines $p^{\text{ièmes}}$ de l'unité ; R_p le groupe multiplicatif des racines $p^{\text{ièmes}}$ de l'unité. Si le polynôme de k[X] $X^p - a$ n'a pas de racine dans k, alors $X^p - a$ est irréductible sur k, et son groupe de Galois sur k est cyclique d'ordre p ; si α est une racine de $X^p - a$, le corps de décomposition est $k[\alpha]$ et les k-automorphismes sont définis par :

$$\alpha \mapsto \xi\,\alpha \qquad \text{où} : \xi \in R_p .$$

0.2.- Théorème. [Bo, § 11, prop. 6] . Soient k un corps commutatif de caractéristique p, K une extension cyclique de k, de degré n non divisible par p, et tel que k contienne le groupe des racines $n^{\text{ièmes}}$ de l'unité. Il existe alors un polynôme irréductible de k[X], de la forme $X^n - a$, tel que K soit engendré par une racine quelconque de ce polynôme.

0.3.- Théorème. [Ja, p. 98, ex. 3] . Soient k un corps commutatif de caractéristique p = 0 et $\mathbb{F}_p$ le corps premier de caractéristique p. Si le polynôme de k[X] $X^p - X - a$ n'a pas de racine dans k, alors $X^p - X - a$ est irréductible sur k, et son groupe de Galois sur k est cyclique d'ordre p ; si α est une racine de $X^p - X - a$, le corps de décomposition est $k[\alpha]$ et les k-automorphismes sont définis par :

$$\alpha \mapsto \alpha + n \qquad \text{où} : n \in \mathbb{F}_p .$$

0.4.- Théorème. [Bo, § 11, ex. 8] . Soit k un corps commutatif de caractéristique p, non nulle. Si K est une extension cyclique de degré p de k, il existe alors un polynôme irréductible de k[X], de la forme $X^p - X - a$, tel que K soit engendré par une racine quelconque de ce polynôme.

(Une démonstration élémentaire du théorème suivant se trouve en appendice).

0.5.- Théorème. ([We, 1.3]). Soient k un corps commutatif, K une extension de k, séparable et de degré fini n, V une variété affine sur K. Il existe une variété affine sur k, $R_{K|k}V$, appelée restriction de Weil de V, telle que :

- $R_{K|k}V$ est isomorphe à V^n sur la clôture galoisienne de K sur k.
- Les points rationnels de V sur K sont en bijection avec les points rationnels de $R_{K|k}V$ sur k.

1.1.- Définition. Soient k un corps commutatif, $\bar{k}$ sa clôture algébrique, I un idéal de $k[X_1,\ldots,X_n]$, et J l'idéal de $\bar{k}[X_1,\ldots,X_n]$ engendré par I. On dit que I est absolument premier si et seulement si J est premier.

1.2.- Définition. Soit V une variété affine sur un corps commutatif k. On dit que V est absolument intègre si et seulement si son idéal (sur k) est absolument premier.

1.3.- Définition. On dit qu'un corps commutatif k est faiblement algébriquement clos si et seulement s'il satisfait l'un des deux énoncés équivalents suivants :

1) Toute variété affine sur k, absolument intègre a un point rationnel sur k.

2) Tout idéal d'un anneau de polynômes sur k à un nombre fini d'indéterminées, absolument premier, a un zéro dans k.

1.4.- Proposition. La classe des corps faiblement algébriquement clos est élémentaire.

Démonstration.- Soit $E = \{f_1(U,X),\ldots,f_n(U,X)\}$ un sous-ensemble fini de $\mathbb{Z}[U_1,\ldots,U_l,X_1,\ldots,X_m]$; il résulte immédiatement de [Va, chap.IV, p. 123, (1.1) et (1.4)] qu'il existe une formule (du premier ordre dans le langage des corps, dont les variables libres sont $u_1,\ldots,u_l$) $\phi'_E(u_1,\ldots,u_l)$ telle que, si K est un corps algébriquement clos et si

$a_1,\dots,a_l$ sont des éléments de K, on a : $K \models \phi'_E(a_1,\dots,a_l)$ si et seulement si l'idéal de $K[X_1,\dots,X_m]$ engendré par $\{f_1(a,X),\dots f_n(a,X)\}$ est premier. La théorie des corps algébriquement clos admettant l'élimination des quantificateurs, soit $\phi_E(u_1,\dots,u_l)$ une formule sans quantificateurs, équivalente dans cette théorie à $\phi'_E(u_1,\dots,u_l)$. Si k est un corps commutatif, et $\bar{k}$ sa clôture algébrique, et si $a_1,\dots,a_l$ sont des éléments de k, on a :

$$k \models \phi_E(a_1,\dots,a_l) \iff \bar{k} \models \phi_E(a_1,\dots,a_l)$$

$\iff$ L'idéal de $\bar{k}[X_1,\dots,X_m]$ engendré par E est premier.

$\iff$ L'idéal de $k[X_1,\dots,X_m]$ engendré par E est absolument premier.

Un corps commutatif est donc faiblement algébriquement clos si et seulement s'il satisfait :

$$\bigcup_{i\in\mathbb{N}} \{\forall u_1,\dots,u_l\ (\phi_E(u_1,\dots,u_l) \rightarrow \exists x (\bigwedge_{f\in E} f(u,x) = 0))\ ;\quad E \in \mathcal{P}_f(\mathbb{Z}[U,X])\}$$

1.5.- Proposition. Soient k un corps faiblement algébriquement clos, I un idéal absolument premier de $k[X_1,\dots,X_n]$, g un élément de $k[X_1,\dots,X_n] - I$. Alors il y a un zéro de I qui n'annule pas g.

Démonstration.- Soit K le corps des fractions de $k[X_1,\dots,X_n]/I = k[\bar{X}_1,\dots,\bar{X}_n]$ (où $\bar{X}_i$ est la classe de X_i). On a :

$$K = k(\bar{X}_1,\dots,\bar{X}_n)$$

Soit :

$$\phi : k[X_1,\dots,X_n,Y] \rightarrow K$$

définie par :

$$\begin{cases} \phi_{|k} = Id_k \\ \text{Pour tout } i = 1,\dots,n,\ \phi(X_i) = \overline{X}_i \\ \phi(Y) = \overline{g}^{\,-1} \qquad (g \notin I \text{ donc } \overline{g} \neq 0) \end{cases}$$

$A = k[X_1,\dots,X_n,Y]_{/\mathrm{Ker}\,\phi}$ est isomorphe à $\mathrm{Im}\,\phi$ qui contient $k[\overline{X}_1,\dots,\overline{X}_n]$ et est inclus dans K ; donc le corps des fractions de A est isomorphe à K. K, corps des fractions de $k[X_1,\dots,X_n]_{/I}$, et est extension régulière de k ([Ax,, p. 244, cor.3]) ; donc, A est absolument intègre (même corollaire). Donc $\mathrm{Ker}\,\phi$ est un idéal absolument premier, qui par conséquent a un zéro dans k. Or, I est inclus dans $\mathrm{Ker}\,\phi$ et $Y g - 1$ est élément de $\mathrm{Ker}\,\phi$, donc ce zéro annule I sans annuler g.

1.6.- Définition. On dit qu'un corps commutatif est pseudo-fini si et seulement s'il est parfait, faiblement algébriquement clos et a, dans une clôture algébrique fixée, exactement une extension de chaque degré fini.

2.1.- Proposition. Toute extension séparable finie d'un corps faiblement algébriquement clos est faiblement algébriquement close.

Démonstration.- Soient k un corps faiblement algébriquement clos, K une extension séparable finie de k, V une variété affine absolument intègre sur K. D'après 0.5, $R_{K|k}V$ est isomorphe à un produit de variétés sur K absolument intègres, donc est absolument intègre sur K, donc sur k ; $R_{K|k}V$ a donc un point rationnel sur k, donc, V a un point rationnel sur K.

2.2.- Corollaire. Toute extension algébrique séparable d'un corps faiblement algébriquement clos est faiblement algébriquement close.

Démonstration.- Soient k un corps faiblement algébriquement clos, K une extension algébrique séparable de k, V une variété affine absolument intègre sur K. Le plus petit corps de définition de V, K' est de dimension finie sur k, donc faiblement algébriquement clos, d'après 2 1. V a donc un point rationnel sur K', donc sur K.

3.0.- Lemme. Soient A un anneau commutatif intègre, et k son corps des fractions ; $f_1(X),\dots,f_n(X) \in A[X]$, unitaires, de degré au moins 1 ; α_i $(i = 1,\dots,n)$ une racine de f_i (dans la clôture algébrique de k). Si :

$$[k(\alpha_1,\dots,\alpha_n) : k] = \prod_{i=1}^{n} d^\circ f_i(X)$$

alors, l'idéal de $A[X_1,\dots,X_n]$ engendré par $\{f_i(X_i) \ ; \ i = 1,\dots,n\}$ est premier,et ne contient aucun élément de A non nul.

Démonstration.- Soient I et J les idéaux respectivement de $A[X_1,\dots,X_n]$ et de $K[X_1,\dots,X_n]$ engendrés par $\{f_i(X_i) \ ; \ i = 1,\dots,n\}$. Soient :

$$\begin{cases} K_0 = K \\ K_{i+1} = K_i(\alpha_i) \end{cases}$$

Comme : $f_{i+1}(X) \in K_i[X]$ et $f_{i+1}(\alpha_i) = 0$

on a : $[K_{i+1} : K_i] \leq d^\circ f_{i+1}(X)$

Pour $i = 1,\dots,n$, $f_i(X)$ est irréductible sur K_{i-1} ; car, s'il existait j tel que f_j ne soit pas irréductible sur K_{j-1}, alors le polynôme minimal de α_j sur K_{j-1} serait de degré strictement inférieur à $d^\circ f_j$; donc :

$$[K_n : K_0] = \prod_{i=0}^{n-1} [K_{i+1} : K_i] < \prod_{i=1}^{n} d^\circ f_i$$

ce qui est contraire à l'hypothèse. Donc :

$$K_{i+1} \simeq K_i[X_{i+1}]/(f_{i+1}(X_{i+1}))$$

$$K_n \simeq K[X_1,\dots,X_n]/_J$$

D'autre part :

$$B = A[X_1,\dots,X_n]/_I \simeq \bigotimes_{i=1}^{n}{}_A \left(A[X_i]/_{(f_i(X_i))}\right)$$

$A[X_i]/_{(f_i(X_i))}$ est un A-module de type fini (car f_i est unitaire), libre (sinon, il y aurait dans l'idéal $(f_i(X_i))$ un polynôme de degré strictement inférieur à $d°f_i$, ce qui n'est pas, car A est intègre), donc B est libre, donc plat. Le plongement canonique : $A \to K$ induit donc un plongement : $A \otimes_A B \to K \otimes_A B$. Or :

$$A \otimes_A B \simeq B$$

$$K \otimes_A B \simeq K[X_1,\dots,X_n]/_J \simeq K_n$$

B se plonge dans un corps, et est donc intègre ; I est donc premier.

Soit : $\phi : B \to K_n$ le plongement obtenu. On voit que, si a est élément de A, alors : $\phi(\bar{a}) = a$. Donc, I ne contient pas d'élément de A non nul.

(Le théorème suivant est probablement connu de nombreux mathématiciens. Cependant, à notre connaissance, il n'a pas été écrit. Aussi, nous en donnons une démonstration).

4.1.- <u>Proposition</u>. Soient p un nombre premier ; k un corps commutatif de caractéristique différente de p, contenant les racines $p^{\text{ièmes}}$ de l'unité ; R_p le groupe multiplicatif des racines $p^{\text{ièmes}}$ de l'unité. Soient α_i une racine de $F_i(X) = X^p - a_i \in k[X]$ (dans une clôture algébrique de k), $i = 1,\dots,n$; si F_1 n'a pas de racine dans k, si, pour tout $i = 2,\dots,n$, F_i n'a pas de racine dans $k[\alpha_1,\dots,\alpha_{i-1}]$, alors $k[\alpha_1,\dots,\alpha_n]$ est galoisien sur k et le groupe de Galois de $k[\alpha_1,\dots,\alpha_n]$ sur k est isomorphe à $R_p^{\,n}$; les k-automorphismes sont définis par :

$$\alpha_i \longmapsto \xi_i \, \alpha_i \qquad \text{où} \quad \xi_i \in R_p \qquad (i = 1,\ldots,n).$$

<u>Démonstration</u>.- Soient :

$$K_o = k$$

$$K_{i+1} = K_i[\alpha_{i+1}] \qquad (i = 0,\ldots,n-1)$$

D'après 0.1, F_{i+1} est irréductible sur K_i, d'où :

$$[K_n : k] = \prod_{i=0}^{n-1} [K_{i+1} : K_i] = p^n$$

α_i est séparable et les conjugués de α_i (c'est-à-dire $\xi \, \alpha_i$ pour $\xi \in R_p$) sont éléments de K_n, donc K_n est galoisien sur k, et $\mathrm{Gal}(K_n|k)$ est d'ordre p^n. Soit :

$$\phi : \mathrm{Gal}(K_n|k) \to R_p^{\,n}$$

$$\sigma \mapsto (\sigma(\alpha_1)\alpha_1^{-1},\ldots,\sigma(\alpha_n)\alpha_n^{-1})$$

ϕ est injective, car $\alpha_1,\ldots,\alpha_n$ engendre K_n sur k, et σ est déterminé par les images de α_i ; $\mathrm{Gal}(K_n|k)$ et $R_p^{\,n}$ ayant même cardinal, ϕ est bijective. De plus, ϕ est un homomorphisme (car $R_p \subset k$), donc un isomorphisme.

4.2.- <u>Proposition</u>. Soient p un nombre premier ; k un corps commtatif de caractéristique différente de p, contenant les racines $p^{\text{ièmes}}$ de l'unité. Soient A un anneau factoriel dont le corps des fractions est k ; $g_i, h_i \in A$ $(i = 1,\ldots,n)$; $q_i \in \mathbb{N}$ $(i = 1,\ldots,n)$; α_i une racine de $f_i(X) = X^p - g_i^{\,q_i}$ (dans une clôture algébrique de k). Si :

Pour tout $i = 1,\ldots,n$, g_i est premier
Pour tout $i = 1,\ldots,n$, $j = 1,\ldots,n$, $i \neq j$, $g_i \neq g_j$
Pour tout $i = 1,\ldots,n$, $j = 1,\ldots,n$, g_i ne divise pas h_j
Pour tout $i = 1,\ldots,n$, p ne divise pas q_i.

alors, $F_1(X)$ n'a pas de racine dans k, et pour $i = 2,\dots,n$, $F_i(X)$ n'a pas de racine dans $k[\alpha_1,\dots,\alpha_{i-1}]$.

Démonstration.- Par récurrence. Soit une racine de $F_1(X)$, élément de k : $\frac{r}{s}$ où r et s sont des éléments de A premiers entre eux :

$$r^p = g_1^{q_1} h_1 s^p$$

Donc s divise r^p, donc r ; comme r et s sont premiers entre eux, s est une unité. L'exposant de g_1 dans la décomposition de $(g_1^{q_1} h_1 s^p)$ est donc q_1, et dans r^p, cet exposant est multiple de p, ce qui est contradictoire. F_1 n'a donc pas de racine dans k.

Supposons maintenant la proposition vraie pour n. Supposons que F_{n+1} ait une racine, que l'on pourra désigner sans inconvénient par α_{n+1}, dans $k[\alpha_1,\dots,\alpha_n]$

$$K' = k[\alpha_1,\dots,\alpha_{n-1},\alpha_{n+1}] \subset k[\alpha_1,\dots,\alpha_{n-1},\alpha_n] = K$$

D'après l'hypothèse de récurrence et 4.1 :

$$[K' : k] = p^n = [K : k]$$

Donc K = K'. Soit ξ une racine primitive $p^{\text{ième}}$ de l'unité. D'après l'hypothèse de récurrence et 4.1, il y a un k-automorphisme de K, σ, défini par :

$$\begin{cases} \text{Si } i = 1,\dots,n-1, \ \sigma(\alpha_i) = \alpha_i \\ \sigma(\alpha_n) = \xi\alpha_n . \end{cases}$$

Soit $N \in \mathbb{N}$, $0 \le N \le p$ tel que :

$$\sigma(\alpha_{n+1}) = \xi^N \alpha_{n+1}$$

et soit :

$$\alpha = \alpha_n{}^{p-N}\ \alpha_{n+1}$$

$$\alpha^p = (\alpha_n^p)^{p-N}\ \alpha_{n+1}{}^p = g_{n+1}{}^{q_{n+1}}\ (g_n{}^{q_n(p-N)}\ h_n{}^{p-N}\ h_{n+1})$$

$$= g_{n+1}{}^{q_{n+1}}\ h$$

α est donc racine de $f(X) = X^p - g_{n+1}{}^{q_{n+1}}\ h$. On voit que les polynômes $f_1,\dots,f_{n-1}$, f satisfont les hypothèses de la proposition, et donc, d'après l'hypothèse de récurrence (comme plus haut) :

$$K = k[\alpha_1,\dots,\alpha_{n-1},\alpha]$$

$$\sigma(\alpha) = \sigma(\alpha_n)^{p-N}\ \sigma(\alpha_{n+1}) = \xi^p\ \alpha_n{}^{p-N}\ \alpha_{n+1} = \alpha$$

Donc σ est l'identité sur K, ce qui est contradictoire ($\sigma(\alpha_n) \neq \alpha_n$).

4.3.- <u>Corollaire</u>. Soient k un corps commutatif ; p un entier naturel premier différent de la caractéristique de k ; $\bar{k}$ la clôture algébrique de k ; $f_i \in k[Y_1,\dots,Y_m]$ $(i = 1,\dots,n)$; $F_i(X) = X^p - f_i \in k[Y_1,\dots,Y_m,X]$. S'il existe $g_i, h_i \in \bar{k}[Y_1,\dots,Y_m]$, $q_i \in \mathbb{N}$ $(i = 1,\dots,n)$ tels que :

Pour tout $i = 1,\dots,n$, $f_i = g_i$
Pour tout $i = 1,\dots,n$, g_i est premier (dans $\bar{k}[Y_1,\dots,Y_m]$)
Pour tout i et $j = 1,\dots,n$, $i \neq j$, $g_i \neq g_j$
Pour tout i et $j = 1$, g_i ne divise pas h_j (dans $\bar{k}[Y_1,\dots,Y_m]$)
Pour tout $i = 1,\dots,n$, p ne divise pas q_i,

alors, l'idéal de $k[Y_1,\dots,Y_m, X_1,\dots,X_n]$ engendré par $\{F_i(X_i)\ ;\ i = 1,\dots,n\}$ est absolument premier, et ne contient aucun élément de $k[Y_1,\dots,Y_m]$ non nul.

<u>Démonstration</u>.- De 4.2, 4.1, et 3.0 $(A = \bar{k}[Y_1,\dots,Y_m])$.

5.1.- <u>Proposition</u>. Soient k un corps commutatif de caractéristique $p \neq 0$, et $\mathbb{F}_p$ le corps premier de caractéristique p. Soient α_i une racine de $F_i(X) = X^p - X - a_i \in k[X]$ (dans une clôture algébrique de k), $i = 1,\dots,n$; si F_1 n'a pas de racine dans k, si pour tout $i = 2,\dots,n$, F_i n'a pas de racine dans $k[\alpha_1,\dots,\alpha_{i-1}]$, alors $k[\alpha_1,\dots,\alpha_n]$ est galoisien sur k, et le groupe de Galois $k[\alpha_1,\dots,\alpha_n]$ sur k est isomorphe à $(\mathbb{F}_p,+)^n$; les k-automorphismes sont définis par

$$\alpha_i \longmapsto \alpha_i + n_i \qquad \text{où : } n_i \in \mathbb{F}_p \quad (i = 1,\dots,n)$$

<u>Démonstration</u>.- Soient

$$\begin{cases} K_o = k \\ K_{i+1} = K_i[\alpha_{i+1}] \qquad (i = 0,\dots,n-1) \end{cases}$$

D'après 0.3, F_i est irréductible sur K_{i-1}, d'où :

$$[K_n : k] = \prod_{i=0}^{n-1} [K_{i+1} : K_i] = p^n$$

α_i est séparable sur k et les conjugués de α_i (c'est-à-dire $\alpha_i + n$ avec $n \in \mathbb{F}_p$) sont éléments de K_n, donc K_n est galoisien sur k, et $\mathrm{Gal}(K_n|k)$ est d'ordre p^n. Soit :

$$\phi : \mathrm{Gal}(K_n|k) \to \mathbb{F}_p^n$$

$$\sigma \longmapsto (\sigma(\alpha_1) - \alpha_1,\dots,\sigma(\alpha_n) - \alpha_n)$$

ϕ est injective, car $\alpha_1,\dots,\alpha_n$ engendre K_n sur k, et σ est déterminé par les images des α_i ; $\mathrm{Gal}(K_n|k)$ et $\mathbb{F}_p^n$ ayant même cardinal, ϕ est bijective. De plus ϕ est un homomorphisme (car : $\mathbb{F}_p \subset k$), donc un isomorphisme.

5.2.- <u>Proposition</u>. Soient k un corps commutatif de caractéristique p et a_i, b_i $(i = 1,\dots,n)$ des éléments de k tels que $\{a_i ; \ i = 1,\dots,n\}$ soit linéairement libre sur $\mathbb{F}_p$; α_i une racine de $F_i(X)=X^p-X-(a_iY+b_i)$

élément de $k[Y][X]$ (dans une clôture algébrique de $k(Y)$). Alors $F_1(X)$ n'a pas de racine dans $k(Y)$, et pour $i = 2,\dots,n$, $F_i(X)$ n'a pas de racine dans $k(Y)[\alpha_1,\dots,\alpha_{i-1}]$.

Démonstration.- Par récurrence. Soit une racine de $F_1(X)$, élément de $k(Y)$: $\frac{r}{s}$ où r et s sont des éléments de $k[Y]$ premiers entre eux :

$$r^p - r\, s^{p-1} - (a_1Y + b_i)\, s^p = 0$$

donc :

$$r(r^{p-1} - s^{p-1}) = (a_1Y + b_i)\, s^p$$

et :

$$s^{p-1}(r + (a_1Y + b_1)s) = r^p$$

s divise donc r^p, et r divise $(a_1Y + b_1)s^p$; comme r et s sont premiers entre eux, s est élément de k et r divise $(a_1Y + b_1)$ qui est premier. Donc, ou bien r est élément de k, et $\frac{r}{s}$ est élément de k :

$$a_1Y + b_1 = (\tfrac{r}{s})^p - \tfrac{r}{s} \in k$$

ce qui n'est pas (les a_i, étant linéairement indépendants sur $\mathbb{F}_p$, ne sont pas nuls).

ou bien il existe $a \in k^*$ tel que :

$$\frac{r}{s} = a(a_1Y + b_1)$$

donc

$$a_1Y + b_1 = (\tfrac{r}{s})^p - \tfrac{r}{s} = a^p\, a_1^{\,p}\, Y^p + a^p\, b_1^{\,p} - a\, a_1\, Y - a\, b_1$$

ce qui n'est pas (car $aa_1 \neq 0$).
En définitive, $F_1(X)$ n'a pas de racine dans $k(Y)$.

Supposons la proposition vraie pour n. Supposons que F_{n+1} ait une racine, que l'on pourra désigner sans inconvénient par α_{n+1}, dans $k[\alpha_1,\ldots,\alpha_n]$.

$$K' = k(Y)[\alpha_1,\ldots,\alpha_{n-1},\alpha_{n+1}] \subset k(Y)[\alpha_1,\ldots,\alpha_n] = K$$

D'après l'hypothèse de récurrence, et 5.1 :

$$[K' : k(Y)] = p^n = [K : k(Y)]$$

Donc K = K'. D'après l'hypothèse de récurrence et 5.1, il y a un k(Y)-automorphisme de K, σ, défini par :

$$\begin{cases} \text{Si } i = 1,\ldots,n-1, \ \sigma(\alpha_i) = \alpha_i \\ \sigma(\alpha_n) = \alpha_n + 1 \end{cases}$$

$N = \sigma(\alpha_{n+1}) - \alpha_{n+1}$ est élément de $\mathbb{F}_p$, car α_{n+1} et $\sigma(\alpha_{n+1})$ sont racines de F_{n+1} ; soit :

$$\alpha = - N\alpha_n + \alpha_{n+1}$$

$$\begin{aligned} \alpha^p - \alpha &= (-N)^p \alpha_n{}^p + \alpha_{n+1}{}^p + N\alpha_n - \alpha_{n+1} \\ &= - N(\alpha_n{}^p - \alpha_n) + (\alpha_{n+1}{}^p - \alpha_{n+1}) \\ &= - N(a_nY + b_n) + (a_{n+1}Y + b_{n+1}) \\ &= (-N a_n + a_{n+1})Y + (-N b_n + b_{n+1}) \end{aligned}$$

Donc α est racine de $F(X) = X^p - X - ((-N a_n + a_{n+1})Y + (-Nb_n + b_{n+1}))$. Comme $a_1,\ldots,a_{n-1}$, $- N a_n + a_{n+1}$ sont linéairement indépendants sur $\mathbb{F}_p$, on voit comme précédemment qu'on a :

$$K = k(Y)[\alpha_1,\ldots,\alpha_{n-1},\alpha]$$

$$\sigma(\alpha) = -N\,\sigma(\alpha_n) + \sigma(\alpha_{n+1}) = - N(\alpha_n + 1) + (\alpha_{n+1} + N) = \alpha$$

Donc, est l'identité sur K, ce qui est contradictoire ($\sigma(\alpha_n) \neq \alpha_n$).

5.3.- Corollaire. Soient $a_i \in k$ $(i = 1,\ldots,n)$ linéairement indépendants sur $\mathbb{F}_p$; $b_i \in k$ $(i = 1,\ldots,n)$. Alors l'idéal de $k[Y,X,\ldots,X_n]$ engendré par $\{X_i^{p^i} - X_i - (a_iY + b_i)\ ;\ i = 1,\ldots,n\}$ est absolument premier.

Démonstration.- De 5.2, 5.1 et 3.30.

6.1.- Définition. Soit T une théorie complète. On dit que $\phi(x_1,\ldots,x_n,u_1,\ldots,u_m)$ a la propriété d'indépendance dans T si et seulement si pour tout ensemble fini I, on a :

$$T \vdash (\exists_{i\in I}\ \bar{u}^i)(\exists_{j\in \mathcal{P}(I)}\ \bar{x}^j)(\bigwedge_{i\in j} \phi(\bar{x}^j,\bar{u}^i) \wedge \bigwedge_{i\notin j} \neg\,\phi(\bar{x}^j,\bar{u}^i))$$

(où "$\bar{u}^i$" abrège "$u_1^i,\ldots,u_m^i$", et "$\bar{x}^j$" abrège "$x_1^j,\ldots,x_n^j$").

On dit que T a la propriété d'indépendance si et seulement s'il existe une formule qui a la propriété d'indépendance dans T.

6.2.- Lemme. Soient K un corps commutatif, k un sous-corps faiblement algébriquement clos de K, et p un nombre premier différent de la caractéristique de K. Soient $\{a_i\ ;\ i \in \mathbb{N}\}$ un ensemble d'éléments de k distincts, I et J deux sous-ensembles finis et disjoints de $\mathbb{N}$. Soit $\phi(y,u)$ la formule :

$$\exists\, x\ (x^p = y + u)$$

Si k contient les racines $p^{\text{ièmes}}$ de l'unité, et qu'il existe un élément δ de k sans racine $p^{\text{ième}}$ dans K, alors K réalise :

$$\{\phi(y,a_i)\ ;\ i \in I\} \cup \{\neg\,\phi(y,a_j)\ ;\ j \in J\}$$

Démonstration.- Soit $\mathfrak{J}$ l'idéal (de l'anneau des polynômes sur k en les indéterminées X_i $(i \in I \cup J)$,Y) engendré par :

$$\{X_i^p - (Y + a_i)\ ;\ i \in I\} \cup \{X_j^p - \delta(Y + a_j)\ ;\ j \in J\}$$

$\mathfrak{J}$ est absolument premier, et ne contient pas $\prod_{j\in J}(Y + a_j)$ (d'après

4.3). $\mathfrak{J}$ a donc un zéro dans k qui n'annule pas $\prod_{j \in J} (Y + a_j)$ (1.5) ; c'est-à-dire qu'il existe $c_i \in k$ $(i \in I \cup J)$, $d \in k$ tels que

Pour tout $i \in I$, $c_i^p - (d + a_i) = 0$

Pour tout $j \in J$, $c_j^p - \delta(d + a_j) = 0$

Pour tout $j \in J$, $d + a_j \neq 0$

Donc, pour $i \in I$, K réalise $\phi(d,a_i)$. Supposons qu'il y ait $j \in J$ tel que :

$$K \models \phi(d,a_j)$$

Il existerait $c'_j \in K^*$ tel que :

$$c'^p_j = d + a_j$$

donc

$$\delta = (c_j c'^{-1}_j)^p$$

ce qui contredit l'hypothèse que δ n'a pas de racine $p^{\text{ième}}$ dans K. Donc pour tout $j \in J$:

$$K \models \neg\ \phi(d,a_j)$$

6.3.- <u>Lemme</u>. Soient K un corps commutatif de caractéristique p, et k un sous-corps faiblement algébriquement clos de K. Soient $\{a_i ; i \in \mathbb{N}\}$ un ensemble d'éléments de k, linéairement indépendants sur le corps premier de k, $\mathbb{F}_p$, I et J deux sous-ensembles finis et disjoints de $\mathbb{N}$. Soit $\psi(y,u)$ la formule :

$$\exists\, x\ (x^p - x = u\, y)$$

S'il existe un élément δ de k sans pseudo-racine dans K, i.e. :

$$(\forall\, x \in K)\ (x^p - x \neq \delta)$$

alors K réalise :

$\{\psi(y,a_i) \;;\; i \in I\} \cup \{\neg\, \psi(y,a_j) \;;\; j \in J\}$

Démonstration.- Soit $\mathfrak{J}$ l'idéal (de l'anneau des polynômes sur k en les indéterminées X_i $(i \in I \cup J)$, Y) engendré par :

$$\{X_i{}^p - X_i - a_iY \,;\, i \in I\} \cup \{X_j{}^p - X_j - (a_jY + \delta) \;;\; j \in J\}$$

$\mathfrak{J}$ est absolument premier d'après 5.3, et donc, a un zéro dans k, c'est-à-dire qu'il existe $c_i \in k$ $(i \in I \cup J)$, $d \in k$ tels que :

Pour tout $i \in I$, $c_i{}^p - c_i - d\,a_i = 0$

Pour tout $j \in J$, $c_j{}^p - c_j - (d\,a_j + \delta) = 0$

Donc, pour tout $i \in I$, K réalise $\psi(d,a_i)$. Supposons qu'il y ait un $j \in J$ tel que :

$$K \models \psi(d,a_j)$$

Il existerait $c'_j \in K$ tel que :

$$c'^p_j - c'_j = d\,a_j$$

donc :

$$\delta = (c_j - c'_j)^p - (c_j - c'_j)$$

ce qui contredit l'hypothèse que δ n'a pas de pseudo-racine dans K.

Donc pour tout $j \in J$:

$$K \models \neg\, \psi(d,a_j)$$

6.4.- Théorème. Soient K un corps commutatif, et k un sous-corps de K faiblement algébriquement clos, non séparablement clos et relativement algébriquement clos dans K. Alors, Th(K) a la propriété d'indépendance.

Démonstration.- Soit l' la clôture galoisienne sur k d'une extension séparable de k (distincte de k) ; soit p un diviseur premier de [l' : k] ; soient un sous-groupe de Gal(l'|k) d'ordre p, et k' son corps d'invariants ; [l' : k'] = p. Puisque l' est séparable sur k', il existe un élément primitif β : l' = k'[β]. Nous distinguons deux cas :

1) p n'est pas la caractéristique de K.

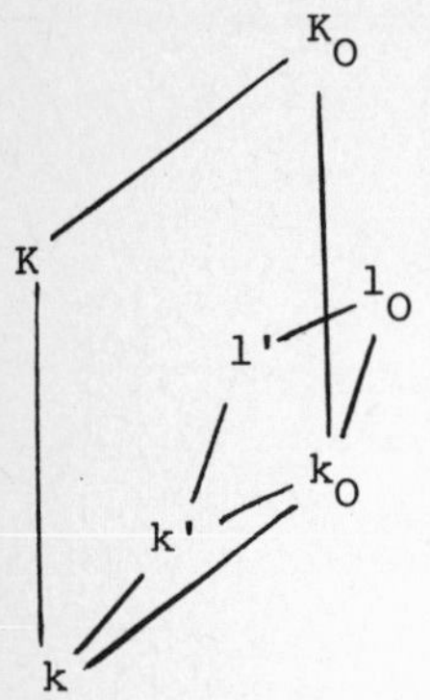

Soit ξ une racine primitive $p^{\text{ième}}$ de l'unité. Soient :

$$k_0 = k'[\xi] \qquad l_0 = l'[\xi]$$

$$l_0 = l'[\xi] = k'[\beta,\xi] = k'[\xi][\beta] = k_0[\beta]$$

Soient F(X) le polynôme minimal de β sur k' ;

$$F(X) \in k'[X] \subset k_0[X]$$

donc :

$$[l_0 : k_0] \leq p$$

D'autre part :

$$[k_0 : k'] \leq p-1$$

$$[l_0 : k'] = [l_0 : l'] \times [l' : k'] = p \times [l_0 : l']$$

$$= [l_0 : k_0] \times [k_0 : k']$$

Donc p divise $[l_0 : k_0] \times [k_0 : k']$, donc $[l_0 : k_0]$ (puisque $[k_0 : k'] \leq p-1$). En définitive :

$$[l_0 : k_0] = p$$

Les conjugués de β sur k_0 sont des conjugués de β sur k' ; puisque l' est galoisien sur k', ils sont éléments de l', donc de

l_0. $l_0 = k_0[\beta]$ est donc une extension galoisienne de k_0, cyclique (puisque de degré premier), et k_0 contient les racines $p^{\text{ièmes}}$ de l'unité. D'après 0.2, il existe $\delta \in k_0$ tel que le polynôme $X^p - \delta \in k_0[X]$ est irréductible, et n'a donc pas de racine dans k_0.

k_0 est séparable sur k, puisque k_0 est séparable sur k' et que k' est séparable sur k ; il existe donc α tel que $k_0 = k[\alpha]$. Soit :

$$K_0 = K[\alpha]$$

k_0 est relativement algébriquement clos dans K_0 (d'après [La,p.59]). k_0 est faiblement algébriquement clos, d'après 2.1.

Soit $\{a_i \; ; \; i \in \mathbb{N}\}$ un ensemble d'éléments de k_0 distincts (un tel ensemble existe, car k_0 est infini, un corps fini n'étant pas faiblement algébriquement clos [Ax,p.245]). Soient I et J deux sous-ensembles finis et disjoints de $\mathbb{N}$. D'après 6.2 et avec les notations de 6.2, K_0 réalise

$$\{\phi(y,a_i) \; ; \; i \in I\} \cup \{\neg\, \phi(y,a_j) \; ; \; j \in J\}$$

Soit $(e_1,\ldots,e_m)$ une base de K_0 sur K, et pour tout $i \in \mathbb{N}$:

$$a_i = \sum_{h=1}^{m} a_i^{\,h}\, e_h$$

Il existe, pour $h = 1,\ldots,m$, $P_h(\vec{V},\vec{X}) \in \mathbb{Z}[\vec{V},\vec{X}]$ ($\vec{V}$ est une suite d'indéterminées de longueur m^2, $\vec{X}$ une suite d'indéterminées de longueur m), et il existe, pour $g = 1,\ldots,m^2$, $b_g \in K$, tels que :

$$(\forall\, x_1,\ldots,x_m \in K) \;\; ((\sum_{h=1}^{m} x_h e_h)^p = \sum_{h=1}^{m} P_h(b,x) e_h)$$

Soit $\phi_1(y_1,\ldots,y_m,u_1,\ldots,u_m,v_1,\ldots,v_{m^2})$ la formule :

$$\exists\, x_1,\ldots,x_m \;\; (\bigwedge_{h=1}^{m} P_n(v,x) = y_h + u_h)$$

On voit que, si $c_h, d_h \in K$, on a :

$$K \models \phi_1(c,d,b) \iff K_0 \models \phi(\sum_{h=1}^{m} c_h e_h, \sum_{h=1}^{m} d_h e_h)$$

ϕ_1 a donc la propriété d'indépendance dans Th(K).

2) p est la caractéristique de K.

D'après 0.4, il existe $\delta \in k'$ tel que le polynôme $X^p - X - \delta \in k'[X]$ soit irréductible, et n'ait donc pas de racine dans k'. Puisque k' est séparable sur k, il existe α' tel qu'on ait : $k' = k[\alpha']$.

Soit : $K' = K[\alpha']$

k' est séparable sur k, donc k' est relativement algébriquement clos dans K' (d'après [La, p.59]).

k' est faiblement algébriquement clos, d'après 2.1.

Soit $\{a_i ; i \in \mathbb{N}\}$ un ensemble d'éléments de k', linéairement indépendants sur le corps premier $\mathbb{F}_p$ (un tel ensemble existe car k' est infini, un corps fini n'étant pas faiblement algébriquement clos [Ax,p.245]) ; soient I et J deux sous-ensembles finis et disjoints de $\mathbb{N}$. D'après 6.3, K' réalise :

$$\{\psi(y,a_i) ; i \in I\} \cup \{\neg\psi(y,a_j) ; j \in J\}$$

Soit $(e_1,\ldots,e_m)$ une base de K' sur K, et pour tout $i \in \mathbb{N}$:

$$a_i = \sum_{h=1}^{m} a_i^h e_h$$

Il existe, pour $h = 1,\ldots,m$, $P_h(\vec{V},\vec{X},\vec{U},\vec{Y}) \in \mathbb{Z}[\vec{V},\vec{X},\vec{U},\vec{Y}]$ ($\vec{V}$ est une suite d'indéterminées de longueur m^2, $\vec{X}$, $\vec{U}$, $\vec{Y}$ des suites d'indéterminées de longueur m), et il existe, pour $g = 1,\ldots,m^2$, $b_g \in K$, tels que :

$$(\forall x_1,\ldots,x_m \in K)\ (\forall y_1,\ldots,y_m \in K)\ (\forall u_1,\ldots,u_m \in K)$$

$$((\sum_{h=1}^{m} x_h e_h)^p - (\sum_{h=1}^{m} x_h e_h) - (\sum_{h=1}^{m} u_h e_h)(\sum_{h=1}^{m} y_h e_h) = P_h(b,x,u,y))$$

Soit $\psi_1(y_1,\ldots,y_m,u_1,\ldots,u_m,v_1,\ldots,v_{m^2})$ la formule :

$$\exists x_1,\ldots,x_m \ (\bigwedge_{h=1}^{m} P_h(v,x,u,y) = 0)$$

On voit que, si $c_h, d_h, f_h \in K$, on a :

$$K \models \psi_1(c_h, d_h, f_h, b_h) \iff K' \models \psi\left(\sum_{h=1}^{m} c_h e_h , \sum_{h=1}^{m} d_h e_h , \sum_{h=1}^{m} f_h e_h\right)$$

ψ_1 a donc la propriété d'indépendance dans Th(K).

6.5.- Corollaire. La théorie d'un corps faiblement algébriquement clos non séparablement clos, ou d'un corps pseudo-fini (notamment d'un ultra-produit de corps finis) a la propriété d'indépendance.

Notation.- Si K est un corps commutatif, on note Abs(K) le sous-corps de K des éléments de K absolument algébriques, c'est-à-dire algébriques sur le corps premier de K.

6.6.- Corollaire. Si K est un corps commutatif de caractéristique non nulle tel que Abs(K) n'est pas fini, ni algébriquement clos, alors Th(K) a la propriété d'indépendance.

Démonstration.- D'après 6.4, car Abs(K) est alors faiblement algébriquement clos, d'après l'hypothèse de Riemann démontrée par Weil (voir [Ax,p.252]).

APPENDICE

Démonstration de 0.5.- Soient $e = (e_1, \ldots, e_n)$ une base de K sur k ; $\sigma_1, \ldots, \sigma_n$ les n k-isomorphismes de K dans une clôture algébrique de k, $\bar{k}$; $(P_1, \ldots, P_m) \in (K[X_1, \ldots, X_\ell])^m$ une base de l'idéal sur K de V. Soient $Y_{j,g}$, $j = 1, \ldots, \ell$, $g = 1, \ldots, n$, $\ell \times n$ indéterminées distinctes, et Q_i le polynôme en les indéterminées $Y_{j,g}$, obtenu en substituant $\sum_{g=1}^{n} Y_{j,g} e_g$ à X_j dans P_i. Il existe n polynômes, éléments de $k[Y_{1,1}, \ldots, Y_{\ell,n}]$, $Q_{i,g}$ tels que

$$Q_i = \sum_{g=1}^{n} Q_{i,g} e_g$$

Soit W_e la variété affine sur k de l'idéal engendré par $\{Q_{i,g}$; $i = 1, \ldots, m$ et $g = 1, \ldots, n\}$. Remarquons que :

1) Si σ est un k-isomorphisme de K, comme $\sigma(Q_{i,g}) = Q_{i,g}$, on a $W_{\sigma(e)} = W_e$.

2) Si $(y_{1,1}, \ldots, y_{\ell,n})$ est un point de W, alors $\left(\sum_{g=1}^{n} y_{1,g} e_g, \ldots, \sum_{g=1}^{n} y_{\ell,g} e_g\right)$ est un point de V.

Nous allons démontrer que W_e et $V^{\sigma_1} \times \ldots \times V^{\sigma_n}$ sont isomorphes sur L, clôture galoisienne de K sur k. Soit ϕ défini sur W_e par :

$$\phi(y_{1,1}, \ldots, y_{1,n}, y_{2,1}, \ldots, y_{\ell,n}) = (x_{1,1}, \ldots, x_{1,\ell}, x_{2,1}, \ldots, x_{n,\ell})$$

où $x_{j,i} = \sum_{g=1}^{n} y_{i,g} \sigma_j(e_g)$ $\qquad (j = 1, \ldots, n,\ i = 1, \ldots, \ell)$

D'après les deux remarques, $(x_{j,1}, \ldots, x_{j,\ell})$ est un point de V^{σ_j} (pour tout $j = 1, \ldots, n$). Il est clair que ϕ est un morphisme de variétés sur L. Soit M la matrice dont le coefficient de la $i^{\text{ème}}$ ligne $j^{\text{ième}}$ colonne est $\sigma_i(e_j)$. On voit qu'on a :

$$\begin{bmatrix} x_{1,i} \\ \cdot \\ \cdot \\ \cdot \\ x_{n,i} \end{bmatrix} = M \begin{bmatrix} y_{i,1} \\ \cdot \\ \cdot \\ \cdot \\ y_{i,n} \end{bmatrix} \qquad (i = 1, \ldots, l)$$

La matrice M est inversible ; en effet, sinon les lignes de M seraient linéairement dépendantes sur $\bar{k}$; il existerait donc des éléments λ_i, $i = 1,\dots,n$, de $\bar{k}$ tels que pour tout $g = 1,\dots,n$, on ait:

$$\sum_{i=1}^{n} \lambda_i \sigma_i(e_g) = 0$$

soit $\sum_{g=1}^{n} a_g e_g$ un élément de K ;

$$\left(\sum_{i=1}^{n} \lambda_i \sigma_i\right)\left(\sum_{g=1}^{n} a_g e_g\right) = \sum_{i=1}^{n} \sum_{g=1}^{n} \lambda_i \sigma_i(a_g)\, \sigma_i(e_g)$$

$$= \sum_{g=1}^{n} a_g \left(\sum_{i=1}^{n} \lambda_i \sigma_i(e_g)\right) \qquad (\text{car } \sigma_i(a_g) = a_g)$$

$$= 0$$

donc les σ_i sont linéairement dépendants sur $\bar{k}$ contrairement au théorème de Dedekind.

Soit ψ défini sur $V^{\sigma_1} \times \dots \times V^{\sigma_n}$ par :

$$\psi(x_{1,1},\dots,x_{n,\ell}) = (y_{1,1},\dots,y_{\ell,n})$$

avec :

$$\begin{bmatrix} y_{i,1} \\ \cdot \\ \cdot \\ \cdot \\ y_{i,n} \end{bmatrix} = M^{-1} \begin{bmatrix} x_{1,i} \\ \\ \\ x_{n,i} \end{bmatrix} \qquad (i = 1,\dots,\ell)$$

Soit pour $j = 1,\dots,m$

$$q_{j,g} = Q_{j,g}(y_{1,1},\dots,y_{\ell,n})$$

Pour tout $i = 1,\dots,n$, $j = 1,\dots,m$:

$$\sum_{g=1}^{n} q_{j,g}\sigma_i(e_g) = \sigma_i(Q_j)(y_{1,1},\ldots,y_{\ell,n}) \quad \text{d'après la remarque 2)}$$

$$= \sigma_i(P_j)(x_{i,1},\ldots,x_{i,\ell}) \quad \text{par définition de } \psi$$

$$= 0 \quad \text{car } (x_{i,1},\ldots,x_{i,\ell}) \in V^{\sigma_i}$$

Pour chaque j, on obtient donc, (en faisant varier i) un système d'équations linéaires dont la matrice est M ; c'est donc un système de Cramer, dont la seule solution est la solution triviale. Donc pour tout $j = 1,\ldots,m$, $i = 1,\ldots,n$, on a :

$$Q_{j,g}(y_{1,1},\ldots,y_{\ell,n}) = 0$$

donc $\psi(x_{1,1},\ldots,x_{n,\ell})$ est un point de W.

Il est clair que ψ est un morphisme de variétés sur L, et que c'est l'inverse de ϕ.

*
* *

BILIOGRAPHIE

[Ax] James AX : "The elementary theory of finite fields" (Ann. of Math. 88 (1968), pp.239-271).

[Al] James AX : "On the undecidability of power series fields" (Proc. Am. Math. Soc. 16^2 (1965), p. 846).

[Bo] Nicolas BOURBAKI : "Algèbre. Chapitre 5. Corps commutatifs" . (Hermann).

[De] Françoise DELON : "Types sur les corps valués henséliens" (à paraître).

[Du] Jean-Louis DURET : "Instabilité des corps formellement réels" Canad. Math. Bull. Vol. 20 (3),1977).

[Er] Ju L. ERSOV : "Undecidability of certain fields" (Doklady 1965. Tom 161,N°1,pp.349-352).

[Fe] Ulrich FELGNER : " $\aleph_1$-kategorische Theorien nicht-kommutativer Ringe" (Fundamenta Mathematicae LXXXII (1975), pp. 331-346).

[Ja] Nathan JACOBSON : "Lectures in abstract algebra. Volume III. Theory of fields and Galois Theory" (D. Van Nostrand company).

[JM] Moshe JARDEN : "Elementary statements over large algebraic fields" (Trans. Am. Math. Soc. Vol. 164, february 1972, pp. 67-91).

[La] Serge LANG : " Introduction to algebraic geometry" (Addison-Wesley publishing company)

[Ma] Angus MACINTYRE : "On ω_1-catagorical theories of fields" (Fundamenta Mathematicae LXXI (1971) p. 1-25).

[Pe] Yu. G. PENZIN : "The undecidibality of fields of rational functions over fields of characteristic 2" (Algebra i Logica, vol. 12, N°2, pp.205-210 March-April 1973. Traduit (en anglais) dans Algebra and Logic).

[RJ] Julia ROBINSON : "Definibality and decision problems in arithmetic" (J.S.L. 14 (1949) pp. 98-114).

[RR] Raphael ROBINSON : " The undecidability of pure transcendental extension of real fields" (Zeischr. f. Math. Logik und Grundlagen d. Math. Bd. 10, S. 275-282 (1964)).

[Sh] Saharon SHELAH : "The lazy model-theorician's guide to stability" (Logique et analyse, septembre-décembre 1975, pp. 241-308)

[Va] L. VAN DEN DRIES : "Model theory of fields. Decidibality, and bounds for polynomial ideals" (Thèse Université d'Utrecht).

[We] André WEIL : "Adèles and algebraic group".

[Wh] William WHEELER : "Model-complete theories of pseudo-algebraically closed fields" (a paraître).

[Wo] Carol WOOD : "Notes on the stability of separably closed fields" (J.S.L. Vol. 44, Nb. 3, Sept. 1979, pp.412-416).

*
* *

HORN-THEORIES OF ABELIAN GROUPS

Ulrich Felgner

Mathematisches Institut der Universität Tübingen

INTRODUCTION. In this paper we classify all abelian groups G whose first-order theory is a HORN-theory. We shall characterize these groups in terms of their SZMIELEW-invariants. We obtain the surprising corollary that the theory of G is a HORN-theory if and only if G and $G \oplus G$ are elementarily equivalent. These results depend on a detailed analysis of the structure of certain reduced products of abelian groups. In the first part of the paper we shall, hence, consider reduced products of abelian groups. The results stated there are slight extensions of results known from the literature. In the second part we apply these results in order to give various characterizations of abelian groups whose first-order theories are HORN-theories.

The results of this paper were obtained in 1976 and were presented the same year at the 5th-Scandinavian Logic Congress at Jyväskylä (Finland) and the 3rd-Conference on Set Theory and Hierarchy Theory in Bierutowice (Poland). I whish to thank Dr. A.Baudisch for some stimulating conversations and suggestions.

§1. REDUCED PRODUCTS.

Let G_i (for $i \in I$) be arbitrary additively written groups and let F be any filter on I. Then

$$\mathcal{N}(F) = \{ f \in \prod_{i \in I} G_i \ ; \ \{ i \in I \ ; \ f(i) = 0 \} \in F \}$$

is a normal subgroup of the complete cartesian product $\prod_{i \in I} G_i$. The factor group $\prod_{i \in I} G_i / \mathcal{N}(F)$ (also written as $\prod_{i \in I} G_i / F$ by most authors) is called the *reduced product* of the groups G_i modulo F. If F is the filter of cofinite subsets of I, $F = \{X \subseteq I ; |I - X| < \aleph_0\}$, then $\mathcal{N}(F) = \bigoplus_{i \in I} G_i$ (= the weak direct product of the G_i). If $F = \{I\}$, then $\mathcal{N}(F) = \{0\}$, and if $F = P(I) = \{X ; X \subseteq I\}$, then $\mathcal{N}(F) = \prod_{i \in I} G_i$

<u>THE SPLITTING LEMMA.</u> *For any filter F on I put $S = \bigcap F$, $T = I - S$ and $F^* = \{X - S ; X \in F\}$. If the G_i (for all $i \in I$) are groups, then*

$$\prod_{i \in I} G_i / \mathcal{N}(F) \cong \left(\prod_{i \in S} G_i \right) \oplus \left(\prod_{i \in T} G_i / \mathcal{N}(F^*) \right).$$

A filter F on I is called *proper*, if $\emptyset \notin F$. A filter F on I is called *ω-incomplete* if F is a proper filter containing sets K_n such that $\emptyset = \bigcap \{K_n ; n \in \omega\}$, where ω is the first infinite ordinal. The following statement is essentially contained in A.Hulanicki [9] and L.Fuchs [5], p.174-175: *If F is any ω-incomplete filter on I and if the G_i are arbitrary abelian groups, then $\prod_{i \in I} G_i / \mathcal{N}(F)$ is algebraically compact.* The algebraically compact groups are by definition the *ω_1-equationally compact* abelian groups. The role of ω-incomplete filters can be explained therefore as follows: *Let F be any free filter on I and let G be an abelian group which is not algebraically compact. Then the reduced power $G^I / \mathcal{N}(F)$ is algebraically compact if and only if F is ω-incomplete.*

'Most' ultrafilters are ω-incomplete. In fact all regular ultrafilters are ω-incomplete and all free ultrafilters on sets whose cardinalities are less than the first measurable cardinal are likewise ω-incomplete. It is, however, easy to encounter ω-complete filters (which are not ultrafilters) on small infinite cardinals (e.g. the filter generated by all closed unbounded subsets of a regular uncountable cardinal). - Let $|X|$ denote the cardinality of X.

<u>DEFINITION.</u> If I is any non-empty set, then

$$F_I = \{ X \subseteq I \ ; \ |I - X| < |I| \}$$

is called the *generalized Fréchet-filter* on I . It is clear that F_I is ω-incomplete if and only if $|I|$ has cofinality ω .

In order to describe the structure of $\prod_{i\in I} G_i/\mathcal{N}(F_I)$ we need a particular form of the Engelking-Karłowicz theorem (cf. [4]) which we shall state below.

If F is a filter on I and if S is a set of subsets of I, then S is called to be *independent modulo* F if, whenever $X_1,..,X_n$ are distinct elements of S and $1\leq m\leq n$, then

$$(I - X_1)\cup\ldots\cup(I - X_m)\cup X_{m+1}\cup\ldots\cup X_n \notin F .$$

LEMMA 1.1. *Let* $J\subseteq I$ *be such that* $|I| = |J| = \kappa \geq \aleph_0$ *. There exists a family* S *of subsets of* J *such that* $|S| = 2^\kappa$ *and* S *is independent modulo the generalized Fréchet-filter* F_I *. Moreover* $I - (X\triangle Y)\notin F_I$ *for any two distinct elements* $X, Y\in S$.

Proof. As usual let A^B denote the set of all functions from B into A and $P_\omega(I) = \{ X\subseteq I ; |X|<\aleph_0 \}$. The set of all ordered pairs (X,f), where $X\in P_\omega(I)$ and $f\in I^{P(X)}$, has cardinality $|I| = \kappa$. Therefore we may use J as an index set of this set,

$$D = \{ (X,f) ; X\in P_\omega(I) \wedge f\in I^P \} = \{ (X_j , f_j) ; j\in J \} .$$

For each $Y\subseteq I$ define a function $h_Y: J\to I$ by $h_Y(j) = f_j(Y\cap X_j)$ and choose $i_o\in I$ arbitrarily. Then $S = \{ h_Y^{-1}(i_o) ; Y\subseteq I \}$ is the required family ($X\triangle Y = (X\cup Y) - (X\cap Y)$ is the symmetric difference).□

We shall use the following group theoretic notation:

$\mathbb{Z}$ = the additive group of all rational integers,

$\mathbb{Q}$ = the additive group of all rational numbers,

$Z(p^\infty)$ = the Prüfer group for the prime p ,

$Z(n)$ = the finite cyclic group of order n ,

$Q_p = \{ \frac{m}{n} ; m,n\in\mathbb{Z} \wedge$ p does not devide $n \wedge n\neq 0 \}$.

If G is any group and λ any cardinal, then $G^{(\lambda)}$ is the weak direct power of G, i.e. $G^{(\lambda)}$ is the set of all functions f from λ into G such that $\{i \in \lambda ; f(i) \neq 0\}$ is finite.

If A is *algebraically compact* then we have a structure theorem for A (cf. Fuchs [5], p.167, Eklof-Fisher [3], §1):

$$(\#) \qquad A \cong D \oplus \prod_p \hat{A}_p ,$$

where D is the maximal divisible subgroup of A and $\hat{A}_p$ is the completion in the p-adic topology of A_p, where p is a prime number,

$$(\#\#) \qquad A_p \cong Q_p^{(\beta_p)} \oplus \bigoplus_{n \in \omega} Z(p^n)^{(\alpha_{p,n})} .$$

The divisible group D can be decomposed as follows:

$$(\#\#\#) \qquad D \cong \mathbb{Q}^{(\delta)} \oplus \bigoplus_p Z(p^\infty)^{(\gamma_p)} .$$

A theorem of Kaplansky states that an algebraically compact group A is uniquely determined (up to isomorphism) by the cardinals $\alpha_{p,n}$, β_p, γ_p and δ, when A is decomposed as in (#), (##) and (###). The following theorem generalizes a result of K. Golema and A. Hulanicki [8]. The cofinality of the cardinal κ is denoted by *cf(* κ *)* and *lim sup* will denote the 'limes superior'.

<u>THEOREM 1.2.</u> *Let* F_κ *be the generalized Fréchet-filter on the cardinal* κ *and for each ordinal* $i < \kappa$ *let* G_i *be an abelian group such that* $2 \leq |G_i| \leq 2^\kappa$. *Assume that* $cf(\kappa) = \omega$. *Then* $A = \prod_{i<\kappa} G_i / \mathcal{N}(F_\kappa)$ *is an algebraically compact group of cardinality* 2^κ, *and if* A *is decomposed as in (#), (##) and (###) then the cardinals* $\alpha_{p,n}$, β_p, γ_p *and* δ *are either 0 or* 2^κ. *Moreover the following holds:*

(1) if $\delta = 0$, *then* $\beta_p = 0$ *and* $\gamma_p = 0$ *for all primes* p,

(2) if $\delta = 0$, *then there exists an integer* m *and a prime* q *such that* $\alpha_{p,n} = 0$ *for all* $n \geq m$ *and all primes* $p \geq q$,

(3) $\beta_p \geq \limsup_{n\to\infty} \alpha_{p,n}$ *for all primes* p, *and*

(4) $\gamma_p \geq \limsup_{n\to\infty} \alpha_{p,n}$ *for all primes* p.

The proof of the theorem is similar to the proof given in Golema and Hulanicki [8] for the case $\kappa = \aleph_0$. In fact, translate all arguments into the language of filters and use lemma1.1 over and over again. This will prove the theorem.

COROLLARY 1.3. *Let A be an abelian group of cardinality 2^κ, where $cf(\kappa) = \omega$. Then the following are equivalent:*

(i) $A \cong A^\kappa / \mathcal{N}(F_\kappa)$,

(ii) there are groups G_i such that $A \cong \prod_{i<\kappa} G_i / \mathcal{N}(F_\kappa)$,

(iii) A is algebraically compact, and when A is decomposed as in (#), (##) and (###), then the cardinals $\alpha_{p,n}$, β_p, γ_p and δ are either 0 or 2^κ and satisfy the relations (1), (2), (3) and (4) of theorem 1.2 ,

Proof. (i)$\Rightarrow$(ii)$\Longrightarrow$(iii) follows from theorem 1.2. It remains to prove (iii)$\Longrightarrow$(i): Put $A^* = A^\kappa/\mathcal{N}(F_\kappa)$ and let $\alpha_{p,n}$, β_p, γ_p and δ be the invariants of A and let $\alpha^*_{p,n}$, β^*_p, γ^*_p and δ^* be the invariants of A^*. If $\delta = 0$, then by hypothesis (iii) A has finite exponent. Then A^* has the same finite exponent, whence $\delta = 0 = \delta^*$. If $\delta = 2^\kappa$, then $\delta^* = 2^\kappa$ is obvious. Thus $\delta = \delta^*$ holds in either case.

Notice that $D = \bigcap \{mA ; m \in \omega\}$ and $D^* = \bigcap\{mA^* ; m \in \omega\}$ are the maximal divisible subgroups of A, resp. of A^* (since both groups are 2^κ-saturated, $\omega_1 \leq 2^\kappa$). Notice further that the canonical homomorphism from A^κ onto A^* maps the torsion subgroup of A^κ onto the torsion subgroup of A^*. Hence, if $\gamma^*_p \neq 0$, then D^* contains a copy of $Z(p^\infty)$. For each n there is, hence, an element $\overline{x}_n \in D^*$ of order p^n. By our remark concerning the torsion subgroups we conclude that also A contains such elements. If these are in D then clearly γ_p is not 0. If these are not in D then condition (4) of theorem 1.2 implies $\gamma_p \neq 0$, hence $\gamma^*_p = 2^\kappa$ implies $\gamma_p = 2^\kappa$. Since the converse is obvious we have that always $\gamma_p = \gamma^*_p$.

If $\beta_p = 0$ then A^κ is generated by $t(A^\kappa) \cup pA^\kappa$, where $t(A^\kappa)$ denotes the torsion subgroup of A^κ. But A^* is a homomorphic image of A^κ, hence $\beta^*_p = \dim(A^* / \langle t(A^*) \cup pA^* \rangle) = 0$. If $\beta_p = 2^\kappa$, then $\beta^*_p = 2^\kappa$ is clear. Hence $\beta_p = \beta^*_p$ always holds.

Since $\alpha_{p,n} = \dim(p^{n-1}A[p] / p^n A[p])$ it follows readily that $\alpha_{p,n}$ is always equal to $\alpha^*_{p,n}$. The corollary is thus proved. □

An abelian group G is algebraically compact if and only if G is a direct summand of *every* abelian group B which contains G as a pure subgroup. It may be interesting to notice that the algebraic compactness of G can be tested by using *one* object only, namely an ultrapower of G. As usual, let $\beta\mathbb{N}$ denote the set of all ultrafilters on $\mathbb{N}$. Thus $\beta\mathbb{N}$ is the Stone-Čech compactification of $\mathbb{N}$. $\beta^*\mathbb{N}$ will denote the set of all free ultrafilters on $\mathbb{N} = \omega$.

LEMMA 1.4. *For any abelian group G the following are equivalent:*

(i) *G is algebraically compact,*

(ii) $\exists F \in \beta^*\mathbb{N}$: *G is a direct summand of* $G^{\mathbb{N}}/\mathcal{N}(F)$,

(iii) $\forall F \in \beta\mathbb{N}$: *G is a direct summand of* $G^{\mathbb{N}}/\mathcal{N}(F)$.

Proof: Of course we have identified G with its canonical isomorphic image in the ultrapower. By Łoś's theorem G is pure in all ultrapowers of it, hence (i) $\Rightarrow$ (iii) holds. (iii) $\Rightarrow$ (ii) is obvious (by the axiom of choice). If F is a free ultrafilter on ω then F is ω-incomplete, and hence $G^\omega/\mathcal{N}(F)$ is algebraically compact. If G is a direct summand of this group, then G is algebraically compact too. Thus (ii) $\Rightarrow$ (i) holds and the lemma is proved. □

DIGRESSION. Also a number of other *finiteness conditions* (algebraic compactness is a finiteness condition) in Group Theory can be expressed by requirements concerning the canonical embedding of G into its ultrapowers. To give a few examples let G be any group which is not necessarily abelian.

(I) G is an *FC-group* $\Longleftrightarrow \forall F \in \beta\omega$: $G \trianglelefteq G^{\omega}/\mathcal{N}(F)$,

(II) G is a *BFC-group* $\Longleftrightarrow \forall F \in \beta\omega$: $G \trianglelefteq G^{\omega}/\mathcal{N}(F) \wedge \left(G^{\omega}/\mathcal{N}(F)\right)/G$ is abelian,

(III) $[G : Z(G)] < \aleph_0 \Longleftrightarrow \forall F \in \beta\omega$: if $A = G^{\omega}/\mathcal{N}(F)$ then $A = G \cdot C_A(G)$,

(we could replace $\forall F \in \beta\omega$ by $\exists F \in \beta^{*}\omega$). Not all finiteness conditions can be expressed in both quantifier forms (consider the Rudin-Keisler pre-ordering on $\mathbb{N}$). This gives rise to a hierarchy and it would be interesting to know whether the corresponding classification of algebraic notions is of algebraic significance.

§ 2 . HORN THEORIES

It was A.Horn who first considered the class of first-order sentences known as *HORN-sentences*. He showed that they are preserved by direct products. Later C.C.Chang and H.J.Keisler have shown that a sentence is preserved by arbitrary reduced products if and only if it is equivalent to a Horn-sentence. Let $\bigwedge_{\nu=1}^{m} \Psi_{\nu}$ denote the conjunction and $\bigvee_{\nu=1}^{m} \Psi_{\nu}$ the disjunction of the formulae $\Psi_1, .., \Psi_m$.

DEFINITION. Let $\mathcal{L}$ be any first-order language.

(i) If Φ is an atomic $\mathcal{L}$-formula, then both, Φ and $\neg\Phi$, are HORN-formulas;

(ii) If $\Phi_0, \Phi_1, .., \Phi_m$ are atomic $\mathcal{L}$-formulae, then $\bigwedge_{\nu=1}^{m} \Phi_{\nu} \Rightarrow \Phi_0$ and $\bigvee_{\nu=1}^{m} \neg\Phi_{\nu}$ are HORN-formulas;

(iii) If Φ and Ψ are HORN-formulas, then $\Phi \wedge \Psi$, $\exists v \Phi$ and $\forall v \Phi$ are HORN-formulas

(iv) Φ is a HORN-formula of $\mathcal{L}$ only if Φ can be obtained by applying the rules (i),(ii) and (iii).

A first-order theory T is called a *HORN-Theory* if T can be axiomatized by a set of HORN-sentences. In the sequel we let $\mathcal{L}$ denote the first-order language of group-theory; the only non-logical symbols of $\mathcal{L}$ are the operation-symbols $+$, $-$ and the constant 0. The set of all $\mathcal{L}$-sentences which are true in the group G is called the *Theory* of G and is denoted by $\mathrm{Th}(G) = \{\Phi ; G \models \Phi\}$.

DEFINITION. A group G is called a ***HORN-group*** if Th(G) is a HORN-theory.

In order to classify all abelian HORN-groups we need the Szmielew invariants.

DEFINITION. An abelian group G is called a ***SZMIELEW-group*** if there are cardinals $\alpha_{p,n}$, β_p , γ_p and δ (for $n \in \omega$ and $p \in$ set of all primes) such that:

(1) $G = \mathbb{Q}^{(\delta)} \oplus \bigoplus_p \left(Z(p^\infty)^{(\gamma_p)} \oplus Q_p^{(\beta_p)} \oplus \bigoplus_{k \in \omega} Z(p^k)^{(\alpha_{p,k})} \right)$;

(2) $\delta \leq 1$, $0 \leq \alpha_{p,k} \leq \aleph_0$, $0 \leq \beta_p \leq \aleph_0$, $0 \leq \gamma_p \leq \aleph_0$ for all p , k ;

(3) If $\limsup_{k \to \infty} \alpha_{p,k} \neq 0$, then $\beta_p = \gamma_p = \aleph_0$;

(4) $\delta = 0 \iff \left(\forall p\colon \beta_p = \gamma_p = 0 \wedge \exists m \in \omega \; \forall p \geq m \; \forall k \geq m\colon \alpha_{p,k} = 0 \right)$.

The importance of the concept of a Szmielew group lies in the following classical theorem due to W.Szmielew (cf. Eklof-Fisher [3]): *every abelian group A is elementarily equivalent to one and only one Szmielew group.* Since we are mainly concerned with reduced products in this paper, we may reformulate W.Szmielew's theorem as follows : *For any abelian group A there is one and only one Szmielew group G such that A and G have isomorphic ultrapowers.*

THEOREM 2.1. *For any abelian group G the following are equivalent:*

(i) G is a HORN-group;

(ii) G and $G \oplus G$ are elementarily equivalent;

(iii) G and $G^\omega/\mathcal{N}(F_\omega)$ are elementarily equivalent;

(iv) G is an elementary substructure of $G^\omega/\mathcal{N}(F_\omega)$;

(v) G is elementarily equivalent to a Szmielew group H such that

the invariants $\alpha_{p,n}$, β_p *and* γ_p *of* H *are* 0 *or* $\aleph_0$.

Proof. Consider first (i) $\Rightarrow$ (ii): If G is a HORN-group then the class of all models of $Th(G)$ is closed under proper reduced products (i.e. under reduced products $\neq \{0\}$). If I is any set of cardinality 2 and $F = \{I\}$, then $G \oplus G \cong G^I/\mathcal{N}(F)$ is hence a model of $Th(G)$.

(ii) $\Rightarrow$ (v): this follows immediately from the Feferman-Vaught theorem.

(v) $\Rightarrow$ (iv): We shall prove first that G and $G^\omega/\mathcal{N}(F_\omega)$ are elementarily equivalent, where $F_\omega = \{X \subseteq \omega \, ; \, |\omega - X| < \aleph_0\}$. Since $G \equiv A$ implies $G^\omega/\mathcal{N}(F_\omega) \equiv A^\omega/\mathcal{N}(F_\omega)$ (cf. Chang-Keisler [1], p.345, theorem 6.3.4.) we may assume without loss of generality that $|G| = \aleph_0$. Let F be any free ultrafilter on ω and put $A = G^\omega/\mathcal{N}(F)$. Since F is ω-incomplete A is algebraically compact (see §1) and $|A| = 2^{\aleph_0}$ (cf. [1], p.201 - p.202, Proposition 4.3.4 and 4.3.7). According to Kaplansky we decompose $A = D \oplus \prod_p \hat{A}_p$, where $D = \mathbb{Q}^{(\sigma)} \oplus \bigoplus_p Z(p^\infty)^{(\tau_p)}$ is the maximal divisible subgroup of A and $\hat{A}_p$ is the completion of

$$A_p = Q_p^{(\rho_p)} \oplus \bigoplus_{n\in\omega} Z(p^n)^{(\zeta_{p,n})} .$$

By the work of Szmielew and Eklof-Fisher [3], $D \oplus \bigoplus_p A_p$ is an elementary subgroup of A. But $A \equiv G$ and the Szmielew invariants of G are known by (v). Hence by Eklof [2], theorem 2.3, and our corollary 1.3 we conclude that $A \cong A^\omega/\mathcal{N}(F_\omega)$. Hence $G \equiv G^\omega/\mathcal{N}(F_\omega)$. Since G is is a pure subgroup of $G^\omega/\mathcal{N}(F_\omega)$ it now follows from Eklof-Fisher [3] Corollary 2.5, that G is an elementary subgroup of $G^\omega/\mathcal{N}(F_\omega)$.

(iv) $\Rightarrow$ (iii) is obvious. (iii) $\Rightarrow$ (i) follows readily from a result of F.Galvin [7], since $P(\omega)/F_\omega$ is an atomless Boolean algebra. The theorem is thus proved. $\square$

<u>COROLLARY 2.2.</u> *Let* T *be any complete* 1st*-order theory of abelian groups. Then* Mod(T) *(= the class of all* T*-models) is closed under proper reduced products if and only if* Mod(T) *is closed under direct products.*

EXAMPLE. Let $\mathbb{F}_\omega = \mathbb{Z}^{(\omega)}$ be the free-abelian group of rank ω. Since $\mathbb{F}_\omega \cong \mathbb{F}_\omega \oplus \mathbb{F}_\omega$ it follows from theorem 2.1 that $Th(\mathbb{F}_\omega)$ can be axiomatized by a set of HORN-sentences.

A characterization of all non-abelian HORN-groups is unknown and similarly there is no classification of all (associative) HORN-rings. As an application of theorem 2.1 we shall prove below that a HORN-ring with unit-element cannot be artinian. This shows that a classification of all HORN-rings would perhaps be rather difficult. By a *ring* will be meant a not necessarily commutative ring (but multiplication is assumed to be associative and distributive on both sides over addition). A *ring* does not need to have a unit element.

LEMMA 2.3. *Among the artinian rings with unit element there is no HORN-ring.*

Proof. Assume, if possible, that there is an artinian HORN-ring R with unit element. Then $R \equiv R \oplus R$, and if R^+ denotes the additive group of R, then similarly $R^+ \equiv R^+ \oplus R^+$. But then $R^+ = \mathbb{Q}^{(\delta)} \oplus B$ for some cardinal δ and some group B of bounded exponent (cf. Fuchs [6], theorem 122.4) since the number of Prüfer-groups occurring in R^+ cannot be finite and non-zero (cf. theorem 2.1). Let $J(R)$ be the Jacobson radical of R. Since R is artinian (i.e. every strictly decreasing sequence of left-ideals is finite) $J(R)$ is nilpotent (cf. [11], p.63 (theorem 14), p.69 (theorem 16) and p.72 (theorem 19). Thus $J(R)^+$, the additive group of $J(R)$, satisfies the minimum condition on subgroups (cf. Fuchs [6], p.296, proposition 122.1). Since R^+ does not contain Prüfer-groups, $J(R)^+$ is finite by a result of Kurosh. $J(R)$ is 1st-order definable in R, $J(R \oplus R) = J(R) \oplus J(R)$ (cf. [11] p.68) and $R \equiv R \oplus R$ then shows that $J(R) = \{0\}$. Thus R is a semi-simple artinian ring with unit element. By Wedderburn's theorem R is hence a direct sum of finitely many matrix rings over skewfields. Looking at the matrix-units it is clear, that R and $R \oplus R$ cannot be elementarily equivalent, a contradiction and the lemma is proved. □

We finally draw the readers attention to a paper by J.Reineke [10] in which ring-theoretic properties are discussed which are preserve under reduced powers.

REFERENCES

[1] C.C.CHANG - H.J.KEISLER: *Model Theory*. Amsterdam 1973.

[2] P.C.EKLOF: *The structure of ultraproducts of abelian groups*. Pacific J.Math.47(1973),pp.67-79.

[3] P.C.EKLOF - E.R.FISHER: *The Elementary Theory of Abelian Groups*. Annals of Math.Logic 4(1972), pp.115-171.

[4] R.ENGELKING - L.KARŁOWICZ: *Some theorems of set theory and their topological consequences*.Fund.Math.57(1965),pp.275-285.

[5] L.FUCHS: *Infinite Abelian Groups, Volume I*,Academic Press, New York 1970.

[6] L.FUCHS: *Infinite Abelian Groups, volume II*.New York 1973.

[7] F.GALVIN: *Horn sentences*. Annals of Math.Logic 1(1970),pp.389-422.

[8] K.GOLEMA - A.HULANICKI: *The structure of the factor group of the unrestricted sum by the restricted sum of Abelian Groups.II*. Fund. Math.53(1964)pp.177-185.

[9] A.HULANICKI: *The structure of the Factor Group of the Unrestricted Sum by the Restricted Sum of Abelian Groups*. Bull.Acad.Polon.Sci. Sêr.sci.math.,astr., et phys., vol.10(1962)pp.77-80.

[10] J.REINEKE: *Reduced Powers of Rings*. Preprint Math.Inst.Techn.Univ. Hannover, Grüne Reihe Nr.16 (1975).

[11] P.RIBENBOIM: *Rings and Modules*. New York 1969 (Interscience Publ.)

ULRICH FELGNER,
MATHEMATISCHES INSTITUT DER UNIVERSITÄT,
Auf der Morgenstelle 10,
74 TÜBINGEN, West Germany.

TWO ORDERINGS OF THE CLASS OF ALL COUNTABLE MODELS OF PEANO ARITHMETIC

Petr Hájek and Pavel Pudlák

Mathematical Institute of the Czechoslovak
Academy of Sciences, Prague, Czechoslovakia

§ 0. Introduction. Consider countable models of Peano arithmetic - PA. Various subrelations of the relation $M_1 \subseteq M_2$ (M_1 is a submodel of M_2) have been studied; for example, $M_1 \subseteq_e M_2$ (M_1 is an initial segment of M_2 or, equivalently, M_2 is an end-extension of M_1), $M_1 \prec M_2$ (M_1 is an elementary submodel of M_2) etc. If $R(M_1,M_2)$ is such a relation, we can extend it by "identifying isomorphic models", i.e. defining

$$\bar{R}(M_1,M_2) \text{ iff } (\exists M_3,M_4)(R(M_3,M_4) \& M_1 \cong M_3 \& M_2 \cong M_4).$$

Let us make a trivial observation that if R is invariant with respect to simultaneous isomorphisms of (M_1,M_2) (i.e. if for each isomorphism $\wp : M_2 \leftrightarrow M_2'$, $R(M_1,M_2)$ implies $R(\wp'' M_1, M_2')$), then the definition of $\bar{R}$ can be slightly simplified:

$$\bar{R}(M_1,M_2) \text{ iff } (\exists M_3)(R(M_3,M_2) \& M_1 \cong M_3).$$

Furthermore, iff R is invariant in this sense, and R is transitive, then $\bar{R}$ is also transitive.

We shall introduce two particular relations $R: M_1 \subseteq_{cd} M_2$ (M_1 is a canonical segment of M_2) and its subrelation $M_1 \subseteq^{o}_{cd} M_2$, both based on a notion of canonical definability of one model in another one (with or without parameters respectively) and study their extensions $\bar{R}$. We obtain two orderings $M_1 \leq_{cd} M_2$ and $M_1 <^{o}_{cd} M_2$ with strikingly different properties. (Strictly speaking, $\leq_{cd}$ is only a preorder, i.e. ordering on the obvious factorization.) We relate our notions to recursive saturatedness. The main technical device is the truth definition for one model definable in another model. In § 4 we obtain some results on symbioticity that might be of independent interest.

§ 1. Notation. We shall consider countable models of Peano Arithmetic here. Thus "model" will be always an abbreviation for "model of PA". Also all the initial segments of the models that we will consider, will be models of PA. The particular choice of a language of arithmetic is not important for the purpose of this paper. Let us take

$0,1,+,\cdot$, let L be the set of formulae in this language, and let L(M) be the formulae of this language extended by constants for elements of M. The standard model will be denoted by N. The Gödel number of a formula φ will be denoted by $\ulcorner\varphi\urcorner$.

§ 2. Canonical definability. We say that a model M_2 is definable in a model M_1 if there are formulae $\varphi(x)$, $\psi_1(x,y,z)$, $\psi_2(x,y,z)$, $\chi_1(x)$, $\chi_2(x)$ of $L(M_1)$ such that $M_2 = \{x \in M_1 : M_1 \models \varphi(x)\}$, and $\psi_1, \psi_2, \chi_1, \chi_2$ define $+, \cdot, 0, 1$ of M_2 in terms of $+, \cdot, 0, 1$ of M_1.

The prominent way to construct some M_2 definable in M_1 is using the Arithmetized Completeness Theorem ACT. This theorem expresses the fact that completeness theorem for the first order logic can be simulated in PA. Namely, if M_1 is a model, T is a theory containing PA (and having the same language), $\tau(x)$ is a numeration of T in $Th(M_1)$ (i.e. $\varphi \in T$ iff $M_1 \models \tau(\ulcorner\varphi\urcorner)$), and $M_1 \models Con(\tau)$, then there is a formula $\sigma(x)$ such that $M_1 \models \forall x(\tau(x) \to \sigma(x))$, and a model M_2 whose domain is a subset of the domain of M_1, and such that σ is a truth definition for M_2 in M_1, i.e., for each $L(M_2)$-sentence φ,

$$M_2 \models \varphi \text{ iff } M_1 \models \sigma(\ulcorner\varphi\urcorner),$$

(and, say, $M_1 \models \neg\sigma(\ulcorner b=b\urcorner)$ for each $b \in M_2 - M_1$). Thus M_2 is a model of T.

If there is a truth definition σ for M_2 in M_1 then we say that M_2 is canonically definable in M_1 (or M_2 is c.d. in M_1). Evidently, if M_2 is c.d. in M_1 then M_2 is definable in M_1, since the definitions of the domain of M_2 and of the operations of M_2 are implicit in the truth definition. The truth definition is in general an $L(M_1)$-formula, i.e. it may contain parameters from M_1. If there is a parameter-free truth definition for M_2 in M_1 (an L-formula) then we say that M_2 is canonically definable in M_1 without parameters (or that M_2 is ^{0}-c.d. in M_1).

If M_2 is canonically definable in M_1 then M_1 is not a submodel of M_2, but it will be useful to represent M_1 as a submodel of M_2. Actually, there is a unique isomorphism ι of M_1 definable in M_1 onto an initial segment of M_2. This isomorphism is given by the formula described informally as follows:

$\iota(x) = y$ iff $\exists z$ (z codes an isomorphism of the interval $[0,x]$ of the universe onto the interval $[0,y]$ of the model).

Definition

$I \subseteq_{cd} M$ if $I \subseteq_e M$ and there exist M_1 and ι such that M is c.d. in M_1, and ι is the isomorphism of M_1 onto I definable in M_1. $I \subseteq^0_{cd} M$ if moreover all the defining formulae are in L, i.e. if there are M_1, ι such that M is ^{o}-c.d. in M_1 and ι is the isomorphism of M_1 and I definable (without parameters) in M_1.

$I \subseteq_{cd} M$ and $I \subseteq^o_{cd} M$ can be read "I is a canonical (^{o}-canonical) segment of M, respectively.

<u>Claim 1</u> (1) The relations "is canonically definable in" and "is ^{o}-canonically definable in" are transitive.

(2) The relations $\subseteq_{cd}$ and $\subseteq^o_{cd}$ are transitive.

<u>Proof</u>: (1) is obvious. (2) Assume $M_1 \subseteq_{cd} M_2 \subseteq_{cd} M_3$. There are M_2' and ι such that M_3 is c.d. in M_2' and $\iota : M_2 \longleftrightarrow M_2'$ is definable in M_2'. Let $M_1' = \iota'' M_1$. Then $M_1' \subseteq_{cd} M_2'$, i.e. there are M_1'' and $\varkappa$ such that M_2 is c.d.in M_1'' and $\varkappa : M_1' \longleftrightarrow M_1''$ is definable in M_1''. Then M_3 is c.d. in M_1'' and the composition of ι and $\varkappa$ is an isomorphism of M_1 and M_1'' definable in M_1''.

<u>Lemma 2</u> If I is strong initial segment of M, then the truth for I is definable in the structure (M,I) - M expanded by the unary predicate I.

<u>Proof</u>: Given I strong in M, and a formula $\varphi = (Q_1(x_1)\dots(Q_n x_n)\ \psi(\underline{x})$ in the prenex normal form, one can find $c_1,\dots,c_n \in M$ such that

(*) $I < c_1 < \dots < c_n$, and for every $i = 1,\dots,n$; $x_1,\dots,x_{i-1} \in I$,

$$(Q_i x_i \in I)(Q_{i+1} x_{i+1} < c_{i+1})\dots(Q_n x_n < c_n)\ \psi(\underline{x}) \Longleftrightarrow$$

$$\Longleftrightarrow (Q_i x_i < c_i)(Q_{i+1} x_{i+1} < c_{i+1})\dots(Q_n x_n < c_n)\ \psi(\underline{x})$$

(see Kirby-Paris [4]). Thus "$I \models \varphi$" is equivalent to the following condition:

$$\exists c_1,\dots,c_n \text{ such that } (*) \text{ and } (Q_1 x_1 < c_1)\dots(Q_n x_n < c_n)\ \psi(\underline{x}).$$

Since truth for bounded formulae is definable in M, the condition can be expressed by a single formula.

<u>Proposition 3</u>

(1) For $k \geq 1$, a model M, there exists a model M' such that $M \subseteq^o_{cd} M'$ and $M \prec_{\Sigma^o_k} M'$.

(2) If $I \subseteq_{cd} M$, then

a) I is regular in M,

b) I is not elementary in M,

c) I is not strong in M.

(3) If $I \subseteq^o_{cd} M$, then $I \not\equiv M$.

(4) If $I \subseteq_{cd} M$, and I is nonstandard, then M is recursively saturated.

Proof: (1) This follows from ACT and from the fact that truth in M for Σ_k-sentences of L(M) can be expressed in M.
(2) a) This is because each coded subset (or partition) of I is definable in I.
(2) b) Let M be c.d. in M_1 along with an isomorphism $M_1 \cong I$. If I were elementary in M, then we would have a truth definition for I in M_1 (= the restriction of the one for M). This composed with the isomorphism would give the truth definition for M_1 in M_1 which is impossible.
(2) c) Using the notation of (2) b), the expanded structure (M,I) is definable in M_1. If I were strong, we would have a truth definition for I in M_1, by Lemma 2. This is impossible, as shown in (2) b).
(3) There is no $\sigma(x) \in L$ such that for some M and each closed $\varphi \in L$, $M \models (\sigma(\ulcorner\varphi\urcorner) \Longleftrightarrow \varphi)$ (by Diagonalization Lemma). If $I \subseteq^{o}_{cd} M$, then we do have $\sigma(x) \in L$ such that for closed $\varphi \in L$, $M \models \varphi \Longleftrightarrow I \models \sigma(\ulcorner\varphi\urcorner)$.
(4) See Smoryński [6].

§ 3. Two orderings.

Definition

a) $M_1 \leq_e M_2$ if $(\exists I \subseteq_e M_2)(I \cong M_1)$.
b) $M_1 \leq_{\delta d} M_2$ if $(\exists I \subseteq_{\delta d} M_2)(I \cong M_1)$.
c) $M_1 <_{cd} M_2$ if $(\exists I \subseteq^{o}_{cd} M_2)(I \cong M_1)$.

Lemma 4 $M_1 \leq_{cd} M_2$ iff M_2 is c.d. in an isomorphic copy of M_1, i.e. iff $(\exists M_1')(M_1' \cong M_1$ & M_2 is c.d. in $M_1')$, similarly for $<^{o}_{cd}$.
Proof: $\Rightarrow$ is obvious. $\Leftarrow$: Let $M_1' \cong M_1$ and let M_2 be c.d. in M_1'. Let ι be the unique isomorphism of M_1' onto a segment I of M_2, ι definable in M_1'. Then $I \subseteq_{cd} M_2$ and $M_1 \cong I$.

For the purpose of the following characterization, we can confine ourselves to models that contain N as an initial segment.

Definition
$SS(M) = \{X; X$ is coded in M, $X \subseteq N\}$,
$\Sigma_1(M) = \{\varphi;\ \varphi$ is closed Σ^{o}_1 and $M \models \varphi$, $\varphi \in L\}$.

Theorem 5
(a) $M_1 \leq_e M_2$ iff $SS(M_1) = SS(M_2)$ and $\Sigma_1(M_1) \subseteq \Sigma_1(M_2)$ (Friedman [1]).
(b) For M_1, M_2 recursively saturated, $M_1 \cong M_2$ iff $SS(M_1) = SS(M_2)$ and

$M_1 \equiv M_2$ (Smoryński [6]).

(c) Let $M_1 \models PA$, let T be a consistent theory in the language L and extending PA. Let $T \in SS(M_1)$ (as a set of Gödel numbers). Then there is an $M_2 \models T$ such that $SS(M_1) = SS(M_2)$ (Friedman [1]).

Lemma 6 For M_1, nonstandard, $M_1 \leq_{cd} M_2$ iff $SS(M_1) = SS(M_2)$, $\Sigma_1(M_1) \subseteq \Sigma_1(M_2)$, and M_2 is recursively saturated.

Proof: The part "only if" was proved above. Let $SS(M_1) = SS(M_2)$, $\Sigma_1(M_1) \subseteq \Sigma_1(M_2)$, and M_2 be recursively saturated. Assume we have some numeration $\{\varphi_n\}_{n=1}^{\infty}$ of closed formulas of L, and some coding of finite sets (definable in PA). Define the following type in M_2:

$$\tau_n(x) \Longleftrightarrow [(\underline{n} \in x \Longleftrightarrow \varphi_n) \,\&\, \mathrm{Con}(x)].$$

Let $c > N$ realize this type. By Theorem 5 (a) we can assume $M_1 \subseteq_e M_2$. Let $d = c \restriction a$, where $N < a \in M_1$, then d also realizes the type. In particular, we have $M_1 \models \mathrm{Con}(d)$. Apply ACT to d, and obtain some M_2' such that $M_1 \leq_{cd} M_2$. By the construction $M_2' \equiv M_2$, and $SS(M_2') = SS(M_2)$, since M_2' contains an initial segment isomorphic to M_1. Now the lemma follows from Theorem 5 (b).

Let us write "$M_1 <_{cd} M_2$" for "$M_1 \leq_{cd} M_2$ and $\neg M_2 \leq_{cd} M_1$".

Theorem 7

(1) $\leq_{cd}$ coincides with the quasiordering $\leq_e$ on the class of recursively saturated models; $M \leq_{cd} M$ iff M is recursively saturated.

(2) No pair of $\leq_{cd}$-noncomparable recursively saturated models has an upper bound.

(3) There are models M, M_1, M_2 such that $M \leq_{cd} M_1$, $M \leq_{cd} M_2$, M_1, M_2 incomparable; M can be taken nonstandard.

(4) There exists a maximal element (i.e. there exists M_1 such that $M_1 <_{cd} M_2$ for no M_2).

(5) Density fails (there exist $M_1 <_{cd} M_2$, M_1, M_2 recursively saturated such that $M_1 <_{cd} M <_{cd} M_2$ for no M).

(6) There exists M recursively saturated such that for each $M_1 \leq_{cd} M$, M_1 recursively saturated, there is M_2 recursively saturated, $M_2 <_{cd} M_1$ (i.e. there is no minimal recursively saturated $M_1 \leq_{cd} M$).

Proof: (1) follows from Theorem 5 (a) and Lemma 6. (2) follows from (1).

(3) It is well-known that there are Σ_1 sentences σ_1, σ_2 such that $\mathrm{Con}(PA + \sigma_1 + \neg\sigma_2)$, $\mathrm{Con}(PA + \neg\sigma_1 + \sigma_2)$. Take for M_1 any model in which these two statements are true, and use ACT to obtain M_2, M_3.

(4) Let S be a maximal set of Σ_1 sentences consistent with PA. Let M_1 be recursively saturated, $M_1 \models PA + S$. Then, by (1) and Theorem 5(a), M_1 is maximal.

(5) Let φ be a Π^0_1-sentence consistent with PA and such that $\neg\varphi$ is Π^0_1-conservative over PA, i.e. whenever $PA + \neg\varphi \vdash \psi$ for a Π^0_1-sentence ψ then $PA \vdash \psi$. For example, take $Con(\pi)$ where π is the usual numeration of PA, for φ (cf. Guaspari [2]). Let S be a maximal set of Σ^0_1-sentences such that $PA + \varphi + S$ is consistent. Then $PA + \neg\varphi + S$ is also consistent because of Π^0_1-conservativity of $\neg\varphi$. Let R be a maximal set of Π^0_1-sentences such that $PA + \neg\varphi + S + R$ is consistent. Let M_1 be a model of $PA + S + \varphi$ such that $S, R \in SS(M_1)$; due to Proposition 3(1), we may assume that M_1 is recursively saturated. By Theorem 5 (c), let $M_2 \models PA + \neg\varphi + S + R$ be such that $SS(M_2) = SS(M_1)$. Then M_2 can be taken such that $M_1 \subseteq_e M_2$. Using Proposition 3(1) again, we may assume that M_2 is recursively saturated. Thus $M_1 \leq_{cd} M_2$, by Lemma 6. Now, if $M_1 \leq_{cd} M \leq_{cd} M_2$, then either $M \models \varphi$ and then $\Sigma_1(M) = S = \Sigma_1(M_1)$ and consequently $M \leq_{cd} M_1$ by Lemma 6, or $M \models \neg\varphi$ and then $\Pi_1(M) = R = \Pi_1(M_2)$ and consequently $M_2 \leq_{cd} M$ by the same lemma.

(6) Call an element $a \in M$, Σ^0_0-definable, if there is a Σ^0_0-formula $\varphi(x) \in L$ (with the single free variable x), such that

$$a = \min\{x \in M;\ M \models \varphi(x)\}.$$

By McAloon [7], there is a model M in which Σ^0_0-definable elements are coinitial with $M - N$ (i.e. for each $b \in M$, $N < b$, there is a Σ^0_0-definable $a \in M$, $N < a < b$). If $I \subseteq_e M$, then $I \models (\exists x)\varphi(x)$ for a Σ^0_0-formula φ iff the element defined by $\varphi(x)$ is in I. It is a well-known fact that initial segments which are models of PA are coinitial with $M - N$ in each model M. As we shall see below, recursively saturated initial segments which are models of PA, are symbiotic with them, thus they are also coinitial. This, together with Lemma 6, gives the conclusion.

Problems:

I. If M_1, M_2 have a lower bound, do they necessarily have a greatest lower bound?

II. If M is not maximal, do there exist M_1, M_2 incomparable, $M \leq_{cd} M_1$, $M \leq_{cd} M_2$?

III. Given M, is there a maximal M_1, $M \leq_{cd} M_1$?

<u>Lemma 8</u> Let $M_1 \subseteq_e M_2$, M_1 Σ^0_k-elementary nonstandard submodel of M_2, let σ be a truth definition for M_3 in M_1, and suppose σ is Σ^0_k.

Then σ is a truth definition for a model M_4 in M_2 such that $M_4 \cong M_3$. Hence, if $\sigma \in L$ then $M_2 <^{o}_{cd} M_3$.

Proof: We have M_3 and M_4 recursively saturated, $SS(M_3) = SS(M_4) = SS(M_1)$. For every sentence $\varphi \in L$ we have $M_1 \models \sigma(\ulcorner\varphi\urcorner)$ iff $M_2 \models \sigma(\ulcorner\varphi\urcorner)$. Thus $M_3 \equiv M_4$, and we can apply Theorem 5 (b).

Theorem 9

(1) $<^{o}_{cd}$ is a strict irreflexive ordering.

(2) There are no $<^{o}_{cd}$-maximal models.

(3) The ordering $<^{o}_{cd}$ is dense on nonstandard models; there are M_1, M_2, M_3, M_4 such that $M_1 <^{o}_{cd} M_3 <^{o}_{cd} M_2$, $M_1 <^{o}_{cd} M_4 <^{o}_{cd} M_2$, M_3 and M_4 are incomparable and $M_3 \not\equiv M_4$.

(4) No pair of $<^{o}_{cd}$-incomparable elements has a g.l.b.

Proof: (1) Transitivity is clear. By Proposition 3(3) we have $M_1 \not\equiv M_2$ if $M_1 <^{o}_{cd} M_2$, whence irreflexivity.

(2) By Proposition 3(1).

(3) Density is an immediate consequence of Lemma 6 and Proposition 3(1).

Concerning the second assertion, we will construct formulae μ_2, μ_3 $\mu_4 \in L$, μ_2 Σ^{o}_{k}-formula, a sentence $\varphi \in L$, and a mapping $\sigma(x) \mapsto \alpha_\sigma$, of all formulae of L with one free variable into sentences of L, such that the following sentences are consistent with PA:

(i) " μ_2, μ_3, μ_4 canonically define models of PA",

(ii) $(\mu_3(\ulcorner\alpha_\sigma\urcorner) \,\&\, \mu_4(\ulcorner\sigma(\ulcorner\neg\alpha_\sigma\urcorner)\urcorner)) \vee (\mu_3(\ulcorner\neg\alpha_\sigma\urcorner) \,\&$ $\&\, \mu_4(\ulcorner\neg\sigma(\ulcorner\neg\alpha_\sigma\urcorner)\urcorner)))$, for each $\sigma \in L$ with one free variable,

(iii) $\psi \Longrightarrow \mu_3(\ulcorner\psi\urcorner) \,\&\, \mu_4(\ulcorner\psi\urcorner)$, for each Σ^{o}_{k}-sentence $\psi \in L$,

(iv) $\mu_3(\ulcorner\varphi\urcorner) \,\&\, \mu_4(\ulcorner\neg\varphi\urcorner)$.

Let μ_2 arbitrary such that $N \models$ " μ_2 canonically defines a model of PA". Let T be PA + Σ^{o}_{k} true sentences + Henkin axioms. Take some sentence $\varphi \in L$ independent of T, and put $K_o = T + \varphi$, $L_o = T + \neg\varphi$. Let $\sigma(x)$ be the n-th formula with one free variable. We will use induction:

(a) n is odd. For $\sigma(x)$ define α_σ as the first formula independent of K_n. If $L_{n-1} + \sigma(\ulcorner\neg\alpha\urcorner)$ is consistent, then define

$$K_n = K_{n-1} + \alpha \ , \ L_n = L_{n-1} + \sigma(\ulcorner\neg\alpha\urcorner).$$

Otherwise define $K_n = K_{n-1} + \neg\alpha$, $L_n = L_{n-1}$.

(b) n is even, $n > 0$. Do the same thing with K_n and L_n interchanged.

Let μ_3 define $\cup K_n$ and μ_4 define $\cup L_n$. Then the statements (i) - (iv) are true in N, thus consistent with PA. Let M_1 be a nonstandard model such that (i) - (iv) is true in it. Then μ_2, μ_3, μ_4 define some models M_2, M_3, M_4 in M_1. By (iii), $M_1 \prec_{\Sigma_k^0} M_3, M_4$. Thus $M_1 <_{cd}^0 M_3$, $M_4 <_{cd}^0 M_2$, by Lemma 8. By (iv), M_3, M_4 are not elementarily equivalent, and by (ii) they are $<_{cd}^0$ incomparable.

Problem:

IV. If a pair has an upper bound in $<_{cd}^0$, does it have a l.u.b.?

§ 4. <u>Loose ends</u>. Let I be an initial segment of M. Denote by $SS_I(M) = \{X \subseteq I;\ X \text{ coded in } M\}$, $\Sigma_{1,I}$ - the set of all Σ_1^0 sentences of L(I)(=containing parameters from I).

<u>Theorem 10</u>

Let $I \subseteq_e M_1, M_2$ (of course, M_1, M_2, I countable models of PA). Assume that $SS_I(M_1) = SS_I(M_2)$ and each $\Sigma_{1,I}$ sentence true in M_1 is also true in M_2. Then there is an isomorphism φ of M_1 onto an initial segment of M_2 such that φ is identical on I.

The <u>proof</u> is a routine modification of the proof of Theorem 5 (a). For the reader's convenience, the proof is spelled out in the Appendix. (This is because in fact Friedman proves a more general theorem than Theorem 5 (a).)

<u>Theorem 11</u>

Let C be a set of proper initial segments $\neq$ N of a model M, let each member of C be a model (of PA), and let C be closed under isomorphisms in the set of all proper initial segments of M. Then C is symbiotic with each of the following two sets

$\{I \in C;\ I \text{ strong in } M\}$,
$\{I \in C;\ I \text{ non-strong in } M\}$.

Remark: A and B are called symbiotic iff for $a<b$, $(\exists I \in A)(a<I<b) \Longleftrightarrow (\exists J \in B)(a<J<b)$.

<u>Proof</u>: Let $I \in C$, $a<I<b$. Take two models $I_1, I_2 \subseteq_e M$, $a<I_1<I<I_2<b$. Find some Σ_1^0-elementary end-extension M_1 of I such that I is strong (non-strong) in M_1 (see Kirby [3], and Proposition 3 respectively). Apply Theorem 10, with $I := I_1$, $M_1 := M_1$, $M_2 := I_2$. Then we have an isomorphism φ such that $I_1 = \varphi'' I_1 < \varphi'' I < \varphi'' M_1 < I_2$. Since $\varphi'' I$ is strong

(non-strong) in $\wp'' M_1$, it is also strong (non-strong) in the whole model.

Corollary 12

(1) Each $I \subseteq_e M$, $I \neq N$ is isomorphic to a $J \subseteq_e M$, J strong in M, and to a $K \subseteq_e M$, K not strong in M; moreover, if $a < I < b$, then J, K can be found such that $a < J$, $K < b$. (Take $C = \{J;\ J$ isomorphic with $I\}$.)
(2) The set $\{I; I \subseteq_{cd} M\}$ is either empty (if M is not recursively saturated) or is not closed under isomorphisms (since, by Proposition 3(2)c) it does not contain any strong segment). In particular, for each M, there is an $I \subseteq_e M$ that is not $I \subseteq_{cd} M$.
(Caution: but if M is recursively saturated, then for each $I \subseteq_e M$, we have $I \leq_{cd} M$, see Theorem 7(1).)

Remark: If M is recursively saturated then there is an $I \subseteq_{cd} M$ such that non $I \subseteq^o_{cd} M$. (We have $M \leq_{cd} M$, thus there is an $I \subseteq_{cd} M$, $I \equiv M$. For this I we have non $I \subseteq^o_{cd} M$.)

Lemma 13 Let T be a recursive extension of PA, and let M be a model such that T is consistent with the set of Σ^o_1 sentences true in M. Then there is an M' recursively saturated, $M \subseteq_e M'$, $M' \models T$ (in fact $M \subseteq_{cd} M'$). If $k \geq 0$, and $M \models T$, we can also require $M \prec_{\Sigma^o_k} M'$.

Proof: By reflexivity of T, we have $T \vdash Con(T \upharpoonright n)$, for each $n \in N$. Let M be nonstandard. Since $\neg Con(T \upharpoonright n)$ is Σ^o_1, our assumption rules out $M \models \neg Con(T \upharpoonright n)$, for $n \in N$. Thus we have $M \models Con(T \upharpoonright n)$, for some $n > N$, and we can apply ACT. If M is standard, apply ACT twice.

Theorem 14

Let T be a recursive extension of PA. Then in each model M, the set of all proper initial segments $\neq N$ that are models of T is symbiotic with its subset consisting of recursively saturated models.
Proof: If $I \models T$, then T is consistent with Σ^o_1 sentences which are true in I. By Lemma 13, I can be extended to some M_1 such that $I \prec_{\Sigma^o_1} M_1$, M_1 recursively saturated, $M_1 \models T$. Now use the trick of Theorem 11.

Remark: Lemma 13 and Theorem 14 have been proved by Murawski, we give alternative proofs here. (See [5],[8].)

Theorem 15
For M recursively saturated, the following two sets are symbiotic

$\{I;\ I \subseteq_{cd} M,\ I \models PA\}$, $\{I;\ I \subseteq_e M,\ I \models PA\}$.

Proof-sketch: Let $I \subseteq_e M$, $a < I < b$. We have to find $J \subseteq_{cd} M$, $a < J < b$. Denote by t_x, $x \in M$, the following term in the sense of M:

$$t_1 = 0 + 1,\ t_{x+1} = (t_x) + 1.$$

Take an $x \in I$ such that for all (standard) formulae $\varphi(u,v)$ with two free variables u, v,

$$M \models \ulcorner\varphi(u,v)\urcorner \in x \Longleftrightarrow \varphi(a,b).$$

Let y be the biggest element such that $y \subseteq_e x$ and

$$M \models Con(\{\varphi(t_a,c);\ \varphi(u,v) \in y\} \cup \{t_x < c;\ x < b\})$$

where $\varphi(u,v)$ are formulae in the sense of M and c is some new constant. Then y is nonstandard, and we have also $I \models Con(T)$ for

$$T = \{\varphi(t_a,c);\ \varphi(u,v) \in y\} \cup \{t_z < c;\ z \in I\}.$$

Let M' be a model of T definable in I (by ACT). Let a', b' be the interpretations of t_a, c respectively. Then the definable isomorphism maps I onto a canonical segment I' of M', such that $a' \in I' < b'$. As we have $(M,a,b) \equiv (M',a',b')$, M, M' recursively saturated, we can find an isomorphism of M onto M' that maps a, b onto a', b' respectively (by an obvious generalization of Theorem 5 (b)). The coimage of I' is the required canonical segment.

APPENDIX Proof of the Theorem 10.

Let $M_1 = \{x_i\}_{i=0}^{\infty}$, $M_2 = \{y_i\}_{i=0}^{\infty}$. We construct the desired isomorphism φ as a union $\bigcup \varphi_n$ of partial isomorphisms. Put $\varphi_0 = \emptyset$; now assume that φ_n has been constructed and has the following properties: Let $dom(\varphi_n) = \{z_0,\dots,z_{h-1}\}$, and put $\varphi_n(z_i) = w_i$ for $i = 0,\dots,h-1$. Then for each Σ_1^0 formula of L(I) with h free variables, $M_1 \models \psi(z_0,\dots,z_{h-1})$ implies $M_2 \models \psi(w_0,\dots,w_{h-1})$, i.e. φ_n is Σ_1^0-preserving. (Note that φ_0 satisfies this since in this case ψ is a sentence.)

Case 1. n is odd, $n = 2m + 1$. If $x_m \in dom(\varphi_n)$ we put $\varphi_{n+1} = \varphi_n$, otherwise we proceed as follows: Put $z_h = x_m$. Let $\omega(u,u_0,\dots,u_h)$ be an universal Σ_1^0 formula of L(I), i.e. for each Σ_1^0 formula of L(I), $\psi(u_0,\dots,u_h)$ with h + 1 free variables there is an $a \in I$ such that

$$(*)\quad PA + diag(I) \vdash \psi(u_0,\dots,u_j) \equiv \omega(a,u_0,\dots,u_h),$$

(diag(I) is the open diagram of I). Such an ω is easily constructed from a Σ_1^0 formula universal for Σ_1^0 formulae with h + 2 free variables observing that, due to pairing, all Σ^0_1 formulae of L(I) are represented by all formulae of the form $\psi(c,u_o,\ldots,u_h)$ where ψ is Σ^0_1 and $c \in I$.

Put $Y = \{a \in I;\ M_1 \models \omega(a,z_o,\ldots,z_{h-1},x_m)\}$.

Then $Y \in SS_I(M_1)$ and, for each $c \in I$, the set

$$Y_c = \{a \leq c;\ M_1 \models \omega(a,z_o,\ldots,z_{h-1}\ x_m)\}$$

is an <u>element</u> of I (via the usual coding of model finite sets; the element of M_1 coding Y_c is less than 2^{c+1} and therefore is an element of I). Let y^* represent Y in M_2. Then, for each $c \in I$, $M_1 \models (\exists v)(\forall a \in Y_c)\,\omega(a,\underline{z},v)$, ($\underline{z}$ is $z_o,\ldots,z_{h-1}$) $M_2 \models (\exists v)(\forall a \in Y_c)\,\omega(a,w,v)$ (the quantifier ($\forall a \ldots$) is bounded!) $M_2 \models (\exists v)(\forall a \in y^* \cap [0,c])\,\omega(a,\underline{w},v)$ and therefore there is a $b \in M_2 - I$ such that

$$M_2 \models (\exists v)(\forall a \in y^* \cap [0,b])\,\omega(a,\underline{w},v) \quad \text{(overspill)}.$$

Take such a v for $w_h = \wp_{n+1}(z_h) = \wp_{n+1}(x_m)$, i.e. define $\wp_{n+1} = \wp_n \cup \{\langle x_m,v\rangle\}$.

Now, if $\psi(u_o,\ldots,u_h)$ is Σ^0_1 formula of L(I)´ and $M_1 \models \psi(z_o,\ldots,z_h)$, and if a satisfies (*), then $M_1 \models \omega(a\ z_o,\ldots,z_h)$, thus $a \in Y_c$ (for any $c > a$), i.e. $M_2 \models a \in y^* \cap [0,b]$, i.e. $M_2 \models \omega(a\ w_o,\ldots,w_h)$; thus $\wp_{n+1}$ preserves Σ^0_1 formulae of L(I). In particular, if $z_h = d$ for some $d \in I$ then $\ulcorner u_h = d \urcorner$ is a Σ^0_1 formula of L(I) satisfied by z_h, in M_1, i.e. it is satisfied by w_n in M_2. This means that $\wp_{n+1}$ is identical on I, whenever defined.

<u>Case 2</u>. n is even, n = 2m + 2. Consider y_m. If $y_m \in rng(\wp_n)$ or if y_m is bigger than all elements of $rng(\wp_n)$ then put $\wp_{n+1} = \wp_n$, otherwise proceed as follows. Let $y_m < w_i$; put $w_h = y_m$ and $Z = \{a \in I;\ M_2 \models \neg\,\omega(a,\underline{w},y_m)\}$. Then $Z \in SS_I(M_2)$ and $Z_c = \{a \in c;\ M_2 \models \neg\,\omega(a,\underline{w},y_m)\}$ is an element of I for each $c \in I$. Consequently, for each $c \in I$ we have the following:

$M_2 \models (\exists v < w_i)(\forall a \in Z_c)\,\neg\,\omega(a,\underline{w},v)$ (this is Π^0_1 formula!),

$M_1 \models (\exists v < z_i)(\forall a \in Z_c)\,\neg\,\omega(a,\underline{w},v)$,

$M_1 \models (\exists v < z_i)(\forall a \in z^* \cap [0,c])\,\neg\,\omega(a,\underline{w},v)$,

where z^* codes Z in M_1; by overspill, there is a $b \in M_1 - I$ such that $M_1 \models (\exists v < z_i)(\forall a \in z^* \cap [0,b])\,\neg\,\omega(a,\underline{w},v)$.

Take such a v for z_h and define $\wp_{n+1} = \wp_n \cup \{\langle v,y_m\rangle\}$, i.e. $\wp_{n+1}(z_h) = y_m$. The proof that $\wp_{n+1}$ preserves Σ^0_1 formulae of L(I) can be safely left to the reader.

Now it is easy to see that $\wp = \bigcup \wp_n$ is an isomorphism of M_1 onto a segment of M_2, identical on I, which concludes the proof.

References

[1] H. Friedman, Countable models of set theories, in Proc. Cambridge Summer School in Logic, Matthias and Rogers eds., Lecture Notes in Mathematics vol. 337, p. 539-573.

[2] D. Guaspari, Partially conservative extensions of arithmetic, Transactions of the AMS.

[3] L.A.S. Kirby, Initial segments of models of arithmetic, thesis, Manchester Univ. 1977.

[4] L.A.S. Kirby, and J.B. Paris, Initial segments of models of Peano's axioms.

[5] R. Murawski: thesis, Warsaw University 1978.

[6] C. Smoryński, Recursively Saturated Nonstandard Models of Arithmetic.

[7] K. McAloon, Diagonal methods and strong cuts in models of arithmetic, in Logic Colloquium 77, eds. A. Macintyre, L. Pacholski, J. Paris, p. 171-182.

[8] L. Kirby, K. McAloon, R. Murawski: Indicators, recursive saturation and expandability, Fund. Math. (to appear).

Ramsey Quantifiers in Arithmetic

Angus Macintyre[(1)]

0. INTRODUCTION

The fundamental work of Kirby and Paris [KP] has inspired a series of researches which have sharpened our understanding of the incompleteness of first-order Peano arithmetic $\mathbb{P}$. We now know that $\mathbb{P}$ fails to capture many of the finite consequences of the infinite version of Ramsey's theorem. The original results were obtained model-theoretically, first by the method of indicators [KP] and then by the strikingly beautiful method in [PH]. Later, Solovay and Ketonen [SK] gave a precise proof-theoretic analysis, relating "Ramsey functions" to a classical hierarchy of provably recursive functions (and using a hierarchy motivated by the so-called subtle hierarchy in set theory). For a very valuable treatment of the whole development, and the connection with Gödel incompleteness, I recommend McAloon [Mc1].

A curious fact in the logical analysis of Ramsey phenomena has been the absence of a vigorous model-theoretic reaction to Jockusch's work [J] relating Ramsey's theorem and the jump. Some of the results of this paper constitute a belated such reaction. But a more important idea in the present context is to relate Ramsey's theorem and the hyperjump.

My paper comes from philosophical reflection on how to respond to incompleteness results. The Gödel phenomena are so pervasive that I can see no way to overcome them. But I do see one sensible way to respond to the Paris-Kirby-Harrington phenomena. Briefly, one should extend $\mathbb{P}$ to include some general version of the infinite Ramsey theorem, and then one easily derives the finite versions. This should eliminate any second generation of incompleteness phenomena relating to finite combinatorics.

The desideratum is to obtain an axiom system $\mathbb{P}^+$, resistant to incompleteness phenomena, and yet having an attractive model theory. To take $\mathbb{P}^+$ as 2nd order arithmetic, with semantics based on unrestricted set quantification, will eliminate all incompleteness, but will also eliminate the model theory (except for questions of definability in the unique model).

My choice of $\mathbb{P}^+$ is a system formulated in the language $L^{<\omega}$ of Magidor-Malitz [MM]. $L^{<\omega}$ has extra quantifiers, often called Ramsey quantifiers, whose expressive power goes far beyond first-order logic. I make a slight modification of the semantics of [MM], to avoid a semantics based on cardinality considerations. I consider only models $\mathfrak{M}$ of $\mathbb{P}$, with ordering $<$, and my semantics is given by:

---

(1) Partially supported by N.S.F. #MCS74-08550, the Polish Academy of Sciences, and the University of Warsaw.

$$\mathfrak{M} \models Q^n v_1 \ldots v_n \psi \Leftrightarrow \text{there is a subset } X \text{ of } \mathfrak{M}, \text{ cofinal for } <, \text{ so that for all distinct } x_1, \ldots, x_n \text{ in } X \ \mathfrak{M} \models \psi(x_1, \ldots, x_n).$$

$\mathbb{P}^+$ (to be called Ind) is just $\mathbb{P}$ + the obvious induction scheme in $L^{<\omega}$ + certain simple axioms for the Q^n. Very general versions of the infinite Ramsey theorem hold in all models of Ind, and the Paris-Kirby-Harrington phenomena disappear.

Due to the second-order semantics, it is not evident that there is a model theory. However, using $\Diamond_{\omega_1}$ and the methods of Magidor-Malitz [MM], I proved a completeness theorem giving many models of power $\aleph_1$. In contrast, I proved that there are no countable nonstandard models (for the intended semantics), a result connected with Lindstrom's theorem [L]. By essential use of an idea of John Schlipf, I proved that there are no models of singular cardinal, and that in uncountable models my semantics agrees with that of Magidor-Malitz. The expressive power of my system can be seen from the existence of a truth-definition for first-order formulas. It must be remarked that the model theory is much reduced from that of first-order logic.

The main results were obtained in the spring of 1978, after hearing a talk by Harvey Friedman at U.C.L.A. The whole approach has acquired respectability only in late 1979, when Schmerl and Simpson [SS] provided both a diamond-free completeness theorem and an illuminating analysis of the proof-theoretic strength of the system. Now Ind can be seen as the restriction of Π^1_1 - CA_0 [F2] to $L^{<\omega}$. I have added some new observations on the model theory, based on the work of Schmerl and Simpson. I am most grateful to Jim Schmerl and Steve Simpson for their interest, and their efforts to clarify the role of the Ramsey quantifiers.

Thanks are due to several friends. Jim Schmerl and Carol Wood steered me towards seeing a serious error in my abstract [Ma]. John Schlipf sent me a simple counterexample which I now use for another purpose, with his kind permission. Harvey Friedman had some effect, not easily analyzed, on my early thoughts on the incompleteness phenomena. Doug Hoover and Ken McAloon helped me in connection with measurability.

There are many points of contact between this paper and Schlipf's [S], though we work in very different systems. Morgenstern [Mo] has found some of the results of this paper independently.

1. PRELIMINARIES

1.1. The first-order logic L for arithmetic has the usual $+, \cdot, 0, 1, <$. $\mathbb{P}$ is the usual axiom system for first-order Peano arithmetic.

For the purposes of this paper, a key property of $\mathbb{P}$ is its ability, observed in Gödel's 1931 paper [Gö], to internalize recursions, by use of the Chinese Remainder Theorem say. This property passes to certain extensions of $\mathbb{P}$ in extensions of L, and I need to outline conditions guaranteeing the property.

So let L' be some logic (not necessarily first-order) extending L. Let $\mathbb{P}'$ be an extension of $\mathbb{P}$ in L', so that $\mathbb{P}'$ includes the obvious L'-induction scheme. Let $\mathfrak{m} \models \mathbb{P}'$, and let F be a partial function from $\mathfrak{m}$ into $\mathfrak{m}$, so that $\mathrm{dom}(F) \subseteq [0,a]$ for some $a \in \mathfrak{m}$. If F is definable (using L' and parameters from $\mathfrak{m}$) one codes it in $\mathfrak{m}$ by the well-defined element $\prod_{j \in \mathrm{dom}(F)} \mathfrak{p}_j^{F(j)+1}$. (The proof of the existence of this element depends on the induction scheme). If F is undefined, it is coded by 1. Then every nonzero element of $\mathfrak{m}$ codes a unique definable partial function with bounded range. So, by coding, a definable partial function with bounded domain is identified with a nonzero element of $\mathfrak{m}$. The notions of <u>compatibility</u> and <u>extension</u> are coded in the obvious way. If G is a definable partial function, and $\alpha \in \mathfrak{m}$, $G \restriction \alpha$ is (the code of) the restriction of G to $\{x \in \mathfrak{m} : x < \alpha\}$.

A recursion problem calls for solving in $\mathfrak{m}$ an equation for G thus:

$$G(\alpha,\_) = \Psi(\alpha,\_,G_{\_} \restriction \alpha)$$

where a) $\_$ stands for some parameters, and $G_{\_}(x) = G(x,\_)$;

b) Ψ is a definable partial function.

The main result is:

<u>Lemma 1.</u> If Ψ is total, there is a unique total definable G solving the above. Given Ψ, the definition of G is uniform for all models of $\mathbb{P}'$.

<u>Proof</u>. Routine. □

2. INDUCTIVE STRUCTURES

2.1. <u>$L^{<\omega}$ and its semantics.</u>

Beginning with an arbitrary first-order logic L one constructs $L^{<\omega}$ by adjoining quantifiers Q^n $(1 \le n < \omega)$ [MM]. The essential formation rule is that Q^n binds n distinct variables, in the form $Q^n v_1 \ldots v_n \psi$.

There are several known semantics for $L^{<\omega}$.

(a) <u>The $\varkappa$-interpretation</u>. Let $\varkappa$ be an infinite cardinal. Let $\mathfrak{m}$ be an L-structure. Then

$\mathfrak{m} \models Q^n v_1 \ldots v_n \psi(v_1,\ldots,v_n) \Leftrightarrow$ there is a subset X of $\mathfrak{m}$, of cardinal $\varkappa$, such that for all distinct $x_1,\ldots,x_n$ in X $\mathfrak{m} \models \psi(x_1,\ldots,x_n)$.

Note that this is sensible only if $\mathrm{card}(\mathfrak{m}) \ge \varkappa$.

(b) <u>The cofinality interpretation</u>. For this one distinguishes a binary $<^{\mathfrak{m}}$ in L, and considers only $\mathfrak{m}$ for which $<^{\mathfrak{m}}$ is a linear order. Then

$\mathfrak{M} \models Q^n v_1 \ldots v_n \psi(v_1,\ldots,v_n) \Leftrightarrow$ there is a subset X of $\mathfrak{M}$, cofinal with respect to $<^{\mathfrak{M}}$, such that for all distinct $x_1,\ldots,x_n$ in X $\mathfrak{M} \models \psi(x_1,\ldots,x_n)$.

To see when the two semantics agree, one must recall:

Definition. Suppose $\mathrm{card}(\mathfrak{M}) = \kappa$, and $<$ is a linear order on $\mathfrak{M}$. $<$ is κ-like if for every $x \in \mathfrak{M}$ $\{y : y < x\}$ has cardinal $< \kappa$.

Then

Lemma 2. Suppose $\mathfrak{M}$ has regular cardinal κ, and $<^{\mathfrak{M}}$ is κ-like. Then the κ-interpretation and the cofinality interpretation agree on $\mathfrak{M}$.

Proof. X cofinal $\Leftrightarrow$ $\mathrm{card}(X) = \kappa$. □

2.2. The connection of these semantics with Ramsey phenomena is evident. I now study the cofinality semantics for models of arithmetic, with $<$ as the usual order. The choice of this over the κ-interpretation is motivated as follows. Consider the standard model $\mathbb{N}$. Then, by Ramsey, one has (in either semantics)

$$\mathbb{N} \models [Q^2xy(x < y \to \psi)] \vee [Q^2xy(x < y \to \neg\psi.)]$$

However, since $\omega_1 \not\to (\omega_1)^2_2$ [ER] one does not anticipate the corresponding result for models of power $\aleph_1$ in the ω_1-interpretation. We also have the heuristic principle "any model of $\mathbb{P}$ thinks it is countable" (see e.g. [Sch]). This principle tends to screen out a semantics based on cardinality. (Partisans of the κ-interpretation need not despair. It will soon be rehabilitated).

So now, we work with the L of 1.1, extend to $L^{<\omega}$, and use the cofinality semantics. Let Ind be the extension of P to the obvious induction scheme in $L^{<\omega}$.

Definition. An inductive structure is a model of Ind.

Lemma 3. There are inductive structures.

Proof. $\mathbb{N}$ is inductive. □

It is far from obvious that there are nonstandard inductive structures. Originally, using $\Diamond_{\omega_1}$, I constructed inductive structures of power $\aleph_1$. Recently Jim Schmerl provided in [ZFC] an elegant proof of the Completeness Theorem for Ind, giving inductive structures of arbitrary uncountable regular cardinal. See [SS].

2.3. I choose now to present the material in reverse order to that of its discovery. So I leave aside the construction of inductive structures, and simply establish the basic properties of these structures. The first result is essentially due to Schlipf. He stated it to me only for countable models, in which case I already had a more complex proof based on considerations from 2.5 below.

Theorem 1. Suppose $\mathfrak{M}$ is inductive, and $a \in \mathfrak{M}$. Then there is no function f such that

(i) $\mathrm{dom}(f) \subseteq [o,a]$;

(ii) $\mathrm{range}(f)$ is cofinal in $\mathfrak{M}$.

Proof. Consider the formula $\Phi(x)$ written below:

$Q^2uv[u$ and v are partial functions with bounded range and u and v are compatible and $\mathrm{dom}(u) \subseteq [o,x]]$.

Suppose, given a, there is a function f satisfying (i) and (ii). Then $\mathfrak{M} \models \Phi(a)$. For let X be the set of all standard finite restrictions of f, and use X to "witness" Q^2 in $\Phi(a)$. Now let α be the least element of $\mathfrak{M}$ satisfying $\Phi(\alpha)$. Evidently $\alpha \neq 0$. For let Y "witness" Q^2 in $\Phi(\alpha)$. Then $\bigcup Y$ is a function defined on a subset of $[o,\alpha]$ with range cofinal in $\mathfrak{M}$. The same argument shows $\Phi(\alpha) \to \Phi(\alpha-1)$ if $\alpha \geq 1$, and we have a contradiction. □

Corollary 1. There are no nonstandard countable inductive structures.

Corollary 2. There are no inductive structures of singular cardinal.

Corollary 3. If κ is regular, and $\mathfrak{M}$ is inductive of power κ, $\mathfrak{M}$ is κ-like.

Proofs. Trivial. □

From Corollary 3 we see that for nonstandard models of power κ the cofinality semantics agrees with the κ-interpretation. This will enable us to bring to bear on our problem the powerful techniques of Magidor-Malitz [MM].

2.4. Now I examine how in inductive structures one witnesses quantifiers Q^n. This is an important point in connection with the Magidor-Malitz method, and for Schmerl's proof.

Theorem 2. Let $\Phi(v_1,\ldots,v_n,\vec{w})$ be an $L^{<\omega}$ formula. Then there is an $L^{<\omega}$ formula $W_\Phi(v,\vec{w})$ such that for all inductive $\mathfrak{M}$ and $\vec{m}$ from $\mathfrak{M}$,

i) for all distinct $x_1,\ldots,x_n$

$$\mathfrak{M} \models \bigwedge_{1\leq j\leq n} W_\Phi(x_j,\vec{m}) \to \Phi(x_1,\ldots,x_n,\vec{m}) ;$$

ii) if $\mathfrak{M} \models Q^n v_1\ldots v_n \Phi(v_1,\ldots,v_n,\vec{m})$ then $\mathfrak{M} \models (\forall x)(\exists y)(y > x \wedge W_\Phi(y,\vec{m}))$.

[Intuitively, $\{x : W_\Phi(x,\vec{m})\}$ "witnesses" $Q^n v_1\ldots v_n \Phi(v_1,\ldots,v_n,\vec{m})$].

Proof. In the interests of intelligibility, I just sketch the idea. The set $\{x:W_\Phi(x,\vec{m})\}$ will be enumerated by an informal recursion which must then be formulated using Lemma 1.

Stage 0. Check if $\mathfrak{m} \models Q^n v_1 \ldots v_n \Phi(v_1,\ldots,v_n,\vec{m})$. If not, let $\alpha_0 = 0$. If yes, we enumerate an element α_0 and construct a formula $\Phi^{(0)}$ as follows. As well as the variables of Φ, $\Phi^{(0)}$ has a new free variable t_0. $\Phi^{(0)}(t_0,\vec{v},\vec{w})$ is $\bigwedge_\sigma \Phi(\sigma(v_1),\ldots,\sigma(v_n),\vec{w})$, where σ runs over all permutations of the variables $\{t_0,v_1,\ldots,v_n\}$. α_0 is the least element of $\mathfrak{m}$ satisfying

i) $\alpha_0 > 0$

ii) $Q^n v_1 \ldots v_n \Phi^{(0)}(\alpha_0,\vec{v},\vec{m})$.

[The existence of α_0 is easily proved from the assumption that $\mathfrak{m} \models Q^n v_1 \ldots v_n \Phi(\vec{v},\vec{m})$].

Stage j+1. If $\alpha_0 = 0$ let $\alpha_{j+1} = 0$. Otherwise, we construct a formula $\Phi^{(j+1)}$ and enumerate an element α_{j+1} as follows.

As well as the variables of $\Phi^{(j)}$, $\Phi^{(j+1)}$ has a new free variable t_{j+1}. $\Phi^{(j+1)}(t_0,\ldots,t_{j+1},\vec{v},\vec{w})$ is $\bigwedge_\sigma \Phi(\sigma(v_1),\ldots, (v_n),\vec{w})$, where σ runs over all permutations of the variables $\{t_0,\ldots,t_{j+1},v_1,\ldots,v_n\}$. α_{j+1} is the least element of $\mathfrak{m}$ satisfying

i) $\alpha_{j+1} > \max_{\ell=j} \alpha_\ell$;

ii) $Q^n v_1 \ldots v_n \Phi^{(j+1)}(\alpha_0,\ldots,\alpha_{j+1},\vec{v},\vec{m})$.

Here I cannot blithely claim the existence of α_{j+1} until I attend to $\Phi^{(j+1)}$ and its semantics. In general, $\Phi^{(j+1)}$ is nonstandard. Moreover, caution is needed in talking about all permutations of the variables, for even though $\{t_0,\ldots,t_{j+1},v_1,\ldots,v_n\}$ is $\mathfrak{m}$-finite not all its permutations will be represented in $\mathfrak{m}$. However, Φ is genuinely finite, and all one needs about the σ is their effect on $v_1,\ldots,v_n$. So it is easy to write down a good definition of the nonstandard formula $\Phi^{(j+1)}$. As to its semantics, $\Phi^{(j+1)}$ is no more complicated, in terms of quantifier complexity, than Φ. It is therefore routine to show that the above "definition" is covered by Lemma 1, so that $n \mapsto \alpha_n$ is total and definable. Then let $\{x:W_\Phi(x,\vec{m})\}$ = range of α. A simple induction shows that this set has the required witnessing properties. □

Remark. The above argument is ultimately model-theoretic. At a later stage we shall look at a proof-theoretic version.

2.5. Definition of satisfaction for first-order formulas. From the proof of Theorem 1 one sees that $L^{<\omega}$ has more expressive power than L, and that in particular $L^{<\omega}$ enables one to simulate in certain situations a single function quantifier. In the standard model $\mathbb{N}$ the satisfaction relation for first-order formulas is definable using a single function quantifier [Sh]. So now I consider the definability of satisfaction for first-order formulas in general inductive structures.

One proceeds uniformly for inductive structures $\mathfrak{M}$. First construct $L_{\mathfrak{M}}$, the extension of L by constants $\bar{m}$ for $m \in \mathfrak{M}$. Then arithmetize $L_{\mathfrak{M}}$ uniformly in $\mathfrak{M}$, in some standard recursive way. (As usual, I never distinguish between an object and its code). In particular one codes such notions as atomic $L_{\mathfrak{M}}$-formula, $L_{\mathfrak{M}}$-formula, sentence, etc. If $\mathfrak{M}$ is nonstandard, the general $L_{\mathfrak{M}}$-formula will be nonstandard, but since L is finite, atomic $L_{\mathfrak{M}}$-formula will have its standard meaning.

The following definition incorporates the most evident requirements on a "generalized truth definition" for arbitrary $L_{\mathfrak{M}}$-sentences. See [K].

Definition. Let T be a set of $L_{\mathfrak{M}}$-sentences. T is a full truth set for $\mathfrak{M}$ if:

(a) T contains all atomic $L_{\mathfrak{M}}$-sentences true in $\mathfrak{M}$;

(b) for each φ, T contains $\neg\varphi$ iff T does not contain φ;

(c) T contains $\varphi_1 \vee \varphi_2$ iff either T contains φ_1 or T contains φ_2;

(d) T contains $\exists v \varphi(v)$ iff for some $m \in \mathfrak{M}$ T contains $\varphi(\bar{m})$.

I outline below a proof of:

Theorem 3. There is an $L^{<\omega}$-definition, uniform for all inductive $\mathfrak{M}$, of a full truth set.

Note that if T is a full truth-definition then the intersection of T with the set of standard $L_{\mathfrak{M}}$-sentences is exactly the set of true standard $L_{\mathfrak{M}}$-sentences.

Proof of Theorem 3. The basic idea is to use Skolem normal form, and the technique (from the proof of Theorem 1) of using Q^2 to simulate a single function quantifier.

Enrich $L_{\mathfrak{M}}$ uniformly to $L_{\mathfrak{M}}^{Sk}$, having lists of new constants and function-symbols of each arity $m \in \mathfrak{M}$, each list recursively in bijection with $\mathfrak{M}$. There is a canonical construction (uniformly first-order definable in $\mathfrak{M}$)

$$\varphi \mapsto Sk(\varphi)$$

assigning to an $L_{\mathfrak{M}}$-sentence φ its Skolem normal form in $L_{\mathfrak{M}}^{Sk}$. $Sk(\varphi)$ has only universal quantifiers. It may be useful to write $Sk(\varphi)$ schematically as

$$\forall\vec{v}\Psi(\vec{v},\vec{c}(\vec{m}),\vec{m},\vec{f}(\vec{v},\vec{m})) \ , \ \ \psi \text{ quantifier-free,}$$

where $\vec{v}$ is the tuple of variables of $Sk(\varphi)$ read from the left, $\vec{c}(\vec{m})$ the corresponding tuple of Skolem constants, $\vec{m}$ the original constants, and $\vec{f}$ the Skolem function symbols. Each tuple may have nonstandard length. Each Skolem function-symbol f occurs always in the same form

$$f(v_{i_1},\ldots,v_{i_k},\vec{m})$$

with fixed $i_1 < \cdots < i_k$. The set $\{i_1,\ldots,i_k\}$ can be called the support of f.

Let $\vec{f} = \langle f_j : j < \alpha\rangle$. I would like to proceed formally and replace $\vec{f}$ by a single g, where

$$f_j(v_{i_1},\ldots,v_{i_k},\vec{m}) = g(2^{j+1}.3^{v_{i_1}+1}\ldots p_{k+1}^{v_{i_k}+1},\vec{m}).$$

[It is of basic importance that the reader observe how unjustified this would be if I was trying to code <u>functions</u> $f_j (j < \alpha)$ by a single <u>function</u> g. The "product"

$$2^{j+1}.3^{v_{i_1}+1}\ldots p_{k+1}^{v_{i_k}+1}$$

has no meaning for general $v_{i_1},\ldots,v_{i_k}$ in $\mathfrak{m}$. Of course if k is <u>finite</u> the product is well-defined].

A special feature of the above is that, uniformly in φ, the map $j \mapsto$ support of f_j is recursive (and so first-order definable).

I want also to replace the $\vec{c}(\vec{m})$ by a single constant. If $\vec{c}(\vec{m}) = \langle c_j(\vec{m}):j < \beta\rangle$, I would put $c_j(\vec{m}) = \gamma(j,\vec{m})$ for some new function symbol γ. [Again, at the level of <u>functions</u>, this is an illegitimate procedure].

Heuristically, I intend that $T(\varphi)$ holds iff $\exists\gamma,g$ such that

$$\mathfrak{m} \models \forall\vec{v}\Psi(\vec{v},\langle\gamma(j,\vec{m}):j < \beta\rangle,\vec{m},\langle g(2^{j+1}.3^{v_{i_1}+1}\ldots p_{k+1}^{v_{i_k}+1},\vec{m}):j < \alpha\rangle).$$

For arbitrary γ,g the notion

$$\mathfrak{m} \models \forall\vec{v}\Psi(\vec{v},\langle\gamma(j,\vec{m}):j < \beta\rangle,\vec{m},\langle g(2^{j+1}.3^{v_{i_1}+1}\ldots p_{k+1}^{v_{i_k}+1},\vec{m}):j < \alpha\rangle)$$

has no sense, but for $L^{<\omega}$-definable γ,g it has an obvious sense (uniform in $\mathfrak{m}$, γ,g) which can be recursively defined. So my second try is: $T(\varphi)$ iff there exist $L^{<\omega}$-definable γ,g such that

$$\mathfrak{m} \models \forall\vec{v}\Psi(\vec{v},\langle\gamma(j,\vec{m}):j < \beta\rangle,\vec{m},\langle g(2^{j+1}.3^{v_{i_1}+1}\ldots p_{k+1}^{v_{i_k}+1},\vec{m}):j < \alpha\rangle).$$

I claim T is uniformly $L^{<\omega}$-definable. I just write down the obvious definition:

$T(\varphi) \Leftrightarrow Q^2 st$ (s and t are finite functions defined on initial segments, and s and t are compatible, and s has domain $\supseteq [0,2\beta]$ and $\mathfrak{m} \models \forall\vec{v}[(\forall j < \alpha)[$ if the support of f_j is $\{i_1,\ldots,i_k\}$, $2.(2^{j+1}.3^{v_{i_1}+1}\ldots p_{k+1}^{v_{i_k}+1}) + 1 \in \mathrm{dom}(s)$ then $\Psi(\vec{v},\langle s(2j):j<\beta\rangle,\vec{m},\langle s(2.(2^{j+1}.3^{v_{i_1}+1}\ldots p_{k+1}^{v_{i_k}+1})+ 1):j < \alpha\rangle)$.

I leave it to the reader to verify that this is a correct $L^{<\omega}$-definition of the T defined earlier.

Now I must prove T is a full truth set. It is evident that T has properties (c) and (d) from the definition of truth set, (a) is trivial. The difficulty is in the verification of (b). The proof of (b) goes by induction on the complexity of Φ (by induction I of course understand internal induction in $\mathfrak{m}$). One must use such equations as

$$Sk(\neg\neg\Phi) = Sk(\Phi),$$

$$Sk((\exists x)\Phi) = Sk(\Phi(x))(x/c), \quad \text{for suitable } c,$$

and

$Sk(\Phi_1 \vee \Phi_2) = Sk(\Phi_1) \vee Sk'(\Phi_2)$, where $Sk'(\Phi_2)$ is got from $Sk(\Phi_2)$ by modifying its Skolem constants and function-symbols to avoid clash with those of $Sk(\Phi_2)$.

This completes my outline of the proof. □

Remark. I know two other very different proofs of the above theorem. The first depends on 3.4 below, and exploits the Paris-Harrington idea of defining truth via indiscemibility for formulas with bounded quantifiers [PH]. The second (due to Simpson) depends on the close connection between my system and Π^1_1 - CA_0. This connection is explained in [SS] and in 2.7 below.

2.6. Consequences of Theorem 3.

In this subsection I present some "general nonsense" consequences of the existence of a full truth set.

The first consequence comes from the comparison of truth and provability.

Theorem 4. The sentence Con($\mathbb{P}$) holds in all inductive structures.

Proof. One shows by internal induction that if Φ is first-order and deducible in first-order logic from $\mathbb{P}$ then $T(\Phi)$. The theorem follows. □

We of course conclude that Th(Inductive structures) is not a conservative extension of P. We must wait a while to see that Th(Inductive structures) is not complete.

The next consequence is pure "general nonsense" of recursive saturation.

Theorem 5. If $\mathfrak{m}$ is a nonstandard inductive structure, $\mathfrak{m}$ is recursively saturated.

Proof. Immediate from the definability of T. □

Of course we do not yet know that Theorem 5 is non-vacuous.

The final consequence is a little less obvious, but still based on very general principles. It involves Theorems 2 and 3.

Theorem 6. Suppose $\mathfrak{m}$ is a nonstandard inductive structure of cardinal κ. Let X be a subset of $\mathfrak{m}$ of cardinal $< \kappa$. Then $\mathfrak{m}$ realizes fewer than κ (1st order) types over X.

Proof. First use Theorem 2 to get an initial segment $[0,\alpha]$ of $\mathfrak{M}$ containing X. Let β be an arbitrary nonstandard element.

For each $\gamma \in \mathfrak{M}$, consider Σ_γ, the set of first-order formulas $\Phi(v)$ such that (the code of) $\Phi(v)$ is less than β, $\Phi(v)$ has constants only from $[0,\alpha]$, and $\Phi(\gamma) \in T$. The sets Σ_γ are uniformly definable. The equivalence relation $\{<\gamma,\delta>:\Sigma_\gamma = \Sigma_\delta\}$ is definable. Clearly the set $\{\Sigma_\gamma : \gamma \in \mathfrak{M}\}$ has a definable injection into $[0,2^{\beta\alpha}]$, and so has cardinality $< \varkappa$ (by Theorem 2). The theorem follows. □

Since nonstandard inductive $\mathfrak{M}$ have the above "Ehrenfeucht-Mostowski property", we expect that they will have large sets of indiscernibles. This will be established after I consider Ramsey phenomena, in 3.4.

2.7. The expressive power of $L^{<\omega}$.

The basic problem is to see how far $L^{<\omega}$ (for inductive structures) simulates set or function quantification.

Suppose f is a function from $\mathbb{N}$ to $\mathbb{N}$. For $x \in \mathbb{N}$, one defines $\bar{f}(x)$ as $\prod_{j<x} p_j^{f(j)+1}$.

Theorem 7. Suppose R is an $L^{<\omega}$-definable relation on $\mathbb{N}$, and S is defined by

$$S(\vec{y}) \Leftrightarrow (\exists f)(\forall x)R(\bar{f}(x),x,\vec{y}).$$

Then S is $L^{<\omega}$-definable on $\mathbb{N}$.

Proof. Evidently

$$\mathbb{N} \models (\exists f)(\forall x)R(\bar{f}(x),x,\vec{y}) \leftrightarrow Q^2uv$$ (u and v are finite functions defined on initial segments, and u and v are compatible and $(\forall x)R(u\restriction[0,x),x,\vec{y})$. □

Corollary. (Simpson). The relations $L^{<\omega}$-definable on $\mathbb{N}$ are exactly those relations on $\mathbb{N}$ recursive in the nth hyperjump, for some $n \in \omega$.

Proof. That the relations recursive in the nth hyperjump are $L^{<\omega}$-definable is evident from Theorem 7. For the converse, it suffices to show that if $R(v_1,\ldots,v_k,\vec{w})$ is recursive in some finite iteration of the hyperjump, then so is $Q^k v_1\ldots v_k R(v_1,\ldots,v_k,\vec{w})$. But

$$Q^k v_1\ldots v_k R(v_1,\ldots,v_k,\vec{w}) \Leftrightarrow (\exists f)(\forall x)[\bar{f}(x) \text{ is monotone increasing} \wedge \text{ for all distinct } v_1,\ldots,v_k < x \;\; R(f(v_1),\ldots,f(v_k),\vec{w})].$$ □

This corollary shows that the Q's act like the hyperjump, and makes all the more remarkable the existence of nonstandard inductive structures. Even now we can see that for nonstandard inductive $\mathfrak{M}$ $L^{<\omega}$ does not allow us to simulate correctly the function quantification involved in hyperjump. Namely

Lemma 4. Suppose $\mathfrak{M}$ is nonstandard inductive. Then there is a subset E of $\mathfrak{M}$, definable in the form

$$y \in E \Leftrightarrow (\exists f)[(\forall x)(f(x+1) = f(x) + 1) \wedge f(0) = 0 \wedge f(y) \neq y] ,$$

but not $L^{<\omega}$ definable.

Proof. Take E as $\mathfrak{M}/\mathbb{N}$. □

Now, one should note that for general $\mathfrak{M}$ and $f:\mathfrak{M} \to \mathfrak{M}$, the definition of $\bar{f}(x)$ given earlier has no sense. But it has a perfectly good sense for inductive $\mathfrak{M}$ and $L^{<\omega}$-definable f. Then the following can be proved just as Theorem 7 is proved.

Theorem 8. Suppose $\mathfrak{M}$ is inductive and R is an $L^{<\omega}$-definable relation on $\mathfrak{M}$. Suppose S is defined by

$$S(\vec{y}) \Leftrightarrow (\exists f)[f \text{ is } L^{<\omega}\text{-definable} \wedge (\forall x)R(\bar{f}(x),x,\vec{y})].$$

Then S is $L^{<\omega}$-definable.

2.8. Q^2 is enough.

Let $L_{-}^{<\omega}$ be the fragment of $L^{<\omega}$ obtained by dropping $Q^j (j \geq 3)$.

Theorem 9. Let $\Phi(\vec{v})$ be an $L^{<\omega}$-formula. Then there is an $L_{-}^{<\omega}$-formula $\Phi_{-}(\vec{v})$ such that for all inductive $\mathfrak{M}$ $\mathfrak{M} \models \forall\vec{v}(\Phi(\vec{v}) \leftrightarrow \Phi_{-}(\vec{v}))$.

Proof. The proof is by induction on the complexity of Φ. The critical case is when $\Phi(\vec{v})$ is $Q^n w_1,\ldots,w_n \psi(\vec{w},\vec{v})$, where ψ is in $L_{-}^{<\omega}$. Use Theorem 2 to get $W_\psi(t,\vec{v})$ witnessing $Q^n w_1 \ldots w_n \psi(\vec{w},\vec{v})$. Let $f_{\vec{v}}$ be the characteristic function of $\{t:W_\psi(t,\vec{v})\}$. Then $f_{\vec{v}}$ is $L^{<\omega}$-definable, and

$$\mathfrak{M} \models (\forall\vec{v})[\Phi(\vec{v}) \leftrightarrow \forall w_1 \ldots w_n (\bigwedge_{i\neq j} w_i \neq w_j \wedge \bigwedge_i f_{\vec{v}}(w_i) = 1 \rightarrow \psi(w_1,\ldots,w_n,\vec{v}))].$$

Whence

$$\mathfrak{M} \models (\forall\vec{v})[\Phi(\vec{v}) \leftrightarrow (\exists f)(f \text{ is } L^{<\omega}\text{-definable} \wedge \{x:f(x) = 1\} \text{ is cofinal} \wedge \forall w_1 \ldots w_n (\bigwedge_{i\neq j} w_i \neq w_j \wedge {}_i\, f(w_i) = 1 \rightarrow \psi(w_1,\ldots,w_n,\vec{v})))].$$

So finally

$$\mathfrak{M} \models (\forall\vec{v})[\Phi(\vec{v}) \leftrightarrow Q^2 uv\ (u \text{ and } v \text{ are functions defined on initial segments} \wedge u \text{ and } v \text{ are compatible} \wedge$$

$$(\forall x)(\mathrm{dom}(u) = [0,x] \to u(x) = 1)$$

$$\wedge \ \forall w_1 \ldots w_n (\bigwedge_{i \neq j} w_i \neq w_j \wedge \bigwedge_i u(w_i) = 1$$

$$\to \psi(w_1,\ldots,w_n,\vec{v}))].$$

The result follows. □

3. VERSIONS OF THE INFINITE RAMSEY THEOREM.

3.1. The work of Jockusch [J] gives clear insight into the complexity of the infinite version of Ramsey's Theorem. Suppose k and r are standard, and R is a function $[w]^k \to r$. Let X be an infinite subset of w homogeneous for R. One seeks a lower bound for the complexity of X in terms of that of R. The interesting result is that this bound depends essentially on k (but not on r). The optimal result is that X can be chosen recursive in $R^{(n)}$.

This semantic result has proof-theoretic consequences. Jockusch proved the above for <u>recursive</u> R. This implies the failure in $\mathbb{P}$ of a certain <u>strong</u> Ramsey Principle which I now explain.

Evidently one has a recursive coding, uniformly in models $\mathfrak{M}$ of $\mathbb{P}$, of the notion x <u>is an</u> α-<u>element set</u> $(x \in [\mathfrak{M}]^\alpha)$.

Consequently one has a uniform recipe which to a formula $\varphi(u,v)$ (possibly with parameters) assigns a formula $P_\varphi(\alpha,r)$ which "says"

$$\{\langle u,v\rangle : \mathfrak{M} \models \varphi(u,v) \wedge u \in [\mathfrak{M}]^\alpha\}$$

is a function $[\mathfrak{M}]^\alpha \to [0,r]$.

Similarly there is a uniform construction giving $H_{\varphi,\psi}(\alpha,r)$ which "says" if $\mathfrak{M} \models P_\varphi(\alpha,r)$ then $\{x \in \mathfrak{M} : \mathfrak{M} \models \psi(x)\}$ is infinite and a homogeneous set for the partition of $[\mathfrak{M}]^\alpha$ given by φ.

When I deal with φ or ψ involving parameters $\vec{\beta}$, I write $\varphi_{\vec{\beta}}(u,v)$ for $\varphi(\vec{\beta},u,v)$, etc.

Let Δ be a set of first-order L-sentences. I say Δ satisfies the <u>strong</u> Ramsey Principle if for every L-formula $\varphi(\vec{\beta},u,v)$ there is an L-formula $\psi(\vec{\beta},x)$ such that $(\forall\vec{\beta})(\forall\alpha)(\forall r)H_{\varphi_{\vec{\beta}},\psi_{\vec{\beta}}}(\alpha,r) \in \Delta$.

<u>Lemma 5</u>. If $\mathbb{N} \models \Delta$ then Δ does not satisfy the <u>strong</u> Ramsey Principle.

<u>Proof</u>. Let $\varphi(\lambda,u,v)$ be a first-order formula which for varying λ enumerates the recursively enumerable subsets of $\mathbb{N} \times \mathbb{N}$. Suppose $\mathbb{N} \models \Delta$ and Δ satisfies the <u>strong</u> Ramsey Principle. Pick $\psi(\lambda,x)$ so that $(\forall\lambda)(\forall\alpha)(\forall r)H_{\varphi_\lambda,\psi_\lambda}(\alpha,r) \in \Delta$.

Now let Π be a recursive partition of $[w]^j$ into r parts. Let λ be an index of the graph of Π, with respect to the enumeration given by Π. Then $\mathbb{N} \models P_{\varphi_\lambda}(j,r)$. Since $\mathbb{N} \models \Delta$,

$$\mathbb{N} \models (\forall\lambda)(\forall\alpha)(\forall r)H_{\varphi_\lambda,\psi_\lambda}(j,r).$$

So $\{x \in \mathbb{N} : \mathbb{N} \models \psi_\lambda(x)\}$ is an infinite homogeneous set for Π. But the Turing degree of this set is bounded by $0^{(n)}$, where n is independent of λ, since ψ is first-order. This contradicts Jockusch's results.□

Remark. Evidently this result is of the same form as Tarski's basic result on undefinability of truth.

3.2. The strong Ramsey Principle would have yielded a schema capturing a "definable version of Ramsey's Theorem". No other version comes to mind for first-order arithmetic. A major advantage of our $L^{<\omega}$ is that in it from induction we can deduce a satisfactory "Ramsey Schema". I now indicate the proof.

One must recall a conventional proof of $\omega \to (\omega)^\alpha_r$. The proof goes by induction on α, roughly as follows. A partition $f:[\omega]^{\alpha+1} \to r$ is given. For each $n \in w$, consider the "section" $f_n:[\omega]^\alpha \to r$, where $f_n(\{x_1,\ldots,x_\alpha\}) = 0$ if $n \in \{x_1,\ldots,x_\alpha\}$, and $= f(\{n,x_1,\ldots,x_\alpha\})$ otherwise.

Stage 0. Consider the least $r_0 < r$ such that for some n and some infinite X, f_n takes the constant value r_0 on $[X]^\alpha$. Given r_0, choose the least such n $(= n_0)$, and select X_0 so that X_0 is infinite and f_{n_0} takes the constant value r_0 on $[X_0]^\alpha$. The existence of r_0 follows by the induction hypothesis $w \to (w)^\alpha_s$ all s.

Stage k+1. Infinite sets $X_0,X_1,\ldots,X_k$ have been defined, and one now defines X_{k+1}. Consider the least $r_k < r$ such that for some $n \in X_k$ and some infinite $X \subseteq X_k$ f_n takes the constant value r_{k+1} on $[X]^\alpha$. Given r_{k+1}, choose the least such n $(= n_{k+1})$ and select X_{k+1} so that X_{k+1} is infinite, $X_{k+1} \subseteq X_k$, and $f_{n_{k+1}}$ takes the constant value r_{k+1} on $[X_{k+1}]^\alpha$. The existence of r_{k+1} follows by the induction hypothesis $w \to (w)^\alpha_s$ all s.

Now, one proves by induction that $r_k \le r_{k+1} < r$, so r_k is eventually constant, say r_∞, at all stages $k \ge k_0$ say. Let $Y_t = Y_{k_0+t}$, and let $g(t) = t^\alpha$ element of Y_t in increasing order. Then $\{g(t):t \in w\}$ is infinite and homogeneous for f.

What is needed to do a "definable version" of this proof for all inductive $\mathfrak{m}$? Evidently, w is replaced by $\mathfrak{m}$, and infinite by cofinal. One supposes one has an $L^{<\omega}$-formula $\varphi(\vec{\beta},\alpha,x,y)$ and specific $\vec{\beta},\alpha,r$ in $\mathfrak{m}$ so that

$$\{\langle u,v\rangle : \mathfrak{m} \models \varphi(\vec{\beta},\alpha,u,v) \wedge u \in [\mathfrak{m}]^\alpha\}$$

is a function $[\mathfrak{m}]^\alpha \to \dot{o},r$.

One wants to define a cofinal $X \subseteq \mathfrak{m}$ homogeneous for the above partition. The essential point is that the $L^{<\omega}$-definition of X should be uniform in α,r.

The first step is to find a "universal" formula of the same complexity as φ. The reason is as follows. Consider the partition $f(\vec{\beta},\alpha)$ of $[\mathfrak{m}]^\alpha$ given by φ as above. In finding homogeneous X for $f(\vec{\beta},\alpha)$ one has to consider the "section partitions" $f_\lambda(\vec{\beta},\alpha)$ $(\lambda \in \mathfrak{m})$ of $\mathfrak{m}^{[\alpha-1]}$ (provided $\alpha \geq 2$). The formula uniformly defining the f_λ is the $\varphi^{sec}(\lambda,\vec{\beta},\alpha,u,v)$ written below:

$$(\lambda \in u \wedge v = 0)$$
$$\wedge\ (\lambda \notin u \wedge \varphi(\vec{\beta},\alpha,\{\lambda\} \cup u,v)).$$

[Of course I here make no distinction between object and code].

An amplified version of $\varphi^{sec}(\lambda,\vec{\beta},\alpha,u,v)$ is of the form

$$(\exists u_1)[(A(\lambda,u,v,u_1) \wedge \varphi(\vec{\beta},\alpha,u_1,v)) \vee B(\lambda,u,v,u_1)]$$

where A,B are L-formulas with bounded quantifiers.

In our intended induction, we shall have to assume that the result is already proved for all instances of φ^{sec}. This makes evident the need for the following simple lemma.

<u>Lemma 6</u>. For each $L^{<\omega}$-formula $\psi(\vec{w}_1,\vec{w}_2,\vec{w}_3)$ there is an $L^{<\omega}$-formula $\psi^*(x,y,\vec{w}_2,\vec{w}_3)$ such that

i) there exists $n \in \omega$ such that for all inductive $\mathfrak{m}$ and all $\vec{m}_1$ in $\mathfrak{m}$ there is μ in $\mathfrak{m}$ such that

$$\mathfrak{m} \models (\forall\vec{w}_2)(\forall\vec{w}_3)[\psi(\vec{m}_1,\vec{w}_2,\vec{w}_3) \leftrightarrow \psi^*(n,\mu,\vec{w}_2,\vec{w}_3)];$$

ii) for all L-formulas $A(v,\vec{w}_2,\vec{w}_3,\vec{w}_4)$, $B(v,\vec{w}_2,\vec{w}_3,\vec{w}_4)$ with bounded quantifiers, for all inductive $\mathfrak{m}$, and for all n,μ,λ in $\mathfrak{m}$ there exists δ in $\mathfrak{m}$ such that

$$\mathfrak{m} \models (\forall\vec{w}_2)(\forall\vec{w}_3)[(\exists\vec{w}_4)((A(\lambda,\vec{w}_2,\vec{w}_3,\vec{w}_4) \wedge \psi^*(n,\mu,\vec{w}_2,\vec{w}_4))$$
$$\vee\ B(\lambda,\vec{w}_2,\vec{w}_3,\vec{w}_4))$$
$$\leftrightarrow\ \psi^*(m,\delta,\vec{w}_2,\vec{w}_3)].$$

<u>Proof</u>. All is clear if one thinks of (the denotation of) ψ^* as the universal predicate recursively enumerable in (the denotation of) ψ. □

The point of this lemma for our proof of a Ramsey Schema is that one replaces $\varphi(\vec{\beta},\alpha,x,y)$ by $\varphi^*(n,\mu,\alpha,x,y)$ for suitable n,μ. Then $\varphi^{sec}(\lambda,\vec{\beta},\alpha,x,y)$ is equivalent to $\varphi^*(m,\delta,\alpha,x,y)$ for suitable m,δ, if $\alpha \geq 1$. So one has the possibility of proving the result for fixed φ^*, by induction on α.

Now I try to formulate in $L^{<\omega}$ a suitable induction hypothesis about φ^*. I first give a naive version formulated in second-order logic, and then force it into $L^{<\omega}$.

<u>Naive Version</u>. For all v,μ, if

$$\{\langle x,y\rangle : \mathfrak{m} \models \varphi^*(v,\mu,\alpha,x,y) \wedge x \in [\mathfrak{m}]^\alpha\}$$

is a function $[X]^\alpha \to [o,r]$, with X cofinal, then there is a cofinal subset Y of X homogeneous for the above partition.

Note first that X above is (uniformly) $L^{<\omega}$-definable from ν,μ,α. To eliminate "$\exists$ cofinal $Y \subseteq X$ ..." one writes:

$(\exists s)[s \leq r \wedge Q^2 tw$ (t and w are finite partial functions with range $\{0\}$ and t and w are compatible and $\mathrm{dom}(t) \subseteq X$ and the partition has constant value s on $[\mathrm{dom}(t)]^{\alpha})$].

So the Naive Version transcribes to an $L^{<\omega}$-version which I write as $H_{\varphi^*}(\alpha)$, and the task now is to prove $H_{\varphi^*}(\alpha)$ by induction on α. I claim that the conventional proof of $\omega \rightarrow (\omega)^{\alpha}_{r}$, outlined earlier, works in all inductive $\mathfrak{m}$. Look back at that proof. Basically, one must show that $k \rightarrow X_k$ is $L^{<\omega}$-definable (uniformly in appropriate parameters). But this is clear by Theorem 2.

This completes my outline of the proof of the following theorem.

Theorem 9. There is a recursive construction which to every $L^{<\omega}$-formula $\varphi(\vec{\beta},\alpha,x,y)$ assigns an $L^{<\omega}$-formula $H_{\varphi}(\vec{\beta},\alpha,x)$ such that for all inductive $\mathfrak{m}$, if $\{\langle u,v\rangle : \mathfrak{m} \models \varphi(\vec{\beta},\alpha,u,v) \wedge u \in [\mathfrak{m}]^{\alpha}\}$ is a function $[X]^{\alpha} \rightarrow [o,r]$ for some cofinal X then $\{x : \mathfrak{m} \models H_{\varphi}(\vec{\beta},\alpha,x)\}$ is a cofinal homogeneous set for this partition.

Remark. This deserves to be called a Ramsey Schema.

3.3. Paris-Harrington in inductive $\mathfrak{m}$.

Suppose $\mathfrak{m}$ is inductive and $f:\mathfrak{m} \rightarrow \mathfrak{m}$ is $L^{<\omega}$-definable (maybe in parameters). PH(f), the Paris-Harrington Principle for f, is the following assertion:

For all α,r,s there exists x_0 such that if $g:[[o,x]]^{\alpha} \rightarrow [o,r]$ is a partition, and $x \geq x_0$, then there is a subset X of $[o,x]$ homogeneous for g, of cardinal $\geq s$, and such that $\mathrm{card}(X) \geq f(\min X)$.

It is well-known [PH] that this statement is first-order in f, and so an $L^{<\omega}$-sentence under our hypothesis on f.

For $f(y) = y$, one knows that PH(f) is true but unprovable in $\mathbb{P}$; [PH]. In inductive structures the situation is much more pleasant.

Theorem 10. $\mathfrak{m} \models PH(f)$, for $\mathfrak{m}$ inductive and f $L^{<\omega}$-definable.

Proof. The proof given in PH adapts perfectly. If $\mathfrak{m} \models \neg PH(f)$ one constructs a definable tree of counterexamples, "of cofinal height". This tree is "finite branching", so Köneg's Lemma applies. So one gets a definable partition, and uses Theorem 9 to get a definable cofinal homogeneous set, and a contradiction in the style of [PH]. □

Remark. Recalling the maneuvres of 3.2, one sees easily that in the obvious sense

$$\mathfrak{m} \models (\forall f)(f \text{ definable} \rightarrow PH(f)).$$

3.4. Indiscernibles.

Now we look back to a remark in 2.6.

Definition. Let $\mathfrak{m}$ be an arbitrary L-structure. Let $<$ be an ordering of $\mathfrak{m}$. Let X be a subset of $\mathfrak{m}$. X is a set of extra-indiscernibles (with respect to $<$) if for each terminal segment Y of X Y is a set of indiscernibles (with respect to $<$) in $\mathfrak{m}_A$, where

$$A = \{m \in \mathfrak{m}: (\forall y \in Y)(m < y)\}.$$

A related notion occurs, unnamed, in Paris-Harrington [PH]. There one has the above notion for L-formulas with bounded quantifiers. In the case of arithmetic, given an infinite such set one can construct a model of P.

Using our Ramsey Schema, we get:

Theorem 11. Suppose $\mathfrak{m}$ is a nonstandard inductive structure. Then $\mathfrak{m}$ has a cofinal $L^{<\omega}$-definable (in parameters) set of first-order extra-indiscernibles with respect to the standard ordering of $\mathfrak{m}$.

Proof. (Outline). Fix a nonstandard α. Using the Ramsey Schema and the Truth Definition, formalize the Ehrenfeucht-Mostouski construction [EM], uniformly in $a \in \mathfrak{m}$, to get $D_{a,\alpha}$ homogeneous for the partitions given by all Φ of code $< \alpha$. Then diagonalize. □

Notes: i) This can be elaborated to take in various stronger notions of indiscernibility corresponding to formulas of bounded complexity in $L^{<\omega}$.

ii) The above method, restricted to formulas with bounded quantifiers, leads to another truth-definition, following Paris-Harrington [PH] (cf 2.5).

iii) Perhaps the main merit of the above is that it helped Schmerl find his beautiful diamond-free proof of the Completeness Theorem, bringing Gaifman's powerful ideas [G1] to the problem.

3.5. A Generalized Ramsey Quantifier.

Formally, the intention is to define quantifiers Q^α where α is an arbitrary member (≥ 1) of $\mathfrak{m}$. There is of course a clear meaning to $Q^\alpha x_1 \ldots x_\alpha \varphi(x_1,\ldots,x_\alpha,\_)$ when α is nonstandard and, say, φ is nonstandard first-order, but this is not what I want.

Let $\varphi(u,v)$ be an $L^{<\omega}$-formula maybe with parameters. Let $\mathfrak{m}$ be inductive, and $\alpha, r \in \mathfrak{m}$, $\alpha \geq 1$. I introduce new formulas $H^{\alpha,r}uv\varphi(u,v)$, with the semantics now described:

$$\mathfrak{m} \models H^{\alpha,r}uv\varphi(u,v) \Leftrightarrow \text{there is a cofinal } X \text{ such that for all } u \in [X]^\alpha \ \mathfrak{m} \models \varphi(u,r).$$

There are two essential points, namely that the "quantifiers" $H^{\alpha,r}$ are uniformly $L^{<\omega}$-definable, and that the Q^n are definable from the $H^{\alpha,r}$.

Lemma 7. For each $L^{<\omega}$-formula $\varphi(u,v,\_\_)$ there is an $L^{<\omega}$-formula $\varphi_H(x,y,\_\_)$ such that for all inductive $\mathfrak{m}$, and all $\alpha,r \in \mathfrak{m}$, $\alpha \geq 1$

$$\mathfrak{m} \models H^{\alpha,r}uv\varphi(u,v,\_\_) \Leftrightarrow \mathfrak{m} \models \varphi_H(\alpha,r,\_\_).$$

Proof. Cf. 3.2., in the transcription of the Naive Version. □

Lemma 8. For each standard $n \geq 1$, and each $L^{<\omega}$-formula $\varphi(v_1,\ldots,v_n,\_\_)$ there is an $L^{<\omega}$-formula $\varphi^H(u,v,\_\_)$ such that for all inductive structures $\mathfrak{m}$

$$\mathfrak{m} \models Q^n v_1 \ldots v_n \varphi(v_1,\ldots,v_n,\_\_) \Leftrightarrow \mathfrak{m} \models H^{n,o}\varphi^H(u,v,\_\_).$$

Proof. Let $\varphi^H(u,v,\_\_)$ "say" that u is an n-element set and $v = 0$ iff $\mathfrak{m} \models \varphi(a_1,\ldots,a_n,\_\_)$ for all $a_1,\ldots,a_n$ such that $u = \{a_1,\ldots,a_n\}$. □

4. EXISTENCE OF NONSTANDARD INDUCTIVE STRUCTURES

4.1. Almost everything proved till now loses interest if there are no nonstandard inductive structures. I now outline a proof that if $\diamondsuit_{\omega_1}$ holds then there are nonstandard inductive structures of power $\aleph_1$. This proof is superceded by [SS], but may be of interest.

From Theorem 1 we know that for nonstandard inductive structures the cofinality semantics agrees with the semantics of Magidor-Malitz [MM]. I now review the main results on the model theory of the latter semantics.

One begins with an arbitrary countable first-order L and forms $L^{<\omega}$ as in 2.1 . With the same kind of uniformity as in predicate calculus, Magidor and Malitz give

(a) a recursively enumerable set of axioms A_L in $L^{<\omega}$;

(b) the usual finitistic rules of proof of first-order logic, extended naturally to $L^{<\omega}$.

From this one has the usual notions of derivability, consistency, etc. One verifies (not without effort, in contrast to the first-order case) that each member of $L^{<\omega}$ is valid for the Magidor-Malitz semantics, and that the rules of proof respect satisfaction, so all theorems are valid. The optimal completeness theorem would be that if Σ is a countable consistent set of $L^{<\omega}$-sentences then Σ has a model. Such a result has not been proved, and the Magidor-Malitz Completeness Theorem is a little more subtle. One must pass from L to an appropriate "free Skolemization". The basic idea is to introduce for each $L^{<\omega}$-formula $Q^n v_1 \ldots v_n \psi(\vec{v},\vec{w})$ a predicate $E(\vec{w},u)$ together with an axiom saying

(a) for all $\vec{w}$, and all distinct $v_1,\ldots,v_n$ in $\{x:E(\vec{w},x)\}$, $\psi(v_1,\ldots,v_n,\vec{w})$;

(b) for all $\vec{w}$, $Q^n v_1 \ldots v_n \psi(\vec{v},\vec{w}) \to Q^1 xE(\vec{w},x)$.

[So, $\{x:E(\vec{w},x)\}$ witnesses $Q^n v_1 \ldots v_n \psi(\vec{v},\vec{w})$].

Beginning with L, one does this freely to get L_1, and then iterates the procedure ω times to get L'. In $(L')^{<\omega}$ one has introduced W_L, the set of all "witnessing axioms" as above. One now observes easily that if Σ is a set of $L^{<\omega}$-sentences which has a model then $\Sigma \cup W_L$ is consistent in the sense of L'. Magidor and Malitz proved the converse for countable Σ, assuming $\Diamond_{\omega_1}$. One deduces a countable compactness theorem. Shelah [Shel] has proved that the latter is not a theorem of ZFC.

4.2. I want to adapt these results to prove a completeness theorem for inductive structures, with the cofinality semantics.

Now L is the usual first-order language for number theory. Ind is the recursively enumerable system in $L^{<\omega}$ with the Magidor-Malitz rules of proof and the following axioms:

$\mathbb{P}$;

the induction scheme in $L^{<\omega}$;

all formulas $Q^1 x\psi(x) \leftrightarrow (\forall x)(\exists y)(y > x \wedge \psi(y))$;

all formulas

$$Q^1 x\psi(x) \wedge \forall v_1 \dots v_k \ [\text{the } v_i \text{ are distinct} \wedge \bigwedge_{j\le k} \psi(v_j) \rightarrow \Theta(v_1,\dots,v_k)] \rightarrow Q^k v_1 \dots v_k \Theta(v_1,\dots,v_k)$$

all formulas

$$Q^k v_1 \dots v_k \Theta(v_1,\dots,v_k,\vec{w}) \rightarrow Q^1 y W_\Theta(y,\vec{w});$$

all formulas

$$(\forall v_1)\dots(\forall v_k)[v_1,\dots,v_k \text{ distinct} \wedge \bigvee_{j\le k} W_\Theta(v_j,\vec{w}) \rightarrow \Theta(v_1,\dots,v_k,\vec{w})].$$

(Here W_Θ is as in 2.4).

Of course, these axioms hold in all inductive structures.

The following is convenient.

<u>Lemma 9</u>. The Magidor-Malitz axioms A_L are theorems of Ind.

<u>Proof</u>. Look at the Magidor-Malitz verification of the <u>validity</u> of A_L. This is proved by transfinite recursions in which nothing strange happens at limit ordinals. Using Gödel's device with the Chinese Remainder Theorem [Gö], these recursions are easily formalized (and proved to work) in Ind. □

Note also:

<u>Lemma 10</u>. $(\forall x) \neg (Q^1 v)(v < x)$ is a theorem of Ind.

<u>Proof</u>. Trivial. □

Now the original completeness theorem for Ind.

Theorem 12. ($\Diamond_{\omega_1}$). Let Σ be a set of $L^{<\omega}$ sentences which is consistent in the sense of Ind. Then Σ has a model of power $\aleph_1$.

Proof. Let $\Sigma_1 = \Sigma \cup \{(\forall x) \neg (Q^1 v)(v < x)\}$. By the preceding lemma Σ_1 is consistent in the sense of Ind. I now consider Σ_1 in the sense of the Magidor-Malitz system. To apply their completeness theorem I need to show that $\Sigma_1 \cup W_L$ is consistent in L' (of course Σ_1 is countable). But this is obvious because of our W_{Θ} axioms. So by Magidor-Malitz, assuming $\Diamond_{\omega_1}$, Σ_1 has a Magidor-Malitz model $\mathfrak{m}$ of power $\aleph_1$. But since $(\forall x) \neg (Q^1 v)(v < x)$ is in Σ_1, $\mathfrak{m}$ is ω_1-like, so the two semantics agree, and the theorem is proved. □

Note: Evidently the result applies to any countable extension of L by constants. But it requires various $\Diamond_\lambda$, and a very deep method of Shelah [She 2], to get by the above method the above result for extensions of L by arbitrary sets of constants.

Another difficulty of the above procedure derives from the complexity of the Magidor-Malitz method. From the above ideas, I made very limited progress in the construction of interesting inductive structures.

Fortunately, Schmerl's proof, in ZFC, of the above theorem, works for arbitrary extensions of L by constants. So it makes e.g. the method of diagrams viable. In Section 5 I will develop some of the basic model theory of Ind, using Schmerl's method (and the necessary ideas of Gaifman [Gl]). This would be a good moment for the reader to consult Schmerl's proof [SS].

5. CONSEQUENCES OF THE COMPLETENESS THEOREM

5.1. Using Gödel's Theorem for $L^{<\omega}$, one sees that there are $2^{\aleph_0}$ complete consistent extensions of Ind. So

Lemma 11. For each regular $\kappa >_0$ there are $2^{\aleph_0}$ inductive structures of power κ which are pairwise inequivalent in the sense of $L^{<\omega}$.

5.2. Now I consider the spectrum problem for an arbitrary complete extension of Ind. That this can be solved completely by someone like me is a tribute to the clarity of Schmerl's construction.

Theorem 13. Let T be a complete extension of Ind. Let κ be a regular cardinal $> \aleph_0$. Then there is a family $\mathfrak{F}_\kappa$ of 2^κ inductive structures of power κ, each a model of T, so that if $\mathfrak{m}$ and $\mathfrak{n}$ are distinct elements of $\mathfrak{F}_\kappa$ then $\mathfrak{m}$ cannot be embedded in $\mathfrak{n}$ as an $L^{<\omega}$-elementary substructure.

Proof. Select a countable weak model $\mathfrak{m}_0$ of T so that $\mathfrak{m}_0$ is pointwise definable in $L^{<\omega}$. (Possible since $T \supseteq$ Ind). Extend L as Schmerl does, by putting in new predicates for the W_Θ, and iterating ω times. This gives L_W say, and $\mathfrak{m}_0$ is to be construed as an L_W-structure. Let F be a Gaifman minimal type operator as in [G1]. Let Λ be a $\varkappa$-like linear order of cardinal $\varkappa$. Let $\mathfrak{m}_0^{F^\Lambda}$ be the result of iterating F along Λ on $\mathfrak{m}_0$. Then $\mathfrak{m}_0 \prec_{L^{<\omega}} \mathfrak{m}_0^{F^\Lambda}$ by Schmerl [SS], where $\mathfrak{m}_0^{F^\Lambda}$ is given the cofinality semantics.

Now let Λ_1, Λ_2 be $\varkappa$-like linear orders of power $\varkappa$. If we have an $L^{<\omega}$ elementary $j:\mathfrak{m}_0^{F^{\Lambda_1}} \to \mathfrak{m}_0^{F^{\Lambda_2}}$, j is necessarily L_W-elementary. Moreover, j must be the identity on $\mathfrak{m}_0$, since $\mathfrak{m}_0$ is pointwise definable. So by [G1] j is induced by an order embedding $\Lambda_1 \to \Lambda_2$. The theorem follows by choosing $2^\varkappa$ incomparable $\varkappa$-like Λ of power $\varkappa$. □

5.3. By the same method, one obtains e.g.:

<u>Theorem 14</u>. (Same hypotheses).

(a) There is a rigid inductive structure $\mathfrak{m}$ of power $\varkappa$ with $\mathfrak{m} \models T$.

(b) If Λ is a $\varkappa$-like linear order of power $\varkappa$, Aut(Λ) acts faithfully on some inductive model of T of power $\varkappa$.

<u>Proof</u>. (b) evident. For (a) (even a sharper version) we simply need rigid $\varkappa$-like Λ, and of course $\varkappa$ is rigid. □

5.3. <u>Upward Löwenheim-Skolem.</u>

Schmerl's proof evidently gives:

<u>Theorem 15</u>. If $\mathfrak{m}$ is inductive and $\varkappa$ is regular with $\varkappa > \operatorname{card}(\mathfrak{m})$, there is an $L^{<\omega}$-elementary end-extension $j:\mathfrak{m} \to \mathfrak{n}$ where $\mathfrak{n}$ is inductive of power $\varkappa$.

5.4. <u>Cofinal Embeddings</u>.

The extensions $j:\mathfrak{m} \to \mathfrak{n}$ most relevant to our semantics are those for which j embeds $\mathfrak{m}$ cofinally in $\mathfrak{n}$. Schmerl's method makes it evident that nontrivial such embeddings exist. (This was not evident via my earlier method). So

<u>Theorem 16</u>. (Same hypotheses as 5.3). There is a diagram $j:\mathfrak{m} \to \mathfrak{n}$ where j is a proper elementary cofinal embedding and $\mathfrak{m}, \mathfrak{n}$ are inductive models of T of power $\varkappa$.

<u>Proof</u>. Use $\Lambda_1 \to \Lambda_2$, a cofinal embedding of $\varkappa$-like orders. □

5.5. Analogues of Some Results of Rabin.

One of the first applications of the overspill method was Rabin's theorem that of $\mathfrak{m}$ is a nonstandard model of $\mathbb{P}$ then $\mathfrak{m}$ has an elementarily equivalent extension which is not a Σ_1-extension [Ra]. The proof has two main ideas:

(a) a first-order property is preserved under extensions of models of T if and only if it is Σ_1-definable;

(b) there is a predicate Σ_1-definable Θ over $\mathfrak{m}$ whose complement is not Σ_1-definable over $\mathfrak{m}$.

From these, a simple use of overspill proves the theorem.

I claim that all this goes through if one replaces first-order logic by $L^{<\omega}$ and $\mathbb{P}$ by Ind. The analogue of (b) is routine, and the analogue of (a) by the method of diagrams for $L^{<\omega}$. So:

Theorem 17. Let $\mathfrak{m}$ be a nonstandard inductive structure. Then there is an embedding $j:\mathfrak{m} \to \mathfrak{n}$ where $\mathfrak{n}$ is inductive, $L^{<\omega}$-equivalent to $\mathfrak{m}$, such that j is not Σ_1.

After considerable effort, Rabin's Theorem was improved by Wilkie [W1], replacing extension by end-extension. I have not been able to extend Wilkie's result to inductive structures. Wilkie's proof has two main ingredients:

(a) Friedman's work on embeddings of countable nonstandard models of $\mathbb{P}$ [F1];

(b) Gaifman's Cofinality Theorem [G2].

In our setting (a) seems completely inapplicable, because of Theorem 1 . I do not know if some analogue of (b) holds for inductive structures. This is certainly a natural question given the semantics of this paper.

I remark, as another example of the utility of the method of diagrams, that Rabin's results (unpublished, but see [MS]) on joint embedding and amalgamation extend from the first-order case to our setting.

5.6. I conclude this section with some remarks about L versus $L^{<\omega}$ in the model theory of arithmetic.

In the $L^{<\omega}$ model theory (not allowing weak models except as auxiliary devices) the Friedman phenomena and indicator theory disappear. I am not aware of any reason why this situation is undesirable. Ind is very resistant to independence results of Ramsey type (cf. 6.1 below). There is an abundance of nonstandard models with the intended semantics. Gaifman's extension operators, overspill and the method of diagrams allow the construction of a variety of interesting models.

It seems interesting to find other fragments of second order arithmetic with nonstandard models in the intended (Cantorian) semantics. Maybe one should consider stationary logic [BMK] in this connection. It would be indeed dramatic to find a Lindstrom Theorem for Cantorian model theories of arithmetic. However, I see no reason to conjecture such.

6. PROOF-THEORETIC STRENGTH OF IND.

6.1. Simpson [SS], on the basis of , made the very satisfying discovery, that Ind has exactly the same first-order consequences as Π^1_1 - CA_0 [F3]. Note that Π^1_1 - CA_0 definitely does not have nonstandard models in the Cantorian semantics.

An important consequence is that the "finitization" of Nash-Williams' Theorem [F2] is provable in Ind, though by FMS independent of the system ATR [ ] .

So the only known independence results for Ind concerning combinatorics are those of Harvey Friedman [F2]. One does not yet discern a pattern in his examples, but it will be important to seek a uniform method for neutralizing them. Simpson's proof actually shows that Ind axiomatizes the $L^{<\omega}$-consequences of Π^1_1 - CA_0. This was pointed out to me by McAloon.

7. COMPLEXITY OF THE COMPLETENESS PROOF.

An interesting theme in the first-order model theory has been the complexity of nonstandard models, and arithmetization of Gödel's Completeness and Incompleteness Theorems. Tennenbaum [T] showed that there is no recursive nonstandard model of $\mathbb{P}$. McAloon [Mc2] made good use of the arithmetization.

As regards Ind, the natural questions seem to be:

(I) If $\mathfrak{m}$ is a Cantorian nonstandard model of Ind with underlying set $\mathbb{R}$, what is the complexity of $+^{\mathfrak{m}}, \cdot^{\mathfrak{m}}$? ;

(II) Is there a uniform "internal" construction which to a nonstandard model $\mathfrak{m}$ of Con(Ind) yields a model $\mathfrak{m}^*$ of Con(Ind) ?

The answer to II is trivially negative, if we require $\mathfrak{m}$ and $\mathfrak{m}^*$ to be Cantorian. For by general nonsense [Mc2] $\mathfrak{m}^*$ would be an end-extension of $\mathfrak{m}$, whence by Theorem 1 $\mathfrak{m}^*$ has cardinal greater than that of $\mathfrak{m}$. But then $\mathfrak{m}^*$ is not interpretable in $\mathfrak{m}$.

If we demand only that $\mathfrak{m}^*$ be a weak model, it is evident that the answer is positive, via a formalized Henkin argument for a two-sorted logic.

1 is a more delicate matter. We know there is an $\mathfrak{m}$ living on $\mathbb{R}$. But Schmerl's construction most naturally involves a well-ordering of $\mathbb{R}$, and so does not give (prima facie) a Borelian structure in the sense of Harvey Friedman [F4]. Whence there arises the fundamental problem of whether there is a Borelian model for Ind on $\mathbb{R}$. (Note that the analogous question for $\mathbb{P}$ has a positive answer, by an unpublished result of Friedman, proved by stretching indiscernibles).

In fact, we have:

Theorem 18. Assume $MA_{2^{\aleph_0}}$. Let $\langle R, \oplus, \odot, \prec \rangle$ be an inductive structure. Then $\prec$ is not Lebesgue measurable.

Proof. (This proof comes from a suggestion of Doug Hoover). It clearly suffices to prove the result with [0,1] replacing $\mathbb{R}$.

Let A be the graph of $\prec$, and suppose A is measurable. The horizontal sections of A have cardinal $< 2^{\aleph_0}$, by Theorem 1 , and so by $MA_{2^{\aleph_0}}$ have measure 0. Then the vertical segments have measure 1. So by Fubini's Theorem, A is not measurable. □

If we assume more about $\prec$, the Fubini argument works without Martin's Axiom.

Theorem 19. Let $\langle R, \oplus, \odot, \prec \rangle$ be an inductive structure. Then neither $\prec$ nor $\oplus$ is Π^1_1.

Proof. If $\prec$ is Π^1_1 it is Δ^1_1. Let A be as in the preceding proof. The horizontal sections are Σ^1_1, and of cardinal $< 2^{\aleph_0}$, so are countable (uncountable Σ^1_1 sets have perfect subsets). So the horizontal sections have measure 0, etc.

If $\oplus$ is Π^1_1 it is Σ^1_1, and so is Σ^1_1, whence Π^1_1. □

Theorem 18 has negative implications for the definability of $\prec$ and $\oplus$.

Example. In Solovay's model [So] for ZFC + GCH + All sets ordinal definable from a real are measurable, neither $\prec$ nor $\oplus$ is ordinal definable from a real. This is clear from Theorem 18 , and the definability of $\prec$ from $\oplus$.

Problem. What can one say about the complexity of $\odot$? All I could show was that the set of powers of a fixed prime has measure 0, if it is measurable. One should not forget, in this context, that neither $\oplus$ nor $\prec$ is second-order definable from $\odot$.

At the opposite extreme from the above, one has:

Theorem 20. If $V = L$, there is a Δ^1_2 inductive structure on $\mathbb{R}$.

Proof. Use Schmerl's construction, together with a Δ^1_2 well-ordering of $\mathbb{R}$. □

In general one sees the close link between definable inductive structures on $\mathbb{R}$, and definable $2^{\aleph_0}$-like orders on $\mathbb{R}$. In particular, the Completeness Theorem does not belong to Borel model theory.

REFERENCES

[BMK] J. Barwise, M. Makkai, M. Kaufmann, Stationary logic, Ann. Math. Logic 13, 171-224.

[EM] A. Ehrenfeucht and A. Mostowski, Models of axiomatic theories admitting automorphisms, Fund. Math. 43, 50-68.

[ER] P. Erdos and R. Rado, A partition calculus in set theory, B.A.M.S. 62 (1956), 427-489.

[F1] H. Friedman, Countable models of set theories, 539-573 in Proceedings of Cambridge Summer School in Mathematical Logic (ed. Mathias) Springer Lecture Notes 337, 1973.

[F2] __________, Systems of second order arithmetic with restricted induction I, II, J.S.L. 41 (1976), 557-557.

[F3] __________, preprints on Borel Diagonalization, Ohio State, 1979.

[F4] __________, Borel structures in mathematics, preprint, Ohio State, 1979.

[FMS] H. Friedman, K. McAloon, S. Simpson, A finite combinatorial principle which is equivalent to the 1-consistency of predicative analysis, in preparation.

[G1] H. Gaifman, Models and types of Peano arithmetic, Annals Math. Logic 9 (1976), 223-306.

[G2] __________, A note on models and submodels of arithmetic, 128-144, in Conference in Mathematical Logic London 1970 (ed. W. Hodges), Springer Lecture Notes 255, 1972.

[Go] K. Godel, Uber formal unentscheidbare Satze der Principia Mathematica and verwandter Systeme, 1, Monatsh. Math. Phys. 38 (1931), 173-198.

[J] C. Jockusch, Ramsey's theorem and recursion theory, J.S.L. 37 (1972), 268-280.

[KP] L. Kirby and J. Paris, Initial segments of models of Peano's axioms, 211-246 in Set Theory and Hierarchy Theory V (ed. A. H. Lachlan, M. Srebrny, Springer Lecture Notes 619, 1977.

[K] S. Krajewski, Nonstandard satisfaction classes, 121-144 in Set Theory and Hierarchy Theory in memory of A. Mostowski (ed. W. Marek), Springer Lecture Notes 537, 1976.

[L] P. Lindstrom, On extensions of elementary logic, Theoria 35 (1969), 1-11.

[MM] M. Magidor and J. Malitz, Compact extensions of L(Q), 1(a), Annals Math. Logic 11 (1977), 217-261.

[Ma] A. Macintyre, Generalized quantifiers in arithmetic, Abstract 755-E11, Notices of A.M.S. 25 (1978), A-386.

[MS] A. Macintyre and H. Simmons, Algebraic properties of number theories, Israel Journal 22 (1975), 7-27.

[Mc 1] K. McAloon, Formes combinatoires du Theoreme d'Incomplétude, 263-277 in Seminaire Bourbaki 1977/78, Springer Lecture Notes 710, 1979.

[Mc 2] __________, Completeness theorems, incompleteness theorems and models of arithmetic, T.A.M.S. 239, 253-277.

[Mo] C. Morgenstern, On generalized quantifiers in arithmetic, to appear in J.S.L.

[PH] J. Paris and L. Harrington, A mathematical incompleteness in Peano arithmetic, 1133-1142 in Handbook of Mathematical Logic (ed. J. Barwise), North Holland, 1977.

[Ra] M. Rabin, Diophantine equations and nonstandard models of arithmetic, 151-158 in Proceedings of 1960 I.C.L.M.S., (ed. Nagell et.al.) Stanford, 1960.

[S] J. Schlipf, Scribblings on papers of Kirby and Paris and Paris and Harrington, Abstract 755-E16, Notices A.M.S. 25 (1978), A-387.

[Sch] J. Schmerl, Extending models of arithmetic, Annals. Math. Logic 14 (1978), 89-109.

[SS] J. Schmerl and S. Simpson, On the role of Ramsey quantifiers in first-order arithmetic, to appear in J.S.L.

[She 1] S. Shelah, unpublished.

[She 2] ________, Omitting Types Theorem for L(Q), notes by W. Hodges, Bedford College, London, 1978.

[Sho] J. Shoenfield, Mathematical Logic, Addison-Wesley 1967.

[So] R. Solovay, A model of set theory in which every set of reals is Lebesgue measurable, Annals Math. 92 (1970), 1-56.

[SK] R. Solovay and J. Ketonen, Rapidly growing Ramsey functions, to appear.

[T] S. Tennenbaum, unpublished.

[W] A. Wilkie, On the theories of end-extensions of models of arithmetic, 305-310 in Set Theory and Hierarchy Theory V (ed. A. H. Lachlan, M. Srebrny, A. Zarach), Springer Lecture Notes 619, 1977.

COMPUTATIONAL COMPLEXITY OF DECISION PROBLEMS IN ELEMENTARY NUMBER THEORY

Kenneth L. Manders

Department of Philosophy and Department of Mathematics
University of Pittsburgh
Pittsburgh, PA 15260 U.S.A.

Introduction

The study of computational complexity of solvable decision problems is usually motivated by considerations of practical computability: As some decidable problems are not primitive recursive, decidability does not even remotely guarantee practality. This has led to the elaboration of a new logical notion, tractability [10], [11], and of a variety of techniques for demonstration of intractability [9], [12], [4].

There is, however, another aspect of the theory of computational complexity, which motivates its appearance here*: the study of definability rather than computability. For to claim, say, that a set of numbers is the set recognized by some Turing machine which halts on every input within some limited number of steps (a function of input size) is to say that the set has a definition of a particular form. The sheer variety of models of computation, and of resource use bounding functions for any given model, allows one to construct definability notions at will, and to specification. For example, nondeterministic and alternating [7], [13] machine models allow one to simulate the effects of quantifiers in definability; running time bounds then give the effect of bounds on the quantifiers. In all cases, our intuitive grasp of the operation of the machine model is a heuristic aid in building the definability theory associated with the complexity notions. And this intuitive grasp is context independent, treating decision problems concerning numbers, strings and graphs on an equal footing.

---

*At Karpacz, the paper was given as the first lecture in a series on weak systems of arithmetic, and served as an introduction to a subsequent lecture by A.Wilkie elsewhere in this volume.

There are closer connections, not yet fully understood, between the definability notions given by linear and polynomial time running bounds and definability and independence questions for weak systems of arithmetic, such as Peano Arithmetic with induction restricted to boundedly quantified formulas. For example, any Turing machine which can be shown in this theory to compute a total function runs in linear time. This follows directly from the model theoretic truncation property possessed by the theory: any initial segment of a model closed under multiplication is again a model. If the Turing machine were not to run in linear time, some model of the theory could be truncated so as to prevent the halting of the computation on some input.

In this paper, we describe the current state of knowledge concerning the time complexity of the most classical of number theoretic problems - primality, factorization, and other problems relating to binary quadratic equations. In section 1 we review the notions of deterministic and nondeterministic polynomial time complexity theory; section 2 applies these notions to the number-theoretic problems. We present only outlines of proofs, with reference to published versions.

1. Computational Complexity: Polynomial Time

Complexity theory is conveniently based on a Turing machine model of computation. A Turing machine is specified by a finite set of quintuples $(a_1,\ldots,a_5)$, each of which specifies the actions (a_3,a_4,a_5) during the current computation step given the current situation (a_1,a_2). For details, see e.g. [11], [21]. Most modifications of the model do not affect the complexity notions considered below, though some care must be exercised in this respect.

We consider Turing machines which start with an input, coded in an alphabet of at least two symbols. Inputs are denoted by 'N' in this section; the input length, or number of symbols in input N, by 'n'. Complexity measures are specified as upper bounds on resource usage, as a function of input length. (Emphasis: input length, not input, even if the input is numerical. The relationship between N and n is exponential.) For any function $t(\cdot)$, a Turing machine M runs in time $t(\cdot)$ if on each input N the machine halts within $t(n)$ steps. M runs in polynomial time if M runs in time t for some polynomial t. P (for polynomial time) is the class of sets (say, of natural numbers) S such that membership in S is decided by some Turing machine which runs in polynomial time.

A Turing machine is nondeterministic if (in opposition to our tacit assumption above and whenever nondererminism is not specifically mentioned) distinct quintuples may share a common (a_1,a_2), allowing distinct actions in the situation specified by such (a_1,a_2). Thus each input gives rise to a tree, each path of which corresponds to a valid computation of the machine. To specify the action of a nondeterministic machine on a decision problem, we assume certain among the halting states are designated, as accepting states. An input is accepted if some valid computation on that input ends in an accepting state; the computation is called an accepting computation. (This definition gives rise to the simulation of the effect of the existential quantifier alluded to the introduction.) A nondeterministic Turing machine (NDTM) accepts a set, e.g. of natural numbers, if it accepts exactly the members of the set as inputs, and no other inputs. An NDTM runs in time $t(\cdot)$ iff on each input N some computation, and on each accepted input some accepting computation, halts within $t(n)$ steps. NP is the class of sets accepted by an NDTM which runs in polynomial time, i.e. in time $t(\cdot)$ for some polynomial $t(\cdot)$. It is sometimes said that such a machine runs in nondeterministic polynomial time.

Because of the existential quantification in the notion of acceptance by NDTM, there is no obvious reason why the complement of a set in NP should again be in NP: the elements of such a complement are the inputs, rejected by all valid computations, and there is no obvious effecient way to check for this condition: nondeterminism does not seem to help, and deterministic search must deal with the entire computation tree, which may have exponential size in the input length. We denote by Co-NP the set of complements of sets in NP. From the definitions, one has the inclusions

$$\underset{\sim}{P} \subseteq \underset{\sim}{NP} \cap \text{Co-}\underset{\sim}{NP} \begin{array}{l} \subsetneq \underset{\sim}{NP} \\ \subsetneq \text{Co-}\underset{\sim}{NP} \end{array};$$

all inclusions are commonly conjectured to be proper, but whether this is so is an open, and apparently difficult problem. (The NP-P-problem.)

As in recursion theory, the major tool in complexity classification is computational reducibility. In studying NP, the most useful notion has been deterministic polynomial time (many-one) reducibility (Karp [12]). For any sets A and B, A is <u>polynomial time reducible to</u> B ($A \leq_p B$) iff some Turing machine M, which runs in polynomial time, computes a function $M(\cdot)$ such that

$$\forall N \in \omega[N \in A \leftrightarrow M(N) \in B]$$

A set X is <u>NP-hard</u> iff

$$\forall Y \in \underset{\sim}{NP}: Y \leq_p X$$

and <u>NP-complete</u> if X is NP-hard and $X \in$ NP.

The prototypical example of an NP-complete set is the set SAT of satisfiable propositional formulas (Levin [20], Cook [9]). That this set is NP is easily seen: Given a formula N, which will contain on the order of n distinct propositional letters, the accepting machine first goes through n nondeterministic steps, composing (on any given path) a table of truth-values for the propositional letters. Then it checks, in some n^2 deterministic steps, whether the valuation in the table satisfies the formula; if so, the path accepts. Now if there is a satisfying valuation, some valid computation will enter it in its table and hence accept; otherwise no computation path accepts.

To show that SAT is NP-hard, let Y be an arbitrary set in NP, and let M accept Y in nondeterministic time n^k. The statement "there is an accepting computation of M on input N" can be reformulated as a statement about a table, with one row per computation step consisting of the contents of the n^k tape locations in principle accessible within n^k steps of Turing machine operation together with indication of tape head position and machine internal state. The machine reducing Y to SAT translates this statement into a propositional formula, introducing propositional letters for basic statements such as "the initial contents of the m-th tape square is '0'", or "the state of M after the k-th step is Q_ℓ".

Many combinatorial problems of practical importance, such as path optimizations through weighted graphs, are NP-complete; a list of hundreds of such problems can be found in Garey and Johnson [11]. The failure of concerted efforts to find (deterministic) polynomial time algorithms for these problems has lead to the conjecture that NP = P, and that NP-complete problems are significantly more difficult than problems in P.

By recursion theoretic methods, the partial order of equivalence classes of sets in NP with respect to polynomial time interreducibility has been studied, assuming NP ≠ P and NP ≠ Co-NP, see Ladner [14]. For example, not all sets properly in NP would be NP-complete: there are infinite strictly descending chains of degrees in NP. Most research, however, has concentrated on classifying the complexity of specific decision problems within NP. The power of the method of computational reduction in showing problems intractable has recently been extended by Adleman and Manders [3], [4], who introduce nondeterministic computational reductions without analogs in ordinary recursion theory. These give rise to (apparently) larger complete degrees in NP, and problems have been shown complete in NP with respect to these reductions which are not known to be NP-complete. (See also Plaisted [28].)

Of course complexity theory comprises studies of many other computational models, other complexity measures (such as space utilized) and other than polynomial bounds, both larger (exponential and beyond) and smaller (log(n) space, linear time, n^2 time). Our selection for presentation of the polynomial time notions represents a particular choice of trade-off between:

a) resolution of complexity classification (improved for smaller bounds, worse for larger), and
b) insensitivity to details of machine model (with opposite dependencies).

This particular choice is so popular and interesting because (a) it still allows a practically and theoretically useful classification of a wide variety of concrete computational problems, including applied combinatorial problems as well as the classical number theoretic ones to be studied below; and (b) it already achieves so much machine independence (e.g. number of tape symbols, number of tapes, simulation by universal machine) that complexity classification can be carried on without explicit reference to the model (see below). These features were already noted by Cobham [8]. An excellent recent exposition of polynomial-time complexity theory as introduced in this section is [11].

2. Number-Theoretic Problems

a. Basic Algorithms

We assume that numerical inputs N are represented in binary notation and then identify n (the number of symbols in N) and $\log_2 N$. It is immediately obvious from the standard algorithms that addition, multiplication and division of integers can be executed in polynomial time. Perhaps the earliest published analysis of running time is that of Lamé (1844, [18]) for the Euclidean GCD algorithm. The algorithm proceeds in stages which require only a fixed number of the elementary arithmetic operations mentioned above; every two stages, the two inputs are replaced by numbers less than half as large (i.e., with one less symbol). Thus the process must terminate after a linear number of stages (in input size) and hence runs in polynomial time. The Euclidean algorithm has a number of further uses. On input a,b with $(a,b) = 1$, it yields a solution to $xa + yb = 1$, and here x is an inverse of a modulo b, i.e., $xa \equiv 1 \bmod b$. This ability to compute inverses leads to a polynomial time realization of the Chinese remainder algorithm, solving

$$x \equiv a_i \bmod m_i, \quad i = 1,\dots,k,$$

for relatively prime moduli m_i.

Much pioneering work on the running time of number theoretic algorithms was done by D. H. Lehmer [19], well before the advent of Complexity Theory. An important algorithm he perfected for machine use and analyzed is the Power Algorithm: to compute $a^N \bmod M$, given a,N,M. We can get a^N by considering the digits N_i of N in binary, starting from the most significant one, N_1, and at <u>stage i</u>,

$i = 1,\dots,n$:

(i) square the previous result (1 if $i = 0$),
(ii) if $N_i = 1$, multiply by a;
(iii) reduce modulo M.

Clearly, there are only n stages; because at each stage we reduce the result modulo M, the numbers involved remain the same size as the input, and the algorithm runs in polynomial time.

b. <u>Primality</u>

The most fascinating of all computational problems in elementary number theory are undoubtedly those of primality testing and factorization. These problems have never ceased attracting attention since the time of Fermat, and motivated the work of major figures such as Gauss.

It is immediate that the set Pr of prime numbers belongs to Co-$\underset{\sim}{NP}$: the composites are recognized by (nondeterministically) "guessing" a nontrivial factorization, which is then verified by multiplication. Pratt [29] noticed that also Pr ε $\underset{\sim}{NP}$. To see this, recall that the primes are the numbers N such that the order of the multiplicative group G modulo N is $N-1$ and, a fortiori, cyclic. The algorithm proceeds by guessing a generator g, $1 < g < N$, for G, and verifying the order of g:

(i) $g^{N-1} \equiv 1 \bmod N$,
(ii) $g^{(n-1)/p} \not\equiv 1 \bmod N$, each prime divisor p of $N-1$.

Here the prime divisors p are obtained nondeterministically; completeness of the system of divisors is verified by repeated division. Primality of the divisors is verified by re-invoking the procedure for each divisor except 2 (which is always present). There are at most n levels of invocation, as at least a factor 2 and hence at least one bit is lost in size at each level. At each level, we perform the verifications for less than n primes p (as the product of all primes at any given level is less than N), and the verifications (i), (ii) run in polynomial time by the power algorithm. So Pr ε $\underset{\sim}{NP}$. Assuming the Extended Riemann Hypothesis (ERH; see below), this can be strengthened to

<u>Theorem</u> (Miller [24]) ERH $\rightarrow$ Pr ε $\underset{\sim}{P}$.

<u>Outline of Proof</u>: The algorithm searches systematically for evidence of non-primality; if none is found, we will eventually be able to conclude that N is prime.

For arbitrary b, $1 < b < N$, we can compute

$$b^{(N-1)/2^k},\ b^{(N-1)/2^{k-1}},\ldots,b^{(N-1)/2},\ b^{N-1} \text{ modulo } N,$$

where 2^k is the highest power of 2 dividing $N-1$, and we check:

(i) $b^{N-1} \equiv 1 \bmod N$,

(ii) if $b^{(N-1)/2^k} \not\equiv 1 \bmod N$, the first residue equal 1 in the computed sequence is preceded by a residue equal -1.

If N is prime, these conditions are always satisfied. Disregarding the easily detected case that N is a proper prime power, let $p \neq q$ be odd prime divisors of N. If the largest powers of 2 dividing the order of b in the multiplicative groups modulo p and q are <u>not</u> the same, say $k_p > k_q$, then b will fail at least one of the tests (i) and (ii). For if b passes Fermat's test (i), we will have: $k \geq k_p > k_q$, $b^{(N-1)/2^{(k-k_p)}} \equiv 1 \bmod N$, and

$$b^{(N-1)/2^{(k-k_p+1)}} \equiv -1 \bmod p$$

$$b^{(N-1)/2^{(k-k_p+1)}} \equiv 1 \bmod q$$

so that this power of b is congruent to neither 1 nor -1 modulo N, i.e., b fails test (ii).

Where is such a b to be found? Depending on whether the power of 2 dividing $p-1$ is greater, equal or less than that for $q-1$, a quadratic nonresidue mod p, resp. any b such that the Jacobi symbol

$$\left(\frac{b}{pq}\right) = \left(\frac{b}{p}\right)\cdot\left(\frac{b}{q}\right) \neq 1 \quad \text{(b quadr. residue modulo exactly one of } p \text{ and } q\text{)},$$

resp. a quadratic nonresidue mod q, will have $k_p \neq k_q$. A theorem of Ankeny [6] guarantees that for a universal constant c, such b always exists with $b < c(\log N)^2$. The formulation of the theorem in Montgomery ([25], p. 120) allows one to specify the use of ERH underlying each case:

$$L(s,\chi) = \sum_n \chi(n)/n^s$$

should have all its nontrivial zeroes s on the line $re(s) = 1/2$, for each character χ given by the Legendre symbol $\chi(n) = (\frac{n}{p})$ for each $p < N$, and for each χ given by the Jacobi symbol $\chi(n) = (\frac{n}{p})\cdot(\frac{n}{q})$ for all distinct odd primes $p,q < N$. P. Weinberger [33] has shown an upper bound $c \leq 4$ for the constant in Ankeny's Theorem.

So Miller's algorithm executes tests (i) and (ii) for all $b < 4(\log N)^2$ and, if no evidence of nonprimality is found, concludes that N is prime. As the testing of (i) and (ii) for each base b can be done in polynomial time, so can the entire test.

c. Factorization and Related Problems

The problem of determining the prime factors of a given number is not itself a decision problem. However, it is polynomial time equivalent to the decision problem F, for arbitrary $a < b$ [24]:

Does b have a nontrivial factor less than a?

Clearly $F \in \underset{\sim}{NP}$, for if there is a factor you guess it and verify. But also $F \in \text{Co-}\underset{\sim}{NP}$, for one can guess a prime factorization of b, and (by Pratt's observation) a verification that it is a prime factorization; then it suffices to check that the smallest prime factor of b is not less than a.

For the equivalence to factorization: given the ability to factor b, F can be decided in (deterministic) polynomial time (as above); given an oracle for F, the factors of b can be located by a binary search for the smallest one: take $a = [b/2], [b/4], \ldots$ until a negative answer is obtained, then work back up through the last interval with decreasing step size until a positive answer is found, then downward again etc.. As the k-th step size is $b/2^k$, this gets the smallest prime factor of b in $\log_2 b$ steps.

A number of classical problems are closely related in complexity to prime factorization. First, the problem of determining whether b is a quadratic residue modulo m. As b is a quadratic residue mod m iff b is a quadratic residue modulo p for each prime p dividing m (e.g., see below), it suffices to guess all such p, confirm their primality by Pratt's method, and check $b^{p-1/2} \equiv 1 \bmod p$ by the power algorithm. So the problem is clearly in $\underset{\sim}{NP} \cap \text{Co-}\underset{\sim}{NP}$.

If b is a quadratic residue mod m, then given the prime factorization of m and a quadratic nonresidue b_p for each prime p dividing m, a solution to $b \equiv x^2 \bmod m$ can be computed in (deterministic) polynomial time (Adleman, Manders and Miller [5]). For using

b_p, one can extract the square root of b_p mod p in polynomial time.

Then if p^α is the highest power of p dividing m, one recursively solves $(x_k)^2 \equiv b \bmod p^k$, $k = 2,3,\ldots,\alpha$ (see [27], Ch. 2.6 or [5]). Finally, the solutions mod p^α are combined by the Chinese remainder algorithm.

Assuming ERH, the needed quadratic nonresidues can be located in polynomial time as in Miller's primality test. Hence, solving $b \equiv x^2 \bmod m$ is polynomial time Turing reducible to factorization of m.

Second, the decision problem for solvability of $x^2 - dy^2 = -1$, the anti-Pellian equation. This famous problem has been studied by many authors. Recently, Lagarias [16] succeeded in showing that the problem can be solved in polynomial time given the prime factorization of d and a nonresidue modulo p for each prime divisor p of d. Thus the problem is in $\underset{\sim}{NP} \cap \text{Co-}\underset{\sim}{NP}$, and, assuming ERH, polynomial time reducible to factorization. Lagarias' result is a considerable improvement on what was available in the literature. The method is to compute a basis for the 2-Sylow subgroup of the class group of binary quadratic forms of determinant d; the problem is reduced to that of deciding equivalence of two forms in this subgroup. The running time analysis requires a systematic analysis of algorithms relating to binary quadratic forms, undertaken in [15].

A number of other number theoretic problems are shown polynomial time equivalent to factorization (often on assumption of ERH) by Miller [24].

The conjectured disparity between the complexity of the power algorithm (and primality testing given ERH, or using randomized techniques not discussed here) on one hand, and factorization on the other hand, is exploited in the Rivest-Shamir-Adleman implementation of a public key cryptosystem [31]. The aspiring recipient of encrypted messages locates two large primes p,q and computes $N = pq$, $\phi(N) = (p-1)(q-1)$, and (using the Euclidean algorithm) a pair s,t of inverses modulo $\phi(N)$. He publishes N and s. An aspiring sender breaks his message up into numerical blocks

$$a_1, a_2, \ldots, 0 < a_i < N$$

and encodes each a_i as: a_i^s modulo $N = x_i$. The recipient of x_i computes

$$(x_i)^t \bmod N = (a_i^s)^t \bmod N = a_i.$$

All these desired computations are polynomial time. However, to break the code one must in effect find t, given s and N. <u>Given the factorization of N</u>, this can be done in polynomial time as above. Conversely, Rabin [30] has shown that N can be factored in polynomial time. Because s.t is a multiple of $\varphi(N)$, breaking the code by finding t given s and N is equivalent to factorization of N (ERH) by Miller [24]. For a variant of the encryption scheme, Rabin has shown that <u>any</u> systematic method for breaking the code is equivalent to factorization. [30]. Given that the asymptotic growth of the running time of the best known algorithms for factorization (see [1] for recent references) is considerably greater than polynomial, one can choose a size for N (≈ 200 digits) which allows quite efficient execution of all desired operations while breaking the code is infeasible.

d. <u>An NP-Complete Binary Quadratic Solvability Problem</u>

<u>Theorem</u> (Manders-Adleman)

1. The set of equations of the form

$$ax^2 + by = c, \quad a,b,c > 0,$$

solvable in natural numbers, is NP-complete.

2. The set of triples a,b,c such that

$$a,b,c > 0, \quad x^2 \equiv a \bmod b \text{ has a solution } x, \quad 0 \leq x < c,$$

is NP-complete.

Via a binary search algorithm, solution of the second problem is deterministic polynomial time equivalent to the problem of finding the least positive solution to a given congruence $x^2 \equiv a$ modulo b, if there is any. In a sense, the theorem is best-possible: integral solvability of single-variable equations can be decided in polynomial time by real root localization techniques (Sturm's Theorem); other simplifications of the problems have been shown in NP $\cap$ Co-NP above.

Wilkie noted [34] that it follows from 1. that if Matijasevic's Theorem (recursively enumerable sets are Diophantine) is provable in $P^- + I\Sigma_0$ (induction for bounded formulas), then NP = Co-NP; this provides prima facie evidence against provability of M's theorem in the system. The author noted that Wilkie's argument works for $P^- + I\Sigma_0^*$, where $I\Sigma_0^*$ is the induction scheme for formulae with bounded quantification of the forms

$$\forall x < 2^{(\log y)^n}; \ \exists x < 2^{(\log y)^n}, \quad n \in \omega.$$

The proof of the second statement is a minor variant of that of the first, which we now discuss. Both problems are clearly in $\underset{\sim}{NP}$, because solutions are always so small that we can afford to write them out nondeterministically and check for correctness.

To show $\underset{\sim}{NP}$-completeness, it suffices to show that some $\underset{\sim}{NP}$-complete problem is deterministic polynomial time reducible to the given problem. The original reduction (Manders-Adleman [23]) is from a variant of SAT; a somewhat shorter reduction can be given [22] from the Set Partition Problem. Here, we only try to disengage the main idea. Consider an attempt at a direct proof. Let M be an arbitrary nondeterministic Turing machine which runs in polynomial time, say n^k. On input N, our reduction must output a Diophantine equation; the equation must be solvable iff M accepts on input N. But M accepts on input N if there exists an appropriate computation history, i.e., an $n^k \times (n^k+1)$ array, one row for each step: the i-th row containing the entries in the first n^k tape squares immediately before the i-th computation step and the machine state and tape position at that moment. To obtain the correspondence with solvability, this array must be encoded in any solution to the Diophantine equation, and the equation must express the constraint on the sequence: that it correspond to an accepting computation of M on N.

We attempt to achieve this (and from here on, I oversimplify) by viewing the array as a sequence of length $n^k.(n^k+1)$, row following row, and encoding this sequence as the sequence of b-ary digits of a number X, for suitably large base b. The major requirement on X, that it corresponds to a computation of M, can be expressed by setting

$$Y = (b^{n^k+1}X) - X;$$

that is, comparing each row in the array with its successor, and requiring that <u>at each position</u> in Y, we find a zero or such changes as would correspond to a state transition of M and corresponding tape changes. Unfortunately, expressing this latter condition would appear to require a universal quantifier. We would be in better shape if we had Y in modular coded form Y', say $K = \prod p_j$; $(Y)_i \equiv Y' \bmod p_i$, $i = 1,\ldots,(n^k+1)^2$. For then we could represent the condition in Diophantine form

$$Y'(Y'-n_1)\ldots(Y'-n_M) \equiv 0 \text{ modulo } K$$

where $n_1,\ldots,n_M$ are the differences which might result by one application of a quintuple from among those defining M.

The crux of the reduction to a Diophantine equation is the provision of a Diophantine expression, computable in polynomial time, for the relationship between a b-ary representation Y and a suitable modular representation Y' of a sequence. A method of constructing these is given in the Conversion Lemma of [23], p. 178. This direct approach yields a rather complicated Diophantine equation. Starting from a suitable NP-complete problem, and incorporating the ideas of the Conversion Lemma directly into the argument, one obtains a binary quadratic.

The need to connect different representations appears to be the expression of a fundamental difficulty in Diophantine definition, recognizable in some form in all published attempts at Diophantine expression of Turing computation or of recursive functions. In each case, it appears that the various constraints to be expressed each have simple Diophantine form in some representation, but no known representation leads to a straightforward simultaneous expression of all constraints.

In the relatively undemanding setting of NP-completeness, we can deal with this by computation of input-dependent parameters. For a <u>uniform</u> Diophantine definition of a given r.e. set, such parameters are not allowed. Then the difficulty can only (so far as now known) be resolved by use of equations with exponentially large minimal solutions. All this suggests that an analysis of the conditions which need to be expressed might lead to further understanding the need for large solutions. Presumably, one must find a way of carrying out the analysis in terms which are representation-neutral (as opposed to the discussion given above).

e. Solvability for Binary Quadratics: Conclusion

Most of the computational problems discussed above can be presented as solvability problems for binary quadratics; including the problem, shown equivalent to factorization above, of whether b has a nontrivial divisor less than a. The complexity of these problems ranges from: in P (as for equations of the form $x^2 = a$), via: in NP $\cap$ Co-NP (as for $x^2 \equiv b \bmod m$), to: NP-complete (nonnegative solvability of $ax^2 + by = c$).

For many of these types, membership in NP is evident by a guess-a-solution-and-check algorithm. But this does not suffice in important cases: the solutions may be too large to verify in polynomial time. Thus, Lagarias [16] shows that for any n, the smallest solution of

$x^2 - 5^{2n+1}y^2 = -1$ has length in binary greater than $\frac{1}{3}5^n$. So in general, more subtle arguments are needed to show that these problems are in NP. Lagarias [17] has shown that the solvability problem (in integers, or in natural numbers) for arbitrary binary quadratics is in NP. Lagarias' arguments rely his analysis of algorithms for binary quadratic forms, and careful elaboration of an idea of Shanks [32] allowing rapid controlled jumping through the principal cycle of forms of a given determinant, using composition and reduction of forms in a manner remniscient of multiplication and reduction in the power algorithm. Together with the NP-completeness result of the previous section, Lagarias' theorem yields:

Theorem. The problem of deciding whether a binary quadratic equation has a solution in natural numbers, is NP-complete.

There are no classes of binary quadratics for which deciding solvability in integers (as opposed to non-negative integers) is known to be NP-complete, though presumably such classes exist. Adleman and Manders have shown completeness in NP of the solvability in integers for various types of binary quadratics [3], [4], with respect to non-deterministic polynomial time reducibilities. Possibly some of these problems are NP-complete.

Curiously, all such presently known intractability results for binary quadratics are for definite or degenerate (see [17]) equations. Lagarias' methods may make it possible to show completeness in NP for indefinite equations. In particular, the (general) Pell equation $x^2 - dy^2 = m$, d nonsquare, is of interest, due to the reduction by Lagrange (see [26]) of solvability in integers of arbitrary indefinite binary quadratics to solvability of this equation (admittedly not exactly in integers).

We conclude with a list of open problems (of, I conjecture, increasing difficulty).

1. Is solvability in integers of $x^2 - dy^2 = m$, d nonsquare, NP-complete?
2. Is factorization polynomial time Turing reducible, perhaps under assumption of ERH, to
   (a) solving $x^2 \equiv b \bmod m$?
   (b) determining solvability in integers of $x^2 - dy^2 = -1$?
3. Is every set in NP expressible by a Diophantine equation with

minimal solutions of length polynomial in the length of the input? (See [2].)

4. Show that factorization is intractable, e.g., complete in some sense.

5. What is the complexity of determining, given b, ℓ and m, whether there is a prime p, $p \equiv \ell$ modulo m, $p < b$?

REFERENCES

1. Adleman, L., A Subexponential Algorithm for the Discrete Logarithm Problem, with Applications to Cryptography. Proc. 20th Annual IEEE Symp. on Foundations of Computer Science, 1979, pp. 55-60.

2. Adleman, L. and Manders, K., Diophantine Complexity. Proc. 17th Annual Symp. on Foundations of Computer Science (IEEE), 1976, pp. 81-88.

3. Adleman, L. and Manders, K., Reducibility, Randomness and Intractability. Proc. 9th Annual ACM Symp. on Theory of Computing, 1977, pp. 151-163.

4. Adleman, L. and Manders, K., Reductions that lie. Proc. 20th Annual IEEE Symp. on Foundations of Computer Science, 1979, pp. 397-410.

5. Adleman, L., Manders, K., and Miller, G.L., On Taking Roots in Finite Fields. Proc. 18th Annual Symp. on Foundations of Computer Science (IEEE), 1977, pp. 175-178.

6. Ankeny, N., The Least Quadratic Nonresidue. Ann. of Math. 55 (1952), pp. 65-72.

7. Chandra, A. and Stockmeyer, L., Alternation. Proc. 17th Annual Symp. on Foundations of Computer Science (IEEE), 1976 , pp. 98-108.

8. Cobham, A., The Intrinsic Computational Difficulty of Functions. In Y. Bar-Hillel (ed.) Proc. 1969 Intern. Congress for Logic, Methodology and Philosophy of Science, North-Holland, Amsterdam, pp. 24-30.

9. Cook, S., The Complexity of Theorem-Proving Procedures. Proc. 3rd Annual ACM Symp. on Theory of Computing, 1971, pp. 151-158.

10. Edmonds, J., Paths, Trees and Flowers. Canadian J. Math. 17 (1965), pp. 449-467.

11. Garey, M. and Johnson, D., Computers and Intractability. W. H. Freeman, 1979.

12. Karp, R., Reducibility Among Combinatorial Problems. In: Complexity of Computer Computation, eds. Miller, R. N. and Thatcher, J., Plenum Press, 1972, pp. 85-104.

13. Kozen, D., On Parallelism in Turing Machines. Proc. 17th Annual Symp. on Foundations of Computer Science (IEEE), 1976, pp. 89-97.

14. Ladner, R., On the Structure of Polynomial Time Reducibility. J. ACM 22 (1975), p. 155-171.

15. Lagarias, J., Worst-Case Complexity Bounds in the Theory of Integral Quadratic Forms, to appear in J. Algorithms.

16. Lagarias, J., On the Computational Complexity of Determining the Solvability or Unsolvability of the Equation $X^2 - DY^2 = -1$, to appear in Trans. AMS.

17. Lagarias, J., Succinct Certificates for the Solvability of Binary Quadratic Diophantine Equations. (Extended Abstract.) Proc. 20th Annual Symp. on Foundations of Computer Science (IEEE), 1979, pp. 47-54.

18. Lamé, M., Note sur la limite du nombre des divisions..., C. R. Acad. Sci. Paris Ser. A-B 19 (1844), pp. 867-869.

19. Lehmer, D.H., Computer Technology Applied to the Theory of Numbers. Studies in Number Theory, M.A.A., 1969, pp. 117-151.

20. Levin, P.A., Universal Sorting Problems (Russian). Problemi Peredaci Informacii IX (1973), pp. 115-116.

21. Machtey, M. and Young, P., An Introduction to the General Theory of Algorithms. North-Holland, 1978.

22. Manders, K., Studies in Applied Logic. Ph.D. Thesis, Berkeley, 1978.

23. Manders, K., and Adleman, L., NP-complete Decision Problems for Binary Quadratics, J. Comp. Sys. Sci. 15 (1978), pp. 168-184. Earlier version: NP-Complete Decision Problems for Quadratic Polynomials. Proc. 8th ACM Conf. on Theory of Computing (1976), pp. 23-29.

24. Miller, G.L., Riemann's Hypothesis and Tests for Primality, J. Comp. Sys. Sci. 13 (1976), pp. 300-317.

25. Montgomery, H., Topics in Multiplicative Number Theory. Springer LNM 227, 1971.

26. Mordell, L., Diophantine Equations, Pure and Applied Mathematics. Vol. 30, Academic Press, 1969.

27. Niven, I., and Zuckerman, H., An Introduction to the Theory of Numbers. J. Wiley, 1960, 1966.

28. Plaisted, D., New NP-Hard and NP-Complete Polynomial and Integer Divisibility Problems. Proc. 18th Annual Symp. on Foundations of Computer Science (IEEE), 1977, pp. 241-253.

29. Pratt, V., Every Prime has a Succinct Certificate. SIAM J. Comput. 4 (1975), pp. 214-220.

30. Rabin, M., Digitalized Signatures and Public Key Functions as Intractable as Factorization. Memo MIT/LCS/TR-212, 1979.

31. Rivest, R., Shamir, A. and Adleman, L., A Method for Obtaining Digital Signatures and Public Key Cryptosystems. Comm. ACM 21 (1978), pp. 120-126.

32. Shanks, D., The Infrastructure of a Real Quadratic Field and Its Applications. Proc. 1972 Number Theory Conf., University of Colorado, Boulder (1972), pp. 217-224.

33. Weinberger, P., Private Communication.

34. Wilkie, A., this volume.

SOME DIOPHANTINE NULLSTELLENSÄTZE

Kenneth McKenna
Harvard University
Department of Mathematics
Cambridge, MA 02138

In this paper we will be concerned with the following problem: Let K be a field and $I \subset K[x_1, \cdots, x_n]$ be an ideal in the polynomial algebra over K. Let $V_K(I)$ be the set of all K-rational points of I. Let $\mathcal{I}(U_K(I))$ be the ideal of all polynomials from $K[x_1, \cdots, x_n]$ which vanish on $V_K(I)$. Our problem is then to describe $\mathcal{I}(U_K(I))$ in some useful way. The prototype for this kind of result is Hilbert's Nullstellensätz which says that if K is algebraically closed then $\mathcal{I}(U_K(I)) = \sqrt{I} = \{f \in K[x_1, \cdots, x_n] \mid$ there exists $m \in \mathbb{N}$ with $f^m \in I\}$. The general form of nullstellensätz in which we will be interested will have the form of a demonstration that, under suitable conditions, $\mathcal{I}(U_K(I)) = \{f \mid f^m_{(x)} D(u_1(x) \ldots u_j(x)) \in I\}$ where the $u_i(x)$ are rational functions and $D(y_1, \cdots, y_j)$ is a polynomial - defined over an appropriate "core field" (such as $\mathbb{Q}$) - with no zero rational over K. An exact statement of the results will be found below.

Nullstellensätze for $\mathbb{R}$ and $\mathbb{Q}_p$ have been demonstrated by Dubois-Reissler [7] and Kochen [3] previously. Our results, we believe, clarify the underlying principals of these Nullstellensätze by both unifying somewhat their proofs and extending the results to obtain Nullstellensätze for other non-algebraically closed fields; notably the pseudo-finite fields investigated first by Ax and separably closed fields.

Our results will deal with concepts derived from model-theoretic algebra. In particular we will lean heavily on the notion of model completeness due to A. Robinson.

§1. Some Model Theoretic Algebra

We will assume some slight familiarity with very basic concepts of mathematical logic [2]. We will consider fields as structures for the (first order) language L which contains the symbols $+, \cdot, -, , 0, 1$ where these denote the ordinary operations and constants. In addition we may wish to introduce constants for certain elements of the field in question: $a_1, a_2, a_3, \cdots$. Thus there is a collection of sentences written in L (denoted by TF) which is such that TF is true in the L-structure K if and only if K is a field. This set TF is just the ordinary axioms of field theory, for example: $\forall x[(x \neq 0) \Rightarrow \exists y(xy = 1)]$, the axiom ensuring inverses for non-zero elements of K.

A consistent collection of L-sentences containing TF is called a theory of fields. For example: If we adjoin to TF the infinite set $\{\varphi_d | d \in \mathbb{N}\}$ where φ_d is the formula $\forall x_0, \cdots, \forall x_{d-1} \exists y[y^d + x_{d-1} y^{d-1} + \cdots + x_0 = 0]$ we obtain the theory of algebraically closed fields (TACF) since any field satisfying φ_d must contain a zero for all of its polynomials of degree at most d. If we adjoin further axioms to fix the characteristic p we obtain the theory of algebraically closed fields of characteristic p, TACF(p).

If T is a theory of fields satisfied by K we write $K \models T$, similarly for an individual sentence, $K \models \varphi$. We will need the important concept (due to Abraham Robinson) of a model complete theory.

Definition: A theory of fields T is said to be model complete if for any two fields $K_i \models T$ with $K_1 \subset K_2$ and for any $\mathcal{C}$ an affine constructible set [8, p. 94] defined over K_1 if $\mathcal{C}$ has point rational over K_2 then $\mathcal{C}$ has a point rational over K_1.

The most obvious example of a model complete theory of fields is TACF, where the conditions of the definition are shown to be true in the course of proving Hilbert's Nullstellensätz. We list some others:

1. Real Closed Fields (TRCF). Here the theory of fields TRCF is such that $K \models$ TRCF if and only if K is real closed in the sense of Artin-Schreier [1], [6].

2. p-adically Closed Fields (TPACF). In the sense of Kochen's beautiful article [3].

3. Let K be a separably closed field (in other words a field with no separable algebraic

extensions) with characteristic p. Suppose further that $[K : K^p] = p^d < \infty$. We may then introduce new constants $t_1, \cdots, t_d$ to distinguish elements of a p-basis in K. We then add to TF axioms ensuring that every separable polynomial has a zero and stipulating that $\{t_1, \cdots, t_d\}$ is a p-basis (for this we make use of the new constants). This theory of fields (with extra constants) we call the theory of separably closed fields of Ershov invariant p^d (where p is the characteristic), TSCF (p^d). This theory is model complete in the expanded language [4] and $K \models$ TSCF (p^d).

4. Let A_p be any <u>infinite algebraic extension of the prime finite field</u> $\mathbb{F}_p$. As Ershov [4] has observed, there are axioms of three types which together determine the totality of all sentences true in A_p: The first set, Σ_1 ensures that every absolutely irreducible variety defined over the field has a point rational over the field. The second, $\Sigma_2(A_p) = \{\exists x\, f(x) = 0 \,|\, f(x) \in \mathbb{F}_p[x]$ where $f(x)$ is a polynomial in one variable with a zero in $A_p\} \cup \{\forall x\, f(x) \neq 0 \,|\, f(x) \in \mathbb{F}_p[x]$ where $f(x)$ is a polynomial in one variable with no zero in $A_p\}$. The third describing how many extensions (either 0 or 1) of each degree A_p has, $\Sigma_3(A_p)$. We then denote $\Sigma_1 \cup \Sigma_2(A_p) \cup \Sigma_3(A_p) \cup$ TF by $T(A_p)$. This $T(A_p)$ is satisfied by A_p and is model complete, [4].

5. Consider the absolute Galois group of any number field, $\mathrm{Gal}(\widetilde{M}/M) = G_M$. This is a compact group with a cononical Haar measure, μ. If $\sigma \in G_M$ and $\widetilde{M}(\sigma)$ is the field fixed by σ then there is a set, Z, of measure one so that the following are truc if $\sigma \in Z$:

a) $\widetilde{M}(\sigma) \models \Sigma_1$, where Σ_1 is as in Example 3 (this was proved by Jarden), [5].

b) $\widetilde{M}(\sigma) \models \Sigma_2$, where $\Sigma_2(\sigma)$ is like $\Sigma_2(A_p)$ in Example 3 except that $\mathbb{F}_p$ is replaced by $\mathbb{Q}$ and A_p by $M(\sigma)$.

c) $\widetilde{M}(\sigma) \models \Sigma_3(\sigma)$ where $\Sigma_3(\sigma)$ ensures the existence of exactly one extension of each degree.

The set $\Sigma_1 \cup \Sigma_2(\sigma) \cup \Sigma_3(\sigma) \cup$ TF completely determines the first order theory of $M(\sigma)$. We denote this theory by $T(\sigma)$. $T(\sigma)$ is model complete. (Thus the family of theories $T(\sigma)$ supplies uncountably many, in fact 2^{ω}, model complete theories of fields.

Below we will determine a specific Nullstellensätz for each of the fields mentioned in Examples 1, 3, 4, and 5.

In each of the cases above the theory of fields mentioned is also complete (as well as model complete). This is to say that for any sentence φ either φ or its negation is formally implied by the theory in question. (For example, if $F(z) \in \mathbb{Z}[z]$ K_1 and K_2 are both models of the same theory (which is chosen from the above list) then F has a zero in K_1 if and only if F has a zero in K_2.)

If T is a theory of fields we may consider $T_{\forall} \subset T$ where $T_{\forall}$ is defined to be all consequences of T which are of the form $\forall x_1, \cdots, \forall x_n\, \varphi(x_1, \cdots, x_n)$ where $\varphi(x_1, \cdots, x_n)$ is quantifier free. A fundamental result due to Tarski then says:

Tarski's First Theorem: If R is any integral domain then there exists a field K which satisfies T and which contains R if and only if $R \models T_{\forall}$.

Another fundamental theorem due to Tarski is the following:

Tarski's Second Theorem: Let R be an integral domain. Adjoin to the language L enough constants from R so that R is generated by these constants, C, over $\mathbb{Z}$ as a ring. Consider the theory of fields in the expanded language obtained by adjoining to TACF (characteristic (R)) the set of all polynomial equations and inequalities from $\mathbb{Z}[C]$ which are true in R (the Diagram of R, $D(R)$). Call this theory TACF (R). Then TACF (R) is complete and model complete in the expanded language.

Let T be a model complete theroy of fields and $K \models T$. Let $I \subset K[x_1, \cdots, x_n]$, I an ideal. We define, with Cherlin [6], the T-radical of I:

$$\sqrt[T]{I} = \cap\, P(I, T)$$

where $P(I, T)$ ranges over all primes of $K[x_1, \cdots, x_n]$ which contain I and for which $K[x_1, \cdots, x_n]/P(I, T)$ may be embedded in a model of T. We then immediately have:

LEMMA 1: $\mathcal{I}(V_K(I)) = \sqrt[T]{I}$.

Proof: Let $f \in \sqrt[T]{I}$. Let $(k) \in V_K(I)$. Let $P(k)$ be the maximal (hence prime) ideal of functions vanishing at (k) . Obviously $K[x_1, \cdots, x_n]/P(k) = K$ so $f \in P(k)$ so $f(k) = 0$. Thus $\sqrt[T]{I} \subset \mathcal{I}(V_K(I))$.

Let $f \notin \sqrt[T]{I}$. Then there is a prime $P \supset I$ together with a field $K_1 \models T$ so $K \subset K[x_1, \cdots, x_n]/P \subset K_1$ and $f \notin P$. But then the K-constructible affine set $\mathcal{C} = V(I) - V(f)$ has a point rational over K_1 . By model completeness $\mathcal{C}$ has a point rational over K . Thus $f \notin \mathcal{I}(V_K(I))$.

Thus we are reduced to describing $\sqrt[T]{I}$. For this we need the following lemma which may be found in [9], but with a more complicated proof.

LEMMA 2: Let T_1 and T_2 be complete, model complete theories. Let $A \models T_1$ $A_1 \models T_1$ and $B \models T_2$. If $A \subset B \subset A_1$, then $T_1 = T_2$.

Proof: We will need a stronger characterization of model completeness. Let T be a model complete theory and K_1 and K_2 satisfy T with $K_1 \subset K_2$. Robinson has shown that under these circumstances if φ is any first order sentence then φ is true in K_1 if and only if φ is true in K_2 . Furthermore, if $K_1 \subset K_2 \subset K_3 \subset \cdots$ is an increasing sequence of models of T then it is true that $K_\infty = \cup K_i$ is a model of T .

We now return to the lemma. Consider the $L(A_1)$ theory consisting of T_2 unioned with the set of all polynomial equalities and inequalities formed over $\mathbb{Z}$ using constants for elements of A_1 which are true in A_1 (the diagram of A_1 , $D(A_1)$). I claim $T_2 \cup D(A_1)$ is consistent. If $T_2 \cup D(A_1)$ is not consistent then there is a finite subset of $D(A_1)$, say $f_1(a_1, \cdots, a_n) = 0$, $f_2(a_1, \cdots, a_n) = 0, \cdots, f_K(a_1, \cdots, a_n) = 0$, $g_1(a_1, \cdots, a_n) \neq 0, \cdots, g_r(a_1, \cdots, a_n) \neq 0$ with $f_i(x_1, \cdots, x_n)$, $g_i(x_1, \cdots, x_n) \in \mathbb{Z}[x_1, \cdots, x_n]$ which is inconsistent with T_2 . But this is impossible because we then have the A-constructible affine set defined by these f_i's and g_i's that has a point rational over A_1 and not A . Thus $T_2 \cup D(A_1)$ is consistent. By

Godel's completeness theorem $T_2 \cup D(A_1)$ has a model B_1 which manifestly is a field containing A_1 and satisfying T_2. So $A \subset B \subset A_1 \subset B_1$.

We may now repeat this argument interchanging T_1 and T_2 alternately to obtain:

$$A \subset B \subset A_1 \subset B_1 \subset A_2 \subset B_2 \subset \cdots$$

where

$$A_i \models T_1$$

$$B_i \models T_2 \quad .$$

But then by what was said at the beginning of the proof $A_\infty = B_\infty$ is a model of both T_1 and T_2. Since T_i is complete $T_1 = T_2$. So we are done.

We recall that we are working with the language of fields, which contains symbols $+, \cdot, -, 0, 1, t_1, t_2, \cdots, t_n, \cdots$, where the presence or absence of $\{t_1, \cdots, t_n\}$ we have treated ambiguously. We will now assume that there are at most finitely many of these t_i and, for convenience, that they always denote a set of independent transcendentals over $\mathbb{Q}$. Thus $\mathbb{Q}(t_1, \cdots, t_n)$ or $\mathbb{F}_p(t_1, \cdots, t_n)$ will be called the core field of the $L(t_1, \cdots, t_n)$ theory of fields under consideration. We assume this theory is complete and model complete and if K is a field presented as a model of T we always assume a determination of $t_1, \cdots, t_n$ has to be made within K. We let the $\mathbb{F}_p(t_1, \cdots, t_n)$ or $\mathbb{Q}(t_1, \cdots, t_n)$ be denoted by $\mathbb{F}(T)$ or just $\mathbb{F}$.

LEMMA 3: Let T be as above and K_1 and K_2 model $\mathbf{T}$. Let A_i be the relative algebraic closure of $\mathbb{F}(T)$ in K_i. Then $A_1 \cong_{\mathbb{F}(T)} A_2$. We denote this unique object by $A(T)$.

<u>Proof:</u> If $f(x) \in \mathbb{F}(T)[x]$ is a one variable polynomial then f has a root in K_1 if and only if f has a root in K_2. A lemma of Ax now implies the desired result.

Now if T is not a theory of algebraically closed fields then $A(T)$ is not algebraically closed. To see that this is true suppose for the sake of contradiction that $A(T)$ were algebraically closed. Then by Tarski's Second Theorem $TACF(\mathbb{Z}[t_1, \cdots, t_n])$ is model complete. But then if $K \models T$ we have:

$$A(T) \subset K \subset \widetilde{K} = \text{algebraic closure of } K \quad .$$

Then by Lemma 2 $K \models TACF$, which is a contradiction.

<u>Definition:</u> Let $N(X) = N(x_1, \dots, x_n) \in \mathbb{F}(T)[x_1, \dots, x_n]$ and let $K \models T$. Then we say N is a <u>T-normic form</u> if it is a homogeneous polynomial with only the trivial zero over K (and so over any $K_1 \supset K$, $K_1 \models T$).

LEMMA 4: Let T be a model complete, complete theory of fields not having algebraically closed models. Then T has normic forms with arbitrarily many variables.

<u>Proof:</u> It is enough to find $N(x,y)$ with two variable, N T-normic, since $N(N(x_1, x_2), N(x_3, x_4))$ has four, etc. By the above lemma $A(T)$ is not algebraically closed. We break the proof into two cases:

<u>Case I:</u> $A(T)$ is not separably closed. Let $A(T)(\alpha)$ be a proper algebraic separable extension of $A(T)$ and E the normal closure of $\mathbb{F}(T)(\alpha)/\mathbb{F}(T)$, let $\mathbb{F}(T) = F$. Note E is not necessarily <u>Galois</u> over F. Then $E = P \otimes_F S$ where P is purely inseparable over F and S is Galois over F. Now $A(T) \not\supset S$ because $A(T)E \supset A(T)(\alpha)$ is not purely inseparable over $A(T)$. Let $S = F(\beta)$, and $N(x,y) = \prod_{\sigma \in Gal(S/F)} (x - \beta^\sigma y)$. $N(x,y)$ is normic over K since $\beta^\sigma \notin A(T)$, and so N is normic over T.

<u>Case II:</u> $A(T)$ is separably closed. Then $A(T)$ is not perfect so there is $a \in A(T)$ with $\alpha = \sqrt[p]{a} \notin A(T)$. Let notation be as in Case I. Then $S \subset A(T)$ since $A(T)$ is separably closed. Thus $A(T)(\alpha) \subseteq A(T)P$. Let $\beta \in P - A(T)$. Then $\beta^{p^k} \in F$ for sufficiently large k, let k be minimal in this regard. Let $N(x,y) = x^{p^k} - y^{p^k} \beta^{p^k} = (x - y\beta)^{p^k}$. Thus N is normic over K and so T-normic. This proves the lemma.

THEOREM 1: Let T be a complete, model complete theory of fields with $K \models T$. Let $\mathbb{F}(T)[X]$ be the polynomial algebra in the infinitely many variarbles

$X = \{x_1, x_2, x_3, \cdots\}$. Let $S(T) \subset \mathbb{F}(T)[X]$ be the set of all polynomials with no zero rational over K. Let $P \subset K[y_1, \cdots, y_n]$, P a prime ideal. Then there is $M \models T$ with $K[y_1, \cdots, y_n]_{/P} \subset M$ if and only if no element of $S(T)$ has a zero rational over the field of fractions of $K[y_1, \cdots, y_n]_{/P}$, which we denote by $K\{P\}$.

Proof: By Tarski's First Theorem we need only show the condition of the theorem is equivalent to $K\{P\} \models T_\forall$. Now if $K\{P\} \models T_\forall$ it is trivially true that no element of $S(T)$ has a zero rational over $K\{P\}$.

Now assume $K\{P\}$ does not satisfy $T_\forall$. Then there is a quantifier free formula, φ, which we may take to be a conjunction of disjunctions of the form $f_1(\vec{x}) = 0 \vee \ldots \vee f_a(\vec{x}) = 0 \vee g_1(\vec{x}) = 0 \vee \ldots \vee g_b(\vec{x}) = 0$ (with f_i, g_i in $\mathbb{F}(T)(X)$) such that $\forall \vec{x}\, \varphi(\vec{x})$ is in $T_\forall$ but $K\{P\} \models \exists \vec{x}\ \varphi(\vec{x})$, that is $K\{P\}$ satisfies a disjunction of conjunctions of type $f_1(\vec{x}) \neq 0 \wedge \ldots \wedge f_a(\vec{x}) \neq 0 \wedge g_1(\vec{x}) = 0 \wedge \ldots \wedge g_b(\vec{x}) = 0$.

We may replace each $f_i(\vec{x}) \neq 0$ by $z_i f_i(\vec{x}) - 1 = 0$ and thus assume that each conjunction in φ is of the form $g_1(\vec{x}) = 0 \wedge \ldots \wedge g_b(\vec{x}) = 0$.

We may also assume K is not algebraically closed and so apply Lemma 4 to obtain a T-normic form $N(w_1, \cdots, w_{b+c})$ with $c \geq 0$. We may thus replace each disjunct in $\varphi(\vec{x})$ by $N(g_1(\vec{x}), \cdots, g_b(\vec{x}), 0, \cdots, 0) = 0$. Taking the product of all of these disjuncts we obtain a polynomial, $D(x_1, \cdots, x_s) \in S(T)$ so that D has a zero rational over $K\{P\}$. This proves the theorem.

Definition: Let T be as above. We call a set $Q(T) \subset \mathbb{F}(T)[X]$ a T-determining set if the following set of conditions are met:

1. The homogenization of no element of Q has a non-trivial zero rational over any model of T, and no element of Q has a zero rational over any such model.
2. $f(x_1, \cdots, x_{j_n}) \in Q \Longrightarrow f(x_{k_1}, \cdots, x_{k_n}) \in Q$.

   That is, Q is closed under interchanging of variables within X.
3. Q is multiplicatively closed.

4. If $K \models T$ and $L \supset K$ is a field extension then $L \models T_\forall$ iff no element of Q has a zero rational over L.

Note that Theorem 1 assures us that there always exists a set of polynomials satisfying 2 through 4 of the definition for any model complete, complete theory.

THEOREM 2: Let T be a complete, model complete theory of fields with $Q \subset \mathbb{F}(T)[X]$ a T-determining set. Let $I \subset K[y_1, \cdots, y_n]$ with $K \models T$. Then $\sqrt[T]{I} = \mathcal{I}(V_K(I)) = \{f \in K[y_1, \cdots, y_n] \mid$ there exists $D(x_1, \cdots, x_s) \in Q$; $u_1(y), \cdots, u_s(y) \in K(y_1, \cdots, y_n)$; and $m \geq 0$, $n \in \mathbb{N}$ so that $f^m(y)\, D(u_1, \cdots, u_s) \in I\}$.

<u>Proof:</u> Call the set in the brackets in the statement of the theorem $R(I)$. Suppose, for the sake of a contradiction, $f \in R(I)$ but $f \notin \mathcal{I}(U_K(I))$ ($= \sqrt[T]{I}$ (by Theorem 1)). Then, by definition of $\sqrt[T]{I}$, there is $P \supset I$ with $K\{P\} \models T_\forall$ and with $f \notin P$. Since $K\{P\} \models T_\forall$ there is $M \models T$ so that $K\{P\} \subset M$. Since T is model complete, K models the first order sentence which maintains that there is a point, k, rational over K for which I vanishes but f does not - that is because M models this sentence with the generic point of P as witness. Let t a new variable and $a = (a_1, \cdots, a_n)$ an n-tuple of new variables. Let $f^m(y)\, D(u_1(y), \cdots, u_s(y)) = A(y) \in K[y_1, \cdots, y_n]$. Then $A(k + ta) \in K[a,t]$ and $A(k + ta) \neq 0$. Specialize a to $\bar{a} \in K^n$ so that $A(k + t\bar{a}) \neq 0$. Then $A(k + t\bar{a}) = R(t) \in K[t]$ and $R(0) = 0$.

Let $D(u_1(k + \bar{a}t), \cdots, u_{s-1}(k + \bar{a}t)) = B(t)$ where $\bar{a}$ is chosen so that the denominator of $\prod_i u_i(k + \bar{a}t)$ is not zero. Then $B(t)$ is a well-defined element of $K(t)$.

Write $u_i(k + \bar{a}t) = c_{-\ell_i} t^{-\ell_i} + \cdots$. Then the point $(u_1(k + \bar{a}t), \cdots, u_s(k + \bar{a}t), 1)$ in projective coordinates is the same as

$$(c_{-\ell_1} + t c_{-\ell_1+1} + \cdots\ t^{\ell_1 - \ell_s} c_{-\ell_s} + \cdots,\ t^{\ell_1}) = P(t) \quad \text{where for convenience we assume}$$

$\ell_1 \geq \ell_i$ $i > 1$. It follows that, since this latter point is well defined at $t = 0$, that if D^* is the homogenization of D that $D^*(P(0))$ is not zero. Therefore, if $D(u_1(k + \bar{a}t), \cdots, u_s(k + \bar{a}t) = b_{-j} t^{-j} + \cdots$ we see $j \geq 0$. Since

$f^m(k + \bar{a}t) B(t) = R(t) \in K[t]$ with $R(0) = 0$ we conclude that $f^m(k + \bar{a}t) = e_\eta t^\eta + \cdots$ with $\eta > 0$. This means $f(k) = 0$ contradicting the choice of k. We conclude $R(I) \subset \sqrt[T]{I}$.

Now suppose $f \notin R(I)$. We must produce a prime $P \supset I$ with $f \notin P$ and $K\{P\} \models T_\forall$ in order to conclude $R(I) \supset \sqrt[T]{I}$.

Consider the set $Q^*(f) \subset K[y_1, \cdots, y_n]$ consisting of all polynomials which also have the form $f^m(y) D(u_1(y), \cdots, u_s(y))$ for some $D \in Q$ and some $u_i(y) \in K(y_1, \cdots, y_n)$. Since Q is T-determining and $f \notin R(I)$ we see: (i) $Q^*(f) \cap I = \emptyset$ and (ii) $Q^*(f)$ is a multiplicatively closed subset of $K[y_1, \cdots y_n]$.

Consider $\mathfrak{R} = K[y_1, \cdots, y_n][Q^*(f)]^{-1}$, the localization of $K[y_1, \cdots, y_n]$ w.r.t. the set $[Q^*(f)]$. This is a ring containing the ideal $I[Q^*(f)]^{-1} = I^*$. If $I^* = \mathfrak{R}$ then $1 = p(y)/f^m D(u_1, \cdots, u_s)$ with $p(y) \in I$. So $I \cap Q^*(f) \neq \emptyset$, contradiction.

Choose $P^* \supset I^*$, P^* a maximal ideal of $\mathfrak{R}$. Let $P = K[y_1, \cdots, y_n] \cap P^*$. $\mathfrak{R}/P^*$ is a field. Suppose $D \in Q$ had a zero rational over $\mathfrak{R}/P^*$. Then $D(u_1, \cdots, u_s, 1)\ p(y)/[f^m(y) D_1(u'_1, \cdots, u'_s)] \in P^*$ where $f^m(y) D_1(u_1, \cdots, u_s) \in Q^*(f)$, $p(y) \in K[y_1, \cdots, y_n]$, $u_i \in \mathfrak{R}$ and $D(u_1, \cdots, u_s) \in P^*$. Hence $p(y) \in P$. Thus $D(u_1, \cdots, u_s) D_1(u'_1, \cdots, u'_s) f^m = p(y) \in P^* \cap Q^*(f)$, which contradicts the choice of P^*. This proves the theorem since $K\{P\} \subset \mathfrak{R}/P^*$ and $\mathfrak{R}/P^*$ satisfies $T_\forall$ by virtue of omitting all zeros of elements of Q.

§2. Some Specific Nullstellensätze

In each case we give a T-determining set, Q.

1. Real Closed Fields: Let Q be generated by substitution of variables and products from $\{1 + x_1^2 + \cdots + x_s^2\}$ then the Dubois-Reissler Nullstellensätz is obtained.

2. P-adically-closed Fields: Kochen has given a Nullstellensätz for TPACF which characterizes $\mathcal{J}(V_{Q_p^k}(I))$ in a manner similar to the one used above. However, instead of

T-determining polynomials, Kochen's Nullstellensätz makes use of rational functions of a special nature. It is of some interest to know that a Nullstellensätz using polynomials as in the above Theorem 2, rather than rational functions, exists - and we now show this to be true. We do not, unfortunately, give any closed form for such a set of TPACF-determining polynomials, but only show they must exist.

Let K be a p-adically closed field and $L \supset K$ a field extension which is not formally p-adic. By Theorem 1 and the fact that TPACF is model complete in the language of pure ring theory we know there is a polynomial $\mathfrak{F}(y_1, \cdots, y_s) \in Q[y_1, \cdots, y_s]$ so that $\mathfrak{F}$ has no zero rational over K but has a zero rational over L. (Unfortunately, $\mathfrak{F}$ may have zeros rational over K at infinity.)

We let $V(\mathfrak{F})$ be the set of zeros of $\mathfrak{F}$ in some large algebraically closed universal domain containing L. We let S(Y) be the singular locus of the algebraic set Y. Then $V(\mathfrak{F}) \supset S(V(\mathfrak{F})) \supset S(S(V(\mathfrak{F}))) \supset S^j(V(\mathfrak{F})) \supset \cdots$ is a finite chain, since $\mathbb{Q}[y_1, \cdots, y_s]$ is a Noetherian ring. The point p then occurs on some maximal $S^j(V(\mathfrak{F}))$. Then p is a simple point on $S^j(V(\mathfrak{F}))$ for some $j \geq 0$ (with $S^0(V(\mathfrak{F})) = V(\mathfrak{F})$). We denote by W' this maximal $S^j(V(\mathfrak{F}))$. Then W' is defined by polynomials with coefficients from $\mathbb{Q}$, W' has no point rational over $\mathbb{Q}_p$, and W' has a simple point rational over L. We let W be the $\mathbb{Q}$-irreducible component on which p lies.

We now appeal to the celebrated Hironaka Resolution of Singularities Theorem - HRST, [10]. Let $\overline{W}$ be the projective closure of W. By HRST we may find X a projective variety defined by polynomials $g_1, \cdots, g_\ell$ with rational coefficients and a rational map $\alpha : X \to \overline{W}$ so that α is a birational isomorphism defined by functions with coefficients from $\mathbb{Q}$ such that α is locally defined and invertible at $\alpha(p)$. Thus $\alpha(p)$ is a point of X rational over L. This may be done by virtue of p being a <u>smooth</u> point of $\overline{W}$.

We now observe that X has no points rational over K. To see this we assume that $K = \mathbb{Q}_p$ and that X has a point q rational over $\mathbb{Q}_p$. Since X is everywhere smooth it has, by virtue of the topological completeness of $\mathbb{Q}_p$, the structure of a $\dim(X)$ - $\mathbb{Q}_p$-manifold locally at q (where "locally" now refers to the p-adic topology). It is thus possible to find - by virtue of X's $\mathbb{Q}$-irreducibility - a point $p = p(\mathfrak{G}) \in \mathfrak{G}$ rational over $\mathbb{Q}_p$ and lying in $\mathfrak{G}$ for every $\mathbb{Q}$-Zariski open set, $\mathfrak{G}$. Since α is invertible on such an $\mathfrak{G}$ and the inverse image $\overline{W} - W$ is contained in a $\mathbb{Q}$-closed set this implies W has points rational over $\mathbb{Q}_p$ - contrary to assumption.

Thus if $K = \mathbb{Q}_p$ we see X has no points rational over K. For a general p-adically closed K the result follows from completeness of TPACF.

We thus have X which is cut out as the zero set of $g_1, \cdots, g_\ell$. Letting g_i have degree d_i and $D_i = \prod_{j \neq i} d_i$ we see X is cut out as the zero set of $g_1^{D_1}, \cdots, g_\ell^{D_\ell}$ and that $\deg(g_j^{D_j}) = \deg(g_k^{D_k})$.

By Krasner's lemma we may choose a simple extension $\mathbb{Q}(T) \supset \mathbb{Q}$ so that $[\mathbb{Q}(T) : \mathbb{Q}] = [\mathbb{Q}_p(T) : \mathbb{Q}_p] = [K(T) : K] = \ell$. Let $N(t_1, \cdots, t_\ell) = \prod_\sigma (t_1 + t_2\alpha + \cdots + t_\ell \alpha^{\ell-1})^\sigma$ where σ ranges over distinct embedding of T into the algebraic closure of $\mathbb{Q}$. $N(t_1, \cdots, t_\ell)$ has only the trivial zero over K. By our above observations $N(g_1^{D_1}, \cdots, g_\ell^{D_\ell})$ is a homogeneous form which has only the trivial zero rational over K but has a non-trivial zero rational over L.

We may now construct our TPACF-determining set easily as follows: Let $P \subset \mathbb{Q}[X]$ be the set of all forms having only the trivial zero rational over $\mathbb{Q}_p$. Let Q be the set of affine polynomials obtained by dehomogenizing the elements of P. It is immediately verified from the above discussion that Q is a TPACF-determining set.

This Q we have chosen above is obviously highly unsatisfactory and a direct construction giving a more usable set of polynomials would be of the utmost interest.

As mentioned earlier Roquette has proved a Nullstellensätz for $GF(q)((t))$ which is totally analogous to Kochen's p-adic Nullstellensätz. It is perhaps of some importance to realize that our methods shed no light on the existence of a Nullstellensätz for

GF(q)((t)) which makes use of polynomials as above.

This darkness may be attributed to two factors: it is unknown whether or not the theory of GF(q)((t)) is model complete in any reasonable language and - perhaps even worse - there is very little information on resolution of singularities for fields of positive characteristic. It is hoped that HRST is true for fields of characteristic $p \neq 0$ however even this would apparently not be enough for our purposes since the resolution we require is absolute, that is we require a non-singular X birationally isomorphic to $\overline{W}$ where everything is defined over the field of definition of $\overline{\mathfrak{W}}$ and where by non-singular we refer to the classical definition in terms of the rank of the Jacobian of X. This is important since it is this classical definition of non-singular that allows us to uniformize X as a local manifold around any point rational over the local field, $\mathbb{Q}_p$. It is well-known that such an absolute resolution of singularities is impossible in non-zero characteristic. (The hoped for characteristic p resolution theorem is formulated in terms of the normality of the local ring of X at each point. These two definitions of a non-singular X agree in characteristic 0 but do not agree over non-algebraically fields in characteristic $p > 0$.)

It would then appear to be of substantial interest just to show that a Nullstellensätz for GF(q)((t)) may be demonstrated for some set of determining polynomials (as opposed to Roquette's rational functions) with coefficients from GF(q)(t).

3. Separably Closed Fields with Ershov Invariant p^n [4]: In this case $[K : K^p] = p^n$. We may thus choose $t_1, \cdots, t_n$ independent transcendental over $\mathbb{F}_p$ so that $\{t_1, \cdots, t_n\}$ is a p-basis for K. Now if $M \supset K$ is a field in which $\{t_1, \cdots, t_n\}$ remains a p-independent then M may be embedded in a model of TSCF(p^n). Thus we are led to consider the polynomial

$\mathcal{J}(X) = \Sigma x^p_{(k_1, \cdots, k_n)} t_1^{k_1}, \cdots, t_n^{k_n}$. Now $\{t_1, \cdots, t_n\}$ remains p-independent in $M \supset K$ if and only if $\mathcal{J}(X)$ has only the trivial zero.

We now consider all polynomials obtained from $\mathcal{J}(X)$ by substitution of variables from $X = \{x_1 x_2 x_3 \cdots\}$ and then closing under products.

This yields the appropriate Q for $TSCF(p^n)$. For the case where $[K : K$

$[K : K^p] = \infty$ see §3.

4. Infinite Algebraic Extensions of $\mathbb{F}_p$, A_p: Let $Q_1 \subset \mathbb{F}_p[x_1]$ where $Q_1 = \{f(x_1) \in \mathbb{F}_p[y_1] \mid f(x_1)$ has only one variable, x_1, and no zero rational over A_p. Then Q_1 is a multiplicative semi-group. Let $Q_i \subset \mathbb{F}_p[x_i]$ be the set of polynomials in x_i obtained by replacing x_1 by x_i in $f(x_1) \in Q_1$. Let $Q = \bigoplus_i Q_i$, that is, let Q be the set obtained by taking arbitrary finite products of the elements of Q_i. Conditions 1-3 of the definition of a T-determining set are automatically fulfilled by Q.

Let $K \models T(A_p)$ and let $P \subset K[y_1, \cdots, y_n]$ be a prime so that no element of Q has a zero rational over $K\{P\}$. Then K is relatively algebraically closed in $K\{P\}$. Since K is perfect $K\{P\}$ is a regular extension of K. We need only now show there is a regular extension M of $K\{P\}$ for which every absolutely irreducible variety defined over $\mathfrak{m}$ has a zero rational over $\mathfrak{m}$, i.e. $\mathfrak{m} \models \Sigma_1$ and such that $\mathfrak{m}$ has the right number of extensions of each degree.

Toward this end we list all of the absolutely irreducible varieties defined over $K(P)$, $V_1, V_2, V_3, \cdots$. We then form a sequence of fields $M_1 \subset M_2 \subset M_3 \subset$ where M_i is obtained from $\mathfrak{m}_{i-1}$ by the adjunction of an $\mathfrak{m}_{i-1}$-generic point for V_i. Then $\mathfrak{m}_i$ is a regular extension of $\mathfrak{m}_{i-1}$. Let $\mathfrak{m}^{(1)} = \bigcup_{i \leq \omega} \mathfrak{m}_i$. We now repeat the process with $\mathfrak{m}^{(1)}$ in place of K and so obtain $\mathfrak{m}^{(2)}$, etc. Thus we obtain $K \subset \mathfrak{m}^{(1)} \subset \mathfrak{m}^{(2)} \subset \cdots$ where every absolutely irreducible variety defined over $\mathfrak{m}^{(i)}$ has a point rational over $\mathfrak{m}^{(i+1)}$. Let $\mathfrak{m}^{(\infty)} = \bigcup \mathfrak{m}^{(i)}$. Then $\mathfrak{m}^{(\infty)}$ is a regular extension of K and every absolutely irreducible variety defined over $\mathfrak{m}^{(\infty)}$ has a point rational over $\mathfrak{m}^{(\infty)}$. Let $\sigma \in Gal(\widetilde{A}_p/A_p)$ be a topological generator of $Gal(\widetilde{A}_p/A_p)$. Let $\mathfrak{m}$ be the field by $\sigma \otimes_{\mathfrak{m}^{(\infty)}} id_{\mathfrak{m}^{(\infty)}}$. Then $\mathfrak{m}$ is an algebraic extension of $\mathfrak{m}^{(\infty)}$ and so, by restriction of scalars, every absolutely irreducible variety defined over $\mathfrak{m}$ has a zero rational over $\mathfrak{m}$.

$Gal(\widetilde{\mathfrak{m}}/\mathfrak{m})$ is isomorphic to $Gal(\widetilde{A}_p/A_p)$ by restriction so the algebraic extensions of $\mathfrak{m}$ are exactly those given by A_p so $\mathfrak{m}$ is a regular extension of K with the correct extensions. Thus $M \models T(A_p)$.

5. $\underline{\widetilde{\mathbb{Q}}(\sigma) \text{ for almost all } \sigma \in Gal(\widetilde{\mathbb{Q}}/\mathbb{Q})}$: Here we form Q as in (4) except over $\mathbb{Q}$ instead of $\mathbb{F}_p$. The proof is then the same as in (4).

6. Pseudo-algebraically closed fields in general: Let K be a pseudo-algebraically closed field. Specifically, we are assuming that K is perfect and every absolutely irreducible variety defined over K has a point rational over K. We may then formulate and prove a Nullstellensätz for K (as opposed to the the theory of K) as follows.

Let $Q_i(K) \subset K[x_i]$ be the set of all polynomials in the one variable x_i with coefficients in K which have no zero in K. Let $Q(K) \subset K[X] = K[x_1, x_2, x_3, \cdots]$ be the set of all finite products of elements from the $Q_i(K)$. Then:

THEOREM 3: Nullstellensätz for a PAC field, K.

$\mathcal{I}(V_K(I)) = \{f \in K[y_1, \cdots, y_n] \mid \exists\, u_j(y) \in K(y)$ and $D(x_1, \cdots, x_r)$ so that $f^m(y)\, D(u_1, \cdots, u_r) \in I\}$.

Theorem 3 may be proved along the lines of the Nullstellensätze for infinite algebraic extensions of $\mathbb{F}_p$ and we will not go into detail here.

We may use Theorem 3 to throw a little light on a question of James Ax - first we need some definitions.

A form $F(x_0, \cdots, x_n)$ of degree d is said to be a C_1 form if $n > d$. A field, K, is said to be a C_1 field if every C_1 form with coefficients in K has a non-trivial zero over K. Ax has asked: Is a p.a.c. field a C_1 field?

We cannot answer this question but we can give a little information on the nature of a minimal counter example - if one exists.

By Theorem 4 below we know that if $F(x_0, \cdots, x_n) \in K[x_0, \cdots, x_n]$ is a form with only the trivial zero over K then

(a) $D(u_1, \cdots, u) = F(1, x_1, \cdots, x_n) G(x_1, \cdots, x_n)$ for some $D \in Q(K)$ and, $u_i \in K(x_1, \cdots, x_n)$ and $G \in K[x_1, \cdots, x_n]$. In this case we may take $D(y_1, \cdots, y_\ell) = \prod_i g_i(y_i)$ where we may take the $g_i(y_i)$ to be irreducible and we allow the possibility that $y_i = y_j$ for $i \neq j$.

Let $u_i = v_i/w_i$ with v_i and w_i being elements of $K[x_1, \cdots, x_n]$. Then (a) becomes (b) $\prod g_i^*(v_i, w_i) = \prod w_i^{\deg(g_i)} F(1, x_1, \cdots, x_n) G(x_1, \cdots, x_n)$ when it is cleared of denominators - where $g_i^*(y_i, z_i) = z_i^{\deg(g_i)} g_i(y_i/g_i)$. Using the irreducibility of the g_i we may extract factors of F from the left of (b) and obtain (c) $\prod_{i \in I} g_i^*(v_i, w_i) = F(1, x_1, \cdots, x_n)$ where $I \subset \{1, \cdots, \ell\}$.

We now suppose F is C_1 with degree d and that all C_1 K-forms of degree less than d have been shown to have non-trivial zeros rational over K.

Under this assumption we may deduce that $F(1, x_1, \cdots, x_n) = g_i(v_i, w_i) = g(v, w)$ for one polynomial g. This is so since a nonsingular general change of coordinates will insure that all variables occur in all factors $g_i(v_i, w_i)$, and a homogenization of each $g_i(v_i, w_i)$ must have a zero by the inductive assumption.

Let $\delta(K)$ be the smallest integer m such that K has an algebraic extension of degree m. Then we may show that $\deg(g) = \delta(K)$.

To see this we assume $\deg(v) \geq \deg(w)$ and we let $v = L(v) + R(v)$ where $L(v)$ is the leading homogeneous term of v. We let $L(w)$ be the term (possibly 0) of w of degree equal to the degree of v.

We let $h(z_1)$ be a polynomial (in one variable) of degree $\delta(K)$ with no root in K. Then $h^*(z_1, z_2)$ has only the trivial zero rational over K.

We may assume all variables from $\{x_1, \cdots, x_n\}$ actually occur in $L(v)$. Therefore, if $\deg(g) > \delta(K)$ $h^*(L(v), L(w))$ is a C_1 form with degree less than d. It would follow that $L(v)$ and $L(w)$ have a common zero, k, rational over K. If this were the case we could expand $F(1, x_1, \cdots, x_n) = g^*(v, w) = g^*(L(v), L(w)) +$

+ [terms of lower order]. It follows that $F(x_0, \cdots, x_n) = g^*(L(v), L(w)) +$ $x_0 H(x_0, \cdots, x_n)$ for some $H \in K[x_0, \cdots, x_n]$. Then $(0, k_1, \cdots, k_n)$ would be a non-trivial zero for $F(x_0, \cdots, x_n)$, contrary to assumption.

We thus conclude $\deg(g) = \delta(K)$.

We can now show that if there is a p.a.c. field which is not C_1 then there is a p.a.c. field, k, which is not C_1 and for which $\delta(L) = 2$.

To see this let K be p.a.c. and not C_1. Assume that the characteristic of K is not 2 for convenience. It is no restriction to assume K is countably (by, say, the Lowenheim-Skolem Theorem). Let $K^* > K$ be a non-principal ultrapower of K.

Let $t \in K^* - K$. Consider the field $K(t^2) = L_0$. L_0 has an extension of degree 2. Let $V_1, V_2, V_3, \cdots$ be a listing of isomorphism types of all absolutely irreducible L_0-varieties. Inductively we define L_i to be obtained from L_{i-1} by adjoining from K^* an L_{i-1} generic point for V_1. This may be done because K^* is w_1-saturated and p.a.c. Let $L_w = \cup L_i$. Let $L_0^{(1)}$ be the perfect closure of L_w. We claim

(1) Every absolutely irreducible variety defined over L_0 has a point rational over $L_0^{(1)}$.

(2) $t \notin L_0^{(1)}$.

We know (1) is true by construction. We know (2) is true because $t \notin L_i$ for any i since L_i is relatively algebraically closed in L_{i+1} and t is algebraic over L_i. Therefore $t \notin L_w$.

Since $[L_w(t) : L_w] = 2 \neq p^r$ for any r we see $t \notin L_0^{(1)}$.

We now repeat the construction with $L_0^{(1)}$ instead of L_0 and obtain $L_0 \subset L_0^{(1)} \subset L_0^{(2)} \subset \cdots$. Let $L = \cup L_0^{(i)}$. By construction we see every absolutely irreducible variety defined over L has a point rational over L. Furthermore, L is obviously perfect - so it is p.a.c.. Since $t \notin L_0^{(i)}$ for any i, $t \notin L$ so L has a quadratic extension.

L is obviously not C_1 because no K-polynomial without a zero rational over K has one rational over L since $L \subset K^*$. We may therefore conclude that if there do exist non-C_1 p.a.c. fields that there exists one for which the minimal counter example has the form $F(x_0, \cdots, x_n) = H(x_0, \cdots, x_n)^2 - \alpha G(x_0, \cdots, x_n)^2$ where $\alpha \in K$ and $\sqrt{\alpha} \notin K$ and H and G have no simultaneous zero rational over K.

§3. Some Further Considerations

In this section we extend some observations due to Kochen and Robinson by combining their observations for specific cases with the general theory developed above.

Again, as before, let T be a complete, model complete theory of fields. Let Q be any T-determining set of polynomials. We are after a characterization of all polynomials from $K[y_1, \cdots, y_n]$ that have no zero rational over K, we call this set $N(K, n) = N$

THEOREM 4: N is the set of all divisors of $Q^*(1)$.

Proof: Clearly $N \supset Q^*(1)$, so N contains all divisors of $Q^*(1)$. Conversely, let $g(y) \in N$. Then $V_K(1) = V_K(g)$ so by Theorem 2 $1^r D(y(y)) \in (g)$ for some $r, D \in Q$ and $u(y)$. Thus $D(u(y)) = g(y)\, m(y)$ for some $m \in K[y]$. This is exactly the claim of the theorem, since $D(u(y)) \in Q^*(1)$.

The following theorem is a reason for developing the theory in terms of T.

THEOREM 5: There exists $D_{(n,d)}(x_1, \cdots, x_m) \in Q$ and an integer $B(n,d)$ so that if $g(y_1, \cdots, y_n)$ has total degree at most d and $g \in N(K,n)$ then there exist $u_1(y), \cdots, u_m(y)$ so that $m < B(d,n)$, $\deg(u_i) < B(d,n)$ so that $g(y) \mid D_{(n,d)}(u(y))$ with $D_{(n,d)}(u(y)) \in Q^*(1)$.

Proof: Suppose the contrary with the pair (n,d) constituting a counter example. Then, by the compactness theorem of first order logic there is a K^* which satisfies T and $g(y_1, \cdots, y_n) \in N(K^*, n)$ so that Theorem 4 fails for g. This contradiction proves

Theorem 5.

The important thing here is that Q is a <u>fixed</u> T-determining set and so is sufficient to represent polynomials with no zeros in <u>any</u> model of T. This is an important difference between the characterizations of, say, Theorem 2 and Theorem 3: There is no finiteness theorem if there is no underlying model complete theory.

Consider the field $\mathbb{F}_q((t))$ of formal Laurent series in t over the finite field, $\mathbb{F}_p$. Roquette [5] has proved a Nullstellensätz for this field and has given a very nice way of representing $N(\mathbb{F}_q((t)), n)$. We will not discuss the specifics of this result except to say that there is no known corresponding finiteness theorem for Roquette's Nullstellensätz. We will show now that the existence of a recursive $B(n, d)$ for $\mathbb{F}_q((t))$ would imply the <u>decidability</u> of $\mathbb{F}_q((t))$.

THEOREM 6: Let T be a theory of fields. Suppose there exists a set, $Q \subset \mathbb{F}\left[\{x_i\}_{i \in \omega}\right]$ so that for any model K of T $N(K, n)$ is the set of divisors of $Q^*(1)$. Then T is model complete.

<u>Proof:</u> As above we deduce the existence of $B(n, d)$ and $D_{(n,d)}$ as in Theorem 5. It therefore follows that we may, for each (n, d), give a specific closed condition for when a polynomial in n variables with degree at most d has a zero in a model of T. This immediately implies model completeness of T.

Thus, in the case of $\mathbb{F}_q((t))$, if $B(n, d)$ exists and is recursive in (n, d) we could easily write down a complete set of axioms for $\mathrm{Th}(\mathbb{F}_q((t)))$.

Finally we note that there are many fields for which the above Nullstellensätze work without the corresponding finiteness theorem. For example: any perfect pseudo-algebraically closed field and separably closed fields of infinite Ershov invariant.

REFERENCES

[1] Jacobson, Lectures in Abstract Algebra III, Princeton, N. J., Van Nostrand.

[2] Chang and Keisler, Model Theory, New York, American Elsevier, 1973.

[3] Kochen, Integer Valued Rational Function over the p-adic Numbers, Proceedings of the Symposia in Pure Mathematics XII, American Mathematical Society, Providence, R. I.

[4] Ershov, Fields with a Solvable Theory, Dokl. Akad. Nauk. SSSR 1967.

[5] Jarden, Elementary Statements over Large Number Fields, Trans. Amer. Math. Soc. 164, 1972, pp. 67-91.

[6] Cherlin, Model Theoretic Algebra, Springer-Verlag Lecture Notes in Mathematics, 521, New York, 1976.

[7] Dubois, D. W.; Efroymson, G., Algebraic Theory of Real Varieties I. Studies and Essays (Presented to Yu-Why Chen on his 60th Birthday, 1 April 1970) pp. 107-135, Math. Fes. Center. Nat. Taiwan Univ., Taipei 1970.

[8] Hartshorne, Algebraic Geometry, Springer-Verlag, New York.

[9] Macintyre, McKenna and vanden Dreis, Elimination of Quantifiers in Algebraic Structures, to appear in Advances in Mathematics.

[10] Hironaka, Resolution of Singularities of an Algebraic Variety Over a Field of Characteristic Zero, Ann. Math. vol. 79 (1964), pp. 109-326.

A Tree Analysis of Unprovable Combinatorial Statements

George Mills*

St. Olaf College, Northfield, MN 55057

Abstract

In this paper we introduce a combinatorial measure of largeness for finite sets of integers (called arboricity) which is a finitary analogue of König's Infinity Lemma. The existence of arboreal sets is shown to be very closely correlated with the Paris-Harrington version of Ramsey's Theorem and is therefore unprovable in Peano arithmetic. Arboricity can also be exactly characterized in terms of ordinal numbers less than ε_0. The ordinal characterization turns out to be only a very slight modification of the notion "α-large" introduced by Ketonen. Thus by "transitivity" this gives a simplified proof of bounds for the Ramsey-Paris-Harrington numbers previously derived by Ketonen and Solovay. Applications of arboreal sets to a selection of other combinatorial problems are also given.

---

*Research supported by NSF grant MCS-7905802

Section 1. Introduction

In this paper we introduce a hierarchy of combinatorial statements which are finitary analogues of König's Infinity Lemma. Thus for each $c \in N$ the statement that a set of integers A is c-arboreal means roughly that any tree built out of A in a certain way must contain either a "long" path or a node with "many" immediate successors. It is possible to iterate the notion of arboreal and thus consider sets which are e-fold c-arboreal. A generalization leads to the notion of e-fold c-f(x)-arboreal where $f: N \to N$.

Arboricity proves to be an effective tool in obtaining sharp numerical estimates in combinatorial problems which otherwise lend themselves to cruder treatment using compactness arguments and König's Infinity Lemma. Several such applications will be given in this paper.

The original motivation for studying arboricity was to obtain information about the Paris-Harrington version of Ramsey's Theorem. Recall that the notation $A \to (*)^e_c$ means that for any partition ("coloring") F of the e-element subsets of A into c classes ("colors"), $F : [A]^e \to c$, there exists a subset $H \subseteq A$ such that

(1) F is constant on $[H]^e$ (H is homogeneous or monochromatic),

(2) $|H| \geq \min H$ (H is relatively large), and

(3) $|H| \geq e + 1$ (H is nontrivial).

Harrington, using ideas of Paris and Kirby-Paris, showed that the variant

$$\forall e \, \forall c \, \exists A \; A \to (*)^e_c$$

of Ramsey's Theorem is not provable in Peano arithmetic, although it is in fact a true statement. Our first goal in this paper will be to establish a very close correlation between the statements "$A \to (*)^{e+1}_c$" and "A is e-fold c-f(x)-arboreal" for suitable functions $f(x)$. In this way we get fairly sharp upper and lower bounds (in terms of arboricity) for the minimal "sizes" of sets A such that $A \to (*)^{e+1}_c$. In other words we have a Ramsey number calculation of sorts. A corollary is that the statement

$$\forall e, c \exists A \text{ A is e-fold c-arboreal,}$$

while true, is not provable in Peano arithmetic. These results will be proved in Sections 2 and 3 of the paper.

In general it is a hopeless task to try to calculate exactly any but the simplest Ramsey numbers. Surprisingly the situation is not so difficult when it comes to arboreal sets. The key to the calculation of "arboreal numbers" is the observation that there is a certain canonical way to (try to) build a tree that is neither too long nor too fat (as determined by the parameters e, c and $f(x)$). We then prove a theorem to the effect that the canonical way is always the most efficient way: If a set A is not e-fold c-f(x)-arboreal, then in fact the canonical construction produces a tree witnessing this fact.

For the proof of this theorem we introduce a small dose of machinery involving ordinal numbers below ε_0. The upshot is an <u>exact</u> characterization of e-fold c-f(x)-arboreal sets A in terms of the ordinal $\omega^c_e = \omega^{\omega^{\cdot^{\cdot^{\cdot^{\omega^c}}}}}$ (a stack of e omegas toped by a c, ordinal exponentiation). For $f(x) = x$ the characterization turns out to be only a very slight modification of the notion of A being ω^c_e-large, introduced by J. Ketonen. (The modification comes in the choice of the fundamental sequences $\{\alpha\}$ for ordinals $\alpha < \varepsilon_0$.) With this modification we have that A is e-fold c-f(x)-arboreal iff $\{\omega_e^c\}(f(a_1),f(a_2),\ldots,f(a_{t-1})) = 0$, where $A = \{a_1,a_2,\ldots,a_t\}$ listed in increasing order, and for $i \geq 1$ $\{\omega_e^c\}(f(a_1),\ldots,f(a_{i+1})) = \{\{\omega_e^c\}(f(a_1),\ldots,f(a_i))\}(f(a_{i+1}))$, an ordinal. For $f(x) = x$, this is notationally essentially Ketonen's definition of A being ω_e^c-large. If we begin with this definition, then our contribution beyond Ketonen and Solovay's work amounts to showing that the ω_e^c-large sets are precisely the e-fold c-arboreal sets. This combinatorial fact together with the generalization to incorporate f(x) considerably increases the versatility of ω_e^c-large sets and reduces the number of technical lemmas needed to establish upper and lower bounds for Ramsey functions. These results are proved in Section 4.

In Section 5 we determine for the record the proof theoretic strength of aboricity. This does not require much further work, as we base our analysis on earlier results of Paris concerning the derivability of the Paris-Harrington partition relation. The conclusion is that for fixed positive integers n and c, Σ_n-induction can prove $\forall a \exists b[a,b]$ is n-fold c-arboreal, but that Σ_n-induction cannot prove $\forall a,c \exists b[a,b]$ is n-fold c-arboreal.

The remaining sections of the paper are devoted to further applications of arboricity. In Section 6 we study the family of "small sets" $T_c^e = \{A \subseteq N \mid A \not\to (*)_c^e\}$. T_c^e forms a countably branching tree when ordered by end-extension: $A \precneq B$ iff $a < b$ for all $a \in A$, $b \in B-A$. Moreover since every infinite subset $X \subseteq \omega$ contains a finite initial segment $A \subset X$ such that $A \to (*)_c^e$, T_c^e has no infinite branches. We can therefore ask what is the ordinal rank of T_c^e. By use of the ordinal characterization of arboricity we calculate the exact value $o(T_c^2) = \omega^c$ and produce reasonably sharp bounds for $o(T_c^e)$ when $e \geq 3$.

In Section 7 we apply arboricity to the problem of determining the strength of iterating the Paris-Harrington partition relation. We place arboreal bounds on the minimal sizes of n-dense sets as originally defined by Paris. Our results here are definitely less than optimal. The main obstacle is our inability to construct "bad" partitions with exponent $e = 2$. Thus the results presented in Section 7 illustrate not only the effectiveness of arboreal methods, but also some of their limitations. We hope publication of these partial results will stimulate further research on these interesting combinatorial problems.

Finally in Section 8 we apply arboricity to derive two combinatorial statements previously studied by H. Friedman and L. Kirby respectively. Friedman formulated a combinatorial statement "B" and showed it to be related to the Ackermann hierarchy. We rederive this result and also obtain more explicit upper bounds than Friedman by showing statement "B" to be a consequence of the existence of 1-fold arboreal sets at the appropriate level. Kirby defined the notion of a finite set A being n-flippable and showed that the existence of n-flippable sets for all n cannot be proved in PA. We show that if A is n-fold $(n+3)$-arboreal, then A is n-flippable. Neither of these results is particularly difficult, but they illustrate that arboricity can be applied outside of the immediate context for which it was designed.

Section 2. Arboricity and exponent two.

In this section we define the notion "A is $c - f(x)$-arboreal" and show that it is intimately connected to the relation $A \rightarrow (*)^2_c$. The letter "A" will always denote a finite subset of $\mathbb{N}$ and we will frequently write A as an indexed set $A = \{a_1, a_2, \ldots, a_t\}$ with the tacit understanding that $a_1 < a_2 < \ldots < a_t$.

2.1 Definition. An A-tree is a tree T whose nodes are all the elements of A and such that aTb implies $a < b$.

For example Figure 1 (i) is not a [2,15]-tree because 3T2 but $3 \not< 2$, and also because 10 does not appear in it. Figure 1 (ii) is a [2,15]-tree.

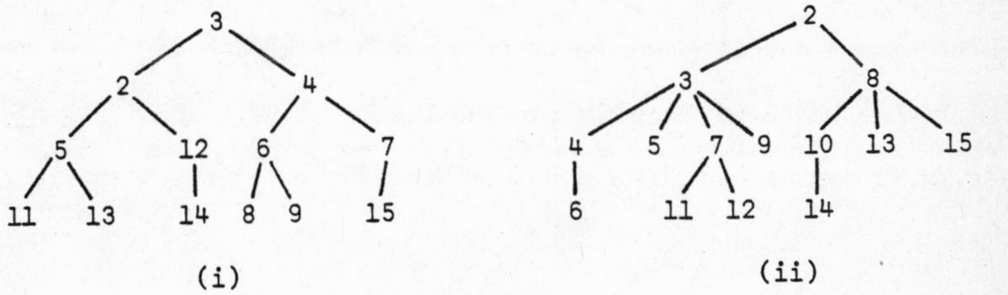

Figure 1

To be a bit more formal, an A-tree T is a transitive, irreflexive binary relation on A such that for all $a,b \in A$

(1) $aTb \rightarrow a < b$

(2) T_a is linearly ordered by T, where $T_a = \{x \in A | xTa\}$, and

(3) there is a unique $r \in A$ with $T_r = \phi$ (r is called the root of T).

In our diagrams we draw our trees with their roots at the top. By (1) the root must always equal minA. For $a \in A$ the rank of a (in T) is defined to be $|T_a|$. Thus the root has rank zero and all other elements have positive rank. We say a is (the) immediate predecessor of b iff aTb and for no c aTcTb. There is a natural 1-1 correspondence between A-trees and "regressive functions" $p: A-\{minA\} \to A$ (regressive meaning $p(a) < a$ for all $a \in A-\{minA\}$). Given T let p(a) = immediate predecessor of a in T. Given p define T by aTb iff $a = p(p(\ldots p(b)\ldots))$ for some (positive) number of applications of p. Most constructions of A-trees will be given by defining the predecessor function p inductively on A-{minA}. We say b is an immediate successor of a (in T) iff a is the immediate predecessor of b. Let $IS_T(a) = \{b | b$ an immediate successor of a in T$\}$. Thus instead of saying "let p(b) = a" we will often say "put b into $IS_T(a)$." It is very natural to think of A-trees as being constructed in a temporal sequence in this way. By a path in T we mean a subset $P \subseteq A$ which is linearly ordered by T. A path P is complete iff no proper superset of P is a path. We refer to $|P|$ as the length of P.

König's Infinity Lemma says that if A is large (infinite) and T doesn't branch too much (each node has only finitely many immediate successors), then T must have a long (i.e., infinite) path. An example of a finite version of this is: If A has 2^n elements and T is 2-branching, then T has a path of length $\geq n + 1$. In this paper we investigate a more general version, of which this turns out to be a special case.

2.2 Definition. An A-tree T is small-branching iff each $a \in A$ has $\leq a$ immediate successors in T ($|IS_T(a)| \leq a$).

For example, Figure 1 (ii) is not small-branching because 3 has four immediate successors. Figure 2 is small branching.

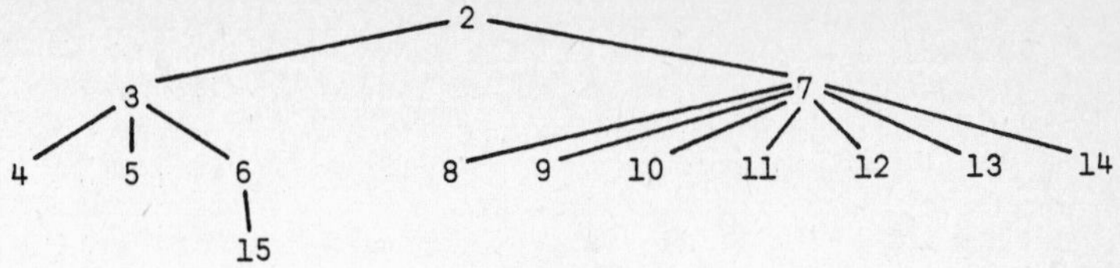

Figure 2

<u>Question</u>: How large must A be to force every small-branching A-tree to have a path of length ≥ 4? By deleting 15 from Figure 2 we can see that [2,14] is not large enough. Analysis shows that [2,15] <u>is</u> large enough: not just Figure 2 but every small-branching [2,15]-tree has a path of length ≥ 4. However, there do exist small-branching [3,16]-trees, in fact a [3,38]-tree with no path of length 4:

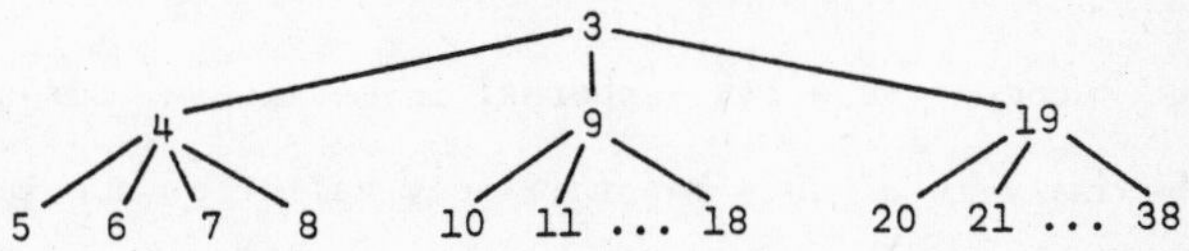

Figure 3

So the answer to our question certainly doesn't just depend on the cardinality of A. Well, if you can't answer a question, make it a definition. Let $c \in \mathbb{N}$.

<u>2.3 Definition</u>. A is <u>c-arboreal</u> iff every small-branching A-tree has a path of length $\geq c + 2$.

The reader should check that $[2, 41 \cdot 2^{39} - 2]$ is not 3-arboreal. It will follow from later results that $[2, 41 \cdot 2^{39} - 1]$ <u>is</u> 3-arboreal, i.e., maximal counterexamples are unique and follow a canonical pattern.

To make the concept of arboricity more versatile we generalize the definition. Suppose $f:N \to N$.

2.4 Definition. An A-tree T is f(x)-small-branching if every $a \in A$ has $\leq f(a)$ immediate successors in T ($|IS_T(a)| \leq f(a)$). A is c - f(x)-arboreal iff every f(x)-small-branching A-tree has a path of length $\geq c + 2$.

Thus c-arboreal is the special case in which f is the identity function. We use a "dummy" variable x in identifying functions. Thus "f(x)" means the same as "f" or "$\lambda x f(x)$."

2.5 Convention. Henceforth "f(x)" will always denote a monotone nondecreasing function $f:N \to N$, i.e., $a \leq b$ implies $f(a) \leq f(b)$.

Though the concept c - f(x)-arboreal is meaningful for arbitrary functions, the analysis of this paper is only valid for monotone nondecreasing ones. As an example take the constant function $f(x) = 2$. Then an f(x)-small branching tree is simply a binary tree. Our earlier remark amounts to saying that if $|A| \geq 2^n$, then A is (n-1)-2-arboreal. (The converse holds too.)

We remark that König's Infinity Lemma can be used to prove that for any c and f(x) and any infinite $I \subseteq N$ there exists a finite $A \subseteq I$ such that A is a c - f(x)-arboreal. A much more precise characterization of c - f(x)-arboreal sets will be given in Section 4.

We now give arboreal upper and lower bounds for the minimal "sizes" of sets A such that $A \to (*)^2_c$. For this purpose define 2^x_y by the equations $2^x_0 = x$ and $2^x_{y+1} = 2^{2^x_y}$. Thus 2^x_y is "a stack of y twos topped by an x." We recall from ordinary Ramsey theory that for all

$a,c \geq 2$, $c^{ca} \to (a)^2_c$. (Using a well known type of Ramsey argument one can easily prove by induction on $k = a_1 + a_2 + \ldots + a_c$ that in fact $c^{k-2c+1} \to (a_1,a_2,\ldots,a_c)^2$.)

2.6 Theorem. Let $c \geq 2$. Let $r(x)$ be a monotone nondecreasing function such that $r(x) \to (x-1)^2_c$ and $2cr(x) \leq 2^x_x$ for all $x \geq c$, $x \geq 4$. ($r(x) = c^{cx-1}$ will do.) Then for every $A \subseteq \mathbb{N}$ with $\min A \geq 4$ each of the following conditions implies the one after it.

(1) A is $(c+3)$-arboreal and $\min A \geq c$.

(2) A is $c - cr(x)$-arboreal.

(3) $A \to (*)^2_c$.

(4) A is $c - (x-2)$-arboreal.

(5) A is $(c-1)$-arboreal.

Note: For $c = 1$, (3) and (4) are both equivalent to $|A| \geq \min A$.

Proof of Theorem. Let $A = \{a_1,a_2,\ldots,a_t\}$.

(4)$\Rightarrow$(5). Assume (4) and let T be a small-branching A-tree. Let T' be obtained from T by removing a_1,a_2 and putting a_{i+2} in place of a_i for $i = 1,2,\ldots,t-2$. The tree T''

T'':

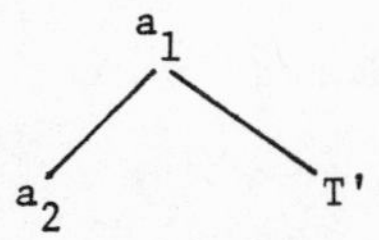

is $(x-2)$-small-branching (since $a_1 \geq 4$ and $a_{i+2} - 2 \geq a_i$), hence by (4) it has a path of length $\geq c + 2$. So T must have a path of length $\geq c + 1$.

(3)$\Rightarrow$(4). Suppose A is not $c - (x-2)$-arboreal. Let T be an $(x-2)$-small-branching A-tree with no path of length $\geq c + 2$. For $a,b \in A$

define $\widehat{ab} = \mathrm{glb}\{a,b\}$ in T. Define $F(a,b)$ = rank of $\widehat{ab}$. Then $F:[A]^2 \to c$ because the rank of $\widehat{ab}$ is one of $0,1,2,\ldots,c-1$. Now suppose $H \subseteq A$ is F-homogeneous. Let $a = \mathrm{glb}H$. Then $a \leq \min H$ and there exist distinct $x,y \in H$ with $\widehat{xy} = a$. Consequently $F(u,v)$ = rank of a, for all distinct $u,v \in H$, so H can contain at most one node at or below each immediate successor of a. Thus even if $a \in H$ we have $|H| \leq 1 + (a-2) < a \leq \min H$. Thus $A \not\to (*)^2_c$.

(2)⇒(3). Assume A is c - cr(x)-arboreal, where $r(x) \to (x-1)^2_c$. To prove $A \to (*)^2_c$, suppose we are given $F:[A]^2 \to c$. We will define a certain A-tree T which by construction cannot have any path of length $\geq c + 2$. Hence T cannot be cr(x)-small-branching. From this fact we will extract a relatively large homogeneous set for F.

We describe T by a construction which takes place in stages. At stage 1 we "place" a_1 at the root of T. Now suppose stage n is completed and we have placed $a_1,a_2,\ldots,a_n$ on the tree (i.e., we have determined the relation $T\restriction\{a_1,\ldots,a_n\}$). For $j \leq n$ we say a_j is <u>available for a_{n+1}</u> (in T) iff for all $i < j$ such that $a_i T a_j$ we have $F(a_i,a_j) = F(a_i,a_{n+1}) \neq F(a_j,a_{n+1})$. Certainly a_1 is always available for a_{n+1}. Pick an a_k of maximal rank in $T\restriction\{a_1,\ldots,a_n\}$ such that a_k as well as a_j is available for a_{n+1} in T for all j such that $a_j T a_k$, and place a_{n+1} as an immediate successor to a_k. This completes stage $n + 1$, and by continuing we define all of T.

By the inequality involved in the definition of "available" we see that no node in T can have rank $c + 1$. (If a_{n+1} had rank $c + 1$ then there would be $c + 1$ a_i's with $a_i T a_{n+1}$. But $a_i T a_j T a_{n+1}$ implies $F(a_i,a_{n+1}) \neq F(a_j,a_{n+1})$, requiring $c + 1$ colors.) Thus T has no path of length $\geq c + 2$, so, since A is $c - cr(x)$-arboreal we have $|IS_T(a)| > cr(a)$ for some $a \in A$. For each $i < c$ let $X_i = \{x \in IS_T(a) | F(a,x) = i\}$. By the pigeonhole principle $|X_t| \geq r(a)$ for some $t < c$. Since $r(a) \to (a-1)^2_c$ we can find $H \subseteq X_t$, H homogeneous for F and $|H| \geq a - 1$. If $F''[H]^2 = \{t\}$, then $\{a\} \cup H$ is homogeneous for F and $|\{a\} \cup H| \geq a = \min\{a\} \cup H$. Since $a \geq 3$ we see that $\{a\} \cup H$ is relatively large and nontrivial.

Suppose $F''[H]^2 = \{s\}$ for some $s \neq t$. Consider any two $x,y \in H$, $x < y$. By construction a_j was available for y for all $a_j T a$ and for $a_j = a$. But since a had maximal rank with this property, x was not available for y. Thus for some $a_i T x$ we have

$$\text{not-}(F(a_i,x) = F(a_i,y) \neq F(x,y)) \qquad (*)$$

Now $a_i \neq a$ (since $F(a,x) = t = F(a,y) \neq s = F(x,y)$), so $a_i T a$. Now since a was available for both x and y we have $F(a_i,a) = F(a_i,x) \neq F(a,x)$ and $F(a_i,a) = F(a_i,y) \neq F(a,y)$. Therefore $F(a_i,x) = F(a_i,y)$. By (*) we must have $F(a_i,y) = F(x,y) = s$. If z is any third element of H then since $a_i T a T z$ and a was available for z, we have $F(a_i,z) = F(a_i,a) = F(a_i,y) = s$. Thus $H \cup \{a_i\}$ is homogeneous to "color" s and is relatively large and nontrivial. This proves (2) $\Rightarrow$ (3).

For the proof of (1) $\Rightarrow$ (2) we require two lemmas. We first make the observation that if A is not $c - f(x)$-arboreal and $B \subseteq A$, then B is not $c - f(x)$-arboreal. Though this is fairly obvious, we will prove it rigorously when we deal with iterated arboricity (Lemma 3.2).

<u>2.7 Lemma</u>. Let $c,d \geq 1$ and suppose A is $(c+d)$-arboreal with $\min A \geq 3$. Suppose for all $x \geq \min A$ $A \cap [x, 2f(x)-1]$ is not d-arboreal. Then A is c-$f(x)$-arboreal.

<u>Proof</u>. Let T be an $f(x)$-small-branching A-tree. We must find a path of length $\geq c+2$ in T. Let $A = \{a_1, a_2, \ldots, a_t\}$, listed in increasing order, and define elements $a_{i_0}, a_{i_1}, \ldots, a_{i_s}$ and subsets $A_1, A_2, \ldots, A_{s+1}$ of A so that $i_0 = 1 < i_1 < i_2 < \ldots < i_s \leq t$, and so that for $j = 0,1,\ldots,s$, $A_{j+1} = \{a_x \mid i_j < x < i_{j+1}\}$ and

(1) if a_j is an endpoint in T, then $A_j = \phi$ (i.e., $i_j = i_{j-1}+1$), and

(2) if a_j is not an endpoint in T, then i_j is the minimal $i > i_{j-1}$ such that $a_i \geq 2f(a_{i_{j-1}})$. (If no such i exists, then i_j is not defined.) Observe that $a_j \leq a_{i_{j-1}}$ for $j \geq 1$.

Let $s = \max\{j \mid i_j$ is defined$\}$. Now for each $j \leq s$ in case (2) we have $A_j \subseteq [a_{i_{j-1}}, 2f(a_{i_{j-1}}) - 1] \cap A$, so by hypothesis A_j is not d-arboreal. Let T_j be a small-branching A_j-tree with no paths of length $d+2$. (For a_j an endpoint, $T_j = \phi$.) Now define T' by substituting a_{i_j} for a_j in T (and deleting a_j for $j > s$. Note that if $a_j T a_k$ and $k \leq s$ then $j \leq s$.) Now define T'' from T' by attaching T_k to a_{i_j} whenever a_{i_k} is an immediate successor of a_{i_j} in T'. (This incorporates T_k for $2 \leq k \leq s$.) Thus a fragment

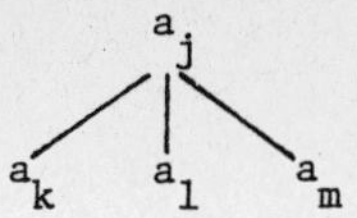

in T becomes

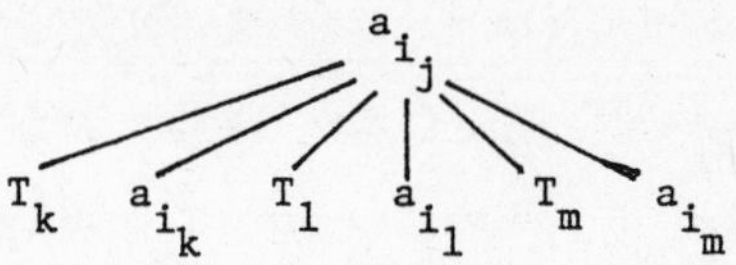

in T''. We finally build T^* from T'' as in this picture:

T^*:

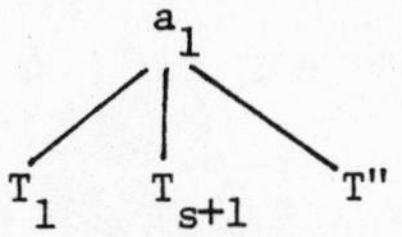

<u>Claim</u>. T^* is a small-branching A-tree.

<u>Proof</u>. Clearly all nodes $a_{i_0}, a_{i_1}, \ldots, a_{i_s}$ and all elements of $A_1, A_2, \ldots, A_{s+1}$ have been incorporated in T^*. We must show for all $a \in A$ that $|IS_{T^*}(a)| \leq a$ and $a < b$ for all $b \in IS_{T^*}(a)$. For $a = a_1$ we have $|IS_{T^*}(a_1)| \leq 3 \leq a_1$ and $a_1 < b$ for all $b \in A - \{a_1\}$. If $a \in A_j$ for some $j \leq s + 1$, then $IS_{T^*}(a) = IS_{T_j}(a)$ so the result is true by choice of T_j. Finally if $a = a_{i_j}$ for some $j \geq 1$ then

$$IS_{T^*}(a_{i_j}) = \{a_{i_k} | a_k \in IS_T(a_j)\} \cup \{\text{root of } T_k | a_k \in IS_T(a_j)\} .$$

Well for such k the root of T_k is $a_{i_{k-1}+1}$. Since $j < k$, we have

$i_j \le i_{k-1}$ and $a_{i_j} \le a_{i_{k-1}} < a_{i_{k-1}+1} \le a_{i_k}$. Also

$|IS_{T*}(a_{i_j})| \le 2|IS_T(a_j)| \le 2f(a_j) \le 2f(a_{i_{j-1}}) \le a_{i_j}$ since $a_j \le a_{i_{j-1}}$

and f is monotone nondecreasing. This proves the claim.

Now let P be a complete path in T* of length $\ge c + d + 2$ (since A is (c+d)-arboreal). The first element of P must be a_1, at most d + 1 elements of P belong to any T_k, and no two elements of P can belong to different T_k's. If no elements of P belong to any T_k, then $P \cap T'$ forms a path in T' of length $\ge c + d + 1 \ge c + 2$. This translates back to a path in T of length $\ge c + 2$.

On the other hand if $P \cap T_k \ne \phi$ for some k, then still $|P \cap T_k| \le d + 1$, so $|P \cap T'| \ge c$. Say $a_{i_j} = \max P \cap T'$, so a_{i_j} is the immediate predecessor of a_{i_k} in T'. By construction this means a_k is not an endpoint in T, say $a_k T a_m$. Let Q be the path in T corresponding to the path $P \cap T'$ in T'. Thus $a_j = \max Q$ and $Q \cup \{a_k, a_m\}$ is a path in T of length $\ge c + 2$. This completes the proof that A is c-f(x)-arboreal, and of the Lemma 2.7.

We remark that Lemma 2.7 is optimal in the sense that "(c+d)-arboreal" cannot be replaced by "(c+d-1)-arboreal." As an example let $A = [a, 2_2^a a]$ where $a \ge 2$. For all $x \ge 2$ $[x, 2 \cdot 2^x - 1]$ is not 2-arboreal and A is 3-arboreal (3=2+2-1), but A is not $2\text{-}2^x$-arboreal. For a verification of the negative claims we refer to the following lemma. The positive claim must await the treatment of canonical trees and calculations of "arboreal numbers."

<u>2.8 Lemma</u>. Suppose for all $x \geq M$, $[x, g(x)]$ is not c-f(x)-arboreal. Then for all $x \geq M$ $[x, g^{f(x)}(x)]$ is not (c+1)-f(x)-arboreal. ($g^i = g \circ g \circ \ldots \circ g$, composition i times).

<u>Proof</u>. Given $x \geq M$, define $x_i = g^i(x)$ for $i = 0,1,\ldots,f(x)$. By hypothesis $[x_i+1, x_{i+1}]$ is not c-f(x)-arboreal, so let T_i be an f(x)-small-branching $[x_i+1, x_{i+1}]$-tree with no path of length $\geq c + 2$. Now form T as pictured

T:

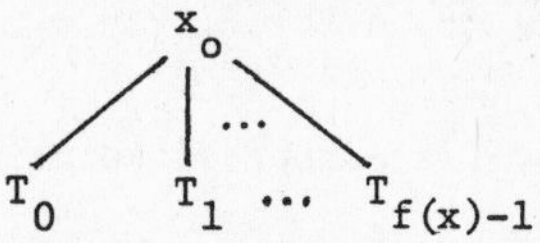

Clearly T is an f(x)-small-branching $[x_0, x_{f(x)}]$-tree with no path of length $\geq c + 3$. But $x_0 = x$ and $x_{f(x)} = g^{f(x)}(x)$, so $[x, g^{f(x)}(x)]$ is not (c+1)-f(x)-arboreal, as required.

Note that it follows from this lemma that for any primitive recursive function $h(x)$ there exists d such that for all $x \geq 2$ $[x, h(x)]$ is not d-arboreal. This is because $[x, 2x]$ is not 1-arboreal, and by considering the sequence $g_0(x) = 2x$, $g_{i+1}(x) = g_i^x(x)$ we obtain a function g_d which dominates h for $x \geq 2$.

We now return to the proof of Theorem 2.6 part (1)$\Rightarrow$(2). Clearly $[x, 2x]$ is never 1-arboreal (easy construction). By Lemma 2.8, $[x, x2^x]$ and hence $[x, 2^x]$ is never 2-arboreal. One more application shows $[x, 2_x^x]$ is never 3-arboreal. Now for $x \geq 4$ and $x \geq c$ we have by hypothesis $2cr(x) \leq 2_x^x$, so $[x, 2cr(x)-1]$ is not 3-arboreal. By Lemma 2.7 then, if A is (c+3)-arboreal and $\min A \geq 4, c$, then A is c-cr(x)-arboreal. QED.

Section 3. Iterated arboricity and higher exponents

Our objective in this section is to extend the results of Theorem 2.6 to higher exponent partition relations. That is we seek (arboreal) bounds for the minimal "sizes" of sets A such that $A \to (*)^e_c$. To carry out this program we define a notion of iterated arboricity. First, for any set of integers X, let $X^o = X - \{\max X\}$.

3.1 Definition. A is 0-fold c-f(x)-arboreal iff $|A^o| \geq c$. A is (e+1)-fold c-f(x)-arboreal iff for every f(x)-small-branching A-tree T there exists a path P in T such that P^o is e-fold c-f(x)-arboreal.

Observe that 1-fold c-f(x)-arboreal coincides with c-f(x)-arboreal, since $|P^{oo}| \geq c$ implies $|P| \geq c + 2$. For example the reader can check that for the constant function $f(x) = 2$ a set A is e-fold c-2-arboreal iff $|A| \geq 2^{c+1}_e$. (We mentioned earlier the special case $e = 1$.)

3.2 Lemma. If $A = \{a_1,\ldots,a_t\}$ is not e-fold c-f(x)-arboreal and $B = \{b_1,\ldots,b_s\}$ with $s \leq t$ and $f(a_i) \leq g(b_i)$ for $i = 1,2,\ldots,s$, (in particular if $B \subseteq A$ and $f = g$), then B is not e-fold c-g(x)-arboreal. (Recall that by convention A and B are listed in increasing order.)

Proof. By induction on e. If $e = 0$ this is immediate. Assume true for given $e \geq 0$, and suppose A is not (e+1)-fold c-f(x)-arboreal. Take an f(x)-small-branching A-tree T with no path P such that P^o is e-fold c-f(x)-arboreal. Form a tree T(B) by substituting b_i for a_i in T, $1 \leq i \leq s$. T(B) is a B-tree since $b_i T(B) b_j \Rightarrow a_i T a_j \Rightarrow a_i < a_j \Rightarrow i < j \Rightarrow b_i < b_j$. In T(B) b_i has $\leq f(a_i) \leq g(b_i)$ immediate successors, so T(B) is g(x)-small branching. Finally if Q is a path in T(B) let P be the corresponding path in T. Then P^o is not e-fold c-f(x)-arboreal. By induction hypothesis Q^o is not e-fold c-g(x)-arboreal, and we are done.

3.3 Lemma. Let $b \geq 2$, $n \geq 1$ be given, let T be any finite tree, $X \subseteq T$, and suppose $|X| \geq b^n$. Then either

(i) there exist distinct $x_1, x_2, \ldots, x_{b+1} \in X$ such that $x_i \neq \widehat{x_i x_j} = \widehat{x_k x_\ell}$ for all $i \neq j$, $k \neq \ell$, or

(ii) there exist distinct $y_1, \ldots, y_{n+1} \in X$ such that $\widehat{y_i y_{n+1}}$ are all distinct for $i = 1, 2, \ldots, n+1$.

Here $\widehat{xy}$ denotes the greatest lower bound of x and y under the ordering of T. In pictures:

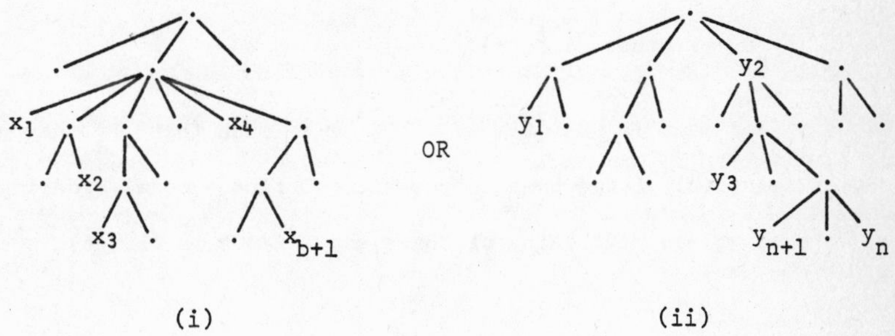

Figure 4

Proof. By induction on n, with b fixed. For $n = 1$ we have $|X| \geq b \geq 2$, and any two distinct elements of X can be indexed so as to satisfy (ii).

Now assume the lemma holds for n and we have $|X| \geq b^{n+1}$. Let $a = \text{glb}X$ and for each $u \in IS_T(a)$, let $X_u = \{x \in X \mid uTx \text{ or } u = x\}$. If there exist distinct $u_1, \ldots, u_{b+1} \in IS_T(a)$ such that $X_{u_i} \neq \phi$ for $i = 1, 2, \ldots, b+1$, then by picking $x_i \in X_{u_i}$ we satisfy (i). So suppose there are $\leq b$ nonempty X_u's.

Now by the pigeonhole principle we have $|X_u| \geq b^n$ for some u. (Otherwise $|X| \leq 1 + \sum_{X_u \neq \phi} |X_u| \leq 1 + b(b^n - 1) < b^{n+1}$). By induction X_u satisfies either (i) or (ii). If (i), then we are done, so suppose $y_1, \ldots, y_{n+1} \in X_u$ satisfy (ii). Let $y_o \in X$ be such that $y_o \widehat{\ } y_{n+1} = a$ (y_o exists since $a = \mathrm{glb}X$ and $y_{n+1} \in X$). Now $y_o \widehat{\ } y_{n+1}$ is distinct from $y_i \widehat{\ } y_{n+1}$ for $i = 1,2,\ldots,n+1$, since aTu and $uTy_i \widehat{\ } y_{n+1}$ or $u = y_i \widehat{\ } y_{n+1}$ for $i = 1,2,\ldots,n+1$. Thus we are done.

<u>3.4 Lemma</u>. (Inductive step for lower bounds) Let $c \geq 1$, $e \geq 2$ and A with $\min A \geq e^{2e}$ be given. Suppose T is an $(x-2)$-small-branching A-tree such that for every path P in T $P^o \not\rightarrow (*)^e_c$. Then

(1) if $e = 2$, then $A \not\rightarrow (*)^3_{2c+2}$.

(2) if $e > 2$, then $A \not\rightarrow (*)^{e+1}_{c+2}$.

<u>Proof</u>. For each $a \in A$ fix $F_a : [T_a]^e \rightarrow c$ such that F_a has no nontrivial relatively large homogeneous set. Define a total ordering $\prec$ on A so that $a \prec b$ iff aTb or there exist $a_1, b_1 \in IS_T(\widehat{ab})$ such that

(i) $a_1 < b_1$

(ii) $a_1 Ta$ or $a = a_1$

(iii) $b_1 Tb$ or $b = b_1$.

Thus $a \prec b$ iff you can get to a by following the path to b and (possibly) forking left (from a bird's eye view). Also for $a \in A - \{\min A\}$ define $p(a)$ = the immediate predecessor of a in T, so $a \in IS_T(p(a))$.

We define the coloring $F : [A]^{e+1} \rightarrow 2c + 2$ (or $c+2$) by cases. Given $\bar{x} = \langle x_o, \ldots, x_e \rangle \in [A]^{e+1}$ define $F(\bar{x})$ by the first case which applies to $\bar{x}$.

<u>Case 1</u>. $x_i \widehat{\ } x_j = x_k \widehat{\ } x_\ell$ for all $i \neq j$, $k \neq \ell$. Let $F(\bar{x}) = 1$.

Case 2. There exists a re-indexing $y_o,\ldots,y_e$ of $\bar{x}$ such that $\min A$, $y_o \hat{} y_e$, $y_1 \hat{} y_e,\ldots,y_e \hat{} y_e$ are all distinct and $y_i \prec y_e$ for $i = 0,1,\ldots,e-1$. Let $F(\bar{x}) = 2 + F_{y_e}(p(y_o \hat{} y_e), p(y_1 \hat{} y_e),\ldots,p(y_{e-1} \hat{} y_e))$.

Case 3. There exists a re-indexing $z_o,\ldots,z_e$ of $\bar{x}$ such that $\min A$, $z_o \hat{} z_e$, $z_1 \hat{} z_e,\ldots,z_e \hat{} z_e$ are all distinct and $z_e \prec z_i$ for $0 \leq i < e$. Let

$$F(\bar{x}) = F_{z_e}(p(z_o \hat{} z_e),\ldots,p(z_{e-1} \hat{} z_e)) + \begin{cases} 2 + c & \text{if } e = 2 \\ 2 & \text{if } e > 2 \end{cases}.$$

Case 4. Otherwise let $F(\bar{x}) = 0$.

Thus we have used $2c + 2$ colors if $e = 2$ and $c + 2$ colors if $e > 2$.

Now suppose $H \subseteq A$ is nontrivial, relatively large and homogeneous for F. Since $|H| \geq \min A \geq e^{2e}$ we have by Lemma 3.3 either

(i) $\exists \bar{x} \in [H]^{e+1}$ with $x_i \neq x_i \hat{} x_j = x_k \hat{} x_l$ for all $0 \leq i < j \leq e$ and all $0 \leq k < l \leq e$, or

(ii) $\exists \bar{w} \in [H]^{2e+1}$ with $w_i \hat{} w_{2e}$ distinct for $i = 0,1,\ldots,2e$.

Suppose (i) happens. Then Case 1 applies to this $\bar{x}$ and $F(\bar{x}) = 1$. Let $a = x_o \hat{} x_1$ and suppose there existed distinct $y, z \in H$ with $\hat{yz} \neq a$. If $\hat{yx_o} = a$, then $F(\langle y,z,x_o,\ldots\rangle) \neq 1$ (where "..." can be anything in H). If $\hat{yx_o} \neq a$, then $F(\langle y,x_o,x_1,\ldots\rangle) \neq 1$. In either case we contradict the homogeneity of H. Therefore $\hat{yz} = a$ for all distinct $y, z \in H$. Thus H contains at most one element at or below each immediate successor of a, so $|H| \leq 1 + |IS_T(a)| < a \leq \min H$, contradicting that H is relatively large. Therefore (i) is impossible.

Now let $\bar{w} \in [H]^{2e+1}$ be as in (ii).

Claim 1. For all $\bar{x} \in [H]^{e+1}$, $\bar{x}$ falls under Case 2 or 3.

Proof. Let $y_o, y_1, \ldots, y_{s-1}$ be those w_i's for which $\widehat{w_i w_{2e}} \neq \min A$ and $w_i \prec w_{2e}$. Let $z_o, \ldots, z_{t-1}$ be those w_i's for which $\widehat{w_i w_{2e}} \neq \min A$ and $w_{2e} \prec w_i$. Since $|\{\min A, y_o, \ldots, y_{s-1}, z_o, \ldots, z_{t-1}, w_{2e}\}| \geq 2e + 1$ we have $s \geq e$ or $t \geq e$. If $s \geq e$, then (re)define y_e to equal w_{2e} and observe that $\langle y_o, \ldots, y_e \rangle$ falls under Case 2 (and not under Case 1). Likewise if $t \geq e$, then (re)define $z_e = w_{2e}$ and observe that $\langle z_o, \ldots, z_e \rangle$ falls under Case 3 (and possibly under Case 2 by reindexing, but not under Case 1). Thus there is at least one $\bar{x} \in [H]^{e+1}$ falling under Cases 2 or 3 and not under Case 1. Since Cases 2 and 3 share no colors with the other cases, all $\bar{x} \in [H]^{e+1}$ must do the same, proving the claim.

Now let $S = \{\widehat{xy} \mid x, y \in H, x \neq y\}$.

Claim 2. S is a path in T.

Proof. Suppose not, say $\widehat{xy} = a$ and $\widehat{zw} = b$ with a,b incomparable. W.l.o.g. $a \prec b$, $x \prec y$ and $z \prec w$. Then since a and b are incomparable we have $x \prec y \prec z \prec w$. Now if $e = 2$ then $\widehat{xw} = \widehat{ab} \neq b = \widehat{zw}$, so $\langle x,z,w \rangle$ falls under Case 2. However $\widehat{xw} = \widehat{ab} = \widehat{yw}$, so $\langle x,y,w \rangle$ cannot fall under Case 2. When $e = 2$ Case 2 shares no colors with other cases, so this contradicts the homogeneity of H. (This is the only place we invoke the nonoverlap of Cases 2 and 3 for $e = 2$.) On the other hand if $e > 2$, look at any $\langle x,y,z,w,\ldots,x_e \rangle \in [H]^{e+1}$. By Claim 1 this must satisfy Case 2 or 3, say Case 2. But $\widehat{xy_e} \neq \widehat{yy_e}$ implies aTy_e while $\widehat{zy_e} \neq \widehat{wy_e}$ implies bTy_e. This is impossible since a and b are incomparable under T. Similarly if Case 3 holds. This proves Claim 2.

Let $m = \max S$ and pick $y_e, z_e \in H$ with $\widehat{y_e z_e} = m$ and $y_e \neq z_e$. (The significance of the subscript will become apparent in due course.)

Claim 3. For all $x \in H - \{y_e, z_e\}$ $\widehat{xy_e} = \widehat{xz_e} \neq m$.

Proof. First suppose $x^\wedge y_e = a \neq b = x^\wedge z_e$. Since $a,b \in S$ we have (w.l.o.g.) aTb by Claim 2. Let us write $u\underline{T}v$ for uTv or $u = v$. Now since $m = \max S$ we have $b\underline{T}m$ by Claim 2. Also $m\underline{T}y_e$, whence $b\underline{T}y_e$. But also $b\underline{T}x$ (since $b = x^\wedge z_e$), so $b\underline{T}x^\wedge y_e$. But $x^\wedge y_e = a$, contradicting aTb. This proves $x^\wedge y_e = x^\wedge z_e$.

Suppose $x^\wedge y_e = x^\wedge z_e = y_e{}^\wedge z_e = m$. Look at $\bar{x} = \langle x,y_e,z_e,\ldots\rangle \in [H]^{e+1}$. This satisfies Case 2 or 3, so there is a $u = x_i$ such that $x^\wedge u$, $y_e{}^\wedge u, z_e{}^\wedge u$ are all distinct. If $u \notin \{y_e,z_e\}$, then $y_e{}^\wedge u = z_e{}^\wedge u$ (we have just proved that). Therefore $u \in \{y_e,z_e\}$, and w.l.o.g. $u = y_e$. Then $x^\wedge u = x^\wedge y_e = m = z_e{}^\wedge y_e = z_e{}^\wedge u$, contradiction. This proves Claim 3.

Now w.l.o.g. $z_e \prec y_e$.

Claim 4. Either for all $\bar{y} \in [H-\{y_e\}]^e$ $\langle\bar{y},y_e\rangle$ falls under Case 2 as indexed, or for all $\bar{z} \in [H-\{z_e\}]^e$ $\langle\bar{z},z_e\rangle$ falls under Case 3 as indexed.

Proof. Look at some $\bar{x} \in [H]^{e+1}$ with $x_0 = y_e$, $x_1 = z_e$. Then $\bar{x}$ falls under Case 2 or 3, say 2. So there is a re-indexing $\langle u_0,\ldots,u_e\rangle$ with $u_0{}^\wedge u_e,\ldots,u_e{}^\wedge u_e$ distinct and $u_i \prec u_e$ for $0 \leq i < e$. If $u_e \notin \{y_e,z_e\}$, then $y_e{}^\wedge u_e = z_e{}^\wedge u_e$, contradiction. Also $u_e \neq z_e$ since $y_e \not\prec z_e$. Therefore $u_e = y_e$, and $x_i \prec y_e$ for all $x_i \neq y_e$. Since the elements of $\bar{x}$ other than y_e,z_e were arbitrary, we have $x \prec y_e$ for all $x \in H-\{y_e\}$. Therefore for all $\langle y_0,\ldots,y_{e-1}\rangle \in H-\{y_e\}$ $\langle y_0,\ldots,y_e\rangle$ cannot fall under Case 3 so it must fall under Case 2. Similar reasoning shows that if $\bar{x}$ falls under Case 3, then for all $\langle z_0,\ldots,z_{e-1}\rangle \in H-\{z_e\}$ $\langle z_0,\ldots,z_e\rangle$ must fall under Case 3. This proves Claim 4.

Now suppose for all $\bar{y} \in [H-\{y_e\}]^e$, $\langle\bar{y},y_e\rangle$ falls under Case 2. Let $H^* = \{p(y^\wedge y_e) \mid y \in H-\{y_e\}\}$. Observe that for $y_0 \neq y_1$ we have $p(y_0{}^\wedge y_e) \neq p(y_1{}^\wedge y_e)$ since $\langle y_0,y_1,\ldots,y_e\rangle$ falls under Case 2 (as indexed).

Therefore $|H^*| = |H| - 1$. Now $H^* \subseteq T_{y_e}$ and H^* is homogeneous for F_{y_e} by the way we have defined F. Therefore either $|H^*| \leq e$ or $|H^*| < \min H^*$. In the former case $|H| \leq e + 1$ and H is trivial. In the latter case $|H| \leq \min H^*$. But $\min H^* \leq p(\hat{yy_e}) < \hat{yy_e} \leq y$ for all $y \in H$, so $\min H^* < \min H$. The upshot: $|H| < \min H$, contradicting that H is relatively large. Thus F has no nontrivial relatively large homogeneous sets.

The case in which $\langle \bar{z}, z_e \rangle$ falls under Case 3 for all $\bar{z} \in [H-\{z_e\}]^e$ is analogous. This completes the proof of Lemma 3.4.

We now prove an inductive lemma for upper bounds. We follow the convention that the binomial coefficient $\binom{a}{n} = 0$ if $a < n$. Given a function $F:[A]^n \to d$ we say $B \subseteq A$ is <u>pre-homogeneous</u> for F iff there is a function $G:[B^o]^{n-1} \to d$ such that $F(x_1,\ldots,x_n) = G(x_1,\ldots,x_{n-1})$ for all $\bar{x} = \langle x_1,\ldots,x_n \rangle \in [B]^n$.

<u>3.5 Lemma</u>. (Inductive step for upper bounds) Let $c,d,e,n \geq 1$. Let A be e-fold c-f(x)-arboreal and $\min A \geq 2$. Suppose $F:[A]^n \to d$ where $d^{\binom{a-1}{n-1}} \leq f(a)$ for all $a \in A$. Then there exists $B \subseteq A$ such that B is pre-homogeneous for F and B^o is (e-1)-fold c-f(x)-arboreal.

(The special case of interest to us will be $n = e + 1$, $d = c \leq \min A$ and $f(x) = 2^x_x$.)

<u>Proof</u>. The argument follows the lines of a well known proof of Ramsey's Theorem. We define an A-tree T in stages. Suppose $a_1,a_2,\ldots,a_m$ have already been placed on T. For $j \leq m$ say a_j is <u>available</u> for

a_{m+1} if for all $\langle x_1,\ldots,x_{n-1}\rangle \in [T_{a_j}]^{n-1}$ $F(x_1,\ldots,x_{n-1},a_j) =$

$F(x_1,\ldots,x_{n-1},a_{m+1})$. Let a_k have maximal rank in $T\restriction\{a_1,\ldots,a_m\}$ such that a_k as well as a_j are available for a_{m+1}, for all $a_j \in T_{a_k}$. Put a_{m+1} into $IS_T(a_k)$.

We show T is $d^{\binom{x-1}{n-1}}$-small-branching and hence $f(x)$-small branching. If $|IS_T(a_k)| \leq 1$ we are ok, since $d^{\binom{a_k-1}{n-1}} \geq d^0 = 1$. Suppose $a, b \in IS_T(a_k)$ with $a < b$. By the maximality of a_k we see a was not available for b. But a_k was available for b, hence for some $\langle x_1,\ldots,x_{n-1}\rangle \in [T_a]^{n-1}$ $F(\bar{x},a) \neq F(\bar{x},b)$. Thus in a canonical way a and b "induce" different functions $[T_a]^{n-1} \to d$. But $a_k = \max T_a$ and $\min T_a \geq 2$, so $|T_a| \leq a_k-1$ and $|[T_a]^{n-1}| \leq \binom{a_k-1}{n-1}$. So there are at most $d^{\binom{a_k-1}{n-1}}$ different functions $[T_a]^{n-1} \to d$, whence $|IS_T(a_k)| \leq d^{\binom{a_k-1}{n-1}} \leq f(a_k)$, as required.

It follows that there is a path B in T such that B^o is $(e-1)$-fold c-$f(x)$-arboreal. Let $b = \max B$ and define $G:[B^o]^{n-1} \to d$ by $G(\bar{x}) = F(\bar{x},b)$. Now for any $\langle x_1,\ldots,x_n\rangle \in [B]^n$ either $x_n = b$ or $x_n T b$ and x_n was available for b in T. In either case $F(x_1,\ldots,x_n) = F(x_1,\ldots,x_{n-1},b) = G(x_1,\ldots,x_{n-1})$. Thus B is pre-homogeneous for F and we are done.

We have laid the groundwork for the following main theorem giving aboreal bounds for the relation $A \to (*)^{e+1}_c$ for $e \geq 2$.

3.6 Theorem. Let $e, c \geq 2$ and A with $\min A \geq e^{2e}, c$ be given.

Then each of the following conditions implies the one after it.

(1) A is e-fold $(c+1)$-arboreal.

(2) A is e-fold c-2^x_x-arboreal.

(3) $A \to (*)^{e+1}_c$.

(4) A is e-fold $([\frac{c}{2}]-e+1)$ - $(x-2)$-arboreal.

(5) A is e-fold $([\frac{c}{2}]-e)$-arboreal.

Here $[\,]$ is the greatest integer function. If $c \leq 2e$ we regard (4) and (5) as vacuously true.

Proof. We prove (2) $\Rightarrow$ (3) $\Rightarrow$ (4) by induction on e. Suppose $e \geq 2$ and the theorem is true for $e - 1$. (Note that for $e - 1 = 1$, (2) $\Rightarrow$ (3) $\Rightarrow$ (4) by Theorem 2.6.)

(2) $\Rightarrow$ (3). Assume (2) and let $F:[A]^{e+1} \to c$. Now $c^{\binom{a-1}{e}} \leq a^{2^a} \leq 2^a_a$ for all $a \geq c, 4$. Hence by the inductive lemma for upper bounds (Lemma 3.5) there is a pre-homogeneous $B \subseteq A$ such that B^o is $(e-1)$-fold c-2^x_x-arboreal. Say $F(x_1,\ldots,x_{e+1}) = G(x_1,\ldots,x_e)$ for all $\langle x_1,\ldots,x_{e+1}\rangle \in [B]^{e+1}$ where $G:[B^o]^e \to c$. By the induction hypothesis $B^o \to (*)^e_c$, so let X be a nontrivial relatively large G-homogeneous subset of B^o. Then $X \cup \{\max B\}$ is F-homogeneous, relatively large and has cardinality $\geq |X| + 1 \geq e + 2$.

(3) $\Rightarrow$ (4). Assume (3). We may assume c is even, $c = 2d$. If $d < e$, (4) is vacuously true, so assume $d \geq e$, i.e., $c \geq 2e$.

Case 1: $e = 2$. Since $A \to (*)^3_{2d}$ we have by the inductive lemma for lower bounds (Lemma 3.4(1)) that for every $(x-2)$-small-branching A-tree T there is a path P in T such that $P^o \to (*)^2_{d-1}$. By

Theorem 2.6, P^{o} is (d-1)-(x-2)-arboreal. Thus A is 2-fold (d-1)-(x-2)-arboreal. But $d - 1 = [\frac{c}{2}] - 1 = [\frac{c}{2}] - e + 1$, as required.

<u>Case 2</u>: $e \geq 3$. Since $A \to (*)_{c}^{e+1}$ we have by the inductive lemma (Lemma 3.4(2)) that every (x-2)-small-branching A-tree has a path P such that $P^{o} \to (*)_{c-2}^{e}$. By the induction hypothesis P^{o} is (e-1)-fold $([\frac{c-2}{2}]-(e-1)+1)$-(x-2)-arboreal. (Observe that since $c \geq 2e$ we have $c - 2 \geq 2(e-1)$). But $[\frac{c-2}{2}]-(e-1)+1 = [\frac{c}{2}]-e+1$. Thus A is e-fold $([\frac{c}{2}]-e+1)$-(x-2)-arboreal, as required.

For the proof of (1)⇒(2) and (4)⇒(5) we require some lemmas.

<u>3.7 Lemma</u>. Let $e \geq 1$, $c \geq 2$ and A be e-fold c-f(x)-arboreal. Suppose $A - \{\min A\} = \bigcup_{i \varepsilon [d]} A_i$ where $1 \leq d \leq f(\min A)$. Then for some i, A_i is e-fold (c-1)-f(x)-arboreal.

<u>Proof</u>. By induction on e. Let e = 1, and suppose no A_i is (c-1)-f(x)-arboreal. For each $i \varepsilon [d]$ let T_i witness this for A_i. Form an A-tree T with root minA and $IS_T(\min A) = \{\text{root } T_i \mid i \varepsilon [d]\}$, $IS_T(a) = IS_{T_i}(a)$ for $a \varepsilon A_i$. Since $|IS_T(\min A)| \leq d \leq f(\min A)$, T is f(x)-small-branching and has no path of length c + 2 (since no T_i has a path of length c + 1). Contradiction.

If $e > 1$, then form f(x)-small-branching A_i-trees T_i so that for any path Q in T_i Q^{o} is not (e-1)-fold (c-1)-f(x)-arboreal. Form T as above. Let P be any path in T. Let $Q = P - \{\min P\} = P - \{\min A\}$. Then Q is a path in some T_i, so Q^{o} is not (e-1)-fold (c-1)-f(x)-arboreal. Applying the induction hypothesis to $P^{o} - \{\min P\} = Q^{o}$ (a union of one set) we see that P^{o} is not (e-1)-fold c-f(x)-arboreal. This contradicts that A is e-fold c-f(x)-arboreal.

3.8 Lemma. Let $e \geq 2$ and $c \geq 1$ be given. Let A be e-fold (c+1)-arboreal with $\min A \geq 3$ and suppose for all $x \geq \min A$, $A \cap [x, 2f(x)]$ is not e-fold c-arboreal. Then A is e-fold c-f(x)-arboreal.

Proof. Let T be any f(x)-small-branching A-tree. Build an A-tree T^* as in the proof of Lemma 2.7 with the appropriate modification, i.e., so that in each subtree T_k, there are no paths P for which P^o is (e-1)-fold c-arboreal. As before T^* is a small-branching A-tree, so we may find a path P in T^* such that P^o is (e-1)-fold (c+1)-arboreal. Now P^o intersects at most one T_k so we may write $P^o - \{\min P^o\}$ as a union of two sets: $P^o \cap T'$ and $P^o \cap T_k$ for some k. By Lemma 3.7 one of these sets must be (e-1)-fold c-arboreal. It cannot be $P^o \cap T_k$, since then $P \cap T_k$ would be a path in T_k for which $(P \cap T_k)^o = P^o \cap T_k$, is (e-1)-fold c-arboreal. Therefore $P^o \cap T'$ is (e-1)-fold c-arboreal.

Now let $X = \{a_j | a_{i_j} \in P^o \cap T'\}$. By construction of T', X is a path in T. Let $a_{j_0} = \max X$. By construction a_{j_0} cannot be an end point in T, so let Q be a path in T such that $Q^o = X$. Suppose Q^o were not (e-1)-fold c-f(x)-arboreal. Then by Lemma 3.2 $P^o \cap T'$ would be not (e-1)-fold c-arboreal, since $f(a_j) \leq a_{i_j}$ for all $a_j \in Q^o$ (see the proof of Lemma 2.7). This contradiction shows that T has a path Q which is (e-1)-fold c-f(x)-arboreal, proving A is e-fold c-f(x)-arboreal. This proves Lemma 3.8.

We now prove (1)$\Rightarrow$(2) in Theorem 3.6. Let A be as in (1). We have already observed that for all x, $[x, 2\cdot 2^x_x]$ is not 3-arboreal. Hence there is a $[x, 2\cdot 2^x_x]$-tree T with no path P of length 5, in particular if $x \geq 4$ no (e-1)-fold c-arboreal path. Thus for all

$x \geq \min A$, $A \cap [x, 2\cdot 2^x_x]$ is not e-fold c-arboreal. By Lemma 3.8 A is e-fold c-2^x_x-arboreal, proving (2).

We now prove (4)$\Rightarrow$(5). Let A be as in (4), $A = \{a_1, a_2, \ldots, a_t\}$. Writing $A - \{\min A\}$ as a union of two sets $\{a_2\}$, $\{a_3, a_4, \ldots, a_t\}$, we see by Lemma 3.7 that $\{a_3, a_4, \ldots, a_t\}$ is e-fold $([\frac{c}{2}]-e)$-$(x-2)$-arboreal. Since $a_i \leq a_{i+2} - 2$ it follows from Lemma 3.2 that $\{a_1, a_2, \ldots, a_{t-2}\}$ is e-fold $([\frac{c}{2}]-e)$-arboreal, whence A is too. This completes the proof of Theorem 3.6.

Section 4. Ordinal characterization of arboreal sets.

Our goal in this section is to provide an exact characterization of e-fold c-f(x)-arboreal sets in terms of ordinal numbers below ε_o. We begin by describing our notation for ordinals. We define $\omega^\alpha_o = \alpha$ and $\omega^\alpha_{e+1} = \omega^{\omega^\alpha_e}$ (ordinal exponentiation, of course). Then $\varepsilon_o = \sup\{\omega^1_e | e \varepsilon N\}$. Given an ordinal $\alpha > 0$ there exist unique positive integers $s, m_1, m_2, \ldots, m_s$ and ordinals $\alpha_1, \alpha_2, \ldots, \alpha_s$ such that $\alpha_1 > \alpha_2 > \ldots > \alpha_s \geq 0$ and

$$\alpha = \omega^{\alpha_1}\cdot m_1 + \omega^{\alpha_2}\cdot m_2 + \ldots + \omega^{\alpha_s}\cdot m_s . \qquad \text{(CNF)}$$

This expression for α is called Cantor normal form (abbreviated CNF). We write $\mathrm{Suc}(\alpha)$ iff $\alpha_s = 0$ (α is a successor ordinal) and $\mathrm{Lim}(\alpha)$ iff $\alpha_s > 0$ (α is a limit ordinal). We note that if $\alpha < \varepsilon_o$ then $\alpha_1 < \alpha$. We write $\alpha >> \beta$ (or $\beta << \alpha$) iff $\beta = \omega^{\delta_1}\cdot n_1 + \ldots + \omega^{\delta_t}\cdot n_t$ in CNF with $\alpha_s \geq \delta_1$.

For each $\alpha < \varepsilon_0$ we define a <u>fundamental sequence</u> $\{\alpha\}$ which is a function $\{\alpha\} : N \to \alpha$ (or $\to 1$ if $\alpha = 0$) given by

$$\{\alpha\}(n) = \begin{cases} 0 & \text{if } \alpha = 0 \\ \omega^{\alpha_1}\cdot m_1 + \ldots + \omega^{\alpha_{s-1}}\cdot m_{s-1} + \omega^{\alpha_s}\cdot(m_s - 1) & \text{if } \mathrm{Suc}(\alpha) \\ \omega^{\alpha_1}\cdot m_1 + \ldots + \omega^{\alpha_{s-1}}\cdot m_{s-1} + \omega^{\alpha_s}\cdot(m_s - 1) + \omega^{\{\alpha_s\}(n)}\cdot n & \text{if } \mathrm{Lim}(\alpha) \end{cases}$$

for α expressed in CNF. (Since $\alpha_s < \alpha$, $\{\alpha_s\}(n)$ has already been defined.) We caution that this choice of fundamental sequences is slightly different from a common alternative choice in which the final coefficient n is replaced by 1 when α_s is a limit ordinal. The present uniform definition for $\mathrm{Suc}(\alpha_s)$ and $\mathrm{Lim}(\alpha_s)$ works out more elegantly for our purposes. We have $\alpha = \sup\{\{\alpha\}(n) \mid n\varepsilon N\}$ whenever $\mathrm{Lim}(\alpha)$. Observe that if $\alpha >> \beta$ then $\{\alpha+\beta\}(n) = \alpha + \{\beta\}(n)$.

The key to the ordinal characterization of arboricity is the observation that there is a certain canonical way to set about trying to build an f(x)-small-branching A-tree which has no paths P for which P^o is (e-1)-fold c-f(x)-arboreal.

<u>4.1 Definition (e ≥ 1)</u>. An A-tree T is <u>e-fold c-f(x)-canonical</u> iff

for every $b\varepsilon A-\{\min A\}$ the immediate predecessor of b in T is the <u>maximal</u> $a\varepsilon A$ such that

(i) $a < b$,

(ii) $T_a \cup \{a\}$ is not (e-1)-fold c-f(x)-arboreal, and

(iii) $|IS_T(a) \cap \{y\varepsilon A \mid y<b\}| < f(a)$.

For example the trees of Figure 3 and of Figure 2 minus node 15 (Section 2) are 2-canonical (i.e., 1-fold 2-x-canonical). The tree of

Figure 1(ii) is not 2-canonical because, e.g., the immediate predecessor of 5 should be 4, not 3. The idea behind canonicity is fairly obvious. One fills in the branches from left to right, using as much length and branching as one can get away with. By placing smaller numbers as far down on the tree as possible one saves the nodes higher up (those with smaller rank) for larger numbers which can do more good since they are allowed to have more immediate successors.

Observe that if in Definition 4.1 the words "the maximal" are weakened to "an", then such a tree T exists iff A is not e-fold c-f(x)-arboreal. Theorem 4.2 shows that remarkably this equivalence holds for canonical trees as well. Since an e-fold c-f(x)-canonical A-tree T is obviously unique if it exists, this means that to test for arboricity of A one need only consider a single A-tree rather than all possible A-trees. As a consequence we are able to provide an exact characterization of arboricity in terms of ordinal numbers below ε_o. We state this as Theorem 4.3. Theorems 4.2 and 4.3 are proved by simultaneous induction.

<u>4.2 Theorem</u> (Canonical characterization). Let $e,c \geq 1$ and A be given. Then the following are equivalent:

(1) A is not e-fold c-f(x)-arboreal.

(2) The e-fold c-f(x)-canonical A-tree T exists.

<u>4.3 Theorem</u> (Ordinal characterization). Let $e \geq 0$, $c,t \geq 1$ and $A = \{a_1,a_2,\ldots,a_t\}$ be given, and define $\alpha_1 = \omega_e^c$, $\alpha_{i+1} = \{\alpha_i\}(f(a_i))$ for $i = 1,2,\ldots,t-1$. Then the following are equivalent:

(1) A is not e-fold c-f(x)-arboreal.

(2) $\alpha_t > 0$.

Proofs. Let Can(e) be the statement of the Canonical Characterization Theorem for the value e, and Ord(e) the statement of the Ordinal Characterization Theorem for e. We will prove Ord(0) and that Ord(e)$\Rightarrow$ Can(e+1) & Ord(e+1) for all $e \geq 0$.

Proof of Ord(0). In this case $\omega_e^c = c$ and $\alpha_i = c - i + 1$ for $i \leq c + 1$, $1 \leq i \leq t$. So $\alpha_t > 0$ iff $t \leq c$ iff $|A| \leq c$ iff $|A^o| < c$ iff A is not 0-fold c-f(x)-arboreal.

Proof that Ord(e)$\Rightarrow$Can(e+1) for $e \geq 0$. Assume Ord(e). We have already observed that (2)$\Rightarrow$(1) in Can(e+1), so now assume A is not (e+1)-fold c-f(x)-arboreal. We must show that the (e+1)-fold c-f(x)-canonical A-tree exists.

Call an A-tree T good if it is f(x)-small-branching and for no path P in T is P^o e-fold c-f(x)-arboreal. Thus we are assuming a good A-tree T exists. For $a,b \in T$ say that a is available for b in T if a satisfies (i), (ii) and (iii) in the definition of (e+1)-fold c-f(x)-canonical; that is,

(i) $a < b$,

(ii) $T_a \cup \{a\}$ is not e-fold c-f(x)-arboreal,

(iii) $|IS_T(a) \cap \{y|y<b\}| < f(a)$.

Call b correctly placed in T if b = minA or the immediate predecessor of b is indeed the maximal $a \in A$ available for b in T. Let $A = \{a_1, a_2, \ldots, a_t\}$. We show by induction on j that there exists a good A-tree T in which $a_1, a_2, \ldots a_j$ are correctly placed. For $j = t$ this implies T is (e+1)-fold c-f(x)-canonical.

For $j = 2$ let T be any good A-tree. Certainly a_1 and a_2 are correctly placed in T. Now suppose $2 < j \leq t$ and T is a good

A-tree in which $a_1, a_2, \ldots, a_{j-1}$ are correctly placed. Supposing a_j is not already correctly placed, we will go through a sequence of modifications of T (T', T'', T^*) ending up with a good A-tree T^* in which $a_1, a_2, \ldots, a_j$ are correctly placed.

Our basic strategy for carrying this out will be to identify the node a_k which occupies the position a_j should occupy $(k > j)$, and to cyclically permute the positions of $a_k, a_{k-1}, \ldots, a_{j+1}, a_j$ on the tree, bringing a_j to its proper place. The only way the resulting tree can fail to be good is if a_j has too many immediate successors or if some path through a_j is too long. By carefully analyzing the situation and by carrying out some preliminary adjustments, we can find an alternative location on the tree to attach the immediate successors out of a_j. This is in accord with our basic intuition as to <u>why</u> the canonical construction is the most efficient way to build an A-tree. In order to identify the suitable alternative location, we will assign ordinals to the nodes in the tree, making use of the induction hypothesis, Ord(e). For each $b \in A$ the assignment is made as follows:

$$\mathrm{ord}_T(b) = \begin{cases} \omega_e^c & \text{if } b = \min A \\ \{\mathrm{ord}_T(a)\}(f(a)) & \text{if } b \in IS_T(a) . \end{cases}$$

Notice that by Ord(e) the condition (ii) in the definition of "available" is equivalent to

(ii') $\mathrm{ord}_T(a) > 0$.

The following lemma deals with sets of ordinals such as might be assigned to the nodes along a path in T.

4.4 Lemma. Suppose $\Omega = \{\alpha_1,\alpha_2,\ldots,\alpha_r\}$ where $\alpha_{i+1} = \{\alpha_i\}(n_i)$ for $i = 1,2,\ldots,r-1$. Then:

(i) If $\beta + \gamma \in \Omega$, $\beta >> \gamma$ and $\beta \geq \alpha_r$, then $\beta \in \Omega$.

(ii) If $m \leq n_i$ for $i = 1,2,\ldots,r-1$ and $\{\alpha_1\}(m) \geq \alpha_r$, then $\{\alpha_1\}(m)\in\Omega$.

Proof. (i) By easy induction on γ using the fact that if $\gamma > 0$, then $\{\beta+\gamma\}(n) = \beta + \{\gamma\}(n)$ and $\beta >> \{\gamma\}(n)$.

(ii) By induction on α_1. Let $\alpha_1 = \omega^{\gamma_1}.m_1 + \ldots + \omega^{\gamma_s}.m_s$ in CNF. If $\gamma_s = 0$, then $\{\alpha_1\}(m) = \{\alpha_1\}(n_1)\in\Omega$. Otherwise

$$\{\alpha_1\}(n_1) = \omega^{\gamma_1}.m_1 + \ldots + \omega^{\gamma_s}.(m_s-1) + \omega^{\{\gamma_s\}(n_1)}.n_1 = \gamma + \omega^{\{\gamma_s\}(n_1)}.n_1$$

and $\{\alpha_1\}(m) = \gamma + \omega^{\{\gamma_s\}(m)}.m$ for the same value of γ. Now using part (i) repeatedly we find that

$$\gamma + \omega^{\{\gamma_s\}(n_1)}\in\Omega$$

$$\gamma + \omega^{\{\gamma_s\}(n_1,n_{i_2})}.n_{i_2}\in\Omega$$

$$\gamma + \omega^{\{\gamma_s\}(n_1,n_{i_2})}\in\Omega$$

$$\gamma + \omega^{\{\gamma_s\}(n_1,n_{i_2},n_{i_3})}.n_{i_3}\in\Omega, \text{ etc.}$$

By the induction assumption we eventually get a sequence $n_1,n_{i_2},\ldots,n_{i_t}$ with $\{\gamma_s\}(m) = \{\gamma_s\}(n_1,n_{i_2},\ldots,n_{i_t})$ and

$$\gamma + \omega^{\{\gamma_s\}(n_1,n_{i_2},\ldots,n_{i_t})}.n_{i_t} = \gamma + \omega^{\{\gamma_s\}(m)}.n_{i_t}\in\Omega.$$

Using part (i) again (since $n_{i_t} \geq m$) we get ${}_{\bullet}\gamma + \omega^{\{\gamma_s\}(m)}.m = \{\alpha_1\}(m)\varepsilon\Omega$. This proves Lemma 4.4.

We resume our analysis of T.

Claim 1. For each $i \leq j$ the set of nodes available for a_i in T forms a path in T (possibly with gaps).

Proof. By induction on i. Let S be the set of nodes available for a_{i-1} in T and assume S forms a path, with maximum element a. Since T is an A-tree, a is also maximal in the ordering of T. Since a_{i-1} is correctly placed in T, $a_{i-1}\varepsilon IS_T(a)$. Now the nodes available for a_i are precisely the elements of S with the possible exception of a and the possible addition of a_{i-1}. This is a path, proving Claim 1.

Now since we are assuming a_j is not correctly placed in T, suppose $a_j \varepsilon IS_T(a_h)$ but a_i is the maximal $a\varepsilon A$ available for a_j in T $(a_i > a_h)$. By Claim 1 we have $a_h T a_i$. Now, if $IS_T(a_i) \cap \{y\varepsilon A | y > a_j\} = \emptyset$, then form T* by moving a_j to $IS_T(a_i)$ and moving a_ℓ to the former position of $a_{\ell-1}$ for all $\ell > j$. By Lemma 3.2 and the fact that f(x) is monotone nondecreasing it is clear that T* is a good A-tree in which $a_1,\ldots,a_j$ are correctly placed. In this case we are done.

Therefore assume now that there does exist $a_k \varepsilon IS_T(a_i)$ with $k > j$. Fix such a k and let $ord_T(a_k) = \beta$.

Claim 2. If $j \leq \ell < k$ and $ord_T(a_\ell) = \beta$, then the tree T' obtained from T by interchanging the two subtrees whose roots are a_k and a_ℓ, respectively, is good and $a_1,\ldots,a_{j-1}$ are correctly placed in it.

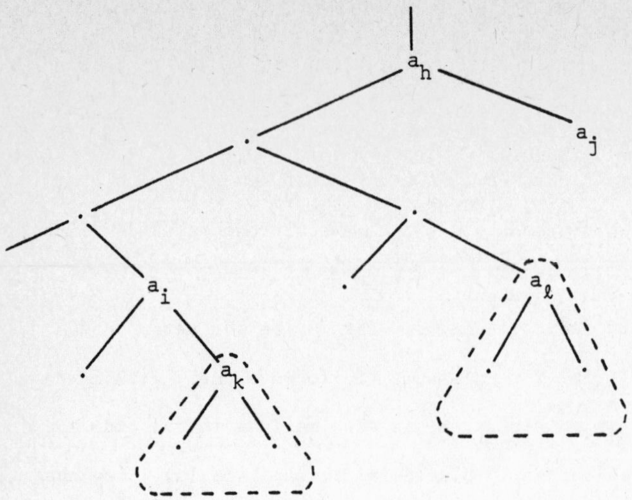

Figure 5

<u>Proof</u>. The last part is obvious, since none of $a_1, \ldots, a_{j-1}$ changed positions. It is also clear that $|IS_T(a)| = |IS_{T'}(a)|$ for all a, so T' is f(x)-small-branching. T' is an A-tree since $a_i < a_j \le a_\ell$ and $p_{T'}(a_k) = p_T(a_\ell) < a_\ell < a_k$. Finally no branches in T' are too long, since the ordinals assigned to the nodes in T' are the same as the ordinals assigned in T. This proves Claim 2.

Now we form a tree T' by applying Claim 2 to the <u>minimal</u> $\ell \ge j$ for which $ord_T(a_\ell) = \beta$. To conserve notation assume w.l.o.g. that k already was minimal, so T = T'. Thus we have $ord_T(a_k) = \beta$ and

$$j \le \ell < k \Rightarrow ord_T(a_\ell) \ne \beta .$$

The argument now breaks into two cases.

Case I. $f(a_j) = f(a_k)$ or $IS_T(a_k) = \phi$. In this case form the tree T^* from T by performing the cyclic permutation $(a_k, a_{k-1}, a_{k-2}, \ldots, a_{j+1}, a_j)$. This brings a_j to its correct place and puts $a_{\ell+1}$ in place of a_ℓ for $\ell = j, j+1, \ldots, k-1$. To show T^* is an A-tree one can show that $a_m T^* a_n$ implies $m < n$ by checking the various cases for m and n ($<j$, $= j$, $\varepsilon(j,k]$, $>k$). Clearly $a_1, \ldots, a_j$ are correctly placed in T^*. Finally T^* is good by Lemma 3.2 since every node in T was replaced by a node with the same or greater f-value ($f(a_{\ell+1}) \geq f(a_\ell)$) with the possible exception of a_j replacing a_k. But either $f(a_j) = f(a_k)$ so again we're safe, or $IS_{T^*}(a_j) = IS_T(a_k) = \phi$ in which case $f(a_j)$ is immaterial. Either way T^* is good, completing the proof for Case I.

Case II. $f(a_j) < f(a_k)$ and $IS_T(a_k) \neq \phi$. In this case let T_1 be the subtree of T whose root node is a_k. T_1 has at least two elements. Since $f(a_j) < f(a_k)$ fix an ℓ with $j \leq \ell < k$ and $f(a_\ell) < f(a_{\ell+1})$. Let T'' be the tree obtained from T by moving the entire subtree T_1 so its root a_k is an immediate successor of a_ℓ rather than of a_i. Formally, for $a \varepsilon A$ we have

$$IS_{T''}(a) = \begin{cases} IS_T(a) \cup \{a_k\} & \text{if } a = a_\ell \\ IS_T(a) - \{a_k\} & \text{if } a = a_i \\ IS_T(a) & \text{otherwise} . \end{cases}$$

Clearly T'' is still an A-tree and the only way it can fail to be f(x)-small-branching is if $|IS_{T''}(a_\ell)| = f(a_\ell) + 1$. We will remedy this later. First we consider the lengths of the paths in T''. This is the crux of the whole argument.

Claim 3. No path in T'' is too long, i.e., for no path P in T'' is P^o e-fold c-f(x)-arboreal.

Proof. Suppose P is a path in T''. We may assume P is complete. If P does not contain a_k then P is also a path in T and the claim follows. So assume $a_k \in P$ and write P as a disjoint union $P = X \cup B \cup Y$, where

$$X = \{a_x \in P | x < j\}$$

$$B = \{a_x \in P | j \leq x < k\}$$

$$Y = \{a_x \in P | k \leq x\}.$$

Let $a_m = \max X$, $a_n = \min B$ and note that $a_\ell = \max B$ and $a_k = \min Y$. Also note that since T is good and since $a_n \varepsilon IS_T(a_m)$, a_m is available for a_n in T. Therefore a_m is available for a_j in T since $m < j \leq n$, and consequently $m \leq i$. Now let

$$C = \{a_x | m < x \text{ and } a_x T a_k\} ,$$

$$Q = X \cup C \cup Y .$$

Observe that Q is a complete path in T and that $X < C < B < Y$ in the sense that $a_x \varepsilon X$ and $a_y \varepsilon C$ implies $a_x < a_y$, etc., because

$$a_x \varepsilon X \text{ implies } x \leq m,$$

$$a_x \varepsilon C \text{ implies } m < x \leq i,$$

$$a_x \varepsilon B \text{ implies } i < j \leq x < k,$$

$$a_x \varepsilon Y \text{ implies } k \leq x.$$

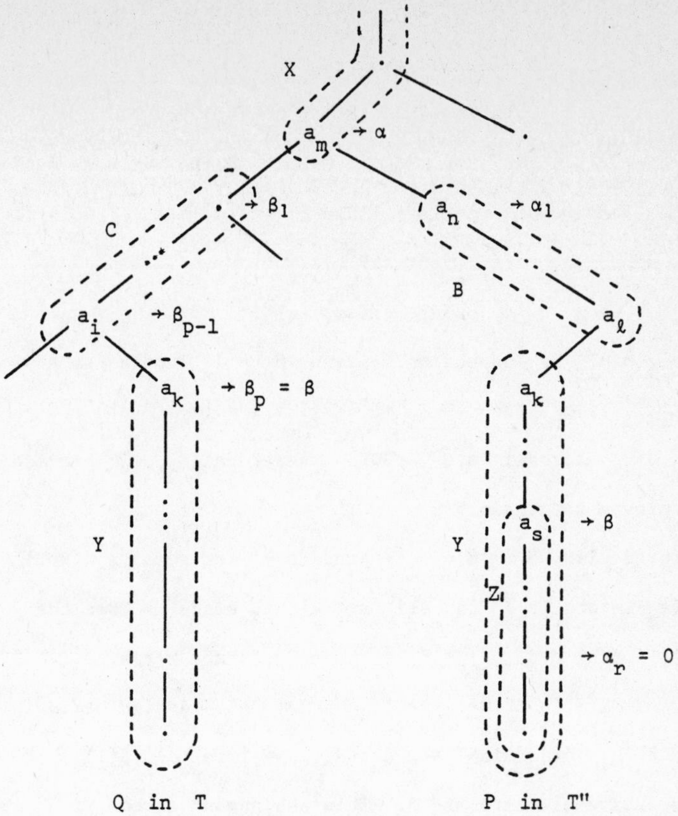

Figure 6

We wish to show that because Q is not too long (being a path in T) therefore P is not too long. This is where ordinals come in. Define an assignment of ordinals to nodes in T'' analogous to the assignment in T. (Only the assignments to nodes in T_1 are changed.) To show P is not too long we must show $ord_{T''}(maxP^o) > 0$.

Suppose $\mathrm{ord}_{T''}(\max P^{o})$ does $= 0$. Let $\alpha = \mathrm{ord}_{T''}(a_m) = \mathrm{ord}_T(a_m)$. Let $\beta_1, \beta_2, \ldots, \beta_p$ be the ordinals assigned in T to the elements of $C \cup \{a_k\}$. Thus $\beta_1 = \{\alpha\}(f(a_m))$, $\beta_2 = \{\beta_1\}(f(\min C))$, and so forth, up to $\beta_p = \beta = \{\beta_{p-1}\}(f(a_i))$. Let $\alpha_1, \alpha_2, \ldots, \alpha_r$ be the ordinals assigned in T'' to the elements of $(B \cup Y)^{o}$. Thus $\alpha_1 = \{\alpha\}(f(a_m))$, $\alpha_2 = \{\alpha_1\}(f(a_n))$, and so forth, up to $\alpha_r = \{\alpha_{r-1}\}(f(\max \dot{P}^{oo})) = 0$.

Claim. For each $q \leq p$ there exists $x \leq r$ with $\beta_q = \alpha_x$.

Proof. $\beta_1 = \alpha_1$ by inspection. Given $\beta_q = \alpha_x$ apply Lemma 4.4(ii) to $\Omega = \{\alpha_x, \alpha_{x+1}, \ldots, \alpha_r\}$ and $\beta_{q+1} = \{\beta_q\}(f(a)) = \{\alpha_x\}(f(a)) \geq 0 = \alpha_r$, since $f(a) \leq f(b)$ for all $a \varepsilon C$, $b \varepsilon B \cup Y$. Therefore $\beta_{q+1} \varepsilon \Omega$ provided $q < p$. This proves the claim.

Now in particular $\beta_p = \beta = \alpha_x = \mathrm{ord}_{T''}(a_s)$ for some $a_s \varepsilon (B \cup Y)^{o}$. Since $\mathrm{ord}_{T''}(a_x) = \mathrm{ord}_T(a_x)$ for all $x < k$ and since k was the minimal $x \geq j$ with $\mathrm{ord}_T(a_x) = \beta$, we conclude that $s \geq k$, so $a_s \varepsilon Y^{o}$.

But now let $Z = \{a_x \varepsilon Y | x \geq s\}$ and $R = X \cup C \cup Z$. Assign ordinals to the elements of R: $\mathrm{ord}_R(\min R) = \omega_e^c$, $\mathrm{ord}_R(b) = \{\mathrm{ord}_R(a)\}(f(a))$ if a, b are consecutive elements of R. This assignment agrees with ord_T on $X \cup C$ and therefore with $\mathrm{ord}_{T''}$ on Z since a_i, a_s are consecutive in R. Therefore $\mathrm{ord}_R(\max R^{o}) = \mathrm{ord}_{T''}(\max R^{o}) = 0$, indicating, by Ord(e), that R^{o} is e-fold c-f(x)-arboreal. This is a contradiction since $R^{o} \subseteq Q^{o}$, and Q^{o} is not e-fold c-f(x)-arboreal. This completes the proof of Claim 3.

We now finish the proof of Ord(e)$\Rightarrow$Can(e+1). We have this tree T'' which is almost good, but a_j is not yet correctly placed. We remedy both these deficiencies at once. Form T^* from T'' by replacing a_x by a_{x+1} for all $x \geq j$ and then put a_j into $IS_{T^*}(a_i)$. Clearly $a_1, \ldots, a_j$ are correctly placed.

Let P be a path in T^*. If $a_j \varepsilon P$, then $Q = (P-\{a_j\})\cup\{a_k\}$ is a path in T'' and $P^o = Q^o$, so P is not too long. If $a_j \notin P$ then P is obtained from some path Q in T'' by possibly increasing some nodes, so by Lemma 3.2 P is not too long. Thus no path in T^* is too long.

Given a node $a_x \varepsilon A$, we show $|IS_{T^*}(a_x)| \leq f(a_x)$. If $x = i$, then $IS_{T^*}(a_i) = IS_{T''}(a_i)\cup\{a_j\} = (IS_T(a_i) - \{a_k\})\cup\{a_j\}$, so $|IS_{T^*}(a_i)| = |IS_T(a_i)| \leq f(a_i)$. If $x = \ell + 1$, then $|IS_{T^*}(a_{\ell+1})| = |IS_{T''}(a_\ell)| = |IS_T(a_\ell)\cup\{a_k\}| = |IS_T(a_\ell)| + 1 \leq f(a_\ell) + 1 \leq f(a_{\ell+1})$. For all other x we have $|IS_{T^*}(a_x)| = |IS_T(a_y)| \leq f(a_y) \leq f(a_x)$ where either $y = x$ or $y = x - 1$. Thus T^* is $f(x)$-small-branching.

We have found a good tree T^* in which $a_1,\ldots,a_j$ are correctly placed. This completes the proof of $\mathrm{Ord}(e)\Rightarrow\mathrm{Can}(e+1)$.

<u>Proof of $\mathrm{Ord}(e)\Rightarrow\mathrm{Ord}(e+1)$</u>. We begin with a definition due essentially to J. Ketonen.

<u>4.5 Definition</u>. A set $A = \{a_1,a_2,\ldots,a_t\}$ is <u>α-$f(x)$-large</u> (α an ordinal $< \varepsilon_o$) if $\alpha_t = 0$, where $\alpha_1 = \alpha$ and $\alpha_{i+1} = \{\alpha_i\}(f(a_i))$ for $i = 1,2,\ldots,t-1$.

Now $\mathrm{Ord}(e)$ simply says that a set is e-fold c-$f(x)$-arboreal iff it is ω_e^c-$f(x)$-large. We are given $\mathrm{Ord}(e)$ and must prove $\mathrm{Ord}(e+1)$. By the previous proof we know $\mathrm{Can}(e+1)$ is true. Therefore it is enough to prove the following: Given A, the $(e+1)$-fold c-$f(x)$-canonical A-tree exists iff A is not ω_{e+1}^c-$f(x)$-large.

Let A be given. Define $\alpha_1,\ldots,\alpha_t$ as in Def. 4.5 with $\alpha_1 = \omega_{e+1}^c$. We may assume $\alpha_{t-1} > 0$. Let $\Omega = \{\alpha_1,\ldots,\alpha_{t-1}\}$ and define a tree structure θ on Ω as follows: The root is α_1. Suppose $j > 1$ and

$\alpha_1,\ldots,\alpha_{j-1}$ have been placed on θ. Say $\alpha_j = \omega^{\beta_1}.s_1 + \ldots + \omega^{\beta_k}.s_k$ in CNF and let q be maximal such that $\gamma = \omega^{\beta_1}.s_1 + \ldots + \omega^{\beta_{k-1}}.s_{k-1} + \omega^{\beta_k}.q \epsilon \Omega$. Now if $\gamma = \alpha_1$ then $q = 1 = s_k$, whence $\alpha_j = \alpha_1$, contradicting $j > 1$. Therefore $\gamma \neq \alpha_1$ so $\gamma = \alpha_{i+1}$ for some $i \geq 1$. Also $i \leq j - 1$ since $\alpha_i > \alpha_{i+1} \geq \alpha_j$. We place α_j as an immediate successor to α_i on θ. Continuing in this way for $j = 2,3,\ldots,t-1$ we define an Ω-tree θ.

Now define an A^o-tree T by: $a_i T a_j$ iff $\alpha_i \theta \alpha_j$. This is an A^o-tree since $a_i T a_j \Rightarrow \alpha_i \theta \alpha_j \Rightarrow \alpha_i > \alpha_j \Rightarrow i < j \Rightarrow a_i < a_j$. For example, suppose $A = \{3,4,\ldots,39\}$, and we want to know whether A is 2-arboreal, i.e., 1-fold 2-x-arboreal. Then $\omega_e^c = \omega^2$, $t = 37$, and Ω, θ and T are as shown in the accompanying Figures.

i	1	2	3	4	5	6	7	8
a_i	3	4	5	6	7	8	9	10
α_i	ω^2	$\omega\cdot 3$	$\omega\cdot 2+4$	$\omega\cdot 2+3$	$\omega\cdot 2+2$	$\omega\cdot 2+1$	$\omega\cdot 2$	$\omega+9$

i	9	...	16	17	18	19	...	35	36	37
a_i	11	...	18	19	20	21	...	37	38	39
α_i	$\omega+8$	...	$\omega+1$	ω	19	18	...	2	1	0

Table 1: Tabulation of Ω for $A = [3,39]$ and $\alpha = \omega^2$.

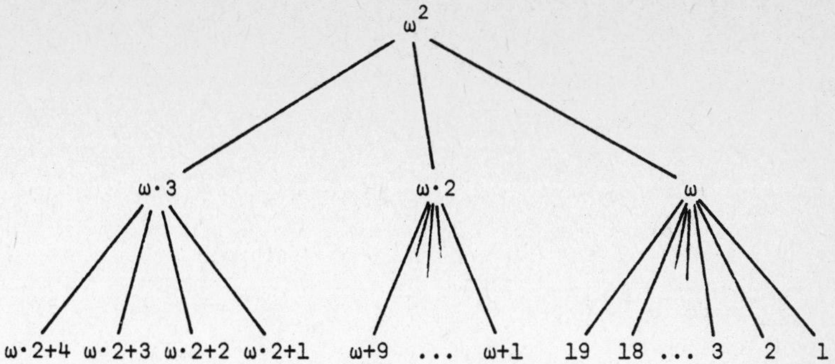

Figure 7: Ω-tree Θ for $A^{\circ} = [3,38]$ and $\alpha = \omega^2$.

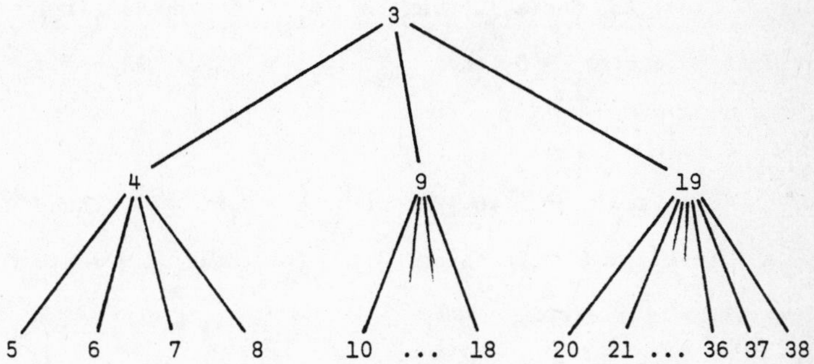

Figure 8: Corresponding A°-tree T.

Our goal is to show that T is (e+1)-fold c-f(x)-canonical, and that if $\alpha_t = 0$ then T is maximal, i.e., there is no node available for a_t in T (as in the example). Also that if $\alpha_t > 0$ then T is not maximal. We proceed via a series of claims. To simplify notation set $n_i = f(a_i)$ for $i = 1,2,\ldots,t$.

<u>Claim 1</u>. If $\alpha_i = \omega^{\gamma_1}.m_1 + \ldots + \omega^{\gamma_s}.m_s$ in CNF with $\gamma_s > 0$, and $\alpha_j \leq \omega^{\gamma_1}.m_1 + \ldots + \omega^{\gamma_s}.(m_s-1) + \omega^{\{\gamma_s\}(n_i)}$ with $j < t$, then $|IS_T(a_i)| = n_i$ and $a_k \leq a_j$ for all $a_k \varepsilon IS_T(a_i)$.

<u>Proof</u>. Let $\beta_x = \omega^{\gamma_1}.m_1 + \ldots + \omega^{\gamma_s}.(m_s-1) + \omega^{\{\gamma_s\}(n_i)}.x$ for $1 \leq x \leq n_i$. Then $\beta_{n_i} = \{\alpha_i\}(n_i)\varepsilon\Omega$ and for each $x < n_i$ $\beta_{n_i} = \beta_x + \gamma_x$ with $\gamma_x \ll \beta_x$. Since $\alpha_j \leq \beta_1 \leq \beta_x$ we have $\beta_x \varepsilon\Omega$ by Lemma 4.4. We have $IS_\theta(\alpha_i) = \{\beta_x | 1 \leq x \leq n_i\}$ and Claim 1 follows.

<u>Claim 2</u>. (a) If $Suc(\alpha_i)$, then $IS_\theta(\alpha_i) = \phi$ and $IS_T(a_i) = \phi$. (b) If $Lim(\alpha_i)$ and $\alpha_t = 0$, then $|IS_\theta(\alpha_i)| = |IS_T(a_i)| = f(a_i)$. In any case $|IS_T(a_i)| \leq f(a_i)$.

<u>Proof</u>. (a) Say $\alpha_i = \omega^\beta.n + m$ with $\beta > 0$, $m > 0$. Then $\alpha_{i+1} = \omega^\beta.n + (m-1)$ and there is no $\gamma\varepsilon\Omega$ for which $m - 1$ is the maximal q with $\omega^\beta.n + q\varepsilon\Omega$.

(b) If $\alpha_t = 0$, then $\alpha_{t-1} = 1$ (since $\alpha_{t-1} \geq 1$), so by Claim 1 with $j = t - 1$ we have $|IS_T(a_i)| = n_i = f(a_i)$. In any case $|IS_T(a_i)| = |IS_\theta(\alpha_i)| = |\{\beta_x | m \leq x \leq n_i\}| \leq n_i$ for some $m \geq 1$.

<u>Claim 3</u>. Let $\theta_\alpha \cup \{\alpha\} = \{\alpha_{i_1}, \alpha_{i_2}, \ldots, \alpha_{i_s}\}$ (the path in θ up to α), let $\beta_1 = \omega_e^c$ and $\beta_{j+1} = \{\beta_j\}(n_{i_j})$ for $j = 1,2,\ldots,s-1$. Then

there exist integers $x_1, x_2, \ldots, x_s \geq 1$ such that for $j = 1,2,\ldots,s$

$$\alpha_{i_j} = \omega^{\beta_1}.(x_1-1) + \ldots + \omega^{\beta_{j-1}}.(x_{j-1}-1) + \omega^{\beta_j}.x_j .$$

Proof. Since $\alpha_{i_1} = \alpha_1 = \omega^c_{e+1} = \omega^{\beta_1}$, we may take $x_1 = 1$. Now assume $1 \leq j < s$ and α_{i_j} is as claimed. Since $\alpha_{i_{j+1}} \varepsilon IS_\theta(\alpha_{i_j})$ we have $Lim(\alpha_{i_j})$ by Claim 2, so $\beta_j > 0$. Hence

$$\{\alpha_{i_j}\}(n_{i_j}) = \alpha_{i_j+1} = \omega^{\beta_1}.(x_1-1) + \ldots + \omega^{\beta_j}.(x_j-1) + \omega^{\{\beta_j\}(n_{i_j})}.n_{i_j} .$$

Since $\beta_{j+1} = \{\beta_j\}(n_{i_j})$ and since $\alpha_{i_{j+1}} \varepsilon IS_\theta(\alpha_{i_j})$, we have

$$\alpha_{i_{j+1}} = \omega^{\beta_1}.(x_1-1) + \ldots + \omega^{\beta_j}.(x_j-1) + \omega^{\beta_{j+1}}.x_{j+1} \quad \text{for some } x_{j+1}$$

with $1 \leq x_{j+1} \leq n_{i_j}$.

Claim 4. For $\alpha_i \varepsilon \Omega$ the following are equivalent:

(a) $T_{a_i} \cup \{a_i\}$ is e-fold c-f(x)-arboreal.

(b) $Suc(\alpha_i)$.

Proof. Let $A' = T_{a_i} \cup \{a_i\} = \{a_{i_1}, a_{i_2}, \ldots, a_{i_s}\}$ and $\Omega' = \{\beta_1, \beta_2, \ldots, \beta_s\}$ with $\beta_1 = \omega^c_e$ and $\beta_{j+1} = \{\beta_j\}(n_{i_j})$ for $j = 1,2,\ldots,s-1$. (Recall $n_{i_j} = f(a_{i_j})$). By Ord(e) we have A' is e-fold c-f(x)-arboreal iff $\beta_s = 0$. But $a_i = a_{i_s}$, so

$\alpha_i = \alpha_{i_s} = \omega^{\beta_1}.(x_1-1) + \ldots + \omega^{\beta_s}.x_s$ with $x_s \geq 1$, by Claim 3. Thus

$\beta_s = 0$ iff $\mathrm{Suc}(\alpha_i)$.

<u>Claim 5</u>. $IS_\theta(\alpha_i) = \phi$ iff $\mathrm{Suc}(\alpha_i)$ or $i = t - 1$.

<u>Proof</u>. If $\mathrm{Lim}(\alpha_i)$ and $i < t - 1$, then $\alpha_{i+1} \varepsilon IS_\theta(\alpha_i)$. Conversely, if $\alpha_j \varepsilon IS_\theta(\alpha_i)$, then $i < j \leq t - 1$ and not $\mathrm{Suc}(\alpha_i)$, by Claim 2.

<u>Claim 6</u>. If $a_j \varepsilon IS_T(a_i)$, then a_i is available for a_j in T.

<u>Proof</u>. We must show (i) $a_i < a_j$; (ii) $T_{a_i} \cup \{a_i\}$ is not e-fold c-f(x)-arboreal; and (iii) $|IS_T(a_i) \cap \{y|y<a_j\}| < f(a_i)$. Well (i) is trivial and (ii) follows immediately by Claims 4 and 5 since $IS_T(a_i) \neq \phi$. (iii) is immediate from Claim 2 since $a_j \varepsilon IS_T(a_i)$, so

$|IS_T(a_i) \cap \{y|y<a_j\}| < |IS_T(a_i)| \leq f(a_i)$.

<u>Claim 7</u>. If $a_j \varepsilon IS_T(a_i)$ and $a_i < a_k$, then a_k is <u>not</u> available for a_j in T.

<u>Proof</u>. Suppose a_k were available. Then $a_k < a_j$ so $\alpha_i > \alpha_k > \alpha_j$ and consequently $\alpha_{i+1} \geq \alpha_k$. Let $\alpha_k = \alpha_j + \beta$ and $\alpha_i = \omega^{\gamma_1}.m_1 + \ldots + \omega^{\gamma_s}.m_s$ in CNF (with $\gamma_s > 0$ by Claim 3). Then

$\alpha_j = \omega^{\gamma_1}.m_1 + \ldots + \omega^{\gamma_s}.(m_s-1) + \omega^{\gamma_{s+1}}.x$ for some $1 \leq x < n_i$, where

$\gamma_{s+1} = \{\gamma_s\}(n_i)$. Also α_{i+1} has the same form, but with $x = n_i$. Now

$\beta \leq \omega^{\gamma_{s+1}}.(n_i-x)$ because $\alpha_j + \omega^{\gamma_{s+1}}.(n_i-x) = \alpha_{i+1} \geq \alpha_k = \alpha_j + \beta$.

Therefore in CNF we have $\alpha_k = \omega^{\gamma_1}.p_1 + \ldots + \omega^{\gamma_u}.p_u$ for some $u \geq s+1$, $\gamma_{s+1} > \gamma_{s+2} > \ldots > \gamma_u$. We have $p_\ell = m_\ell$ for $\ell = 1,2,\ldots,s-1$. Also $p_s = m_s - 1$ and $x \leq p_{s+1} \leq n_i$. If $p_{s+1} = x$ then $u \geq s+2$.

Case 1: $\gamma_u = 0$. Then $\mathrm{Suc}(\alpha_k)$, so by Claim 4 $T_{a_k} \cup \{a_k\}$ is e-fold c-f(x)-arboreal and a_k is not available for anything. Contradiction.

Case 2: $\gamma_u > 0$. For $z = 1,2,\ldots,n_k$ let

$$\alpha_{k,z} = \omega^{\gamma_1}.p_1 + \ldots + \omega^{\gamma_{u-1}}.p_{u-1} + \omega^{\gamma_u}.(p_u-1) + \omega^{\{\gamma_u\}(n_k)}.z\ .$$

Since $\alpha_j < \alpha_{k,z}$ for all z we have $\alpha_{k,z} \varepsilon \Omega$ by Lemma 4.4(i). Therefore $IS_\theta(\alpha_k) = \{\alpha_{k,z} | 1 \leq z \leq n_k\}$. Letting $\alpha_{i_z} = \alpha_{k,z}$ we have

$IS_T(a_k) = \{a_{i_z} | 1 \leq z \leq n_k\}$ and $a_{i_z} < a_j$ (because $\alpha_{k,z} > \alpha_j$). Therefore $|IS_T(a_k) \cap \{y | y < a_j\}| = n_k = f(a_k)$, so a_k is not available for a_j. Contradiction.

This completes the proof of Claim 7.

We have reached our goal of proving that T is (e+1)-fold c-f(x)-canonical, by Claims 6 and 7. To finish the proof we need to show T is maximal.

Claim 8. If $\alpha_t = 0$, then there is no node $a_i \varepsilon A^o$ available for a_t in T.

Proof. If $\mathrm{Lim}(\alpha_i)$, then by Claim 2 $|IS_T(a_i)| = f(a_i)$. Since $a \varepsilon A^o \Rightarrow a < a_t$, we have $|IS_T(a_i)| = |IS_T(a_i) \cap \{y | y < a_t\}| = f(a_i)$ so a_i is not available for a_t. If $\mathrm{Suc}(\alpha_i)$, then by Claim 4 a_i is not available for anything. This proves Claim 8.

It follows from Claim 8 that if $\alpha_t = 0$, then the (e+1)-fold c-f(x)-canonical A-tree does not exist. By Can(e+1) this implies A is (e+1)-fold c-f(x)-arboreal.

Now suppose $\alpha_t > 0$. Then let $a_{t+1} = a_t + 1$ and repeat the whole argument with $B = A \cup \{a_{t+1}\}$. We have shown that the (e+1)-fold c-f(x)-canonical B^o-tree exists. But $B^o = A$, so A is not (e+1)-fold c-f(x)-arboreal. This completes the proof of Theorems 4.2 and 4.3.

Section 5. The proof theoretic strength of arboricity.

In this section we simply record for the sake of completeness the proof theoretic strength of arboricity in terms of various natural fragments of Peano arithmetic. These results are based on Paris's work concerning the derivability of the Paris-Harrington partition relation. If Γ is a class of formulas in the language of PA let $I\Gamma$ denote the fragment of PA in which the induction schema is allowed only for formulas in Γ:

$$[\theta(0) \wedge \forall x(\theta(x) \rightarrow \theta(x+1)) \rightarrow \forall x \theta(x)] \text{ for } \theta \varepsilon \Gamma .$$

$I\Gamma$ includes also the usual axioms recursively defining $+$, $\cdot$, and the properties of zero and successor. Let $\Pi_1(PA)$ denote the set of Π_1 formulas which are theorems of PA. Given $e \geq 1$ Paris has proved the following:

$$\text{(i)} \quad I\Sigma_e + \Pi_1(PA) \nvdash \forall z\, \forall x\, \exists y\, [x,y] \rightarrow (*)_z^{e+1}$$

but for each fixed $c \geq 1$

$$\text{(ii)} \quad I\Sigma_e + \Pi_1(PA) \vdash \forall x\, \exists y\, [x,y] \rightarrow (*)_c^{e+1} .$$

Now all of the combinatorial arguments used in the proofs of Theorems 2.6 and 3.6 can certainly be carried out in $I\Sigma_1$ (and probably even $I\Sigma_o$) since they deal only with given finite sets. It follows that for each $e \geq 1$

$$\text{(i)} \quad I\Sigma_e + \Pi_1(PA) \nvdash \forall z\, \forall x\, \exists y\, [x,y] \text{ is e-fold z-arboreal,}$$

but for each fixed $c \geq 1$

$$\text{(ii)} \quad I\Sigma_e + \Pi_1(PA) \vdash \forall x\, \exists y\, [x,y] \text{ is e-fold c-arboreal.}$$

and of course it follows immediately from (i) that

$$\text{(iii)} \quad PA \nvdash \forall w\, \forall z\, \forall x\, \exists y\, [x,y] \text{ is w-fold z-arboreal.}$$

Remark on rates of growth: Define $A_c^e(k)$ = least n such that $[k,n]$ is e-fold c-arboreal. Let $A_c(k) = A_c^1(k)$. The following exact values follow directly from Theorems 4.2 and 4.3.

$$A_o(k) = \kappa + 1$$

$$A_1(k) = 2k + 1$$

$$A_2(k) = (k+2)2^k - 1$$

and in general

$$A_{c+1}(k) = A_c(A_c(\ldots(A_c(k+1))\ldots)),\ k \text{ applications of } A_c.$$

Thus for each fixed c, the function f(k) = "least n s.t. [k,n] is c-arboreal" is primitive recursive. However even for k = 2, the function g(c) = "least n s.t. [2,n] is c-arboreal" grows more rapidly than any primitive recursive function. It is essentially Ackermann's function. This use of trees provides a quick and graphic definition of Ackermann's function for those unfamiliar or uncomfortable with definitions by recursion.

Ackermann's function is relatively tame compared with the rates of growth encountered for iterated arboricity. In general (verify) $A_1^{e+1}(k) = A_{k+1}^{e}(k)$, so A_1^{e+1} dominates all functions A_c^e. Thus A_1^2 is, again, a variant of Ackermann's function, and the hierarchy continues up from there. For fixed $c \geq 1$ $I\Sigma_e + \Pi_1(PA)$ can prove that A_c^e is totally defined, but not that A_1^{e+1} is totally defined. In fact A_1^{e+1} dominates (is eventually greater than) any recursive function which $I\Sigma_e + \Pi_1(PA)$ can prove to be total. Finally A, as a function of three variables, dominates any recursive function which PA can prove to be total.

Section 6. Ordinal of the tree of "small sets".

In this section we use the ordinal characterization of arboricity to place fairly strict bounds on a certain ordinal associated in a natural way with the Paris-Harrington partition relation. This seems to give another reasonable measure of the complexity of that relation. Some earlier results on this measure were obtained independently by Peter Aczel.

<u>6.1 Definition</u>. $T_c^e = \{A \subseteq N \mid A \not\to (*)_c^e\}$.

We consider T_c^e to be an ω-branching tree under the relation of end extension. That is $A \prec B$ (in the sense of T_c^e) iff $a < b$ for all $a \in A$, $b \in B - A$. (Of course T_c^e is a different sort of tree than we have been considering up to now.) Since every infinite $X \subseteq N$ contains a finite initial segment $A \subseteq X$ such that $A \to (*)_c^e$, it follows that T_c^e contains no infinite paths, i.e., T_c^e is well-founded. There is a traditional way in logic to associate an ordinal with any well-founded tree. We will describe it in terms of a game.

The game $G_c^e(\alpha)$ is played by two players, One and Two, who alternate turns. Player One names ordinals while Two names positive integers. The record of a play of $G_c^e(\alpha)$ through t moves can be written thus:

One	Two
α_1	a_1
α_2	a_2
α_3	a_3
.	.
.	.
.	.
α_t	a_t

One's ordinals must satisfy $\alpha \geq \alpha_1 > \alpha_2 > \alpha_3 > \ldots > \alpha_t$, and Two's integers must satisfy $a_1 < a_2 < \ldots < a_t$. The game ends when One has no legal move, i.e., when $\alpha_t = 0$. At that point Two is deemed the winner iff $\{a_1, a_2, \ldots, a_t\} \varepsilon T_c^e$, that is, iff $\{a_1, a_2, \ldots, a_t\} \nrightarrow (*)_c^e$. In other words One tries, by choosing large ordinals, to prolong the game until Two has named such a large set that it $\rightarrow (*)_c^e$.

<u>6.2 Definition</u>. $o(T_c^e)$ = the minimum α such that One has a winning strategy in the game $G_c^e(\alpha)$.

<u>6.3 Theorem</u>. (1) $o(T_c^1) = \omega \cdot c$.

(2) $o(T_c^2) = \omega^c$.

(3) For $e \geq 2$ $\omega_e^d \leq o(T_c^{e+1}) \leq \omega_e^c$ where $d = [\frac{c}{2}] - e + 1$.

For the proof we first need to calculate the ordinal of a closely related game. Define a game $H(\alpha, \beta, f(x))$ exactly as $G_c^e(\alpha)$ except that Two's winning criterion is that $\{a_1, \ldots, a_t\}$ must be not β-$f(x)$-large.

<u>6.4 Lemma</u>. If $f(x)$ is unbounded, then One has a winning strategy for $H(\alpha, \beta, f(x))$ iff $\alpha \geq \beta$.

<u>Proof</u>. If $\alpha \geq \beta$ then One's strategy is to play $\alpha_1 = \beta$ and $\alpha_{i+1} = \{\alpha_i\}(f(a_i))$. When $\alpha_t = 0$ the set $\{a_1, \ldots, a_t\}$ is, by definition, β-$f(x)$-large. One has won.

If $\alpha < \beta$, Two's winning strategy is to define auxiliary ordinals $\beta_1 > \beta_2 > \ldots > \beta_t$ so that $\beta_1 = \beta$ and given $\beta_i > \alpha_i$ (as $\beta_1 > \alpha_1$), to choose a_i so large that $\{\beta_i\}(f(a_i)) \geq \alpha_i$. Then set

$\beta_{i+1} = \{\beta_i\}(f(a_i))$, so $\beta_{i+1} \geq \alpha_i > \alpha_{i+1}$. In the end $\beta_t > 0 = \alpha_t$ and the β's show that $\{a_1,\ldots,a_t\}$ is not β-$f(x)$-large. Two has won. This proves the lemma.

We remark that if $f(x)$ is bounded, then One has a winning strategy in $H(\alpha,\beta,f(x))$ for some finite ordinal α.

6.5 Lemma. For $c \geq 1$ the following are equivalent:

(1) $A \to (*)^1_c$

(2) A is $\omega\cdot c$-$(x-2)$-large.

Proof. If $A = \{a_1,\ldots,a_t\}$ is not $\omega.c$-$(x-2)$-large, define $\alpha_1 = \omega.c$, $\alpha_{i+1} = \{\alpha_i\}(a_i-2)$, so that $\alpha_t > 0$. Define $F:A \to c$ by

$$F(a_i) = \min\{r \mid \alpha_i \leq \omega.(r+1)\} .$$

Now if $a_i = \min F^{-1}(r)$, then $\alpha_i = \omega.(r+1)$ by Lemma 4.4. Thus $\alpha_{i+1} = \omega.r + (a_i-2)$, $\alpha_{i+a_i-1} = \omega.r$, and

$F^{-1}(r) = \{a_i,a_{i+1},\ldots,a_{i+a_i-2}\}$, not relatively large. Thus $A \not\to (*)^1_c$.

Conversely if $A \not\to (*)^1_c$ call a partition $F:A \to c$ good if $|F^{-1}(i)| < \min F^{-1}(i)$ for all i, and call $a \in A$ initial if $a = \min F^{-1}(i)$ for some i. The following is easy to verify: If F is good, a is initial and if $b < a < d$ for some b,d with $F(b) = F(d)$, then the partition F' obtained by "interchanging a and d" is good. Thus there exists a good partition F with $F^{-1}(0) < F^{-1}(1) < \ldots < F^{-1}(c-1)$. We may further assume $F^{-1}(i)$ has maximal cardinality for each $i < c - 1$; thus $|F^{-1}(i)| = (\min F^{-1}(i)) - 1$. If we now assign ordinals

$\alpha_1, \alpha_2, \ldots, \alpha_t$ testing for $\omega.c-(x-2)$-largeness, it turns out that α_i is a successor ordinal iff a_i is not initial. The c initial points must be "used up" to account for the c limit ordinals $\leq \omega.c$ hence α_t is a successor, not 0, so A is not $\omega.c-(x-2)$-large.

Proof of Theorem 6.3. (1) By Lemma 6.5 $G_c^1(\alpha)$ and $H(\alpha,\omega.c,x-2)$ are in fact the same game. Hence by Lemma 6.4, $o(T_c^1) = \omega.c$.

(2) If $\alpha \geq \omega^c$, let One play $G_c^2(\alpha)$ using his/her winning strategy for $H(\alpha,\omega^c,c^{cx})$. The resulting set $\{a_1,\ldots,a_t\} = A$ is ω^c-c^{cx}-large, hence c-c^{cx}-arboreal by Theorem 4.3. By Theorem 2.6 $A \rightarrow (*)_c^2$, provided $\min A \geq 4$. To ensure this, correct One's strategy by having One add four to each of Two's moves before calculating a response. Then $\{a_1+4, a_2+4,\ldots,a_t+4\} \rightarrow (*)_c^2$ by the argument above. A fortiori $A \rightarrow (*)_c^2$.

If $\alpha < \omega^c$, let Two play $G_c^2(\alpha)$ by using his/her winning strategy for $H(\alpha,\omega^c,x-2)$ and adding four. Then A is not ω^c-$(x-2)$-large, hence not c-(x-2)-arboreal, hence $A \not\rightarrow (*)_c^2$, by Theorems 4.3 and 2.6.

(3) Similar argument using Theorem 3.6.

Section 7. Iteration of the Paris-Harrington relation

In this section we use the machinery of arboreal sets to measure the complexity of iterated partition relations.

7.1 Definition. A is 0-dense(e,c) iff $|A| \geq \min A$ and $|A| > e$. A is (n+1)-dense(e,c) iff for every $F:[A]^e \rightarrow c$ there exists a homogeneous $B \subseteq A$ such that B is n-dense(e,c).

Thus A is 1-dense(e,c) iff $A \to (*)^e_c$. Recall that one of the first independent combinatorial statements obtained by Paris was the statement "For all n there exists an n-dense(3,2) set." It is still an open question whether the statement "For all n there exists an n-dense(2,2) set" is provable in PA.

We obtain an arboreal upper bound for the size of n-dense(3,2) sets via the following lemma. Recall that 2^x_y denotes "a stack of y twos, topped by an x".

7.2 Lemma. Suppose $1 \leq n \leq e$, $2 \leq c$ and $\min A = a \geq 2$. Suppose A is e-fold c-2^x_x-arboreal and $F:[A]^n \to 2^a_{a-2}$. Then there exists a homogeneous $B \subseteq A$ which is (e-n+1)-fold $(c-1)$-2^x_x-arboreal.

Proof. Let $d = 2^a_{a-2}$ and note that $d^{\binom{x-1}{m-1}} \leq d^{2^x} \leq 2^x_x$ for all $x \in A$, $m \leq n$. Thus we may iterate Lemma 3.5 n-1 times to obtain a subset $H \subseteq A$ and partition $G:H^{oo\ldots o} \to d$ (n-1 o's) such that $H^{oo\ldots o}$ is (e-(n-1))-fold c-2^x_x-arboreal and for all $\langle x_1,\ldots,x_n\rangle \in [H]^n$, $F(\bar{x}) = G(x_1)$. Let $C = H^{oo\ldots o}$ and write $C = \bigcup_{i<d} G^{-1}(i)$. Since $d < 2^{\min C}_{\min C}$ and $e - n \geq 1$ we have by Lemma 3.7 that some $G^{-1}(i)$ is (e-n+1)-fold $(c-1)$-2^x_x-arboreal. That $G^{-1}(i)$ is certainly homogeneous for F.

7.3 Theorem. Suppose $\min A \geq e + 2$ and A is 2e-fold (e+2)-arboreal. Then A is e-dense(3,2).

Proof. We will actually prove by induction on e that if A is 2e-fold $(e+1)$-2^x_x-arboreal, then A is e-dense(3,2). The theorem then follows immediately from Lemma 3.8.

For $e = 1$ A is 2-fold $2\text{-}2^x_x$-arboreal, so $A \to (*)^3_2$ by (the proof of) Theorem 3.6 ($\min A \geq e^{2e}$ was not used), i.e., A is 1-dense(3,2). Now assume the hypothesis holds for a given $e \geq 1$ and A is (2e+2)-fold $(e+2)\text{-}2^x_x$-arboreal. Let $F:[A]^3 \to 2$. By Lemma 7.2 there is a homogeneous $B \subseteq A$ which is ((2e+2)-3+1)-fold $(e+2-1)\text{-}2^x_x$-arboreal. Therefore B is e-dense(3,2) by the induction hypothesis, proving A is (e+1)-dense(3,2). This proves Theorem 7.3.

Similar upper bounds can be derived for n-dense(e,c) for other values of n,e,c. We conjecture that these upper bounds are close to the actual values, i.e., no degree of (2e-1)-fold arboricity is an upper bound for e-dense(3,2). However the currently known lower bounds fall far short of showing this. The reason is that we do not have an effective construction of bad partitions of exponent two. What is lacking is a version of Lemma 3.4 for $e = 1$. The following bound is a rough indication of the best that can be obtained using the current methods contained in Lemma 3.4.

<u>7.4 Theorem</u>. Suppose $n \geq 1$, $\min A \geq 16$ and A is n-dense(3,4). Then A is (n+1)-fold 1-(x-2)-arboreal.

<u>Remark</u>. We conjecture that A must in fact be 2n-fold 1-(x-2)-arboreal.

<u>Proof</u>. By induction on n. We use the fact that $\min A \geq 2^{2\cdot 2}$. For $n = 1$, we have $A \to (*)^3_4$, so A is 2-fold $([\frac{4}{2}]-2+1)$-(x-2)-arboreal by Theorem 3.6, as required. Now assume the theorem holds for a given $n \geq 1$ and A is (n+1)-dense(3,4) but <u>not</u> (n+2)-fold 1-(x-2)-arboreal.

Let T be an (x-2)-small branching A-tree with no path P such that P^o is (n+1)-fold 1-(x-2)-arboreal. For each $a \in A$, take $F_a:[T_a]^2 \to 1$ to be the constant function and define $F:[A]^3 \to 4$ $(=2\cdot 1+2)$ as in the proof of Lemma 3.4(1).

By assumption there is an n-dense(3,4) homogeneous subset $H \subseteq A$. Now H is certainly relatively large, so the reasoning of the proof remains valid up through the definition of the set H*. Let $P = H^* \cup \{p(y_e), y_e\}$. Now P is a path in T and P^o is n-dense(3,4) since it is obtained from H by replacing each y_i by the smaller $p(y_i \frown y_e)$ (the usual sort of monotonicity argument). By the induction hypothesis P^o is (n+1)-fold 1-(x-2)-arboreal, contradiction. This completes the proof.

Clearly this proof seems to have wasted an exponent, since no use was made of the potentially nasty partitions $F_a:[T_a]^2 \to c$. By how much can an exponent two partition reduce the size of its homogeneous sets? Lemma 7.2 provides a lower bound for the guaranteed size of a homogeneous set when the parent set is at least 2-fold-arboreal. The following companion lemma holds for 1-fold-arboreal sets.

<u>7.5 Theorem.</u> $\omega^{nc+3} \to (\omega.n)^2_c$ in the sense that if A is ω^{nc+3}-large (nc+3-arboreal) with $\min A \geq nc, 3$ and if $F:[A]^2 \to c$, then there exists an $\omega\cdot n$-large F-homogeneous subset $B \subseteq A$.

<u>Proof.</u> The construction is based on the same underlying idea as was used in the proof of (2) ⇒ (3) of Theorem 2.6. In fact the construction is essentially the same for the case n = 1, though the formal description is quite different, making the two proofs appear unrelated. The technique may be called "collapsing of subtrees" and is an effective way to utilize the properties of arboreal sets.

Let A and F be given. Begin by constructing the "ordinary Ramsey tree" T for this partition; that is, T is a c-branching A-tree with the property that aTbTd implies $F(a,b) = F(a,d)$. We now assign a vector of ordinals to each node in T. Set $\mathrm{ord}(\min A) = \langle\omega\cdot n, \omega\cdot n, \ldots, \omega\cdot n\rangle$, c components. If $b \in IS_T(a)$ and $\mathrm{ord}(a) = \langle\alpha_1, \alpha_2, \ldots, \alpha_c\rangle$, then set $\mathrm{ord}(b) = \langle\beta_1, \beta_2, \ldots, \beta_c\rangle$ where

$$\beta_j = \begin{cases} \{\alpha_j\}(a) & \text{if } j = F(a,b) \\ \alpha_j & \text{if } j \neq F(a,b) . \end{cases}$$

Thus only one component is decreased, corresponding to the color $F(a,b)$. (We identify c with $[1,c]$.) Note (inductively) that each β_j has the form $\beta_j = \omega\cdot x + y$ with $x \leq n$ and $y \leq a$. We use the notation $\mathrm{ord}_j(b)$ to denote the jth component of $\mathrm{ord}(b)$.

Suppose for the moment that for some i and some b, $\mathrm{ord}_i(b) = 0$. Let $a_1, a_2, \ldots, a_r$ enumerate $\{a \mid aTb\}$, and let $j_1, j_2, \ldots, j_{t-1}$ be those j's for which $F(a_j, b) = i$. Set $\alpha_1 = \omega\cdot n$ and $\alpha_{x+1} = \{\alpha_x\}(a_{j_x})$ for $x = 1,2,\ldots,t-1$. It is easy to check that in fact $\alpha_x = \mathrm{ord}_i(a_{j_x})$, whence $\alpha_t = \mathrm{ord}_i(b) = 0$. Thus $B = \{a_{j_1}, a_{j_2}, \ldots, a_{j_{t-1}}, b\}$ is $\omega\cdot n$-large. Since B is homogeneous for F (to color i) we are done.

Thus it suffices to prove that for some i and some b, $\mathrm{ord}_i(b) = 0$. Suppose that is not the case. For each $a \in A$, define $L(a) = n_1 + n_2 + \ldots + n_c$, where $\mathrm{ord}_i(a) = \omega\cdot(n-n_i) + m_i$. ($L(a)$ measures how many limit ordinals $\mathrm{ord}(a)$ has passed.) Observe that $L(a) = 0$ iff $a = \min A$ and that L satisfies $L(a) \leq L(b) \leq L(a) + 1$ for $b \in IS_T(a)$. We define a "collapsed tree" T* in terms of its predecessor function:

$$\mathrm{pred}_{T*}(b) = \max\{a\varepsilon A \mid aTb \quad \text{and} \quad L(b) = L(a) + 1\} \ .$$

Now $L(a)$ = rank of a in $T*$. Since $L(a) \leq nc$ for all a, we see $T*$ has no path of length $nc + 2$. But A is $(nc+3)$-arboreal, hence nc-2^x_x-arboreal by Lemma 2.7. Therefore $T*$ is not 2^x_x-small-branching.

Pick an $a\varepsilon A$ with $|IS_{T*}(a)| > 2^a_a$. Let $S = \{b\varepsilon a \mid aTb$ and $L(b) = L(a) + 1\}$. Thus $IS_{T*}(a) \subseteq S$, so $|S| > 2^a_a$. But $S \cup \{a\}$ is a subtree of T, whence a rough calculation shows that $|S| \leq kc^k$ where k is the maximal length of any path in S under the ordering of T. Suppose $b_1, b_2, \ldots, b_k$ is such a maximal path. Say $ord_i(b_1) = \omega \cdot n_i + m_i$ for $i = 1,2,\ldots,c$. Recall $m_i \leq a$. Now for any $j \geq 1$, since $L(b_j) = L(b_1)$ we have $ord_i(b_j) = \omega \cdot x + y$ with $x = n_i$ and $y \leq m_i$. Since each of the c components of $ord(b_j)$ is greater than zero and one of them decreases at each step (from j to $j + 1$), we must have $k \leq c(a+1)$. Therefore $|S| \leq (a+1)c^{c(a+1)+1} \leq$ $(a+1)^{(a+1)^2} < 2^a_3 \leq 2^a_a$. This is a contradiction, and the proof is complete.

<u>Remark</u>. We conjecture that for some (small) c there is no d such that $\omega^d \to (\omega^2)^2_c$ in the sense of Theorem 7.5.

<u>Section 8. Applications to other partition statements</u>

We conclude with two applications of arboricity to combinatorial statements previously studied by H. Friedman and L. Kirby, respectively. We include these applications primarily to illustrate further the use of arboricity in combinatorial arguments. Both statements studied were motivated by the Paris-Harrington result, and, like it, were devised by finitizing established infinitary combinatorial statements. We will begin with Friedman's property.

8.1 Definition. (Friedman) $H_k(m)$ = least n such that the following holds: Let $g_1,\ldots,g_k:[1,n] \to [1,n]$ be such that each $g_i(j) \leq j$. Then there is an increasing sequence $b_1 < b_2 < \ldots < b_m \leq n$ such that for each $i \in [1,k]$, either g_i is constant on $b_1,\ldots,b_m$ or each $b_j < g_i(b_{j+1})$.

Friedman defines an Ackermann hierarchy thus: $F_1(m) = 2^{m-1} + 1$, $F_{k+1}(m) = F_k(\ldots F_k(1)\ldots)$, m iterations. He then proves the following theorem, giving a combinatorial proof for the lower bound and a metamathematical argument for the upper bound.

8.2 Theorem. (Friedman) $\forall k \exists q \forall m > 3 \quad F_k(m) < H_k(m) < F_q(m)$.

Using arboricity we obtain the following stronger theorem with explicit upper bounds.

8.3 Theorem. Let A be $(k+3)$-arboreal with $\min A \geq m^k$, $m > 3$. Then for any functions $g_1,\ldots,g_k:A \to N$ there exists an increasing sequence $b_1 < b_2 < \ldots < b_m$ from A such that for each $i \in [1,k]$, either g_i is constant on $b_1,\ldots,b_m$ or each $b_j < g_i(b_{j+1})$ and $m^k < g_i(b_1)$. Thus $H_k(m) \leq A_{k+3}(m^k)$.

Proof. Let $A = \{a_1,\ldots,a_t\}$ and $g_1,\ldots,g_k$ be given. Build an

A-tree T in stages. Having placed $a_1,\dots,a_j$ on T, say that a_p is available for a_{j+1} iff either $p = 1$ or, letting $a_p \in IS_T(a_q)$, we have

$$(1)\quad \forall i\in[1,k] \quad g_i(a_p) \leq a_q \Rightarrow g_i(a_{j+1}) = g_i(a_p), \text{ and}$$

$$(2)\quad \exists i\in[1,k] \quad g_i(a_p) > a_q \text{ and } g_i(a_{j+1}) \leq a_p .$$

Choose as immediate predecessor for a_{j+1} the maximal $b\in\{a_1,\dots,a_j\}$ such that all $a\in T_b \cup \{b\}$ are available for a_{j+1}. This defines T.

Now if $a\in IS_T(b)$ define $L(a) = |\{i|g_i(a)>b\}|$. Then by (2) if $c\in IS_T(a)$ and $a\in IS_T(b)$, then $L(c) < L(a)$. Since $0 \leq L(a) \leq k$ for all $a\in A-\{\min A\}$, T has no path of length $\geq k + 3$. But by Lemma 2.7 A is $(k+1)$-$(x+1)^{k+1}$-arboreal, since $[x,2(x+1)^{k+1}]$ is never 2-arboreal for $x \geq m^k \geq \min A$. Therefore T is not $(x+1)^{k+1}$-small-branching. Fix $a\in A$ with $|IS_T(a)| > (a+1)^{k+1}$. For each $b\in IS_T(a)$ define $\sigma_b:[1,k] \to [1,a+1]$ by

$$\sigma_b(i) = \begin{cases} g_i(b) & \text{if } g_i(b) \leq a \\ a + 1 & \text{if } g_i(b) > a \end{cases}$$

As there are at most $(a+1)^k$ possibilities for σ_b there must exist distinct $b_1,b_2,\dots,b_{a+1}\in IS_T(a)$ with $\sigma_{b_p} = \sigma_{b_q}$ for $1 \leq p < q \leq a + 1$.

Let $B = \{b_1,\dots,b_m\}$. Now for each i we consider two possibilities.

Case 1. $g_i(b_j) \leq a$ for some j. Then $a \geq g_i(b_j) = \sigma_{b_j}(i) = \sigma_{b_h}(i) = g_i(b_h)$ for all h, so g_i is constant on B.

Case 2. $g_i(b_j) > a$ for all j. Then for each j since b_j was not available for b_{j+1} we have either

(1') not $\forall \ell \varepsilon [1,k]$ $g_\ell(b_j) \leq a \Rightarrow g_\ell(b_{j+1}) = g_\ell(b_j)$, or

(2') not $\exists \ell \varepsilon [1,k]$ $g_\ell(b_j) > a$ and $g_\ell(b_{j+1}) \leq b_j$.

But (1') contradicts Case 1, so by (2') $g_i(b_j) > a \Rightarrow g_i(b_{j+1}) > b_j$. Since $m^k \leq a < g_i(b_1)$, the theorem follows. In particular $[1, A_{k+3}(m^k)]$ satisfies the combinatorial property, so $H_k(m) \leq A_{k+3}(m^k)$.

We now turn to finitary flipping properties, defined and studied by L. Kirby. Suppose $\langle A_i : i<m \rangle$ is a sequence of subsets of A. We define a sequence of elements $\langle x_i : i<r \rangle$ for some $r \leq m$, by defining inductively for as long as possible:

$$x_o = \min A_o$$

$$x_{n+1} = \min\{x | x>x_n \text{ and } x \varepsilon \bigcap \{A_i : i<x_n\}\} ,$$

Note that x_n is used as the bound for the subscripts of the sets used to define x_{n+1}. The ladder of $\langle A_i : i<m \rangle$ is the set $\{x_i : i<r\}$. Given a sequence $\langle A_i : i<m \rangle$, by a flip of this sequence we mean any sequence $\langle A'_i : i<m \rangle$ such that for each $i < m$

$$A'_i = A_i \text{ or } A'_i = A - A_i .$$

8.4 Definition. (Kirby) A is 0-flippable if $|A| \geq 2$.

A is (n+1)-flippable iff any sequence $\langle A_i : i < \max A \rangle$ of subsets of A has a flip whose ladder is n-flippable.

By studying flippability in the context of nonstandard models of arithmetic, Kirby has shown that the statement

$$\forall n \; \exists A \quad A \text{ is n-flippable}$$

cannot be proved in PA although it is a true statement. In fact the function $W(a,b) = \max\{n \mid [a,b] \text{ is n-flippable}\}$ is an indicator for strong initial segments of models of PA.

The (proof of the) following theorem shows that flippability and arboricity are essentially the same concept in different guises.

8.5 Theorem. Let A be n-fold $(n+1)$-2^x-arboreal, $\min A > 1$. Then A is n-flippable.

Proof. By induction on n. For $n = 0$ A is 0-fold 1-2^x-arboreal, i.e., $|A^o| \geq 1$, so $|A| \geq 2$, A is 0-flippable. For $n > 0$ let $\langle A_i : i < m \rangle$ be a sequence of subsets of A with $m = \max A$. Let $A^+ = A \cup \{1\}$ and define an A^+-tree in stages: $\text{root}T = 1$. Given $b \in A$, if all $a \in A^+$, $a < b$, are already placed on T, say a is available for b if $a = 1$ or $a \in IS_T(c)$ and for all $i < c$ $a \in A_i \Leftrightarrow b \in A_i$. Let the immediate predecessor of b be the maximal $a < b$ such that all $x \in T_a \cup \{a\}$ are available for b.

If some node c had $> 2^c$ immediate successors, then for some $a, b \in IS_T(c)$, $a < b$, we would have $\{i < c \mid a \in A_i\} = \{i < c \mid b \in A_i\}$. Thus a would be available for b and b not correctly placed. Thus T is

2^x-small-branching. Let P be a path in T such that P^o is $(n-1)$-fold $(n+1)$-2^x-arboreal. Let $b = \max P$ and define the flip $\langle A_i' : i<m \rangle$ by $A_i' = A_i$ iff $b \in A_i$. As to the ladder of this flip we have

$$\begin{aligned} x_o = \min A_0' &= \min\{a \in A \mid a \in A_o \Leftrightarrow b \in A_o\} \\ &= \min\{a \in A \mid \forall i<1 \quad a \in A_i \Leftrightarrow b \in A_i\} \\ &= \text{the element of } P \cap IS_T(1) . \end{aligned}$$

Similarly x_{n+1} = the element of $P \cap IS_T(x_n)$. Thus the ladder of the flip is precisely $P - \{1\}$, which is $(n-1)$-fold n-2^x-arboreal since P is $(n-1)$-fold $(n+1)$-2^x-arboreal. By induction $P - \{1\}$ is $(n-1)$-flippable and we are done.

Note that if x_o were defined to be minA rather than $\min A_o$ there would be no need to fuss with first inserting and then deleting 1, and we would have for all A, if A is n-fold 1-2^x-arboreal then A is n-flippable.

<u>8.6 Definition</u>. (Kirby) A is <u>0-weakly-flippable</u> iff $|A| \geq 2$.

A is <u>(n+1)-weakly-flippable</u> iff whenever $\langle A_i : i<m \rangle$ is a sequence of subsets of A, either $\{a \in A \mid a<m\}$ is n-weakly-flippable or there is a flip $\langle A_i' : i<m \rangle$ such that $\bigcap_{i<m} A_i'$ is n-weakly-flippable.

Kirby has shown that weak-flippability gives rise to an indicator for regular initial segments. The following theorem confirms this.

<u>8.7. Theorem</u>. For all A each of the following conditions implies the one after it:

(1) A is $2n$-2^x-arboreal,

(2) A is n-weakly-flippable,

(3) A is n-2^x-arboreal.

Proof. Exercise. (Hint: A is $(n+1)$-$f(x)$-arboreal iff whenever $A - \{\min A\} = \bigcup_{i<m} B_i$ where $m \leq f(\min A)$, then some B_i is n-$f(x)$-arboreal.)

BIBLIOGRAPHY

1. Aczel, Peter, "A Generalization of Ramsey's Theorem," handwritten notes.

2. -------, "The Ordinal Height of a Density," handwritten notes.

3. Erdős, Paul and George Mills, "Some Bounds for the Ramsey-Paris-Harrington Numbers," J. Combinatorial Theory, Series A, to appear.

4. Friedman, Harvey, "A Combinatorial Theorem Related to the Ackermann Hierarchy," manuscript dated April, 1978

5. Ketonen, Jussi and Robert Solovay, "Rapidly Growing Ramsey Functions," to appear.

6. Kirby, Laurie, "Initial Segments of Models of Arithmetic," Ph.D. dissertation, Univ. of Manchester, England, 1977.

7. -------, "Flipping Properties in Arithmetic," to appear.

8. ------- and Jeff Paris, "Initial Segments of Models of Peano's Axioms," in Proceedings of the Bierutowice Conference 1976, Springer-Verlag (Lecture Notes in Mathematics #619), 211-226.

9. Mills, George, "Extensions of Models of Peano Arithmetic," Ph.D. dissertation, Univ. of California, Berkeley, USA, 1977.

10. Paris, Jeff, "Some Independence Results for Peano Arithmetic," J. Symbolic Logic, 43 (1978), 725-731.

11. -------, "A Combinatorial Hierarchy in Models of Arithmetic," in these proceedings.

12. ------- and Leo Harrington, "A Mathematical Incompleteness in Peano Arithmetic," in Handbook of Mathematical Logic (J. Barwise, ed.), North-Holland, Amsterdam, 1977, 1133-1142.

A HIERARCHY OF CUTS IN MODELS OF ARITHMETIC

J.B. Paris
(Manchester)

Abstract In this paper we show that it is possible to classify most of the natural families of cuts considered to date in terms of a single hierarchy. This classification gives conservation and independence results for fragments of arithmetic.

Acknowledgment Over the past three years I have been in frequent and completely open communication with George Mills and Laurie Kirby concerning the problems discussed in this paper. Obviously I wish to express my sincere thanks to them for their considerable help with ideas and suggestions.

I also wish to thank Peter Aczel for generously providing a key lemma in the proof of the main theorem.

Preliminaries Let P be Peano's 1st order axioms for arithmetic and let $I\Sigma_n$ be the same set of axioms but with the induction shema restricted to Σ_n formulas. Throughout this paper M is a countable non-standard model of $I\Sigma_1$ and N is the standard model of P which, as usual, we identify with an initial segment of M. As far as possible we use n, m, k for elements of N, a, b, c, d, e for elements of M and $\alpha, \beta, \gamma, \delta$ for ordinals. For $a \in M$ we identify a with $\{x | x \in M \ \& \ x < a\}$, so a is also a subset of M.

Let $A \subseteq M$, $a \in M$. We write $A \gtrless a$ if every element of A is greater than (resp. less than) a. $I \subseteq M$ is a cut in M, denoted $I \subset_e M$, if $2 \in I$, I is closed under multiplication, $I \neq M$, and $a < b \in I \Rightarrow a \in I$.

Let $I \subset_e M$. If $A \subseteq I$ we say A is coded in M if there is some $b \in M$ such that for $a \in I$.

$$a \in A \Longleftrightarrow \text{the } a\text{'th prime divides } b.$$

Similiarly we can talk of $f: I \to M$ being coded etc.

Henceforth all subsets of I or functions on I etc. which are mentioned are assumed to be coded in M unless otherwise stated.

Suppose $f: I \to a$ is coded by e in M. Then by overspill there is a $b > I$ such that e codes a map $f': b \to a$ with $f' \restriction I = f$. Hence we have a continuation of f a little way above I. This justifies the use of expressions like f(c) for c close to, but exceeding, I.

Let X, V be sets of cuts in M. We say that X and V are symbiotic if for all $a, b \in M$,

$$\exists J \in X, a \in J < b \Longleftrightarrow \exists J \in V, a \in J < b.$$

A Σ_1 formula $Y(x, y) = z$ is an indicator for X in M if

(i) $M \models \forall x, y \exists ! z \; Y(x, y) = z$, (we denote this function by Y^M or simply by Y when no confusion can arise),

(ii) for $a, b \in M$,

$Y^M(a, b) > N \iff \exists J \in X, a \in J < b$

(iii) for $a_1, a_2, b_1, b_2 \in M$, if $a_1 \leq a_2$ and $b_2 \leq b_1$ then $Y^M(a_2, b_2) \leq Y^M(a_1, b_1)$.

We recall the following theorem concerning indicators.

Theorem 0. Let T be a rec. theory in the language of arithmetic and let $T \supseteq I\Sigma_1$. Suppose that $Y(x, y) = z$ is an indicator for $\{I \subseteq_e M | I \models T\}$ in any countable $M \models T$. Then

(i) $T \nvdash \forall x, z \exists y \; Y(x, y) \geq z$,

(ii) $\forall n \in N, T \vdash \forall x \exists y \; Y(x, y) \geq \underline{n}$,

(iii) if $N \models T$ then $\forall x, z \exists y \; Y(x, y) \geq z$ is independent of T,

(iv) if $f(x) = y$ is Σ_1 and $T \vdash \forall x \exists ! y \; f(x) = y$ then $\exists n \in N$ such that $T \vdash \forall x (f(x) < g_n(x))$, where $g_n(x) = \mu y : Y(x, y) \geq n$.

The proof of this result for $T = P$ is contained in [6], indeed the proof there shows that we can replace T by $T + R$ for any Π_1 set of sentences R consistent with T. The general result is proved similarly using for (ii) that if M is a countable recursively saturated model of $I\Sigma_1$ then M is isomorphic to a proper initial segment of itself. (See [1].)

Let $I \subseteq_e M$. We write $M \prec_I K$ if $M \prec_{\Sigma_0} K$, $I \subseteq_e K$ and for some $\pi \in K$, $I < \pi < (M-I)$. Say I is 1-extendible (in M) if $\exists K, M \prec_I K$ and I is $(n+1)$-extendible $(n > 0)$ if $\exists K, M \prec_I K$ and I is n-extendible in K. Write $I \models I\Sigma_n^*$ if for every Σ_n^0 formula $\theta(x, X_1, \ldots, X_m)$ in the second order language of arithmetic and $A_1, \ldots, A_m \subseteq I$ (coded by our convention)

$$I \models \exists x \theta(x, A_1, \ldots, A_m) \rightarrow \exists x (\theta(x, A_1, \ldots, A_m) \wedge \forall y < x \; \neg \; \theta(x, A_1, \ldots, A_m)).$$

Write $I \models B\Sigma_n^*$ if for all such Σ_n^0 formulae $\theta(x, y, X_1, \ldots, X_m)$ and $A_1, \ldots, A_m$ and $a \in I$,

$$I \models \forall x < a \exists y \theta(x, y, A_1, \ldots, A_m) \rightarrow \exists z \forall x < a \exists y < z \theta(x, y, A_1, \ldots, A_m).$$

$I \models I\Sigma_n$, $I \models B\Sigma_n$ denote the same properties for first order θ.

Let $[I]^k = \{<a_1, \ldots, a_k> | a_1, \ldots, a_k \in I \; \& \; a_1 < a_2 < \ldots < a_k\}$ (=I if $k = 1$). By using the standard pairing function we can code subsets of $[I]^k$ in M. $A \subseteq I$, not necessarily coded, is unbounded (in I) if $\forall a \in I \exists b > a, b \in A$. For $k \geq 1$, $A \subseteq [I]^{k+1}$, not necessarily coded, is unbounded (in $[I]^{k+1}$) if $\{a_0 | \{<a_1, \ldots, a_k> | <a_0, a_1, \ldots, a_k> \in A\}$ is unbounded$\}$ is unbounded. It is easy to see that if $1 \leq s < k+1$ then $A \subseteq [I]^{k+1}$ is unbounded if and only if $\{<a_0, \ldots, a_{s-1}> | \{<a_s, \ldots, a_k> | <a_0, \ldots, a_k> \in A\}$ is unbounded in $[I]^{k-s+1}\}$ is unbounded in $[I]^s$. I is k-Ramsey $(k \geq 1)$ if for all coded $f: [I]^k \rightarrow a$ with $a \in I$,

$b < a$ such that $f^{-1}\{b\}$ is unbounded.
Hence Regular = 1-Ramsey.

<u>Preamble.</u> Early work by the Laurie Kirby, George Mills and the author (see [3], [4], [5]) had shown that for $I \subseteq_e M$,

$$I \text{ is regular} \iff I \text{ is 1-extendible} \iff I \models B\Sigma_2^*$$

$$\text{and} \qquad I \text{ is semi-regular} \iff I \models I\Sigma_1^* .$$

Whilst semi-regular cuts need not be regular, Leo Harrington showed that regular cuts have a primitive recursive indicator from which it followed that the cuts satisfying any one of these properties or the properties $I \models I\Sigma_1$ or $I \models B\Sigma_2$ are all symbiotic. This raises the question as to whether for $k > 1$ there are combinatorially defined cuts (like regular) which are symbiotic with $\{I \subseteq_e M | I \models I\Sigma_k\}$ etc. A number of properties suggested themselves, the most reasonable being k-Ramsey and k-extendible. This result was eventually obtained for $k = 2$ and then, upon the arrival of an elegant framework divised by R. Solovay and J. Ketonen, for $k = 3$. Finally Peter Aczel generously provided a general proof theoretic result which enabled the result to be proved for all k.
We now state the main theorem.

<u>Theorem 1.</u> For $1 \leq k \in N$ the cuts I satisfying any of the following are symbiotic:-

(1) I is k-extendible.

(2) I is k-Ramsey.

(3) $I \models B\Sigma_{k+1}^*$.

(4) $I \models I\Sigma_k^*$.

(5) $I \models B\Sigma_{k+1}$.

(6) $I \models I\Sigma_k$.

We shall prove theorem 1 via a series of lemmas. Clearly (3) ⇒ (5) and (4) ⇒ (6). It is shown in [8] that (5) => (6) and the proof that (3) => (4) is similar. Thus it is enough to show that (1) ⇒ (2), (2) ⇒ (3) and cuts satisfying (1) are dense in the cuts satisfying (6). These will be proved in Propositions 2, 3, and 13 and 15 respectively. Propositions 2 & 3 also appear in [3].

<u>Proposition 2</u> (with L. Kirby). Let I be k-extendible, $k \geq 1$. Then I is k-Ramsey.

<u>Proof.</u> For $k = 1$ this result is proved in [4], so suppose $k > 1$ and that the result has been proved for $k - 1$. Let $M \prec_I K$ where I is (k-1)-extendible in K and let $f: [I]^k \to a \in I$ in M. Pick $c \in K$ such that $I < c < M-I$ and, in K, define $h: [I]^{k-1} \to a$ by $h(a_1,\ldots, a_{k-1}) = f(a_1,\ldots, a_{k-1}, c)$. [$f(a_1,\ldots, a_{k-1}, c)$ is defined according to the convention introduced in the earlier section on notation.]

Since I is (k-1)-Ramsey in K we can pick $b < a$ such that $h^{-1}\{b\}$ is unbounded in $[I]^{k-1}$. Let $\langle a_1,\ldots, a_{k-1}\rangle \in [I]^{k-1}$ and $h(a_1,\ldots, a_{k-1}) = b$. Then for $d \in I$, $f(a_1,\ldots, a_{k-1}, c) = b$ & $d < c$. In K let p be minimal such that $f(a_1,\ldots, a_{k-1}, p) = b$ & $d < p$. Since $d, b, a_1,\ldots, a_{k-1} \in M$ and $M \prec_{\Sigma_0} K$, $p \in M$ and since $p \leqslant c$, $p \in I$. Thus $\{x \in I | f(a_1,\ldots, a_{k-1}, x) = b\}$ is unbounded in I and hence $f^{-1}\{b\}$ is unbounded in $[I]^k$ as required.

<u>Proposition 3.</u> (with L. Kirby) Let I be k-Ramsey, $k \geqslant 1$. Then $I \models B\Sigma^*_{k+1}$.

<u>Proof</u>. Assume the result for k-1, so $I \models B\Sigma^*_k$. (An easy overspill argument shows that $I \models B\Sigma^*_1$.) Let $\theta(x, y, x_1,\ldots, x_k, x_{k+1}, X)$ be a Σ^0_0 formula in the 2nd order language of arithmetic, let $B \subseteq I$, $a, b \in I$ and suppose

$$I \models \forall x < a \exists y \forall x_1 \exists x_2 \ldots \theta(x, y, x_1,\ldots, x_k, b, B).$$

Let $I < e$ and $B' \subseteq e$ be such that $B = B' \cap I$. For $\langle a_1,\ldots, a_k\rangle \in [I]^k$ and $x < a$ define $g(x, a_1,\ldots, a_k)$ to be the least y such that $\forall x_1 < a_1 \exists x_2 < a_2 \ldots \theta(x, y, x_1,\ldots, x_k, b, B')$ if such exists, 0 otherwise. Define $f(a_1,\ldots, a_k)$ to be the (least) $x < a$ such that $g(x, a_1,\ldots, a_k)$ is maximal.

Since I is k-Ramsey we can pick $x_0 < a$ such that $f^{-1}\{x_0\}$ is unbounded in $[I]^k$. Pick $y_0 \in I$ such that

$$I \models \forall x_1 \exists x_2 \ldots \theta(x_0, y_0, x_1,\ldots, x_k, b, B).$$

We claim that $I \models \forall x < a \exists y \leqslant y_0 \forall x_1 \exists x_2 \ldots \theta(x, y, x_1,\ldots, x_k, b, B)$. Suppose not and pick $z < a$ such that

$$I \models \forall y \leqslant y_0 \exists x_1 \forall x_2 \ldots \theta(z, y, x_1,\ldots, x_k, b, B).$$

Pick $t \in I$ such that

$$I \models \forall x_1 \exists x_2 \ldots \theta(z, t, x_1,\ldots, x_k, b, B).$$

Then since $I \models B\Sigma^*_k$ and $f^{-1}\{x_0\}$ is unbounded we can find $\langle a_1,\ldots, a_k\rangle \in f^{-1}\{x_0\}$ such that

$$I \models \forall x_1 < a_1 \exists x_2 < a_2 \ldots \theta(x_0, y_0, x_1,\ldots, x_k, b, B)$$
$$I \models \forall y \leqslant y_0 \exists x_1 < a_1 \forall x_2 < a_2 \ldots \neg\theta(z, y, x_1,\ldots, x_k, b, B)$$
$$I \models \forall x_1 < a_1 \exists x_2 < a_2 \ldots \theta(z, t, x_1,\ldots, x_k, b, B).$$

Since these formulae are bounded by elements of I replacing B by B' and I by M has no effect. Hence

$$g(x_0, a_1,\ldots, a_k) \leqslant y_0 < g(z, a_1,\ldots, a_k) \leqslant t.$$

But then $f(a_1,\ldots, a_k) \neq x_0$, a contradiction, and the claim and lemma are proved.

Before completing the proof of theorem 1 we need to recall some ideas and results developed by R. Solovay and J. Ketonen.

In what follows $\alpha, \beta, \gamma, \delta$ are ordinals less than ε_0 which we assume are represented in Cantor normal form. We write $\beta >> \alpha$ if, in Cantor normal form. $\beta = \omega^{\gamma_1}n_1 + \ldots + \omega^{\gamma_i}n_i$, $\alpha = \omega^{\delta_1}m_1 + \ldots + \omega^{\delta_j}m_j$ with $\gamma_1 > \gamma_2 > \ldots > \gamma_i \geq \delta_1 > \delta_2 > \ldots > \delta_j$. For $\alpha < \varepsilon_0$, $n \in \omega$ define $\{\alpha\}(n)$ as follows:-

$$\{\beta+1\}(n) = \beta, \qquad \{\omega^{\gamma+1}(\beta+1)\}(n) = \omega^{\gamma+1}\beta + \omega^{\gamma}n,$$

$$\{0\}(n) = 0, \qquad \{\omega^{\delta}(\beta+1)\}(n) = \omega^{\delta}\beta + \omega^{\{\delta\}(n)} \quad \text{for limit } \delta.$$

Write $\alpha \underset{n}{\rightarrow} \beta$ if there is a finite sequence $\alpha_0, \ldots, \alpha_k$ such that $\alpha_0 = \alpha$, $\alpha_k = \beta$ and for $m < k$ $\alpha_{m+1} = \{\alpha_m\}(j_m)$ some $j_m \leq n$. Write $\alpha \underset{n}{\Rightarrow} \beta$ if $\alpha \underset{n}{\rightarrow} \beta$ with each $j_m = n$.

We now produce a series of lemmas (4-12) involving ordinals. For convenience we shall state and prove these for the standard model. However these can all be carried out in M assuming that $M \models I\Sigma_1$. To see this set

$$O(0, c) = \{0,1,\ldots, c\},$$

$$O(n+1, c) = \{\omega^{\alpha_1}a_1 + \ldots + \omega^{\alpha_m}a_m \mid a_1, \ldots, a_m \leq c$$

$$\alpha_1, \ldots, \alpha_m \text{ are distinct elts of } O(n, c)\}.$$

Then in M, for $n \in N$,

(a) if $\alpha \in O(n, c)$, $j \leq c$ then $\{\alpha\}(j) \in O(n, c)$,

(b) $\forall\, S \subseteq O(n, c)\ \exists$ order preserving $f: S \twoheadrightarrow |S|$,

(c) $\forall\, S \subseteq O(n, c)$, $S \neq \emptyset \Rightarrow S$ has a least elt.

The next lemma is well known and straightforward to prove.

<u>Lemma 4.</u> (i) If $\beta >> \alpha$ then $\{\beta+\alpha\}(n) = \beta+\{\alpha\}(n)$.

(ii) If $\beta >> \alpha$ and $\alpha \underset{n}{\Rightarrow} \gamma$ then $\beta+\alpha \underset{n}{\Rightarrow} \beta+\gamma$.

(iii) If $\alpha \underset{n}{\Rightarrow} \beta$ and $1 \leq n$ then $\omega^{\alpha} \underset{n}{\Rightarrow} \omega^{\beta}$.

(iv) $\{\alpha\}(j) \underset{n}{\Rightarrow} \{\alpha\}(j-1)$ for $0 < j \leq n$.

(v) $\alpha \underset{n}{\rightarrow} \beta$ if and only if $\alpha \underset{n}{\Rightarrow} \beta$.

The next Lemma is also well known but for convenience we include a proof.

<u>Lemma 5.</u> For $\alpha > 0$, $n > 0$, $\alpha \underset{n+1}{\rightarrow} \{\alpha\}(n)+1$.

<u>Proof.</u> By induction on α. If α is a successor the result is clear.

If $\alpha = \omega^{\beta+1}(\delta+1)$ then $\{\alpha\}(n+1) = \omega^{\beta+1}\delta + \omega^{\beta}n + \omega^{\beta} \underset{n+1}{\Longrightarrow} \omega^{\beta+1}\delta + \omega^{\beta}n + 1 = \{\alpha\}(n)+1$, using lemma 4(i) since $\omega^{\beta} \underset{n+1}{\Longrightarrow} 1$. If $\alpha = \omega^{\gamma}(\delta+1)$ with γ a limit then $\{\alpha\}(n+1) = \omega^{\gamma}\delta + \omega^{\{\gamma\}(n+1)}$, $\{\alpha\}(n)+1 = \omega^{\gamma}\delta + \omega^{\{\gamma\}(n)} + 1$. By inductive hypothesis and lemma 4(iii) $\omega^{\{\gamma\}(n+1)} \underset{n+1}{\Longrightarrow} \omega^{\{\gamma\}(n)+1}$. Hence $\omega^{\gamma}\delta + \omega^{\{\gamma\}(n+1)} \underset{n+1}{\longrightarrow} \omega^{\gamma}\delta +$ $\omega^{\{\gamma\}(n)+1} \underset{n+1}{\longrightarrow} \omega^{\gamma}\delta + \omega^{\{\gamma\}(n)} + \omega^{\{\gamma\}(n)} n \underset{n+1}{\longrightarrow} \omega^{\gamma}\delta + \omega^{\{\gamma\}(n)} + 1$ since $\omega^{\{\gamma\}(n)} n \underset{n+1}{\longrightarrow} 1$.

We now introduce the Solovay-Ketonen notion of a bounded, coded subset of M being η-large where η is a ordinal in the sense of M (see [2]). For convenience we introduce the definitions and lemmas in the special case M = N, the necessary generalisation being evident.

Definition. For finite $X \subseteq N$, X is 1-large if $|X| \geq 2$.
X is α-large if $\{n \in X | X-n$ is $\{\alpha\}(n)$-large$\}$ is 1-large.
We use Γ, Λ for finite subsets of N and denote by $\Lambda_0, \Lambda_1, \ldots, \Lambda_{|\Lambda|-1}$ the elements of Λ in increasing order.

The following lemmas 6-11 are due to R. Solovay and J. Ketonen, see [2]. For the sake of completeness we include sketch proofs.

Lemma 6. If $\Gamma \supseteq \Lambda$ and Λ is α-large then Γ is α-large.

Proof. By induction on α.

Lemma 7. If $\alpha \underset{n}{\longrightarrow} \beta, n \leq \Lambda_0$ and Λ is α-large then Λ is β-large.

Proof. By induction on α it is enough to show that Λ is $\{\alpha\}(n_0)$-large, given that Λ is α-large. So let $\Lambda_i < \Lambda_j$ be such that $\Lambda-\Lambda_i$ is $\{\alpha\}(\Lambda_i)$-large and $\Lambda-\Lambda_j$ is $\{\alpha\}(\Lambda_j)$-large. Then by inductive hypothesis, since $\{\alpha\}(\Lambda_i) \underset{\Lambda_i}{\longrightarrow} \{\alpha\}(n_0) \underset{\Lambda_i}{\longrightarrow} \{\{\alpha\}(n_0)\}(\Lambda_i)$, $\Lambda-\Lambda_i$ is $\{\{\alpha\}(n_0)\}(\Lambda_i)$-large and $\Lambda-\Lambda_j$ is $\{\{\alpha\}(n_0)\}(\Lambda_j)$-large and the result follows.

Remark. Suppose $\Lambda-\Lambda_i$ is $\{\alpha\}(\Lambda_i)$-large and $k < i$. Then $\Lambda-\Lambda_i$ is $\{\alpha\}(\Lambda_k)$-large so $\Lambda-\Lambda_k$ is $\{\alpha\}(\Lambda_k)$-large. Hence

$$\Lambda \text{ is } \alpha\text{-large} \iff \Lambda-\Lambda_0 \text{ is } \{\alpha\}(\Lambda_0)\text{-large and } \Lambda-\Lambda_1 \text{ is } \{\alpha\}(\Lambda_1)\text{-large}$$
$$\iff \Lambda-\Lambda_1 \text{ is } \{\alpha\}(\Lambda_1)\text{-large.}$$

Lemma 8. Let $\alpha >> \beta$. Then Λ is $(\alpha+\beta)$-large if and only if $\{\Lambda_i | \Lambda-\Lambda_i$ is α-large$\}$ is β-large.

Proof. Λ is $(\alpha+\beta)$-large $\iff \{\Lambda_i | \Lambda-\Lambda_i$ is $\alpha+\{\beta\}(\Lambda_i)$-large$\}$ is 1-large

$\iff \{\Lambda_i | \{\Lambda_j | (\Lambda-\Lambda_i)-\Lambda_j$ is α-large & $i \leq j\}$ is $\{\beta\}(\Lambda_i)$-large$\}$ is 1-large

$\iff \{\Lambda_i | \{\Lambda_j | \Lambda-\Lambda_j$ is α-large$\} - \Lambda_i$ is $\{\beta\}(\Lambda_i)$-large$\}$ is 1-large

$\iff \{\{\Lambda_j | \Lambda-\Lambda_j\}$ is α-large$\}$ is β-large.

Lemma 9. If $\alpha_n >> \alpha_{n-1} >> \dots >> \alpha_1$ and Λ is $(\alpha_n + \alpha_{n-1} + \dots + \alpha_1)$-large then $\exists\, i_0 < i_1 < \dots < i_n$ such that $\Lambda \cap [\Lambda_{i_{j-1}}, \Lambda_{i_j}]$ is α_j-large for $1 \leq j \leq n$.

Proof. Suppose Λ is $(\alpha_2 + \alpha_1)$-large. Let $i_0 = \Lambda_0$ and let i_1 be maximal such that $\Lambda - \Lambda_i$ is α_2-large. Then by lemma 8, $\Lambda \cap [\Lambda_{i_0}, \Lambda_{i_1}]$ is α_1-large and $\Lambda \cap [\Lambda_{i_1}, \Lambda_{|\Lambda|-1}]$ is α_2-large.

Lemma 10. Suppose Λ is ω^α-large $(\alpha \geq 2)$, $0 < \Lambda_0$, and let $m_0 \leq m_1 \leq \dots \leq m_{\Lambda_1}$, $m_0 \leq \Lambda_0$, $\Lambda_{|\Lambda|-1} \leq m_{\Lambda_1}$. Thenfor some $1 \leq i \leq \Lambda_1$, $\Lambda \cap [m_{i-1}, m_i]$ is $\omega^{\{\alpha\}(\Lambda_0)}$-large.

Proof. If $\alpha = \beta+1$ then $\Lambda - \Lambda_1$ is $\omega^\beta \Lambda_1$ - large so there are $i_0 < i_1 < \dots < i_{\Lambda_1}$ such that $(\Lambda - \Lambda_1) \cap [\Lambda_{i_{j-1}}, \Lambda_{i_j}]$ is ω^β-large for $1 \leq j \leq \Lambda_1$. It follows that $\Lambda \cap [m_{i-1}, m_i]$ is ω^β-large for some $1 \leq i \leq \Lambda_1$ and since $\omega^{\{\alpha\}(\Lambda_0)} = \omega^\beta$ the result follows. If α is a limit then $\Lambda - \Lambda_1$ is $\omega^{\{\alpha\}(\Lambda_1)}$-large. Therefore, since by lemma 5 $\{\alpha\}(\Lambda_1) \xrightarrow[\Lambda_1]{} \{\alpha\}(\Lambda_0+1) \xrightarrow[\Lambda_0+1]{} \{\alpha\}(\Lambda_0)+1$, $\Lambda - \Lambda_1$ is $\omega^{\{\alpha\}(\Lambda_0)+1}$-large. Hence $\Lambda - \Lambda_1$ is $\omega^{\{\alpha\}(\Lambda_0)} \Lambda_1$-large and the result follows as for the successor case.

Lemma 11. Let $\alpha \geq 1$ and $|\Lambda| \geq 2$. Then Λ is α-large if and only if

$$\{\dots\{\{\{\alpha\}(\Lambda_1)\}(\Lambda_2)\}\dots\}(\Lambda_{|\Lambda|-1}) = 0.$$

Proof. Write $\{\alpha\}(\Lambda_1, \Lambda_2, \dots, \Lambda_{|\Lambda|-1})$ for $\{\dots\{\{\{\alpha\}(\Lambda_1)\}(\Lambda_2)\}\dots\}(\Lambda_{|\Lambda|-1})$. The proof is by induction on α. If $\alpha = 1$ the result is obvious. Assume true for $\beta < \alpha$. Then

$$\Lambda \text{ is } \alpha\text{-large} \iff \Lambda - \Lambda_1 \text{ is } \{\alpha\}(\Lambda_1)\text{-large}$$
$$\iff \{\alpha\}(\Lambda_1, \Lambda_2, \dots, \Lambda_{|\Lambda|-1}) = 0.$$

Notation. Set $\omega_0^\alpha = \alpha$, $\omega_{k+1}^\alpha = \omega^{\omega_k^\alpha}$. Hence by lemma 4(iii) $\omega_k^{m+1} \xrightarrow[1]{} \omega_k^m$ and for limit α, $\{\omega_k^\alpha\}(n) = \omega_k^{\{\alpha\}(n)}$.

Lemma 12. Suppose Λ is ω_k^{m+1}-large with $m, k > 0$ and $\Lambda_0 \geq 2$. Then $\{\Lambda_1, \Lambda_2, \dots, \Lambda_{|\Lambda|-2}\}$ is ω_k^m-large.

Proof. By lemma 10 one of $\{\Lambda_0, \Lambda_1\}$, $\{\Lambda_1, \Lambda_2, \dots, \Lambda_{|\Lambda|-2}\}$, $\{\Lambda_{|\Lambda|-2}, \Lambda_{|\Lambda|-1}\}$ must be $\{\omega_k^{m+1}\}(\Lambda_0)$-large. Clearly it must be the second of these sets. Finally, since $\omega^m \Lambda_0 \xrightarrow[1]{} \omega^m$, lemmas 4 (iii) and 7 imply that $\{\Lambda_1, \Lambda_2, \dots, \Lambda_{|\Lambda|-2}\}$ is ω_k^m-large.

At this point we move back to M to establish a connection between largeness and extendibility.

Proposition 13. Suppose $a, b \in M$, $k > 0$, and, in the sense of M, $[a, b]$ is

ω_k^η-large for some $\eta > N$. Then there is a k-extendible cut I such that $a \in I < b$.

Proof. For simplicity suppose $k = 3$. Fix $d \in M$ so that d is very large with respect to b, hereafter expressed $d >> b$. By treating $+, \cdot$ as relations we can, and do, treat d $(= \{x \in M | x < d\})$ as a substructure of M.

Before proceeding on the proof we set up some apparatus. Since $M \models I\Sigma_1$ we can find, as a coded subset of M, a satisfaction relation "$d \models \theta(e_1, \ldots, e_k)$" for the structure d which is defined whenever $\ulcorner\theta\urcorner, k, e_1, \ldots, e_k \leq d$ and here satisfies Tarski's satisfaction conditions (and so is an extension of the standard satisfaction relation). With such a satisfaction relation in mind and $g \leq c << d$ set

$$Q(g, c) = \{51x | x < \tfrac{g}{3}\} \cup \{2^{4x} | x < \tfrac{g}{3}\} \cup$$

$$\cup \{x \leq d | d \models \forall z(\theta(z, c, \vec{e}) \longleftrightarrow z = x)$$

$$\text{for some formula } \theta \text{ with } \langle\ulcorner\theta\urcorner, \vec{e}\rangle < \tfrac{g}{3}\}.$$

Notice that $|Q(g, c)| \leq g$ and $Q(g, c)$ is a subset of d coded in M.

Call a strictly decreasing sequence of intervals $[g_j, h_j]$, $j \leq q$ a c-sequence if $h_0 \leq c$ and for all $j < q$

$$(g_{j+1}, h_{j+1}) \cap Q(g_j, c) = \emptyset .$$

Suppose we are given such a c-sequence with $c << d$. We claim that if $i < q-54$ then $2^{g_i} \leq g_{i+1}$. For suppose $g_i \leq g_{i+1} < 2^{g_i}$. Then since $2^{4x} \in Q(g_i, c)$ for $x < \frac{g_i}{3}$, $[g_{i+1}, h_{i+1}] \subseteq [2^{4x}, 2^{4x+4}]$ for some x. Hence $g_{i+1} \leq g_{i+2} \leq 16g_{i+1}$. Again since $51x \in Q(g_{i+1}, c)$ for $x < g_{i+1}/3$, $[g_{i+2}, h_{i+2}] \subseteq [51x, 51x + 51]$ for some x. Hence since the intervals are strictly decreasing we must have $q \leq i + 54$.

In view of this if $J \subseteq_e M$, $J < q$, $c << d$ then with the obvious notation $I \subseteq \bigcup_{i\in J} Q(g_i, c) \prec_I d$ where $I = \bigcup_{i \in J} g_i$.

Returning now to the proof of this proposition, by using lemma 10 repeatedly we can construct in M a b-sequence $\{[g_j, h_j] | j \leq q\}$ such that $[g_0, h_0] = [a, b]$, $\{\omega_2^\eta\}(\{g_j | j \leq q\}) < 2$, and for $j < q$ $[g_j, h_j]$ is $\omega^{\{\omega_2^\eta\}(\{g_r | r<j\})}$-large. By lemma 11 it follows that $\{a -1, g_q+1\} \cup \{g_j | j \leq q\}$ is ω_2^η -large and hence, by lemma 12, $\{g_j | j \leq q\}$ is ω_2^ν-large for some $\nu > N$.

Again using lemma 10 repeatedly we can find $\langle b, g_q\rangle$-sequence $\{[g'_j, h'_j] | j \leq s\}$ such that $g'_j, h'_j \in \{g_t | t \leq q\}$ for $j \leq s$ and $\{\omega^\nu\}(\{g'_j | j \leq s\}) < 2$. Hence $\{g'_j | j \leq s\}$ is ω^μ-large some $\mu > N$. Finally using lemma 10 again we can find $\langle b, g_q, g'_s\rangle$-sequence $\{[g''_j, h''_j] | j \leq r\}$ such that $g''_j, h''_j \in \{g'_t | t \leq s\}$ and $\{\mu\}(\{g''_j | j \leq r\}) < 2$. Clearly $r > N$. Let $I = \bigcup_{n\in N} g''_n$. Then

$$\bigcup_{g_i \in I} Q(g_i, b) \prec_I \bigcup_{g'_i \in I} Q(g'_i, \langle b, g_q\rangle) \prec_I \bigcup_{g''_i \in I} Q(g''_i, \langle b, g_q, g'_s\rangle) \prec_I d$$

since for example $I < g_q < \bigcup_{g_i \in I} Q(g_i, b) - I$. It follows that in $\bigcup_{g_i \in I} Q(g_i, b)$, I is 3-extendible and proposition 13 now follows from the next proposition.

Proposition 14. (The Transfer Lemma). Suppose $K \prec d \in M$, $I \subseteq_e K$ and $a \in I < b \in K$ where $d >> b$. Then if I is 3-extendible in K (in the obvious sense) then $\exists J \subseteq_e M$ such that $a \in J < b$ and J is 3-extendible in M.

Proof. The proof is fairly standard but for completeness we include it. Given $e < b$, $e \in K$ consider the 2 player game G_e in K. On the c'th go $(c \leq e)$ player A plays $f_c: [a, b]^i \to b$, $i = 1, 2$ or 3 and player B plays $Y_c \subseteq [a, b]^3$. Assuming B has not already lost he maintains this state if

(i) $\bigcap_{q \leq c} Y_q \neq \emptyset$.

(ii) If $f_c: [a, b]^3 \to r$ then $Y_c = [a, r]^3$ or $Y_c = f^{-1}\{q\}$ some $q < r$.

(iii) If $f_c: [a, b]^2 \to b$ then either $Y_c \subseteq \{\langle x, y, z\rangle \in [a, b]^3 | x < f(y, z)|$

or $Y_c \subseteq \{\langle x, y, z\rangle \in [a, b]^3 | f(y, z) = q\}$ some $q \leq \min Y_c$.

(iv) If $f_c: [a, b] \to b$ then either $Y_c \subseteq \{\langle x, y, z\rangle \in [a, b]^3 | y < f(z)\}$

or $Y_c \subseteq \{\langle x, y, z\rangle \in [a, b]^3 | f(z) = q\}$ some $q \leq \min Y_c$.

G_e goes on for e goes and B wins G_e if he has not lost by the last go.

Now suppose $K \prec_I K_1 \prec_I K_2 \prec_I K_3$ and pick $\pi_i \in K_i$, $i = 1, 2, 3$ such that $I < \pi_1 < (K-I)$, $I < \pi_2 < (K_1-I)$, $I < \pi_3 < (K_2-I)$. Then for $n \in N$ player B can beat any strategy for player A by playing as follows. If A plays $f: [a, b]^3 \to r$, B plays $[a, r]^3$ if $I < r$ and plays $f^{-1}\{f(\pi_3, \pi_2, \pi_1)\}$ (as computed in K_3) if $r < I$. If A plays $f: [a, b]^2 \to b$, B plays $\{\langle x, y, z\rangle \in [a, b]^3 | f(y, z) > x\}$ if $f(\pi_2, \pi_1) > \pi_3$ and plays $\{\langle x, y, z\rangle \in [a, b]^3 | f(y, z) = f(\pi_2, \pi_1)\} - f(\pi_2, \pi_1)$ if $f(\pi_2, \pi_1) \leq \pi_3$ (in which case $f(\pi_2, \pi_1) \in I$). Finally if A plays $f: [a, b] \to b$, B plays $\{\langle x, y, z\rangle \in [a, b]^3 | f(z) > y\}$ if $f(\pi_1) > \pi_2$ and $\{\langle x, y, z\rangle \in [a,b]^3 | f(z) = f(\pi_1)\} - f(\pi_1)$ if $f(\pi_1) \leq \pi_2$.

Hence B has a winning strategy for G_n for $n \in N$. [We are assuming here that d is much larger than b so all this can be expressed in K.] Thus B has a winning strategy for G_q in K some $N < q$. Hence B has a winning strategy for G_q in d and hence M. Since M is countable we can produce a "strategy" for A (in the outside world) which plays all relevant functions f in the first N goes. If B uses his winning strategy against this strategy for A then in N goes B produces a non-principal ultrafilter V on the subsets of $[a, b]^3$ which are coded in M such that if $J = \{c | [a, c]^3 \in' V\}$ then:-

(i) If $f: [a, b]^3 \to c$ and $c \in J$ then $f^{-1}\{e\} \in V$ some $e < c$.

(ii) If $f: [a, b]^2 \to M$ then either $\{\langle x, y, z\rangle \in [a, b]^3 | x < f(y, z)\} \in V$

or $\exists c \in J$ such that $\{\langle x, y, z\rangle \in [a, b]^3 | f(y, z) = c\} \in V$.

(iii) If $f: [a, b] \to M$ then either $\{\langle x, y, z\rangle \in [a, b]^3 | y < f(z)\} \in V$

or $\exists c \in J$ such that $\{\langle x, y, z\rangle \in [a, b]^3 | f(z) = c\} \in V$.

Now let X_3 be the set of maps $f: [a, b]^3 \to M$ (f coded in M) let
$X_2 = \{f \in X_3 | \exists h: [a, b]^2 \to M, h(y, z) = f(x, y, z) \forall \langle x, y, z\rangle \in [a, b]^3\}$ and let
$X_1 = \{f \in X_3 | \exists h: [a, b] \to M, h(z) = f(x, y, z) \forall \langle x, y, z\rangle \in [a, b]^3\}$.
Let M_i $(i = 1, 2, 3)$ be the ultrapower of M with respect to V using the maps in X_i, and let $\sigma_1, \sigma_2, \sigma_3$ be the elements of M_3 corresponding respectively to the projections $f_1(x, y, z) = x$, $f_2(x, y, z) = y$, $f_3(x, y, z) = z$. Then $M \prec M_1 \prec M_2 \prec M_3$, J is an initial segment of M_3, $\sigma_3 \in M_1$, $\sigma_2 \in M_2$, $\sigma_1 \in M_3$ and $J < \sigma_3 < (M-J)$, $J < \sigma_2 < (M_1-J)$, $J < \sigma_3 < (M_2-J)$. Hence J is 3-extendible in M and since $a \in J < b$ the result is proved

Remark. Proposition 14 is a special case of the following more general result:-

Let T be a recursive theory in the language of second order arithmetic. Then there is a Σ_0 formula $W(x, y, z) = t$ such that for any countable $M \models I\Sigma_0$,

(i) $M \models \forall x, y, z \exists !t\, W(x, y, z) = t$,

(ii) for $a, b, e \in M$,

$W^M(a, b, e) > N \iff I \subset_e M$, $a \in I < b$ and

$\langle I$, subsets of I coded in $M\rangle \models T$ and, in the sense of M,

$e \geqslant b^{b^{\cdot^{\cdot^{\cdot^{b}}}}}\}$10 times.

The proof of this result is essentially contained in [3].

To complete the proof of theorem 1 it is enough to show the following result which was generously proved for us by Peter Aczel.

Proposition 15. For all $n \in N$, $I\Sigma_{k+2} \vdash \forall x \exists y [x, y]$ is ω^n_{k+2}-large. To see that theorem 1 follows from this lemma suppose $I \models I\Sigma_{k+2}$, $a \in I$. Then since $I \models I\Sigma_2$ there is $a < b \in I$ and $\eta > N$ such that $I \models [a, b]$ is ω^η_{k+2}-large. Hence $M \models [a, b]$ is ω^η_{k+2}-large and the result follows from proposition 13. Notice that this result (together with results in [1] for $k = 1$) implies that for $k \geqslant 1$ W defined by

$W(a, b) = \max.c$ such that $[a, b]$ is ω^c_k-large

(= 0 if no such c exists)

is an indicator for cuts satisfying $I\Sigma_k$.

To prove Propositiom 15 we first introduce some apparatus. As usual we do this with respect to the standard model, the transfer to M being transparent.

Let T_0 be the Σ_1 subsets of $N \times N$. Since we can handle elements of T_0 via their codes we can think of T_0 as a set of natural numbers. Similarly we can treat T^i_{k+1} as a set of natural numbers where T^i_{k+1} is the set of primitive recursive maps from T_i to T_i.

For $f_i \in T_i$, $m \leqslant i \leqslant k$ let $(f_k, f_{k-1}, \dots, f_m)$ denote

$$(\dots((f_k(f_{k-2}))(f_{k-3})\dots)(f_m).$$

For $g_n \varepsilon T_0$, $n \varepsilon N$, let

$$\mathcal{L}_n g_n = \{<x, y> | \{z | x \leq z \leq y \ \& <z, y> \varepsilon g_z\} \text{ is 1-large}\}.$$

For $e, f \varepsilon T_0$ let $ef = \{<x, z> | \exists y(<x, y> \varepsilon e \ \& <y, z> \varepsilon f)\}$.
Pick, primitively recursively, $z_k \varepsilon T_k$ such that

$$z_0 = \{<x, y> | [x, y] \text{ is 1-large}\}$$

$$(z_{k+1}, f_k, \ldots, f_0) = \mathcal{L}_n(f_k^n, f_{k-1}, \ldots, f_0) \text{ for } f_i \varepsilon T_i, 0 \leq i \leq k,$$

where f_k^n is the n-fold iteration of f_k.

Since z_{k+1} is, in effect, a map from codes to codes we may assume that we have chosen the z_i so that distinct terms built up from the z_i actually give distinct functions. This will be useful, although not essential, in what follows.

Our first aim is to show that $(z_{k+1}^n, z_k, \ldots, z_0) = \{<x, y> | [x, y]$ is ω_{k+1}^n-large$\}$. To this end we define $S_k \subseteq T_k$ and with each $f \varepsilon S_k$ associate an ordinal $O(f)$ and a sequence $\{f\}(n)$, $(n \varepsilon N)$, of elements of T_k as follows.

(i) $z_k \varepsilon S_k$, $O(z_k) = 1$, $(\{z_k\}(n))(f) = f^n$ for $k > 0$, $f \varepsilon T_{k-1}$,

$$\{z_0\}(n) = \{<x, x> | x \varepsilon N\}.$$

(ii) If $t, s \varepsilon S_k$ and $O(t) << O(s)$ then $ts \varepsilon S_k$, $O(ts) = O(t) + O(s)$ and $\{ts\}(n) = \{t\}(n)s$.

(iii) If $t \varepsilon S_{k+1}$ then $t(z_k) \varepsilon S_k$, $O(t(z_k)) = \omega^{O(t)}$ and $\{t(z_k)\}(n) = (\{t\}(n))(z_k)$.

Notice any element of S_k can be represented either in the form $(ts, z_{p-1}, \ldots, z_k)$ with $t, s \varepsilon S_p$, $O(t) << O(s)$ and $p \geq k$ or in the form $(z_p, z_{p-1}, \ldots, z_k)$ with $p \geq k$. Also if $(t, z_i) \varepsilon S_i$ then, after simplification, $\{(t, z_i)\}(n) \varepsilon S_i$.

We also now give a series of lemmas after which we can prove proposition 15. Again these lemmas are stated and proved within the standard model. In view of the statement of proposition 15 it is enough to notice that they also hold in any $M \models I\Sigma_2$.

<u>Lemma 16.</u> Let $s \varepsilon S_{k+1}$, $t = (s, z_k)$. Then $\{O(t)\}(n) = O(\{t\}(n))$.

<u>Proof</u>. By induction on $O(t)$ and then by the following cases:-

(i) $S = (z_p, z_{p-1}, \ldots, z_{k+1})$, $p > k$.

(ii) $S = (z_p^m f, z_{p-1}, \ldots, z_{k+1})$, $p > k$, $m > 0$, $f \varepsilon S_p$ and $O(z_p^m) << O(f)$

(iii) $S = (gf, z_{p-1}, \ldots, z_{k+1})$, $p > k$, $g, f \varepsilon S_p$, $O(g) << O(f)$ and

$g = (h, z_p)$ some $h \varepsilon S_{p+1}$ (so $O(g)$ is a limit.)

Lemma 17. Let $t \in S_{k+1}$. Then for $f_i \in T_i$, $0 \leq i \leq k$,

$$(t, f_k, \ldots, f_0) = \mathcal{L}_n (\{t\}(n), f_k, \ldots, f_0).$$

Proof. By induction on $O(t)$. If $t = (z_p, \ldots, z_{k+1})$, $p > k$ the result follows easily. If $t = (gh, z_{p-1}, \ldots, z_{k+1})$, $p > k$ with $g, h \in S_p$ and $O(g) \ll O(h)$ then

$$\begin{aligned}(t, f_k, \ldots, f_0) &= (gh, z_{p-1}, \ldots, z_{k+1}, f_k, \ldots, f_0)\\ &= (g, h(z_{p-1}), z_{p-2}, \ldots, z_{k+1}, f_k, \ldots, f_0)\\ &= \mathcal{L}_n (\{g\}(n), h(z_{p-1}), z_{p-2}, \ldots, z_{k+1}, f_k, \ldots, f_0)\\ &= \mathcal{L}_n (\{g\}(n)h, z_{p-1}, \ldots, z_{k+1}, f_k, \ldots, f_0)\\ &= \mathcal{L}_n (\{t\}(n), f_k, \ldots, f_0).\end{aligned}$$

Lemma 18. For $t \in S_0$, $t = \{<x, y> | [x, y]$ is $O(t)$-large$\}$.

Proof. By induction on $O(t)$ and then by the following cases

(i) $t = gf$ with $g, f \in S_0$, $O(g) \ll O(f)$

(ii) $t = (z_1, z_0)$

(iii) $t = (z_1^m f, z_0)$ with $m > 0$, $f \in S_1$ and $O(z_1^m) \ll O(f)$

(iv) $t = (gf, z_0)$ with $g, f \in S_1$, $O(g) \ll O(f)$, $g = (h, z_1)$ some $h \in S_2$.

In case (i) we invoke lemma 8. Cases (ii), (iii), (iv) use lemmas 16 and 17.

Proof of proposition 15.

As an immediate corollary of this last lemma we have that $(z_{k+2}^n, z_{k+1}, \ldots, z_0) = \{<x, y> | [x, y]$ is ω_{k+2}^n-large$\}$, so in order to prove the lemma it is enough to show that for $k, n \in N$

$$I\Sigma_{k+2} \vdash G_0((z_{k+2}^n, z_{k+1}, \ldots, z_0))$$

where $G_0(f)$ is $\forall x \exists y <x, y> \in f \wedge \forall x, y, z (<x, y> \in f \wedge y \leq z \rightarrow <x, z> \in f)$.

To this end let $G_{k+1}(f)$ be $\forall y(G_k(y) \rightarrow G_k(f(y)))$, so $G_k(x)$ is Π_{k+2}. Clearly $I\Sigma_{k+2} \vdash G_{k+2}(z_{k+2}) \wedge \cdots \wedge G_0(z_0) \rightarrow G_0((z_{k+2}^n, z_{k+1}, \ldots, z_0))$ so since $I\Sigma_2 \vdash G_0(z_0) \wedge G_1(z_1)$ it is enough to show that

$$I\Sigma_{k+2} \vdash G_{k+2}(z_{k+2}).$$

Clearly $I\Sigma_2 \vdash \forall x(G_{k+1}(x) \rightarrow \forall y G_{k+1}(x^y)) \rightarrow G_{k+2}(z_{k+2})$ so it is enough to show that

$$I\Sigma_{k+2} \vdash \forall x(G_{k+1}(x) \rightarrow \forall y\, G_{k+1}(x^y))$$

- or equivalently

$$I\Sigma_{k+2} \vdash \forall x, z(G_{k+1}(x) \wedge G_k(z) \rightarrow \forall y G_k(x^y(z)))$$

But since $G_k(x^y(z))$ is Π_{k+2} this follows by $I\Sigma_{k+2}$ and proposition 15 is proved.

We now have three sections dealing with some of the applications of theorem 1. Throughout M is a countable non-standard model of $I\Sigma_1$.

<u>The combinatorial relation $\overset{\rightarrow}{*}$.</u>

Recall that for $n \in N$ $A, c \in M$, $A \overset{\rightarrow}{*} (n+1)^n_c$ stands for:-

$\forall f: [A]^n \rightarrow c\ \exists B \subseteq A$ such that B is homogeneous for f (i.e. f is constant on $[B]^n$) and $n+1, \min(B) \leq |B|$.

<u>Theorem 22.</u> For $a, b \in M$, $a \leq b$ and $0 < n \in N$,

$$Y(a, b) = \max c \text{ such that } [a, b] \overset{\rightarrow}{*} (n+2)^{n+1}_c$$

is an indicator for n-extendible cuts.

<u>Proof.</u> We first show that if $[a, b] \overset{\rightarrow}{*} (n+2)^{n+1}_c$, $c > N$, then there is an n-extendible cut I such that $a \in I < b$. For simplicity assume that $n = 3$, and that $[a, b] \overset{\rightarrow}{*} (5)^4_c$ for some $c > N$. By results in [7] it follows that there is a $p \in N$ such that for all $f: [a, b]^{n+1} \rightarrow c/p$ there is a set $A \subseteq [a, b]$ homogeneous for f and such that $N < 2^{\min(A)} \leq |A|$. It what follows we shall use this stronger version.

Fix $d \in M$, d much larger than b (hereafter written as $d >> b$) and for $a \leq a_1 < a_2 < a_3 \leq b$ set

$$|a_1, a_2, a_3|_0 = a$$

$$|a_1, a_2, a_3|_{q+1} = \max(\mathrm{def}_{q+1}(d, \{a_2, a_3\} \cup |a_1, a_2, a_3|_q) \cap [a, a_1]),$$

where $\mathrm{def}_q(d, X)$ is the set of elements of d definable in d by a formula with code at most q and parameters from X.
Set $\chi(a_1, a_2, a_3)$ to be the max. q such that

$$[a_1, a_2] \cap \mathrm{def}_q(d, \{a_3\} \cup |a_1, a_2, a_3|_q) = \emptyset$$
$$\text{and } [a_1, a_3] \cap \mathrm{def}_q(d, |a_1, a_2, a_3|_q) = \emptyset.$$

If $\chi(a_1, a_2, a_3) > N$ then these conditions ensure that if $I \subsetneq_e M$ is a limit point of the $|a_1, a_2, a_3|_q$ for $q < \chi(a_1, a_2, a_3)$ then with the obvious notation

$$\mathrm{def}_N(d, I) \precneq_I \mathrm{def}_N(d, \{a_3\} \cup I) \precneq_I \mathrm{def}_N(d, \{a_2, a_3\} \cup I) \precneq_I d$$

and by proposition 14 the result follows.

Hence to show the result it is sufficient to show that $\exists\ a_1, a_2, a_3$ such that $\chi(a_1, a_2, a_3) > N$ and $a \leq a_1 < a_2 < a_3 \leq b$.

To show this define $H : [a, b]^4 \to {}^c/2p$ as follows. Let $<x, y, z, t> \in [a, b]^4$. If for some $q < {}^c/8p$,

$$|y, z, t|_q \geq x$$

set $H(x, y, z, t)$ to be the least such q. If not but for some $q < {}^c/8p$,

$$[y, z] \cap \mathrm{def}_q(d, \{t\} \cup |y, z, t|_q) \neq \emptyset$$

set $H(x, y, z, t)$ to be the least such q plus ${}^c/8p$. If both these fail but for some $q < {}^c/8p$,

$$[y, t] \cap \mathrm{def}_q(d, |y, z, t|_q) \neq \emptyset$$

set $H(x, y, z, t)$ to be the least such q plus ${}^c/4p$. Otherwise set $H(x, y, z, t) = {}^{3c}/8p{+}1$.

Now pick a homogeneous set $B \subseteq [a, b]$ for H with $N < 2^{\min(B)} \leq |B|$. Say $B = \{a_i | i < |B|\}$ in increasing order. We claim that H must take value ${}^{3c}/8p{+}1$ on $[B]^4$ and hence that $\chi(a_1, a_2, a_3) \geq {}^c/8p$.

To prove this claim suppose that H takes value q on $[B]^4$ with $q < {}^c/8p$. Clearly $q > 0$ so for $i < j < k < m < |B|$,

$$[a_i, a_j] \cap \mathrm{def}_q(d, |a_j, a_k, a_m|_{q-1} \cup \{a_k, a_m\}) \neq \emptyset$$

whilst, by homogeneity, $|a_j, a_k, a_m|_{q-1} < a_0$. Thus $[a_i, a_j] \cap \mathrm{def}_q(d, a_0 \cup \{a_e, a_{e+1}\}) \neq \emptyset$ for all $i < j < |B| - 2$ where $e = |B| - 2$. But this is impossible since

$$|\mathrm{def}_q(d, a_0 \cup \{a_e, a_{e+1}\})| \leq (a_0+2)^{q+1} < |\{[a_{2i}, a_{2i+1}] \,|\, 2i+3 < |B|\}|.$$

A very similar argument shows that H cannot take a value less than ${}^{3c}/8p$ on $[B]^4$ and the claim is proved.

To complete the proof of the theorem we must show that if I is n-extendible $(n > 0)$ and $a \in I < b$ then $[a, b] \underset{*}{\to} (n+2)_k^{n+1}$ for all $k \in N$. In fact we shall prove by induction on n the ostensibly stronger result:-

"If I is n-extendible, $a \in I < b$, $f: [a, b]^{n+1} \to k$, $k \in N$, $s \in I$ and C is unbounded in I then $\exists\ A \subseteq C$, A a bounded subset of I such that A is homogeneous for f and s, $\min(A) \leq |A|$."

To this end assume the hypothesis of this result and let $M \underset{I}{\prec} K$ where I is (n-1)-extendible in K for $n > 1$. Let $\pi \in K$, $I < \pi < (M-I)$ and $\pi \in C$ (in the sense of K). In K define an increasing sequence b_p as follows. Let b_0, $b_1, \ldots, b_{n-1}$ be the first n elements of C. Now suppose b_q found and $b_q < \pi$ for $q < p$, $(p \geq n)$. Define b_p to be minimal such that $b_p \in C$ and for all $q_1 < \ldots < q_n < p$,

$$f(b_{q_1},\ldots, b_{q_n}, b_p) = f(b_{q_1},\ldots, b_{q_n}, \pi).$$

The sequence stops when $b_p = \pi$. Notice if $q \in I$ then the map $h: [q]^n \to k$ defined by $h(q_1,\ldots, q_n) = f(b_{q_1},\ldots, b_{q_n}, \pi)$ is coded in M. From h we can successively define b_r for $r < q$ in M. Hence, since $b_r \leqslant \pi$, $b_r \in I$ for $r < q$. It follows that if $q \in I$ then $b_q \in I$ and hence that the sequence of b_r's is unbounded in I.

We now consider two cases. If $n = 1$ let $f(b_{q_1}, \pi),\ldots, f(b_{q_m}, \pi)$, with $s \leqslant q_1 < \ldots < q_m \in I$ be all the distinct value of $f(b_j, \pi)$ for $s \leqslant j \in I$. [It is here that we require $k \in N$.] For $1 \leqslant j \leqslant m$ let

$$A_j = \{b_q | f(b_q, \pi) = f(b_{q_j}, \pi) \ \& \ s \leqslant q \leqslant s+mb_{q_m}\} \subseteq C.$$

Then it is easy to see that A_j is coded in M, A_j is homogeneous for f and one of these A_j's satisfies s, $\min(A_j) \leqslant |A_j|$ as required.

Finally if $n > 1$ then by inductive hypothesis we can find $B \subseteq \{b_q | q \in I\}$ such that B is a bounded subset of I (so B is coded in M) B is homogeneous for the map $g: [a, \pi-1]^n \to k$ defined by $g(x_1,\ldots, x_n) = f(x_1,\ldots, x_n, \pi)$ and s, $\min(B) \leqslant |B|$. It follows that B is homogeneous for f and the result follows.

Whilst this theorem is fresh in our minds we will state some immediate improvements which follow from the proof.

<u>Definition</u>. Let $b \in I \subseteq_e M$. We say that $r(I) > b$ if whenever $I = \bigcup_{j<b} A_j$ is coded in M then for some $j < b$, A_j in unbounded in I.

Clearly I is regular just if $r(I) > b \ \forall b \in I$.

<u>Corollary 23.</u> Let $I \subseteq_e K_0 \models I\Sigma_1$, $c \in I$ and suppose $K_1, K_2,\ldots, K_n$ are such that $K_0 \prec_I K_1 \prec_I K_2 \prec_I \ldots \prec_I K_n$ and in K_n $r(I) > c$. Then for $A \in K_0$ unbounded in I $A \underset{*}{\to} (n+2)^{n+1}_c$ holds in K_0.

<u>Proof.</u> If $r(I) > c$ in K_n we can replace the finite k in the second part of the proof of theorem 22 by c. The only new problem that arises is that the A_j's, $1 \leqslant j \leqslant m$ (with m now possibly infinite) may be unbounded in I. However since $m \leqslant c$ the condition $r(I) > c$ ensures that this does not occur.

<u>Lemma 24</u>. Let $A, b \in M$ with $|A| > b^\nu$ some $\nu > N$. Then $\exists I \subseteq_e M$ such that A is unbounded in I and $r(I) > b$.

<u>Proof.</u> Enumerate as $\bigcup_{j<b} Q_{ij}$, $i \in N$ all coded partitions of A into b parts. Now pick a decreasing sequence A_i, $i \in N$ of subsets of A such that

$$A_0 = A$$

$$|A_{i+1}| \geqslant |A|.b^{-i}$$

and $A_{i+1} \subseteq A_i \cap Q_{ij}$ some $j < b$.

It is easy to see that if $I = \bigcup_{i \varepsilon N} \min(A_i)$ then $r(I) > b$.

Corollary 25. Let $A, c, s, \varepsilon M$ and $n \varepsilon N$ be such that $A \underset{*}{\to} (n+2)_c^{n+1}$ and $1 < s^\nu < c$ some $\nu > N$. Then $\exists I \subseteq_e M$ and $M_1, M_2, \ldots, M_n$ such that

$M \underset{I}{\prec} M_1 \underset{I}{\prec} M_2 \underset{I}{\prec} \ldots \underset{I}{\prec} M_n$, A is unbounded in I and, in M_n, $r(I) > s$.

Proof. By the proof of the first part of theorem 22 with A in place of $[a, b]$ and $<d, A>$ in place of d etc. we can obtain $<a_1, a_2, a_3> \varepsilon [A]^3$ such that $\chi(a_1, a_2, a_3) \geq {}^c/8p$. By lemma 24 we can take $I \subseteq_e M$ to be a limit point of $\{|a_1, a_2, a_3|_q \; |q < {}^c/8p\}$ such that $r(I) > s$. Finally since we have replaced d by $<d, A>$, A will be unbounded in I.

Corollary 26. Let $A, H \varepsilon M$, $A \underset{*}{\to} (n+1)_b^n$, $H:[A]^m \to e$ where $2, m \leq n$, and b, $e < \min(A)$ and $b > N$. Then for any s such that $s^i < b, \forall i \varepsilon N \; \exists B \subseteq A$, $B \varepsilon M$ such that B is homogeneous for H and $B \underset{*}{\to} (n-m+2)_s^{n-m+1}$.

Proof. By Corollary 25 find $I \subseteq_e M$ and $M_1, \ldots, M_{n-1}$ such that

$$M \underset{I}{\prec} M_1 \underset{I}{\prec} M_2 \underset{I}{\prec} \ldots \underset{I}{\prec} M_{n-1}$$

and $r(I) > s$ in M_{n-1}. As in Corollary 23 there is a set $E \varepsilon M_{m-1}$ such that E is unbounded in I, $E \subseteq A$ and E is homogeneous for H. By Corollary 23 again, in M_{m-1}, $E \underset{*}{\to} ((n-1)-(m-1)+2)_s^{(n-1)-(m-1)+1}$. We can now pull back into M a sufficiently large initial segment of E to give the required set B.

Corollary 27. There is an $i \varepsilon N$ such that for $2, m \leq n$ the following is provable in $I\Sigma_1$:-

For all sufficiently large b if $A \underset{*}{\to} (n+1)_{b^i}^n$, $F:[A]^m \to e$ and $e, b < \min(A)$ then $\exists B \subseteq A$ such that B is homogeneous for F and $B \underset{*}{\to} (n-m+2)_b^{n-m+1}$.
With a little more work b^i here could be replaced by ib. However for more precise bounds we refer the reader to the paper by George Mills in this volume.

Combining theorems 1 & 22 we see that

$$Y(a, b) = \max.c \text{ such that } [a, b] \underset{*}{\to} (n+2)_c^{n+1}$$

is an indicator for cuts satisfying $I\Sigma_n$ in any countable $M \models I\Sigma_1$. Hence by theorem 0,

Corollary 28

(i) $I\Sigma_n \not\vdash \forall x, z \; \exists y, \; [x, y] \underset{*}{\to} (n+2)_z^{n+1}$,

(ii) $\forall m \in N$, $I\Sigma_n \vdash \forall x \exists y, [x, y] \overset{\to}{*} (n+2)^{n+1}_m$,

(iii) $\forall x, z \exists y\ [x, y] \overset{\to}{*} (n+2)^{n+1}_z$ is independent of $I\Sigma_n$,

(iv) if $f(x) = y$ is Σ_1 and $I\Sigma_n \vdash \forall x \exists! y\ f(x) = y$ then $\exists\, m \in N$ such that $I\Sigma_n \vdash \forall x(f(x) < g_m(x))$ where $g_m(x) = \mu y : [x, y] \overset{\to}{*} (n+2)^{n+1}_m$.

Combining this with the following lemma we obtain reasonable estimates on the proof complexity of "finite" versions of Leo Harringtons independent sentence $\forall x, z, t, w \exists y\ [x, y] \overset{\to}{*} (t)^z_w$.

<u>Lemma 29</u>. $I\Sigma_1 \vdash [\forall x \exists y, [x, y] \overset{\to}{*} (n+2)^{n+1}_2] \to [\forall x, z \exists y, [x, y] \overset{\to}{*} (n+1)^n_z]$.

<u>Proof</u>. Let $c, a \in M$ and suppose that

$$M \models \forall x \exists y, [x, y] \overset{\to}{*} (n+2)^{n+1}_2 .$$

Pick $d > c, a$ such that in the usual notation of Ramsey's theorem $d \to (n+1)^n_c$. Pick b such that $[d, b] \overset{\to}{*} (n+2)^{n+1}_2$. Now given $f: [a, b]^n \to c$ define $H: [d, b]^{n+1} \to 2$ by

$$H(\vec{x}) = 0 \iff \vec{x} \text{ is homogeneous for } f.$$

Let $B \subseteq [d, b]$ be homogeneous for H with $|B| \geq \min(B) \geq d$. Then $B \to (n+1)^n_c$ so for some, and hence all, $\langle\vec{x}\rangle \in [B]^{n+1}$, $H(\vec{x}) = 0$. Hence by lemma 2.7 of [7] B is homogeneous for f as required.

<u>Corollary 30</u>. The cuts in M satisfy $I\Sigma_n$, $I\Sigma_{n+1}$ are not symbiotic. Hence the hierarchy of theorem 1 is non-trivial.

<u>Proof.</u> For $n = 0$, $N < a \in M$, $\bigcup_{n \in N} a^n \models I\Sigma_0$ but, since this cut is not closed under exponentiation, it is not a limit of cuts satisfying $I\Sigma_1$.

Now suppose $n > 0$ and that the corollary is false. Let $N < b \in M$ and pick d maximal such that $[d, b] \overset{\to}{*} (n+3)^{n+2}_2$. By overspill $N < d$ since $N \models I\Sigma_{n+1}$. By the proof of lemma 29, $[d, b] \overset{\to}{*} (n+2)^{n+1}_c$ for some $N < c$. Hence by our assumption and theorem 22 $\exists\, J \subseteq_e M$, $J \models I\Sigma_{n+1}$ and $d \in J < b$. By corollary 28 (ii), $[e, f] \overset{\to}{*} (n+3)^{n+2}_2$ some $d < e, f \in J$. But then $[e, b] \overset{\to}{*} (n+3)^{n+2}_2$ contradicting the choice of d.

<u>Conservation results.</u>

The following lemma is well known:-

<u>Lemma 31</u>. Suppose X, Y are symbiotic sets of cuts in M. Then for θ a Π_2 sentence, $\forall J \in X, J \models \theta \iff \forall J \in Y, J \models \theta$.

Proof. Suppose $\forall J \in X, J \models \theta$ where θ is $\forall x \exists y\ \psi(x, y)$ with ψ bounded. Let $I \in Y$ and $a \in I$. Then since there are $J \in X$ arbitrarily close to I, the least c such that $\psi(a, c)$ must lie in I. Hence $I \models \theta$.

Hence, by theorem 1.

Corollary 32. For $\theta \in \Pi_2$, $n > 0$,
$I\Sigma_n \vdash \theta \iff$ n-extendible $J \subseteq_e M \models I\Sigma_1$, $J \models \theta$.

A potential value of this result is that it may allow us a backdoor method of showing $I\Sigma_n \vdash \theta$ for combinatorial θ. Again by theorem 1 a similar result holds with $B\Sigma_{n+1} + I\Sigma_0$ in place of $I\Sigma_n$. (We need to add in $I\Sigma_0$ here to ensure that $B\Sigma_{n+1} + I\Sigma_0 \vdash I\Sigma_1$.) Since $I\Sigma_n$ is strictly weaker than $B\Sigma_{n+1} + I\Sigma_0$ (see [8]) this gives a conservation result. In fact our methods yield a much better conservation result:-

Theorem 33. (Also, independently, by H. Friedman.)
$I\Sigma_n$ and $B\Sigma_{n+1} + I\Sigma_0$ have the same Π_{n+2} consequences.

Before giving the proof of this theorem we prove a lemma. This lemma follows easily from an early result of Leo Harrington on indicators for regular cuts. However the proof we give below generalises to higher levels.

Lemma 34. Let M be a countable non-standard model of $I\Sigma_1$ and suppose $A \in M$ and $\exists I \subseteq_e M$, A is unbounded in I and I semi-regular. Then $\exists J \subseteq_e M$, A is unbounded in J and J is regular.

Proof. Pick $A \ll d$, and let A be unbounded in I, I semi-regular. Adopting the notation of lemma 13 for each $n \in N$ there is an $\{A\}$-sequence

$$[0, d] = [a_0\ b_0] \supset [a_1, b_1] \supset \ldots \supset [a_n, b_n]$$

with $a_i, b_i \in A$ for $i \leqslant n$. Since, to produce such a sequence, pick for each $0 \leqslant i \leqslant n$ $a_i, b_i \in A$ such that $a_i \in I < b_i$ and $\mathrm{Def}_{a_{i-1}}(d, \{A\} \cup a_{i-1}) \cap [a_i, b_i] = \emptyset$, possible since $\mathrm{Def}_{a_{i-1}}(d, \{A\} \cup a_{i-1})$ cannot be unbounded in I. It follows that there is an $\{A\}$ - sequence $[a_0, b_0] \supset [a_1, b_1] \supset \ldots \supset [a_\nu, b_\nu]$ with $a_i, b_i \in A$ and and $\nu > N$. Let $K = \mathrm{Def}_J(d, \{A\} \cup J)$ where $J = \bigcup_{i \in N} a_i$. Then J is regular in K and, in K, A is unbounded in J. As in the transfer lemma, proposition 14, such a regular cut exists in M.

Proof of theorem 33.

We first consider the case $n = 0$. So suppose $J \models I\Sigma_0$, J countable. Let K be the full ultrapower of J so J is not confinal in K. Let $\bar{J} = \{x \in K \mid \exists y \in J, x \leqslant y\}$. Then, since $K \models I\Sigma_0$, $\bar{J} \models B\Sigma_1$, J is cofinal in and $J \prec_{\Sigma_1} \bar{J}$. It follows that any Π_2 sentence true in $\bar{J}$ must be true in J and this yields the required result.

Now suppose $n > 0$ and

$$I\Sigma_n \nvdash \forall x \exists y \forall z \ \theta(x, y, z)$$

where $\theta \in \Sigma_{n-1}$. We shall show that

$$B\Sigma_{n+1} + I\Sigma_0 \nvdash \forall x \exists y \forall z \ \theta(x, y, z).$$

Pick countable non-standard $M \models I\Sigma_n + \forall y \exists z \neg\theta(a, y, z)$, where $a \in M$. Define $H: M \to M$ by

$$H(x) = y \iff \neg\theta(a, x, y) \wedge \forall t < y \ \theta(a, x, t),$$

so $H \in \Delta_n$. Let $\exists y \ \psi(x, y)$ be the standard complete Σ_{n-1} formula with $\psi \in \Pi_{n-2}$. (If $n = 1$ omit this step.) Define $F: M \to M$ by

$$F(s) = t \iff \forall x \leq s \ (\exists y \ \psi(x, y) \to \exists y \leq t \ \psi(x, y)) \wedge$$
$$\exists x \leq s \ (\psi(x, t) \wedge \forall z < t \ \neg\psi(x, z),$$

so $F \in \Delta_n$. Notice that if $a \in J \subseteq_e M$ and J is closed under F and H then $J \prec_{\Sigma_{n-1}} M$ and $J \models \forall x \exists y \neg\theta(a, x, y)$.

Now define

$$G_0(x) = F(x) + H(x) + x + 1$$
$$G_{m+1}(x) = G_m^{x+2}(x).$$

Since $M \models I\Sigma_n$, each G_m is Σ_n and is total for $m \in N$. Also, by $I\Sigma_n$,

$$\{\eta \mid G_\eta(a) \text{ is defined}\}$$

must contain an element $\eta > N$. Pick such an η. As in [4] $\exists a \in I \subseteq_e M$ such that I is semi-regular and $A = \{G_0^k(a) \mid G_0^k(a) \leq G_\eta(a)\}$ is unbounded in I. By the lemma there is a regular $J \subseteq_e M$ such that $a \in J$ and A is unbounded in J. By earlier remarks it follows that $J \prec_{\Sigma_{n-1}} M$ and $J \models \forall x \exists y \neg\theta(a, x, y)$.

It remains to show that $J \models B\Sigma_{n+1}$. To this end suppose $J \models \forall x < b \exists y \forall z \ \phi(x, y, z)$ where ϕ is Σ_{n-1}. Let $J < h \in M$. Then since $M \models I\Sigma_n$ there is a coded subset B of h such that

$$M \models \forall x, y, z \leq h(\phi(x, y, z) \longleftrightarrow \langle x, y, z\rangle \in B).$$

Therefore since $J \prec_{\Sigma_{n-1}} M$,

$$J \models \forall x, y, z \ (\phi(x, y, z) \longleftrightarrow \langle x, y, z\rangle \in B). \quad \ldots \quad \text{(i)}$$

Now pick $e \gg B$ and by the usual 1-extendibility property of regular cuts find

$e \underset{J}{\prec} e'$ with $J < d < e-J$. Then since

$$J \models \forall x < b\ \exists y\ \forall z\ \langle x, y, z\rangle \in B,$$

$$e' \models \forall x < b\ \exists y < p\ \forall z < d\ \langle x, y, z\rangle \in B$$

for any $J < p$. The least such p must be in J. Hence $J \models \forall x < b\ \exists y < p\ \forall z\ \langle x, y, z\rangle \in B$ and $J \models B\Sigma_{n+1}$ follows by (i). Finally since $J \models \exists x\ \forall y\ \exists z\ \neg\theta(x, y, z)$ it follows that $B\Sigma_{n+1} + I\Sigma_0 \nvdash \forall x\ \exists y\ \forall z\ \theta(x, y, z)$.

Remarks. This result is optimal since hidden in the proof of proposition 7 (with n in place of n-1) of [8] is a Σ_{n+2} sentence provable from $B\Sigma_{n+1} + I\Sigma_0$ but not from $I\Sigma_n$.

We would conjecture that 2 could be replaced by n+2 in corollary 32. However the most we have at present is an involved proof that 2 can be replaced 3. Two related results are:-

(i) For $\theta \in \Pi_{n+2}$, $I\Sigma_n^* \vdash \theta \leftrightarrow B\Sigma_{n+1}^* \vdash \theta$.

(ii) For any θ, $I\Sigma_1 \vdash \theta \leftrightarrow I\Sigma_1^* \vdash \theta$.

Here $I\Sigma_n^* \vdash \theta$ stands for "for all countable $M \models I\Sigma_1$, if $I \subseteq_e M$ and $I \models I\Sigma_n^*$ then $I \models \theta$".

The next corollary sheds more light on the relationship between $I\Sigma_n$ and $B\Sigma_{n+1} + I\Sigma_0$.

Corollary 35. Let $M \models I\Sigma_n$. Then M has a confinal extension $\bar{M}$ such that $M \prec_{\Sigma_{n+1}} \bar{M}$ and $\bar{M} \models B\Sigma_{n+1} + I\Sigma_0$.

Proof. The result for $n = 0$ has been proved in the proof of the theorem so suppose $n > 0$ and let $M \models I\Sigma_n$. By the theorem we can obtain, by the completeness theorem, $M \prec_{\Sigma_{n+1}} K \models B\Sigma_{n+1} + I\Sigma_0$. Let $\bar{M} = \{x \in K \mid \exists y \in M,\ x \leq y\}$. We first show that $\bar{M} \prec_{\Sigma_n} K$. To this end suppose $a \in \bar{M}$, say $a < b \in M$ and $K \models \exists x\ \theta(a, x)$ with $\theta \in \Pi_{n-1}$. Then

$$K \models \forall z < b\ \exists x[\theta(z, x) \vee (\neg\exists y\ \theta(z, y) \wedge x = 0)].$$

Therefore since $K \models B\Sigma_{n+1}$,

$$K \models \exists t \forall z < b\ \exists x < t[\theta(x, z) \vee (\neg\exists y\ \theta(z, y) \wedge x = 0)].$$

Hence $$K \models \exists t \forall z < b\ [\exists x\ \theta(z, x) \rightarrow \exists x < t\ \theta(z, x)].$$

Since $M \prec_{\Sigma_{n+1}} K$ and $M \models B\Sigma_n$ this also holds in M. Hence for some $t \in M$,

$$K \models \forall z < b[\exists x\ \theta(z, x) \rightarrow \exists x < t\ \theta(z, x)].$$

In particular there is an $x < t$ (hence $x \in \bar{M}$) such that $K \models \theta(a, x)$ and $\bar{M} \prec_{\Sigma_n} K$ follows by Tarski's Criterion.

From $\bar{M} \prec_{\Sigma_n} K$ it is easy to show that $M \prec_{\Sigma_{n+1}} \bar{M}$. Finally to show that $\bar{M} \models B\Sigma_{n+1}$ suppose that $\bar{M} \models \forall x < b \, \exists y \, \theta(x, y)$ with $\theta \in \Pi_n$. Then as $\bar{M} \prec_{\Sigma_n} K$, $K \models \forall x < b \, \exists y < p \, \theta(x, y)$ for any $\bar{M} < p \in K$. [If no such p exists there is nothing to prove.] Since $K \models I\Sigma_n$ there is a least such p which must be in $\bar{M}$ and the result follows.

[An alternative proof of this corollary and the preceding theorem using ultrapowers will appear in a forthcoming paper by the author.]

Reflection Principles

Recall that in [7] the standard independent combinatorial sentence for P was shown to be equivalent to the Σ_1 - reflection principle or equivalently $\mathrm{Con}(P + T_1)$ where T_1 is the set of Π_1 sentences which are true according to the standard complete Π_1 formula. In this section we show that a similar result holds for $I\Sigma_n$, $(n > 0)$.

By earlier results the following functions are indicators for cuts satisfying $I\Sigma_n$ $(n > 0)$

$$Y(a, b) = \max. c \text{ such that } [a, b] \underset{*}{\rightarrow} (n+2)_c^{n+1}$$

$$W(a, b) = \max. c \text{ such that } [a, b] \text{ is } \omega_n^c\text{-large.}$$

Theorem 36 For $n > 0$ the following are equivalent in $I\Sigma_1$:-

(i) $\mathrm{Con}(I\Sigma_n + T_1)$,

(ii) $\forall x, z \, \exists y \; Y(x, y) \geq z$,

(iii) $\forall x, z \, \exists y \; W(x, y) \geq z$.

We shall prove this result in the next three lemmas.

Lemma 37. $I\Sigma_1 \vdash \mathrm{Con}(I\Sigma_n + T_1) \rightarrow \forall x, z \, \exists y W(x, y) \geq z$.

Proof. Let $M \models I\Sigma_1$ and work in M. Assume $\mathrm{Con}(I\Sigma_n + T_1)$ but that for some c, $d \in M$, $\neg \exists y \; W(c, y) \geq d$. Then with the notation of proposition 15,

$$\mathrm{Con}(I\Sigma_n + G_n(z_n^{\underline{d}}) + \neg \exists y \; W(\underline{c}, y) \geq \underline{d})$$

since $$I\Sigma_n + G_n(z_n^{\underline{d}}) \vdash \forall x \, \exists y \; W(x, y) \geq \underline{d}.$$

[Here $\underline{d}$ is the numeral of d.]

Let e be minimal such that

$$\mathrm{Con}(I\Sigma_n + G_n(z_n^{\underline{e}}) + \neg \exists y \; W(\underline{c}, y) \geq \underline{d}).$$

Then $e > 0$ since $\mathrm{Con}(I\Sigma_n + T_1)$ and

$$I\Sigma_n + T_1 \vdash G_n(z_n^{\overset{o}{-}}) + \neg \exists y\, W(\underline{c}, y) \geq \underline{d}.$$

But since $$I\Sigma_n \vdash G_n(z_n^{\underline{e-1}}) \to G_n(z_n^{\underline{e}})$$

we obtain $$\neg \mathrm{Con}(I\Sigma_n + G_n(z_n^{\underline{e-1}}) + \neg \exists y\, W(\underline{c}, y) \geq \underline{d})$$

- a contradiction.

Lemma 38. $I\Sigma_1 \vdash \forall x, z\ \exists y\ W(x, y) \geq z \to \forall x, z \exists y\ \underline{Y}(x, y) \geq z.$

Proof. The proof is essentially a re-run of the proof of Proposition 13. We outline the proof for the case $n = 2$. Let M be a non-standard countable model of $I\Sigma_1 + \forall x, z \exists y\ W(x, y) \geq z$. Let $a, c \in M$. Working in M let e be non-standard, $e >> c$, and pick b such that $[a, b]$ is ω_2^e - large. We shall show that $[a, b] \underset{*}{\to} (4)_c^3$.

Let $f: [a, b]^3 \to c$ and fix $d >> b$. Using the notation of proposition 13 construct an $\{f, b\}$ - sequence $\{[g_j, h_j] | j \leq q\}$ where $[g_0, h_0] = [a, b]$.

By Lemma 12 we can arrange that $\{g_j | j \leq q\}$ is ω^{e-1}-large. Similarly we can find an $\{f, b, g_q\}$ - sequence $\{[g_j', h_j'] | j \leq s\}$ such that g_j, $h_j \in \{g_i | i \leq q\}$ for $j \leq s$ and $\{g_j' | j \leq s\}$ is (e-2) - large. It follows that $s \geq e-2$ and hence, since e is much larger than c, that $s > (c+2)^{\nu}$ for some $\nu > N$.

By lemma 24 let $I \subseteq_e M$ such that $\{g_j' | j \leq s\}$ is unbounded in A and $r(I) > c$. Then I is a proper cut and, as in proposition 13, I is closed under exponentiation.

Let
$$K_0 = \mathrm{Def}_I(d, \{f, b\} \cup I),$$
$$K_1 = \mathrm{Def}_I(d, \{f, b, g_q\} \cup I),$$
$$K_2 = d.$$

Then $$K_0 \underset{I}{\prec} K_1 \underset{I}{\prec} K_2, \quad a, b, f \in K_0 \text{ and } r(I) > c \text{ in } K_2.$$

By corollary 23 there is coded subset B of $[a, b]$ in K_0 which is large and homogeneous for f. The result follows.

Preamble. Before concluding the proof of the theorem we recall some results. Let T be a Π_1 theory in the language of arithmetic. Let L^* be the language obtained by Skolemizing the language of arithmetic. So for each formula $\exists w\ \theta(w, \vec{u})$ of the language of arithmetic there is a function symbol $F_{\exists w\ \theta(w, \vec{u})}$ of length $\vec{u}$ arguments. Now let

$$\forall x_1 \exists x_2 \forall x_3 \exists x_4 \ldots \phi(x_1, x_2, x_3, x_4, \ldots)$$

be an axiom of T (assumed in PNF) with ϕ open. A slice of this axiom is an open sentence of L^* of the form

$$\phi(t_1,\ F_\lambda(t_1),\ t_2,\ F_\psi(t_1,\ F_\lambda(t_1),\ t_2),\ldots)$$

where $t_1, t_2, \ldots$ are closed terms of L^* and

$$\lambda(x_1) = \exists x_2\ \forall x_3\ \exists x_4 \ldots \phi(x_1, x_2, x_3, x_4 \ldots)$$

$$\psi(x_1, x_2, x_3) = \exists x_4 \ldots \phi(x_1, x_2, x_3, x_4 \ldots) \text{ etc.}$$

Let T^* consist of all slices of axioms of T together with the equality axioms

$$t = t$$

$$\vec{t} = \vec{s} \to F(\vec{t}) = F(\vec{s}),$$

$$t_1 = t_2 \to t_2 = t_1,$$

$$(t_1 = t_2 \wedge t_2 = t_3) \to t_1 = t_3,$$

where t_1, t_2 etc. are closed terms of L^*.

By Herbrands theorem if T is inconsistent then T^* is inconsistent in the sentential calculus. Furthermore this reduction can be carried out in $I\Sigma_1$.

Now suppose that with each function symbol F of L^* we associate a total function $\bar{F}$ in such a way that the map sending F to the Σ_1-code for $\bar{F}$ is recursive. For t a closed term of L^* define

$V(t) = b \iff$ replacing each F in t by $\bar{F}$ and evaluating gives b. (Give $+, \cdot, 0, '$, their standard interpretation.)

Using $I\Sigma_1$, V is Σ_1 and total on closed terms. Let Γ be the natural Σ_1 formula such that for $\theta \in \Sigma_0$,

$$\theta \leftrightarrow \Gamma(\ulcorner\theta\urcorner)$$

holds in any model of $I\Sigma_1$. For $\theta(x_1,\ldots, x_n)$ an open formula in the language of arithmetic and $t_1,\ldots, t_n$ closed terms of L^* define

$$\Vdash \theta(t_1,\ldots, t_n) \Leftrightarrow \exists \vec{b}\,[V(\vec{t}) = \vec{b} \wedge \Gamma(\ulcorner\theta(\underline{\vec{b}})\urcorner)].$$

Now suppose T was inconsistent. Then there would be a proof p in the sentential calculus of $\neg(\lambda_1 \wedge \ldots \wedge \lambda_q)$ for some $\lambda_1,\ldots, \lambda_q \in T^*$. Then by induction on the length of p (and we only need $I\Sigma_1$ here) we can show $\Vdash \neg(\lambda_1 \wedge \ldots \wedge \lambda_q)$ and hence, by standard properties of Γ, $\Vdash \neg\lambda_i$ some $1 \leq i \leq q$.

Summing up then we have, in $I\Sigma_1$, that if for every finite subset S of T^* there is an assignement $F \longmapsto \bar{F}$ such that $\Vdash \lambda$ for all $\lambda \in S$ then $Con(T)$. We are now ready to complete the proof of the theorem.

<u>Lemma 39</u>. $I\Sigma_1 \vdash \forall x, z\ \exists y\ Y(x, y) \geq z \to Con(I\Sigma_n + T_1)$.

<u>Proof.</u> For convenience assume $n = 3$. Let M be a non-standard model of $I\Sigma_1 + \forall x, z\ \exists y[x, y] \not\geq (5)_z^4$. Working in M let S be a finite subset of

$(I\Sigma_3 + T_1)^*$ where we assume that the sentences in $I\Sigma_3 + T_1$ are in PNF. Let $\nu > N$ be much larger than (the code for) S, denoted as usual by $S << \nu$. Pick b such that $|0, b| \underset{*}{\to} (5)^4_\nu \nu$. Let $d >> b$ and, as in the proof of corollary 25 (with $A = [0, b]$) find $<\beta, \gamma, \delta> \varepsilon [0, b]^3$ such that if

$$a_0 = 0$$

$$a_{q+1} = \max(\mathrm{def}_{q+1}(d, \{\gamma, \delta\} \cup a_q) \cap \beta)$$

then for $q < \nu$

$$[a_{q+1}, \delta] \cap \mathrm{def}_{q+1}(d, a_q) = \phi \quad \ldots \text{(i)}$$

$$[a_{q+1}, \gamma] \cap \mathrm{def}_{q+1}(d, a_q \cup \{\delta\}) = \phi \quad \ldots \text{(ii)}$$

$$[a_{q+1}, \beta] \cap \mathrm{def}_{q+1}(d, a_q \cup \{\gamma, \delta\}) = \phi \quad \ldots \text{(iii)}$$

Fix $N < k$, $S << k << \nu$. Let $\exists x \forall y \exists z\, \theta(x, y, z, \vec{u}) \varepsilon \Sigma_3$ with θ bounded, $\ulcorner\theta\urcorner < k$. Then by (i), (ii), (iii), for $k < i < j < q \leq \nu$, $\vec{b} \leq a_{i-1}$

$$\exists x \leq \beta \forall y \leq \gamma \exists z \leq \delta\, \theta(x, y, z, \vec{b})$$

$$\Longleftrightarrow \exists x \leq a_i \forall y \leq \gamma \exists z \leq \delta \theta(x, y, z, \vec{b})$$

$$\Longleftrightarrow \exists x \leq a_i \forall y \leq a_j \exists z \leq \delta \theta(x, y, z, \vec{b})$$

$$\Longleftrightarrow \exists x \leq a_i \forall y \leq a_j \exists z \leq a_q \theta(x, y, z, \vec{b}) \quad \ldots \text{(iv)}$$

We now define the map $F \to \bar{F}$ for F a function symbol of L*. For $\theta(\vec{u})$ a Σ-formula (in the sense of M) in PNF, say

$$\theta = \exists x_1 \forall x_2 \exists x_3 \ldots \forall x_p\, \phi(\vec{x}, \vec{u})$$

and $\vec{b} \varepsilon M$ define $\bar{F}_\theta(\vec{b})$ as follow. If $e \leq \nu$ is minimal such that $\vec{b} \leq a_e$ and $e + p \leq \nu$ set

$\bar{F}_\theta(\vec{b})$ = least $x_1 \leq a_{e+1}$ such that
$\Gamma(\ulcorner \forall x_2 \leq a_{e+2} \exists x_3 \leq a_{e+3} \ldots \forall x_p \leq a_{e+p}\, \phi(\underline{x}_1, x_2, \ldots, x_p, \underline{\vec{b}})\urcorner)$
if such an x_1 exists.

In all other cases set $\bar{F}_\theta(\vec{b}) = 0$. Notice that if $\vec{b} \leq a_e$ then $\bar{F}_\theta(\vec{b}) \leq a_{e+1}$. Hence for s a term appearing in some $\phi \varepsilon S$, $V(s) < a_k$. We shall show that with this assignment, $\Vdash \phi$ for all $\phi \varepsilon S$.

First suppose $\lambda(t_1, \ldots, t_p) \varepsilon S$ is a slice of

$$\forall x_1 \exists x_2 \forall x_3 \ldots \forall x_p\, \lambda(x_1, \ldots, x_p) \varepsilon T_1$$

where $\exists x_2 \forall x_3 \ldots \forall x_p\, \lambda(x_1, \ldots, x_p)$ is the PNF of a bounded formula. Then $\Gamma(\ulcorner \exists x_2 \forall x_3 \ldots \forall x_p\, \lambda(V(t_1), x_2, \ldots, x_p)\urcorner)$ and by induction on the number of quantifiers $\Gamma(\ulcorner\lambda(V(t_1),\ldots, V(t_p))\urcorner)$ as required.

It remains to consider a slice in S of an instance of $I\Sigma_3$. So suppose θ is bounded,

$$\forall x_1 [\exists x_2 \forall x_3 \exists x_4\ \theta(x_1,x_2,x_3,x_4) \vee \exists x_5 \leq x_1 [\neg \exists x_6 \forall x_7 \exists x_8\ \theta(x_5,x_6,x_7,x_8) \wedge$$
$$\wedge \forall x_9 < x_5 \exists x_{10} \forall x_{11} \exists x_{12} \theta(x_9,x_{10},x_{11},x_{12})]]$$

is an instance of an axiom of $I\Sigma_3$ which has (as indicated by the variables) a PNF

$$\forall x_1 \exists x_2 \ldots \exists x_{12} \forall x_{13} \ldots \exists x_p\ \lambda(x_1,\ldots, x_p)$$

where $\forall x_{13} \ldots \exists x_p\ \lambda(x_1,\ldots, x_p)$ is the PNF of a bounded formula and $\lambda(s_1,\ldots, s_p) \in S$. Let e be minimal such that $V(s_1) \leq a_e$. There are now two cases. If $\Gamma(\ulcorner \exists x_2 \leq a_{e+1} \forall x_3 \leq a_{e+2} \exists x_4 \leq a_{e+3}\ \theta(V(s_1), x_2, x_3, x_4)\urcorner)$ then $V(s_2) \leq a_{e+1}$ and $\Gamma(\ulcorner\forall x_3 \leq a_{e+2} \exists x_4 \leq a_{e+3}\ \theta(V(s_1), V(s_2), x_3, x_4)\urcorner)$ Let f be minimal such that $V(s_3), V(s_1), V(s_2) \leq a_f$. Then by (iv) $\Gamma(\ulcorner\forall x_3 \leq a_{f+1} \exists x_4 \leq a_{f+2}\ \theta(V(s_1), V(s_2), x_3, x_4)\urcorner)$ so $\Gamma(\ulcorner \exists x_4 \leq a_{f+2}\ \theta(V(s_1), V(s_2), V(s_3), x_4)\urcorner)$. [In this argument we are repeatedly using the fact that $S << k << \nu$.] A similar argument now shows that $\Gamma(\ulcorner\theta(V(s_1),\ldots, V(s_4))\urcorner)$ and hence, as above, $\Gamma(\ulcorner\lambda(V(s_1),\ldots, V(s_p))\urcorner)$.

Now suppose $\neg\Gamma(\ulcorner\exists x_2 \leq a_{e+1} \forall x_3 \leq a_{e+2} \exists x_4 \leq a_{e+3}\ \theta(V(s_1), x_2, x_3, x_4)\urcorner)$ Let $c \leq V(s_1)$ be minimal such that

$$\neg\Gamma(\ulcorner \exists x_2 \leq a_{e+1} \forall x_3 \leq a_{e+2} \exists x_4 \leq a_{e+3}\ \theta(c, x_2, x_3, x_4)\urcorner).$$

Then by (iv) for any $e < j < q < r \leq \nu$, c is minimal such that $\neg\Gamma(\ulcorner\exists x_2 \leq a_j \forall x_3 \leq a_q \exists x_4 \leq a_r\ \theta(c, x_2, x_3, x_4)\urcorner)$ so $V(s_5) = c$ and by arguments as in the first case

$$\Gamma(\ulcorner\neg\theta(V(s_5), V(s_6), V(s_7), V(s_8)) \wedge \theta(V(s_9), V(s_{10}), V(s_{11}), V(s_{12}))\urcorner).$$

Hence $\Gamma(\ulcorner\lambda(V(s_1),\ldots, V(s_p))\urcorner)$ as required. This completes the proof of proposition 39 and theorem 36.

<u>Corollary 40</u>. For $n > 0$, $I\Sigma_{n+1} \vdash Con(I\Sigma_n + T_1)$.

<u>Proof</u>. By theorem 36, Corollary 28 and lemma 29.

<u>Remark.</u> For the case $n = 1$ theorem 36 also holds for the indicator for 1-extendible cuts defined by

$$Y(a, b) = \min c : g_c(a) \geq b,$$

where $g_0(x) = x + 2$, $g_{n+1}(x) = g_n^{x+2}(x)$. (For a proof that this is an indicator see [4].) Hence with the natural coding

$$I\Sigma_1 \vdash Con(I\Sigma_1 + T_1) \leftrightarrow \forall z, g_z \text{ total} \leftrightarrow \forall z, g_z(z) \text{ defined.}$$

Notice $g_z(z)$ is (essentially) Ackermans function.

References

[1] C. Dimitracopoulos, Doctorial thesis, Manchester. To appear.

[2] J. Ketonen & R. Solovay, "Rapidly growing Ramsey functions", to appear.

[3] L. Kirby, "Initial segments of models of arithmetic", Doctorial thesis, Manchester, 1977.

[4] L. Kirby & J. Paris, "Initial segments of models of Peano's axioms". Springer-Verlag lecture notes in mathematics, Vol. 619.

[5] G. Mills, "Extensions of models of Peano arithmetic" Doctorial thesis, Berkeley, 1977.

[6] J. Paris, "Some independence results for Peano arithmetic", J.S.L. 43 (1978), pp. 725-731.

[7] J. Paris & L. Harrington, "An incompleteness in Peano arithmetic". Handbook for Mathematical Logic, (ed. J. Barwise.), North Holland, 1976, pp.1133-1142.

[8] J. Paris & L. Kirby, "Σ_n-collection schemas in arithmetic". Logic Colloquium '77, North Holland 1978, pp.199-209.

COFINAL EXTENSION PRESERVES RECURSIVE SATURATION

C. Smoryński and J. Stavi

0. Introduction. In this note we show that elementary cofinal extensions of recursively saturated models of arithmetic are again recursively saturated. The proof runs as follows: We produce some special types and show them to i. be realized arbitrarily high in recursively saturated models, whence in their cofinal extensions, and ii. induce, upon their realization in a model, a degree of recursive saturation below their realizing elements. The mechanism by which ii is accomplished is quite simple: The element a realizing a special type serves as a parameter to encode enough information to reduce the complexities of all formulae in a great many recursive types. By Overspill and the existence of partial truth definitions, this guarantees the types will be realized.

Our immediate goals are to establish some notation and recall two useful lemmas. Following this, in section 1 we define our first special type and prove the main theorem. In section 2 we introduce another type which can be used to prove this theorem. The inclusion of this second type is not necessary to the purpose of this note; but we find the type amusing and feel it worthy of display.

On to our immediate goals: By arithmetic we mean any extension of Peano arithmetic by the addition of finitely many new relation symbols and the full schema of induction in the augmented language. We denote models of arithmetic by gothic capitals, $\mathfrak{M}, \mathfrak{N}$, with $|\mathfrak{M}|$ denoting the domain of $\mathfrak{M}$. Thus, $\mathfrak{M} = (|\mathfrak{M}|;0,+,\cdot,',<,\dots)$. Natural numbers are denoted by x,y,z,... and elements of $|\mathfrak{M}|$ by a,b,c,.... The constant naming an object a is denoted $\bar{a}$.

Given two models $\mathfrak{M} \subseteq \mathfrak{N}$ of arithmetic, we say that $\mathfrak{N}$ is a cofinal extension of $\mathfrak{M}$, written $\mathfrak{M} \subseteq_c \mathfrak{N}$, if every element of $|\mathfrak{N}|$ lies below an element of $|\mathfrak{M}|$:

$$\forall a \in |\mathfrak{N}|\ \exists b \in |\mathfrak{M}| (\mathfrak{N} \models \bar{a} < \bar{b}).$$

We shall only concern ourselves with elementary cofinal extensions-- not really a severe restriction: In Gaifman 1972 it is shown that, for the usual arithmetic language, cofinal extensions are automatically elementary.

Fix a model $\mathfrak{M}$ of arithmetic. A set

$$\tau v v_0 \dots v_{n-1} = \{ \varphi_0 v v_0 \dots v_{n-1}, \varphi_1 v v_0 \dots v_{n-1}, \dots \}$$

of formulae in the free variables indicated (and possessing no non-numerical constants) and a list $a_0,\dots,a_{n-1} \in |\mathfrak{M}|$ determine a type $\tau v \bar{a}_0 \dots \bar{a}_{n-1}$ over $\mathfrak{M}$ if, for each $k \in \omega$,

$$\mathfrak{M} \models \exists v \bigwedge_{i \le k} \varphi_i v\bar{a}_0 \ldots \bar{a}_{n-1}.$$

A type $\tau v\bar{a}_0 \ldots \bar{a}_{n-1}$ arising from $\tau v v_0 \ldots v_{n-1}$ is <u>recursive</u> if $\{\ulcorner\varphi\urcorner : \varphi \in \tau v v_0 \ldots v_{n-1}\}$ is recursive. For $k \in \omega$, $\tau v\bar{a}_0 \ldots \bar{a}_{n-1}$ is a <u>Σ_k-type</u> if every $\varphi \in \tau$ is Σ_k. The basic fact about recursive Σ_k-types is the following lemma from <u>Robinson 1963</u> and <u>Friedman 1973</u>:

<u>0.1. Lemma</u>. Let $\mathfrak{M}$ be nonstandard, $a_0, \ldots, a_{n-1} \in |\mathfrak{M}|$, and (for some $k \in \omega$) $\tau v\bar{a}_0 \ldots \bar{a}_{n-1}$ a recursive Σ_k-type over $\mathfrak{M}$. Then $\tau v\bar{a}_0 \ldots \bar{a}_{n-1}$ is realized in $\mathfrak{M}$, i.e. there is an $a \in |\mathfrak{M}|$ such that $\mathfrak{M} \models \varphi \bar{a}\bar{a}_0 \ldots \bar{a}_{n-1}$ for all $\varphi \in \tau$.

The proof is a simple appeal to the existence of a Σ_k truth definition for Σ_k sentences and Overspill. Informally: For every recursive τ, for $x \in \omega$,

$\mathfrak{M} \models \exists v$ "v satisfies the first $\bar{x}$ formulae of τ ".

By Overspill, for some nonstandard b,

$\mathfrak{M} \models \exists v$ "v satisfies the first $\bar{b}$ formulae of τ ". (*)

If a witnesses this sentence, i.e. if

$\mathfrak{M} \models$ "$\bar{a}$ satisfies the first $\bar{b}$ formulae of τ ",

then a realizes τ.

As the proof of the lemma relies so heavily on the truth definition (without which (*) is inexpressible), it cannot be applied to the case of unbounded types, i.e. types with formulae of unbounded complexity. Indeed, many nonstandard models omit many recursive types. Models in which all recursive types are realized are called <u>recursively saturated</u>.

A slight generalization of the notion of recursive saturation is that of short recursive saturation. A type $\tau v\bar{a}_0 \ldots \bar{a}_{n-1}$ is <u>short</u> if it contains the formula $v < \bar{a}_0$. A model $\mathfrak{M}$ is <u>short recursively saturated</u> if it realizes all of its short types. The second lemma we wish to recall (from, e.g., <u>Lesan 1978</u> and <u>Smoryński A</u>) asserts that, in a tall model, we can shorten types. To state this properly, we write $b \gg a_0, \ldots, a_{n-1}$ if b is larger than $Fa_0 \ldots a_{n-1}$ for any parameter-free Skolem function F.

<u>0.2. Lemma</u>. Let $a_0, \ldots, a_{n-1} \in |\mathfrak{M}|$, $\tau v\bar{a}_0 \ldots \bar{a}_{n-1}$ a type over $\mathfrak{M}$, and $b \gg a_0, \ldots, a_{n-1}$ for some $b \in |\mathfrak{M}|$. Then

$$\tau' v\bar{b}\bar{a}_0 \ldots \bar{a}_{n-1} = \tau v\bar{a}_0 \ldots \bar{a}_{n-1} \cup \{v < \bar{b}\}$$

is a type over $\mathfrak{M}$.

We define $\mathfrak{M}$ to be <u>tall</u> if for every $a_0, \ldots, a_{n-1} \in |\mathfrak{M}|$ there is a $b \in |\mathfrak{M}|$ with $b \gg a_0, \ldots, a_{n-1}$. Noting that tallness is just the realization of a particular recursive type, Lemma 0.2 has the following immediate corollary:

<u>0.3. Corollary</u>. Let $\mathfrak{M}$ be nonstandard. $\mathfrak{M}$ is recursively saturated iff $\mathfrak{M}$ is tall and short recursively saturated.

In view of the corollary, we shall establish the main theorem of the paper by showing short recursive saturation to be preserved under elementary cofinal extension.

(That tallness is so preserved also requires a small argument-- which we leave to the reader.)

One last bit of notation: Recall the canonical indexing of finite sets:

$$D_x = \begin{cases} \emptyset, & x = 0 \\ \{x_0,\ldots,x_{m-1}\}, & x = 2^{x_0} + \ldots + 2^{x_{m-1}} \text{ and } x_0 < \ldots < x_{m-1}. \end{cases}$$

There is a Σ_1 formula, which we denote by $v_0 \in D_{v_1}$, which defines this indexing in arithmetic and for which Separation is provable: For any φ,

$$\vdash \forall v_0 \exists v_1 \forall v_2 [v_2 \in D_{v_1} \leftrightarrow v_2 \le v_0 \wedge \varphi v_2].$$

1. The main theorem. Without further ado, we prove the following

1.1. Theorem. Let $\mathfrak{M}$ be (short) recursively saturated and let $\mathfrak{N}$ be an elementary cofinal extension of $\mathfrak{M}$. Then $\mathfrak{N}$ is (short) recursively saturated.

Proof: Let $\mathfrak{M}, \mathfrak{N}$ be given as stated and let $\sigma v\bar{a}_0\ldots\bar{a}_{n-1}$ be a short recursive type over $\mathfrak{N}$. Pick $a \in |\mathfrak{M}|$ nonstandard such that $a > a_i$ for each i.

We shall define a special type $\tau v\bar{a}$ over $\mathfrak{M}$ which will be realized in $\mathfrak{M}$, whence in $\mathfrak{N}$, and which in $\mathfrak{N}$ will reduce the complexity of $\sigma v\bar{a}_0\ldots\bar{a}_{n-1}$. We simply let τvv_0 consist of all formulae of the form (for $\varphi v_1\ldots v_{n+1}$ containing only $v_1,\ldots,v_{n+1}$ free):

$$\forall v_1\ldots v_{n+1} < v_0[\varphi v_1\ldots v_{n+1} \leftrightarrow \langle v_1,\ldots,v_{n+1},\ulcorner\varphi\urcorner\rangle \in D_v]$$

(where $\langle\cdot,\ldots,\cdot\rangle$ is the usual tupling function). By arithmetic separation one easily sees $\tau v\bar{a}$ to be a recursive type. To realize it in $\mathfrak{M}$ when $\mathfrak{M}$ is only <u>short</u> recursively saturated, we must provide a bound $c \in |\mathfrak{M}|$ and prove that the set

$$\tau' v\overline{ca} = \tau v\bar{a} \cup \{v < \bar{c}\}$$

is a recursive type. We choose

$$c = a^{n+2}\cdot 2^{\langle a,\ldots,a\rangle},$$

where the tuple exhibited is an n+2-tuple.

Obviously $\tau' v\overline{ca}$ is recursive and we need only show it to be a type to realize it. To this end, let $\varphi_0,\ldots,\varphi_{m-1}$ be given and apply separation to obtain $d \in |\mathfrak{M}|$ such that

$$\mathfrak{M} \models \forall v_1\ldots v_{n+2}[\langle v_1,\ldots,v_{n+2}\rangle \in D_{\bar{d}} \leftrightarrow \bigvee_{i=0}^{m-1} (v_{n+2} = \ulcorner\varphi_i\urcorner \wedge \bigwedge_{j=1}^{n+1} v_j < \bar{a} \wedge \varphi_i v_1\ldots v_{n+1})].$$

d satisfies the axioms

$$\forall v_1\ldots v_{n+1} < \bar{a}[\langle v_1,\ldots,v_{n+1},\ulcorner\varphi_i\urcorner\rangle \in D_v \leftrightarrow \varphi_i v_1\ldots v_{n+1}]$$

of $\tau v\bar{a}$ and it suffices to show that $d < c$. But

$$d = \sum_{i=0}^{m-1} \sum_{d_1<a} \ldots \sum_{d_{n+1}<a} \{ 2^{\langle d_1,\ldots,d_{n+1},\ulcorner\varphi_i\urcorner\rangle} : \mathfrak{M} \models \varphi_i \bar{d}_1\ldots\bar{d}_{n+1} \}$$

$$< m \cdot a^{n+1} \cdot 2^{\langle a,\ldots,a\rangle} < a^{n+2} \cdot 2^{\langle a,\ldots,a\rangle} = c,$$

since $a > m, \ulcorner \varphi_i \urcorner$ by the assumption that a is nonstandard.

We can now let $b \in |\mathfrak{M}|$ realize $\tau' v\overline{ca}$, whence $\tau v\bar{a}$. Note that b still realizes $\tau v\bar{a}$ in $\mathfrak{N}$; in particular, for $\varphi v_0 \ldots v_n$ with the indicated free variables,

$$\mathfrak{N} \models \forall v{<}\bar{a}[\varphi v\bar{a}_0 \ldots \bar{a}_{n-1} \leftrightarrow \langle v,\bar{a}_0,\ldots,\bar{a}_{n-1}, \ulcorner \varphi v_1 \ldots v_{n-1} \urcorner \rangle \in D_{\bar{b}}], \qquad (*)$$

where $a_0,\ldots,a_{n-1} < a$ are the parameters from the short type $\sigma v\bar{a}_0 \ldots \bar{a}_{n-1}$ over $\mathfrak{N}$ introduced earlier.

We define a new set $\sigma_0 v v_0 \ldots v_{n+1}$ by:

$\psi \in \sigma_0$ iff for some $\varphi \in \sigma$, ψ is: $\langle v,v_1,\ldots,v_{n+1}, \ulcorner \varphi v_1 \ldots v_{n+1} \urcorner \rangle \in D_{v_0}$.

By (*), an element $c \in |\mathfrak{N}|$ realizes $\sigma_0 v\overline{ba}_0 \ldots \bar{a}_{n-1}$ (or a finite subset thereof) iff it realizes $\sigma v\bar{a}_0 \ldots \bar{a}_{n-1}$ (respectively, the corresponding finite subset). It follows that $\sigma_0 v\overline{ba}_0 \ldots \bar{a}_{n-1}$ is a recursive Σ_1-type, hence (by Lemma 0.1) it is realized in $\mathfrak{N}$. Thus $\sigma v\bar{a}_0 \ldots \bar{a}_{n-1}$ is realized in $\mathfrak{N}$. QED

We confess to being rather pleased with ourselves over this proof. The interplay between arithmetic encoding and recursive saturation is at the aesthetically proper level-- neither so shallow as to trivialize the proof nor so deep as to obscure it. Happily, we can show that some such interplay was necessary: The result fails for a fragment of arithmetic too weak to perform any encoding.

1.2. Example. Elementary cofinal extensions of recursively saturated models of Th(ω;0,<,') need not be recursively saturated. The counterexample is very simple. Define models M_1, M_2 by

$$M_1\colon \omega + (\omega^* + \omega)(\eta + \eta)$$
$$M_2\colon \omega + (\omega^* + \omega)(\eta + 2 + \eta),$$

where we use the customary notation for order types. M_1 is recursively saturated, M_2 is not, and M_1 easily elementarily cofinally embedds in M_2.

A possibly interesting open problem is whether or not the theorem holds for Presburger-Skolem arithmetic, i.e. Th(ω;0,+,'). Although the theory itself has little expressive power, its recursively saturated models are at least capable of encoding classes of reals with decent closure properties.

In celebrating the interplay between arithmetic encodability and recursive saturation, we should also offer some emphasis to the role of the latter concept.

1.3. Example. Let Γ be a class of sets of formulae closed under relative recursiveness; and define $\mathfrak{M}$ to be Γ-saturated if $\mathfrak{M}$ realizes every type in Γ over $\mathfrak{M}$. Among models of arithmetic, Γ-saturation is preserved under elementary cofinal extension. To see this, let $\mathfrak{M}$ be Γ-saturated, $\mathfrak{N}$ an elementary cofinal extension of $\mathfrak{M}$, $\sigma v v_0 \ldots v_{n-1} \in \Gamma$, and $a_0,\ldots,a_{n-1} \in |\mathfrak{N}|$ such that $\sigma v\bar{a}_0 \ldots \bar{a}_{n-1}$ is a type over $\mathfrak{N}$. By the Γ-saturation of $\mathfrak{M}$ and the closure of Γ under relative recursiveness, one easily finds an element $b \in |\mathfrak{M}|$ such that, for all appropriate $\varphi v v_0 \ldots v_{n-1}$,

$$\mathfrak{M} \models \ulcorner\varphi\urcorner \in D_{\bar{b}} \quad \text{iff} \quad \varphi \in \sigma v v_0 \ldots v_{n-1}.$$

Clearly $\sigma v \bar{a}_0 \ldots \bar{a}_{n-1}$ is equivalent in $\mathfrak{M}$ to the recursive (hence: realized) type $\sigma_1 v \bar{a}_0 \ldots \bar{a}_{n-1} \bar{b}$ defined by choosing $\sigma_1 v v_0 \ldots v_n$ to consist of all formulae,

$$\varphi v v_0 \ldots v_{n-1} \leftrightarrow \ulcorner\varphi\urcorner \in D_{v_n},$$

where φ has only the free variables indicated.

In particular, $\aleph_0$-saturation-- the realization in a model of all types over it-- is preserved. Kotlarski has recently shown that higher saturations, in particular $\aleph_1$-saturation,are not similarly preserved. The extension of the above proof breaks down in trying to establish an analogue to Lemma 0.1 in the presence of infinitely many constants.

We can further refer to the literature for an example of the failure of the preservation of a property lying somewhere in strength between full saturation and recursive saturation:

1.4. Example. Resplendence is not preserved under elementary cofinal extension among models of arithmetic. (The reader unfamiliar with resplendent models is referred to Barwise and Schlipf 1976, wherein it is remarked that) resplendent models are not two-cardinal models, i.e. if $\mathfrak{M}$ is resplendent, then any infinite definable subset of $|\mathfrak{M}|$ has the same cardinality as $|\mathfrak{M}|$. But Paris and Mills 1979 have shown any countable nonstandard $\mathfrak{M}$ to have an uncountable two-cardinal elementary cofinal extension, hence a non-resplendent such extension. Starting with $\mathfrak{M}$ resplendent yields the counterexample.

2. Another type. A moment's thought and a minor calculation show that the type $\tau v \bar{a}$, or better, the type $\tau_0 v \bar{a}$ defined by

$$\tau_0 v \bar{a}: \quad \forall v_1 < v_0 [\varphi v_1 \leftrightarrow \langle v_1, \ulcorner\varphi v_1\urcorner\rangle \in D_v],$$

is interesting independently of the cofinal extension which led us to consider it. Let, for $a \in |\mathfrak{M}|$ nonstandard,

$$I_a = \{ b \in |\mathfrak{M}| : P(b) < a \text{ for all polynomials } P(X) \in \mathbb{N}[X] \},$$

be the largest initial segment below a closed under all the n-tupling functions (for $n \in \omega$). Our proof of Theorem 1.1 essentially contained the following fact:

2.1. Fact. Let $\mathfrak{M}$ be a model of arithmetic, $a, b \in |\mathfrak{M}|$ nonstandard, and let b realize $\tau_0 v \bar{a}$. Then every short recursive type $\sigma v \bar{a}_0 \ldots \bar{a}_{n-1}$ over $\mathfrak{M}$, with $a_0, \ldots, a_{n-1} \in I_a$, is realized in $\mathfrak{M}$.

Via this Fact, our proof of Theorem 1.1 can be restated: In a (short) recursively saturated model of arithmetic-- hence in any elementary cofinal extension thereof-- the type $\tau_0 v \bar{a}$ is realized for arbitrarily large a. But, if the a's go arbitrarily high, so do the I_a's-- whence the arbitrarily high realization of short types $\tau_0 v \bar{a}$ entails (short) recursive saturation.

The type schema $\tau_0 v\bar{a}$ is not unique. We now exhibit another type whose realization entails a degree of recursive saturation in an initial segment. The idea of this type is not to encode information, but to encode growth. To define this new type, say $\tau_1 v\bar{a}$, we recall that natural numbers, besides coding finite sets, also encode finite sequences. The usual notation for the decoding (i.e. projection) function is

$$(x)_y = \begin{cases} \text{the y-th element of the finite sequence } (x_0,\ldots,x_{lh(x)\dot{-}1}) \\ \text{encoded by x, if } y < lh(x) \\ \\ 0, \quad \text{if } y \geq lh(x), \end{cases}$$

where $lh(x)$ denotes the length of the sequence encoded by x. These functions, being decently representable in arithmetic, will be assumed given by terms, $(v_0)_{v_1}$ and $lh(v_0)$, respectively.

With this notation, we define the type $\tau_1 v\bar{a}$ by

$$\tau_1 vv_0: \ (v)_{\bar{0}} = v_0, \text{ and}$$

$$\forall v_1 < lh(v)\dot{-}2[\exists v_2 \varphi((v)_{v_1}, v_2) \rightarrow \exists v_2 < (v)_{v_1+1} \varphi((v)_{v_1}, v_2)].$$

Using the notation, $\ll$, of the introduction, the second schema asserts

$$\forall v_1 < lh(v)\dot{-}2[(v)_{v_1} \ll (v)_{v_1+1}].$$

Thus, if b realizes $\tau_1 v\bar{a}$ in $\mathfrak{M}$, we have

$$a = (b)_0 \ll (b)_1 \ll \ldots,$$

each $(b)_{x+1}$ being unreachable from $(b)_x$ via parameter-free Skolem functions.

For $a,b \in |\mathfrak{M}|$, b realizing $\tau_1 v\bar{a}$ over $\mathfrak{M}$, we define

$$I^{a,b} = \{ c \in |\mathfrak{M}| : \ \exists x \in \omega \, (c < (b)_x) \},$$

and

$$\mathfrak{J}^{a,b} = (I^{a,b}; 0, +, \cdot, ', <, \ldots);$$

i.e. $\mathfrak{J}^{a,b}$ is simply the structure obtained by restricting the operations and atomic relations of $\mathfrak{M}$ to $I^{a,b}$.

<u>2.2. Theorem</u>. Let $a,b \in |\mathfrak{M}|$ and b realize $\tau_1 v\bar{a}$ over $\mathfrak{M}$. Then $\mathfrak{J}^{a,b}$ is a recursively saturated elementary initial segment of $\mathfrak{M}$.

Proof sketch: The definitions of $\tau_1 v\bar{a}$ and $I^{a,b}$ guarantee that $I^{a,b}$ is closed under Skolem functions (!) and hence that $\mathfrak{J}^{a,b}$ is an elementary initial segment of $\mathfrak{M}$.

Recursive saturation is again proven by reducing the complexities of formulae in types. The trick that accomplishes this is familiar from indicator theory-- specifically, the proof that strong segments model Peano's axioms. (cf. e.g. <u>Kirby and Paris 1977</u>.)

Let $\sigma v\bar{a}_0 \ldots \bar{a}_{n-1}$ be a recursive type over $\mathfrak{J}^{a,b}$. Pick x such that each $a_i \leq (b)_x$ and appeal to Lemma 0.2 to shorten σ by bounding v by $(b)_{x+1}$. Notationally, we thus might as well assume $\sigma va_0 \ldots$ to already be short and $a_0 = (b)_x$ to be the largest of the a_i's.

Given a formula $\varphi \in \sigma v\bar{a}_0 \ldots \bar{a}_{n-1}$, say

$$\forall v_0 \exists v_1 \forall v_2 \psi v v_0 v_1 v_2 \bar{a}_0 \ldots \bar{a}_{n-1},$$

with $\psi \in \Delta_0$, we associate with it the formula $\varphi^* \in \Delta_1$:

$$\forall v_0 < (\bar{b})_{\overline{x+1}} \exists v_1 < (\bar{b})_{\overline{x+2}} \forall v_2 < (\bar{b})_{\overline{x+3}} \psi v v_0 v_1 v_2 \bar{a}_0 \ldots \bar{a}_{n-1}$$

and claim

$$\mathcal{J}^{a,b} \models \forall v < (\bar{b})_{\bar{x}} (\varphi v \leftrightarrow \varphi^* v). \qquad (*)$$

Given this, the effectiveness of the transformation $\varphi \mapsto \varphi^*$ effectively reduces σ to a Σ_1-type whose realization is guaranteed by Lemma 0.1.

Modulo the proof of the equivalence (*), which we leave the more ambitious reader to crib from the above cited paper, we have completed the proof of the Theorem.

QED

To use $\tau_1 v\bar{a}$ to re-establish Theorem 1.1, it suffices to realize $\tau_1 v\bar{a}$ for arbitrarily large $a \in |\mathfrak{M}|$ for recursively saturated $\mathfrak{M}$. To do this, it suffices to note that $\tau_1 v\bar{a}$ is a recursive type over $\mathfrak{M}$. Note that we do not say anything about <u>short</u> recursive saturation-- for, though, $\tau_1 v\bar{a}$ is a type, it is not a short type and cannot be shortened by anything definable from a. Should there be any $c \in |\mathfrak{M}|$ such that $a \ll c$, then, by Lemma 0.2, we can shorten $\tau_1 v\bar{a}$ and so realize it in $\mathfrak{M}$ even in the short case.

Incidently, this last remark allows us to give a quick proof of a basic result of <u>Kotlarski A</u> which we can state simply by letting $\mathfrak{N} \prec_e \mathfrak{M}$ denote the relation of proper elementary end extension.

<u>2.3. Corollary</u>. Let $\mathfrak{M}$ be a countable recursively saturated model of arithmetic and define

$$R = \{\mathfrak{N} : \mathfrak{N} \prec_e \mathfrak{M} \text{ and } \mathfrak{N} \text{ is recursively saturated}\}.$$

Then: $\prec_e$ orders R in the type of the reals.

Proof: We must show that

i. R is non-empty, with neither first nor last element

ii. R is dense

iii. R is complete (every non-trivial cut determines an element)

iv. R has a countable dense subset.

i. Pick $a \in |\mathfrak{M}|$ and let b realize $\tau_1 v\bar{a}$. Since $\mathcal{J}^{a,b} \in R$, R is non-empty.

If $\mathfrak{N} \in R$, choosing $a \notin |\mathfrak{N}|$ will yield $\mathfrak{N} \prec_e \mathcal{J}^{a,b}$. Choosing $a,b \in |\mathfrak{N}|$ will yield $\mathcal{J}^{a,b} \prec_e \mathfrak{N}$.

ii. Let $\mathfrak{N}_0, \mathfrak{N}_1 \in R$, $\mathfrak{N}_0 \prec_e \mathfrak{N}_1$. As in i, choose $a \in |\mathfrak{N}_1| - |\mathfrak{N}_0|$ and $b \in |\mathfrak{N}_1|$. Then

$$\mathfrak{N}_0 \prec_e \mathcal{J}^{a,b} \prec_e \mathfrak{N}_1.$$

iii. This is nothing more than the fact that recursive saturation is preserved under the union of elementary chains.

iv. By ii, the set $\{\mathcal{J}^{a,b}$: b realizes $\tau_1 v\bar{a}\}$ is dense. Since $|\mathfrak{M}|$ is countable, so is it. QED

References

J. Barwise and J. Schlipf

1976 An introduction to recursively saturated and resplendent models, JSL 41, pp. 531-536.

H. Friedman

1973 Countable models of set theories, in: A.R.D. Mathias and H. Rogers, eds., Cambridge Summer School in Mathematical Logic, Springer-Verlag, Heidelberg.

H. Gaifman

1972 A note on models and submodels of arithmetic, in: W. Hodges, ed., Conference in Mathematical Logic-- London '70, Springer-Verlag, Heidelberg.

L. Kirby and J. Paris

1977 Initial segments of models of Peano's axioms, in: A. Lachlan, M. Srebrny, and A. Zarach, eds., Set Theory and Hierarchy Theory V, Springer-Verlag, Heidelberg.

H. Kotlarski

A On elementary recursively saturated cuts in models of Peano arithmetic, to appear.

H. Lesan

1978 Models of arithmetic, dissertation, Manchester.

J. Paris and G. Mills

1979 Closure properties of countable non-standard integers, Fund. Math. 103, pp. 205-215.

A. Robinson

1963 On languages which are based on nonstandard integers, Nagoya Math. J. 22, pp. 83-117.

C. Smoryński

A Recursively saturated nonstandard models of arithmetic, to appear in JSL.

BAR-ILAN UNIVERSITY

SOME MODEL THEORY AND NUMBER THEORY FOR MODELS OF WEAK SYSTEMS OF ARITHMETIC

Lou van den Dries

Introduction

Various mathematicians have been interested in the number theoretic strength of certain fragments of the set of 1st order Peano axioms. (This set of axioms seems to imply all of classical number theory, whether algebraic, analytic, diophantine or transcendental.)

A first "independence" result is due to J. C. Shepherdson: he constructed a model of open induction, i.e. a discretely ordered domain satisfying the induction axioms for open formulas, in which the equations

$$x^2 = 2y^2 \qquad xy \neq 0$$

$$x^3 + y^3 = z^3, \qquad xyz \neq 0$$

have a solution, thus answering questions of Skolem and Kreisel respectively, cf. [Sh].

The problem then arose: which diophantine equations are solvable in models of open induction? In this direction A. Wilkie proved (cf. [Wi,3.3]):

A polynomial $f(X_1,\ldots X_n) \in \mathbb{Z}[X_1,\ldots,X_n]$ has a zero in a model of open induction if and only if there is an ideal $I \subset \mathbb{Z}[X_1,\ldots,X_n]$ containing $f(X_1,\ldots,X_n)$ such that:

(i) I has for each prime number p a p-adic integral zero,

(ii) the ring $\mathbb{Z}[X_1,\ldots,X_n]/I$ can be discretely ordered.

In [v. d. D. 2] I used the theory of algebraic functions in one variable to obtain from Wilkies theorem a decision method for $n = 2$. For larger n, a decision method (for solvability in models of open induction) is not yet known, but I strongly suspect there is one.

This paper is first of all intended as a survey of the above development. Because we give proofs of the results mentioned which are somewhat simpler or more systematic than the original ones, the treatment we give is reasonably selfcontained and hopefully useful to some readers.

We also pay attention to how the results mentioned above have to be adapted if the theory of open induction is augmented by certain axioms. For instance, although $\mathbb{Z}$ is normal, i.e. integrally closed in its fraction field (a consequence of unique factorization), models of open induction do not always have this property. Therefore we are particularly interested in the normal models of open induction, which of course preserve more of the arithmetic structure of $\mathbb{Z}$ than arbitrary models of open induction: e.g. the "irrationality of $\sqrt{2}$" is a (trivial) consequence of the theory of normal models of open induction. Far less obvious is that Hilbert's Irreducibility Theorem is true in all normal discretely ordered domains (the open induction axioms are not even needed), a fact proved in §2.

The reader of that section may find it amusing to prove Liouville's Theorem for an arbitrary discretely ordered domain Z: if α is algebraic of degree $n > 1$ over the fraction field Q of Z, then there is $C > 0$ in Z such that $|\alpha - p/q| > 1/Cq^n$ for all p,q in Z, $q > 0$. (See [Ba, pp. 1-2] for a proof of Liouville's Theorem for $\mathbb{Z}$ which can be adapted to our case, if one keeps in mind that the absolute value function, defined on the algebraic closure of Q, takes its values in a fixed real closure of Q.)

A reason why fragments of Peano may have some number theoretic interest can be illustrated as follows: suppose an irreducible polynomial $f(X,Y) \in \mathbb{Z}[X,Y]$ has infinitely many zeros; this is the same as saying that $f(X,Y)$ has a non-standard zero in an elementary extension of $\mathbb{Z}$, or that $\mathbb{Z}[x,y] \overset{\text{def.}}{=} \mathbb{Z}[X,Y]/(f(X,Y))$ can be embedded in an elementary extension of $\mathbb{Z}$. So $\mathbb{Z}[x,y]$ inherits (by restriction) much of the rich arithmetic structure of such an elementary extension. Now, independently of whether the assumption on $f(X,Y)$ is satisfied or not, one may study the question when $\mathbb{Z}[x,y]$ admits certain of these structures and obtain in this way necessary conditions for the existence of infinitely many integral zeros of $f(X,Y)$. This is in fact what happens in the non-standard proof of the Siegel-Mahler Theorem in [R,R], where the arithmetic divisor theory of an elementary extension of $\mathbb{Z}$ is restricted to a divisor theory on (the fraction field of) $\mathbb{Z}[x,y]$, thus leading naturally to Weil's distributions on the function field of $f(X,Y) = 0$.

In §§3, 4 we are led to consider embeddings of $\mathbb{Z}[x,y]$ in models of certain fragments of Peano and find necessary and sufficient conditions on $\mathbb{Z}[x,y]$ to allow such an embedding. A crucial difference with nonstandard arithmetic is that in our case the fragments considered are recursively axiomatized, while in [R,R] the underlying theory is certainly not recursively axiomatizable, although A. Robinson indicated how to replace it by one which has an axiom system recursive relative to the bounds provided by Roth's Theorem. Of course this amounts to saying that the Siegel-Mahler Theorem is not yet known to be "effective", but is known to be effective relative to Roth's Theorem; see also [Ma, §7] for a discussion of this point.

In §1 we make some conventions and define the algebraic notions which are used later on. Sections 3 and 4 give a systematic treatment of (normal) models of open induction, in particular with respect to solvability of polynomial equations. Section 5 concludes again with remarks and suggests some topics which in my opinion are worth the attention of those mathematical logicians who wonder what models of arithmetic have to do with arithmetic.

§1. Conventions and Definitions

All rings are commutative with identity 1. $\mathbb{N}$, $\mathbb{Z}$, $\mathbb{Q}$, $\mathbb{R}$, $\mathbb{C}$ will have their usual meaning:

$$\mathbb{N} = \{1,2,3,\ldots\}, \quad \text{etc.}$$

On the other hand Z will denote an arbitrary discretely ordered ring, that is, a linearly ordered ring in which 1 is the smallest (strictly) positive element. Q will always denote the (ordered) fraction field of the domain Z, R a fixed real closure of Q, and C the algebraic closure of Q given by $C = R(i)$, $i^2 = -1$. $\mathbb{Z}$ will always be considered as an ordered subring of Z, $\mathbb{Q}$ as an ordered subfield of Q. L is the language of ordered rings: $L = \{+,\cdot,-,0,1,<\}$.

A $\mathbb{Z}$-ring is a discretely ordered ring Z such that for all $n \in \mathbb{N}$:

$$Z/nZ \simeq \mathbb{Z}/n\mathbb{Z},$$

or equivalently: for all primes p and $n \in \mathbb{N}$:

$$Z/p^nZ \simeq \mathbb{Z}/p^n\mathbb{Z}.$$

A model of open induction is a discretely ordered ring satisfying all induction axioms for open L-formulas $\phi(\bar{x},y)$:
$[\phi(\bar{x},0) \wedge \forall y \geq 0(\phi(\bar{x},y) \rightarrow \phi(\bar{x},y+1))] \rightarrow \forall y \geq 0\ \phi(\bar{x},y)$.

A domain D is called normal if each root in the fraction field Q(D) of D of a monic polynomial over D is already in D. The normalization of a domain D is the smallest normal subdomain of the fraction field of D which includes D.

Finally, we will freely use the basic facts on real fields and fields of power series; in particular, if K is a field, then $K((X^{\frac{1}{\infty}}))$ denotes the union of the powerseries fields $K((X^{\frac{1}{m}}))$, $m \in \mathbb{N}$ and so is algebraically (real) closed iff K is.

§2. An Example: Hilbert's Irreducibility Theorem

(2.1) Hilbert's irreducibility theorem states:

<u>if</u> $f(T,X) \in \mathbb{Q}(T)[X]$ <u>is irreducible, then there are infinitely many</u> $t \in \mathbb{Z}$ <u>such that</u> $f(t,X)$ <u>is defined, and irreducible in</u> $\mathbb{Q}[X]$.

A quick proof using only elementary calculus is given in [La,Ch.8]. Despite its simplicity this proof <u>seems</u> to depend rather essentially on the structure of $\mathbb{Z}$, in particular on the fact that $\mathbb{Z} \subset \mathbb{R}$. However, model theoretic transfer principles enable us to show that this "same" proof works under rather general assumptions.

(2.2) <u>Theorem</u> <u>Let</u> Z <u>be any discretely ordered normal domain and</u> $f = f(T,X)$ <u>an irreducible polynomial in</u> $Q(T)[X]$.

<u>Then there are infinitely many</u> $t \in Z$ <u>such that</u> $f(t,X)$ <u>is defined, and irreducible in</u> $Q[X]$.

Proof Let $f(T,X) = a_n(T)X^n + a_{n-1}(T)X^{n-1} + \ldots + a_0(T)$, $a_n(T) \neq 0$, $n > 1$. Expanding the roots of f as power series at infinity, we get:

$$f = a_n(T)(X-x_1(T)) \ldots (X-x_n(T)),$$

where $x_1(T),\ldots, x_n(T)$ are in the field of (descending) Puiseux series $C((T^{-\frac{1}{\infty}})) = \bigcup_{m\geq 1} C((T^{-\frac{1}{m}}))$, which is an algebraically closed extension of the field $Q(T)$ of coefficients of f.

Our assumption that f is irreducible means that each of its 2^n-2 factors $\prod_{i\in S} (X-x_i(T))$, $\phi \neq S \subsetneq \{1,2,\ldots,n\}$, has one of its coefficients not in $Q(T)$. Taking such a coefficient for each of the 2^n-2 factors we obtain a family $y_1(T),\ldots,y_N(T)$, $N = 2^n-2$, of series in $C((T^{-\frac{1}{m}}))$ for some $m \geq 1$, which are algebraic over $Q(T)$, but do not belong to $Q(T)$.

Classically, i.e. in case $Z = \mathbb{Z}$, $Q = \mathbb{Q}$, the argument proceeds now as follows: for some real number $c > 0$ the n series $x_1(T),\ldots, x_n(T)$, and hence the N series $y_1(T),\ldots, y_N(T)$ converge for all real values t of T with $t > c$, where $T^{-\frac{1}{m}}$ is of course evaluated as the positive real $t^{-\frac{1}{m}}$. If t is any integer larger than c such that none of $y_1(T),\ldots,y_N(T)$ assumes a rational value at $T = t$, then $f(t,X)$ is clearly irreducible in $\mathbb{Q}[X]$ (assuming that $a_n(t)$ is defined). To prove that there are infinitely many such $t \in \mathbb{Z}$, $t > c$, we multiply the series $y_1(T),\ldots, y_N(T)$ with some $q(T) \in \mathbb{Z}[T]\setminus\{0\}$ such that $z_1(T) = q(T)\cdot y_1(T),\ldots,z_N(T) = q(T)\cdot y_N(T)$ are all integral over $\mathbb{Z}[T]$. These $z_i(T)$ are still outside $\mathbb{Q}(T)$, and if $z_i(t)$ is rational, $c < t \in Z$, then $z_i(t)$ is in fact an integer. (Here we use the crucial fact that $\mathbb{Z}$ is normal.) So we only have to prove that there are infinitely many integers $t > c$ such that $z_1(t),\ldots, z_N(t)$ are not integers. In fact, elementary calculus (involving higher derivatives of the z_i's, and a mean value theorem) shows that, if c is large enough, the integers $t > c$ such that some $z_i(t) \in \mathbb{Z}$ become more widely spaced as t increases, in a certain technical sense, from which the result follows easily. See [La,Ch.8] for details.

In the general case we want the real closure R of Q to replace the field of reals, and similarly we take C instead of $\mathbb{C}$. To carry through the proof described above for $\mathbb{Z}$ we are faced with the problem to assign values in C to our "algebraic functions" $y_i(T)$ and $z_i(T)$ at all points of an interval $(c, \infty) \subset R$, in such a way that elementary calculus applies to the functions so defined. Now, for quite different purposes, this task was essentially carried out in [v. d. D., R] (for arbitrary real closed fields R and $C = R(i)$, $i^2 = -1$). The main tools used in [v. d. D., R] are transfer principles due to Tarski, Ax and Kochen.

(2.3) <u>Remarks</u> (a) The usual corollaries hold: the symmetric and alternating groups S_n and A_n, $n \geq 1$, can be realized as Galois groups over Q, if Z is normal.

(b) The assumption that Z is normal cannot be omitted from the theorem. There are in fact Z's whose fraction field Q is real closed, see (3.5), and for such Z there are no irreducible polynomials of degree > 2 in Q[X], while there are many such in Q(T)[X].

(c) If we define for a polynomial $f = f(T,X) \in Z[T,X]$ $\ell(f)$ as the maximum plus 1 of the absolute values of the coefficients of f, then, given d and n in $\mathbb{N}$, there is $M = M(d,n) \in \mathbb{N}$, such that for each normal Z and each $f \in Z[T,X]$ of total degree $\leq d$ and irreducible in Q(T)[X], there are at least n elements t in Z such that $|t| \leq \ell(f)^M$ and f(t,X) is irreducible in Q[X]. Moreover M can be taken as a recursive function of (d,n).

This follows from the theorem by a familiar model theoretic compactness argument, noting that if Z is normal and $1 < a \in Z$, then $\{b \in Z \mid |b| \leq a^M$ for some $M \in \mathbb{N}\}$ is a discretely ordered normal domain contained in Z.

§3. <u>Models of open induction</u>

The first result says that models of open induction are those discretely ordered domains for which an "integral part" operator is defined on their real closure.

(3.1) <u>Theorem</u> (Shepherdson, [Sh]).

<u>Z is a model of open induction if and only if for each</u> $r \in R$ <u>there is</u> $a \in Z$ <u>with</u> $a \leq r < a + 1$.

<u>Proof</u> Let Z be a model of open induction and $r \in R$. To find a (necessarily unique) $a \in Z$ with $a \leq r < a + 1$, we may assume that $r > 0$. Let r be a root of $f(X) \in Z[X] \setminus \{0\}$. We first assume that r is the smallest positive root of f(X) in R. The case $r \leq 1$ is trivial, so assume $r > 1$. Let $\phi(y)$ be an open L(Z)-formula expressing in R that f has a zero x with $0 < x < y$. Then $Z \models \neg \phi(0)$, but Z does not satisfy $\forall y \geq 0 \neg \phi(y)$, so for some $0 \leq a \in Z$ we must have $Z \models \neg \phi(a) \wedge \phi(a+1)$. Then clearly $a \leq r < a + 1$. To handle similarly the second positive root of f, if there is one, if suffices to consider the case that this root is $> a + 1$, so this root is the smallest positive root of $f(X + a + 1)$, and the previous technique applies. In this way we can handle all roots of f(X).

Suppose conversely that each $r \in R$ has an "integral part" $[r] \in Z$, with $[r] \leq r < [r] + 1$. Then the following facts are easily established:

(i) each open $L(Z)$-formula $\phi(y)$ is equivalent, w. r. t. Z, to a boolean combination of formulas $f(y) > 0$, with $f(X) \in Z[X]$;

(ii) each formula $f(y) > 0$ $(f(X) \in Z[X])$ is equivalent, w. r. t. Z and R, to a boolean combination of formulas $y > r$, where r is a root of $f(X)$ in R.

(iii) each formula $y > r$ $(r \in R)$ is equivalent, w. r. t. Z, to the formula $y \geq [r] + 1$.

By (i), (ii) and (iii) we are reduced to showing that Z satisfies the induction axioms for boolean combinations of formulas $y > a$, with $a \in Z$. The subsets X of Z defined by such boolean combinations are so simple that one easily establishes that if $0 \in X$ and for all $a \geq 0$ in Z $(a \in X \Longrightarrow a + 1 \in X)$, then X contains all nonnegative elements of Z. ∎

(3.2) Remarks

(a) Two easy consequences are: if Z is a model of open induction, then Z is a $\mathbb{Z}$-ring and its fraction field Q is dense in its real closure R. (As for the proof of the first property: let $a \in Z$ and $n \in \mathbb{N}$, and write $a = n \cdot [\frac{a}{n}] + i$, $0 \leq i < n$.)

(b) The last part of the proof of the theorem can be adapted to show that a model of open induction Z satisfies an, at first sight stronger, form of induction, namely: if $\phi(y)$ is any open $L(Z)$-formula, then

$Z \models [\phi(0) \wedge \forall y \geq 0\{(\forall 0 \leq x \leq y\ \phi(x)) \rightarrow \phi(y+1)\}] \rightarrow \forall y \geq 0\ \phi(y)$.

(3.3) Shepherdson used his results to indicate concrete non-standard models of open induction. Generalizing his construction, we define for any Z its discretely ordered extension $Z\langle T\rangle$:

$$Z\langle T\rangle = \{a_n T^{r_n} + a_{n-1} T^{r_{n-1}} + \ldots + a_1 T^{r_1} + a_0 \mid r_n > r_{n-1} > \ldots > r_1 > 0,\ r_i \in \mathbb{Q},$$

$$a_i \in R \text{ for } 1 \leq i \leq n,\ a_0 \in Z\},$$

which we consider as an ordered subring of the real closed field of Puiseux series $R((T^{-\frac{1}{\infty}})) = \bigcup_{m\geq 1} R((T^{-\frac{1}{m}}))$ (in descending powers of T). So in the notation above we have, if $a_n \neq 0$: $a_n T^{r_n} + \ldots + a_0 > 0 \Longleftrightarrow a_n > 0$. Note that $Z[T] \subset Z\langle T\rangle$.

(3.4) Proposition If Z is a model of open induction, then $Z\langle T\rangle$ is also a model of open induction.

Proof Let $r = a_n T^{r_n} + \ldots + a_0 + \ldots$ be an element of the real closed extension $R((T^{-\frac{1}{\infty}}))$ of $Z\langle T\rangle$; then putting $a = a_n T^{r_n} + \ldots + [a_0]$, we have $a \in Z\langle T\rangle$ and $a - 1 < r < a + 1$. So the result follows from (3.1). ∎

(3.5) Shepherdson used the model $\mathbb{Z}\langle T\rangle$ to show that the irrationality of $\sqrt{2}$ is independent of open induction: $(\sqrt{2}\cdot T, T)$ is a solution in $\mathbb{Z}\langle T\rangle$ of $(\frac{X}{Y})^2 = 2$, $Y \neq 0$. He also posed the problem to characterize the diophantine equations which are solvable in models of open induction, see problem 1 in [Wi]. One trivial observation is that a homogeneous polynomial $f(X_0,\ldots,X_n) \in \mathbb{Z}[X_0,\ldots,X_n]$ has a non-trivial zero in a model of open induction iff it has a non-trivial real zero: for in that case it has a non-trivial real algebraic zero $(r_0,\ldots, r_n)$, whence $(r_0 T,\ldots, r_n T)$ is a non-trivial zero of $f(X_0,\ldots,X_n)$ in $\mathbb{Z}\langle T\rangle$.

In fact, we have the stronger result that every model Z of open induction can be embedded in a model of open induction which has real closed fraction field: let $Z_0 = Z$, $Z_{n+1} = Z_n\langle T_{n+1}\rangle \supset Z_n$, and put $Z_\infty = \bigcup_{n\geq 0} Z_n$. Then Z_{n+1} contains the real closure of Z_n, hence Z_∞ is a model of open induction with real closed fraction field.

(This, by the way, shows that the fraction fields of models of open induction, or of discretely ordered domains, do not form an elementary class: some real closed fields belong to the class, but R does not.)

(3.6) One way to prevent this undesirable destruction of arithmetic structure is to add the axioms of normality to those of open induction; in the previous section we have seen that this can have good effect. We will see in (3.19) and (3.20) what consequence it has on the solvability of homogeneous diophantine equations. For ordinary diophantine equations Wilkie gave the following algebraic criterion for their solvability in models of open induction, cf. [Wi, Th. 3.3].

(3.7) Theorem (Wilkie) *Let* $f = f(X) \in \mathbb{Z}[X]$, $X = (X_1,\ldots,X_n)$.

Then f *has a zero in a model of open induction if and only if there is an ideal* $I \subset \mathbb{Z}[X]$ *containing* f, *such that*

(i) $\mathbb{Z}[X]/I$ *can be discretely ordered, and*

(ii) I *has for each* $m \in \mathbb{N}$ *a zero modulo* m *in* $\mathbb{Z}$ (*or equivalently*: I *has for each prime number* p *a* p-*adic integral zero*).

(3.8) Using Ax' theorem in [Ax] one can certainly decide for any ideal $I \subset \mathbb{Z}[X]$ given by a finite set of generators, whether (ii) of (3.7) holds. If one could similarly give a decision procedure for (i), then, by Gödel's completeness theorem, we would have a positive solution of problem 1 of [Wi]: in that case the set of $f \in \mathbb{Z}[X]$ having a zero in a model of open induction is recursive. However, this has only been done up till now for $n = 2$ in [v. d. D. 2], see also the next section, ($n = 1$ is a trivial case).

We will indicate a proof of (3.7), emphasizing the algebraic constructions involved in Wilkies proof. This will enable us to study the effect of adding the normality axioms to those of open induction and to prove the analogue of Wilkies theorem for the so extended theory. See (3.15) and (3.10).

(3.9) The first construction is the following.

Let Z be a $\mathbb{Z}$-ring, K a real closed field extending Z and $\beta \in K$ such that $f(\beta)$ is infinite, i.e. $|f(\beta)| > n$, $\forall n \in \mathbb{N}$, for each non-constant $f \in Z[T]$. Then

$$Z[\tfrac{\beta}{\infty}] \overset{\text{def.}}{=} \bigcup_{n \in \mathbb{N}} Z[\beta/n]$$

is again a $\mathbb{Z}$-ring containing β. Note that

$$Q(Z[\tfrac{\beta}{\infty}]) = Q(\beta).$$

(3.10) Lemma If Z is normal, then $Z[\frac{\beta}{n}]$ is normal.

Proof. β is transcendental over Z, hence $Z[\frac{\beta}{n}] \simeq Z[T]$ is normal if Z is, and normality is clearly preserved under taking directed unions. ∎

The following is Wilkies lemma 3.1 in [Wi]. It enables us to fill in "gaps" in $\mathbb{Z}$-rings.

(3.11) Lemma Let Z be a $\mathbb{Z}$-ring, K a $\mathrm{card}(Z)^+$-saturated real closed extension of Z, and $r \in K$ such that there is no $a \in Z$ with $a \leq r \leq a+1$. Then there is is $\beta \in K$ such that $\beta \leq r < \beta + 1$ and $f(\beta)$ is infinite for all $f(T) \in Z[T] \setminus Z$.

By iterating the $Z \mapsto Z[\frac{\beta}{\infty}]$ construction and taking unions of ascending chains, the previous lemma and Shepherdson's theorem enable us to embed every $\mathbb{Z}$-ring in a model of open induction. In fact, we have a more precise result by (3.10).

(3.12) Lemma Let Z be a $\mathbb{Z}$-ring. Then Z has an extension $\overline{Z}$ which is a model of open induction such that $Q(\overline{Z})$ is a purely transcendental extension of Q. If moreover Z is normal, then $\overline{Z}$ is normal.

(3.13) Hence, to find the subrings of models of open induction we have only to determine the discretely ordered rings which can be embedded in $\mathbb{Z}$-rings. Note first that for each $\mathbb{Z}$-ring Z and prime number p there is a unique ring morphism $Z \to \mathbb{Z}_p$, as is clear from the commuting diagram of ring morphisms:

$$\begin{array}{ccccc} Z & \longrightarrow & Z/p^{n+1}Z & \simeq & \mathbb{Z}/(p^{n+1}) \\ & \searrow & \downarrow & & \downarrow \\ & & Z/p^nZ & \simeq & \mathbb{Z}/(p^n) \end{array}$$

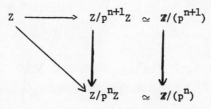

Conversely, <u>suppose that for each prime</u> p <u>a ring morphism</u> $\phi_p : Z \to \mathbb{Z}_p$ <u>is given</u> (Z not necessarily a $\mathbb{Z}$-ring). These morphisms define canonically an embedding of Z into a $\mathbb{Z}$-ring:
define a binary relation div on $\mathbb{N} \times Z$ by:

$$n \text{ div } a \iff n \text{ divides } \phi_p(a) \text{ in } \mathbb{Z}_p \text{ for each prime } p;$$

let $$Z_{div} = \{a/n \mid n \text{ div } a,\ n \in \mathbb{N},\ a \in \mathbb{N}\}.$$

(3.14) <u>Proposition</u> Z_{div} <u>is a $\mathbb{Z}$-ring.</u>

<u>If</u> Z <u>is normal, then</u> Z_{div} <u>is normal.</u>

<u>Proof</u> It is left to the reader to check that the ordering on Z_{div} is discrete. Let p be a prime number.

The map $a/n \to \phi_p(a)/n$ is clearly the unique extension of ϕ_p to a ring morphism $Z_{div} \to \mathbb{Z}_p$ (which we will also denote by ϕ_p). To see that Z_{div} is a $\mathbb{Z}$-ring one applies the following general (and easily proved) result:
if $\phi : A \to \mathbb{Z}_p$ is a ring morphism, then $A/p^nA \simeq \mathbb{Z}/(p^n)$ for all $n \in \mathbb{N}$ if and only if $\phi^{-1}(p\mathbb{Z}_p) = pA$.

Suppose now that Z is normal. Z_{div} is included in the localization $\mathbb{N}^{-1}Z$ which is normal, so we have only to show that an element a/n, $a \in Z$, $n \in \mathbb{N}$, which is integral over Z_{div} belongs to Z_{div}. So let $(a/n)^k + b_1(a/n)^{k-1} +\ldots+ b_k = 0$, $b_i \in Z_{div}$, hence $a^k + (b_1n)a^{k-1} +\ldots+ b_k n^k = 0$. Applying ϕ_p to this equation and dividing by n^k gives: $\phi_p(a)/n$ is integral over $\mathbb{Z}_p$, so belongs to $\mathbb{Z}_p$; because this holds for every prime p, we have n div a, i.e $a/n \in Z_{div}$. ∎

(3.15) Combining (3.12) and (3.14) we obtain:

Z <u>can be embedded in a model of open induction if and only if there exists for each prime</u> p <u>a ring morphism</u> $Z \to \mathbb{Z}_p$.

(Note that in this condition the particular discrete ordering on Z plays no role.) This leads easily to explicit universal axioms characterizing the substructures of models of open induction, and to a proof of (3.7). Restricting ourselves to normal models of open induction we have the following analogue of (3.7).

(3.16) <u>Theorem Let</u> $f = f(X) \in \mathbb{Z}[X]$, $X = (X_1,\ldots,X_n)$. <u>Then</u> f <u>has a zero in a normal model of open induction if and only if there exists an integer</u> $m \geq 0$, <u>and an ideal</u> $I \subset \mathbb{Z}[X,Y]$, $Y = (Y_1,\ldots,Y_m)$, <u>containing</u> f <u>and for each</u> $1 \leq i \leq m$ <u>a polynomial in</u> $\mathbb{Z}[X,Y_i]$ <u>which is monic in</u> Y_i, <u>such that</u>

(i) $\mathbb{Z}[X,Y]/I$ <u>can be discretely ordered and is normal, and</u>

(ii) I <u>has for each prime</u> p <u>a</u> p-<u>adic integral zero</u>.

Proof Suppose $f(a_1,\ldots,a_n) = 0$, $a_1,\ldots,a_n$ in a normal model of open induction Z. Then Z contains the normalization $\mathbb{Z}[a_1,\ldots,a_n, b_1,\ldots,b_m]$ of $\mathbb{Z}[a_1,\ldots,a_n]$. (Recall that the normalization of a finitely generated domain is a finitely generated domain.) Let $I = \{p \in \mathbb{Z}[X,Y] \mid p(a,b) = 0\}$, where $(a,b) = (a_1,\ldots,a_n,b_1,\ldots,b_m)$. Then I clearly satisfies the requirements of the theorem.

To prove the other direction, we write $a_i = X_i + I$, $b_i = Y_i + I$. Then by (i) the normal domain $\mathbb{Z}[a,b] = \mathbb{Z}[a_1,\ldots,a_n,b_1,\ldots,b_m]$ can be discretely ordered, and for each prime p we have a ringmorphism $\mathbb{Z}[a,b] \to \mathbb{Z}_p$ by (ii), so $\mathbb{Z}[a,b]$ endowed with any of its discrete orderings can be embedded in a normal model of open induction, by (3.12), (3.13) and (3.14). So f has a zero in this model, because $f(a) = 0$. ■

(3.17) In the next section we shall see how this result leads to a decision procedure for $n = 2$, the case of plane curves. Here we will determine the homogeneous diophantine equations which have a non-trivial solution in a normal model of open induction. This is clearly equivalent with determining the polynomials $f(X) \in \mathbb{Z}[X]$, $X = (X_1,\ldots,X_n)$, which have a zero in the fraction field of a normal model of open induction.

(3.18) Lemma A field K can be embedded in the fraction field of a normal model of open induction if and only if K is a (formally) real field in which the prime field $\mathbb{Q}$ is algebraically closed.

Proof Suppose $K \subset Q(Z)$ where Z is a normal model of open induction and let $\alpha \in K$ be algebraic over $\mathbb{Q}$. Then $n\,\alpha$ is integral over $\mathbb{Z}$ for some $n \in \mathbb{N}$, but Z is normal, so $n\,\alpha \in Z$. Also $|n\alpha| <$ some integer, so $n\alpha \in \mathbb{Z}$, hence $\alpha \in \mathbb{Q}$. We have shown that $\mathbb{Q}$ is algebraically closed in K.

Conversely, let K be an ordered field in which $\mathbb{Q}$ is algebraically closed. Then we define

$$K[T;\mathbb{Z}] = \{f \in K[T] \mid f(0) \in \mathbb{Z}\} \subset K[T]$$

ordered by putting $T > k$ for all $k \in K$.

Then $K[T;\mathbb{Z}]$ is a discretely ordered normal domain: if $f \in K(T)$ is integral over $K[T;\mathbb{Z}]$, then f is integral over $K[T]$, hence belongs to $K[T]$. Let $f^n + p_1 f^{n-1} + \ldots + p_n = 0$, $p_i \in K[T;\mathbb{Z}]$. Then $f(0)^n + p_1(0)\, f(0)^{n-1} + \ldots + p_n(0) = 0$, all $p_i(0) \in \mathbb{Z}$. So $f(0)$ is integral over $\mathbb{Z}$, $f(0) \in K$; hence $f(0) \in \mathbb{Z}$, because of our assumption that $\mathbb{Q}$ is algebraically closed in K. So $f \in K[T;\mathbb{Z}]$.

The map $f \mapsto f(0)$ defines for each prime p a morphism $K[T;\mathbb{Z}] \to \mathbb{Z}_p$. Using (3.12), (3.13) and (3.14) we obtain that $K[T;\mathbb{Z}]$ can be embedded in a normal model of open induction; then K is embedded in the fraction field of such a model. □

(3.19) Corollary Let $f \in \mathbb{Z}[X]$, $X = (X_1, \ldots, X_n)$. Then f has a zero in the fraction field of a normal model of open induction if and only if there is a prime $p \subset \mathbb{Q}[X]$ containing f such that

(i) p is a real prime, i.e. the domain $\mathbb{Q}[X]/p$ is formally real, and

(ii) p is absolutely prime, i.e. $p \cdot \widetilde{\mathbb{Q}}[X]$ is a prime ideal in $\widetilde{\mathbb{Q}}[X]$, where $\widetilde{\mathbb{Q}}$ is is the algebraic closure of $\mathbb{Q}$.

Proof. Suppose $f(x) = 0$, $x = (x_1, \ldots, x_n) \in \mathbf{Q}^n$, where $\mathbf{Q}$ is the fraction field of a normal model of open induction. Put $p = \{g \in \mathbb{Q}[X] \mid g(x) = 0\}$.

Then by the preceding lemma the field $\mathbb{Q}$ is algebraically closed in the fraction field of $\mathbb{Q}[X]/p \simeq \mathbb{Q}[x_1, \ldots, x_n]$ which is formally real. So p satisfies (i) and (ii).

Conversely, if p is a prime ideal of $\mathbb{Q}[X]$ containing f and satisfying (i) and (ii), then $\mathbb{Q}$ is algebraically closed in the fraction field of the real domain $\mathbb{Q}[X]/p$; putting $x_i = X_i + p$, we have: $f(x_1, \ldots, x_n) = 0$ and $\mathbb{Q}[x_1, \ldots, x_n] \simeq \mathbb{Q}[X]/p$ can be embedded in the fraction field of a normal model of open induction, by the preceding lemma. ■

(3.20) We now have an algorithm for deciding whether a given polynomial $f \in \mathbb{Z}[X_1, \ldots, X_n]$ has a zero in the fraction field of a normal model of open induction: by Gödel's completeness theorem the set of f's having no such zero is recursively enumerable, and the above corollary implies that the set of f's having such a zero is also recursively enumerable. Hence the set of such f's is recursive.

Of course, for this argument to be valid we need algorithms to decide whether an ideal of $\mathbb{Q}[X]$, given by a finite set of generators, is an absolutely real prime ideal. For a principal ideal $g \cdot \mathbb{Q}[X]$ this algorithm exists because $g \cdot \mathbb{Q}[X] (\neq 0)$ is an absolutely real prime ideal iff g is absolutely irreducible and g assumes positive as well as negative values for real arguments. The case of an arbitrary ideal can be effectively reduced to the case of a principal ideal, see [v. d. D. 1, Ch. IV, §3]

(3.21) We have seen that fraction fields of normal models of open induction preserve some of the arithmetic structure of $\mathbb{Q}$: irreducible polynomials over $\mathbb{Q}$ remain irreducible over the fraction field of a normal model of open induction, by (3.18).

The p-adic valuations determine another important part of the arithmetic structure of $\mathbb{Z}$ and $\mathbb{Q}$. In an attempt to preserve their properties we might proceed as follows: let p be a prime number; define a p-valuation ring of Z as a valuation ring $\mathcal{O}$ of Q, such that $\mathcal{O} \supset Z$, $p\mathcal{O}$ = maximal ideal of $\mathcal{O}$, and

$O/pO \simeq \mathbb{F}_p = \mathbb{Z}/(p)$; it follows easily that **there is a ring morphism** $O \to Z$

A <u>valued</u> discretely ordered domain is now defined as a structure $(Z, (O_p)_{p \text{ prime}})$, where for each prime p O_p is a p-valuation ring of Z.

Along the lines of (3.11)-(3.14) one can show that each valued discretely ordered (normal) domain can be embedded in a valued (normal) model of open induction. Analogues of (3.7) and (3.16) are easy consequences. Let us state one of them explicitly.

<u>A polynomial</u> $f \in \mathbb{Z}[X]$, $X = (X_1,\dots,X_n)$, <u>has a zero in a valued model of open induction if and only if there is an ideal</u> I <u>of</u> $\mathbb{Z}[X]$ <u>containing</u> f <u>such</u>

(i) $\mathbb{Z}[X]/I$ <u>can be discretely ordered, and</u>

(ii) I <u>has for each prime</u> p <u>a non-singular</u> p-<u>adic integral zero</u> (non-singular as a point on the Zariski-closed subset of $\mathbb{A}^n(\tilde{\mathbb{Q}}_p)$ defined by I).

(3.22) Concluding this section we mention **the following example of failure of joint embedding, due to A. Wilkie:**

(1) Models of open induction do not have the joint embedding property: **the domains** $\mathbb{Z}[X, X\sqrt{2}, \sqrt{3X^2+1}\,]$ and $\mathbb{Z}[Y, Y\sqrt{3}, \sqrt{2Y^2+1}\,]$, each of which can be embedded in a model of open induction, cannot be embedded in one and the same discretely ordered ring. In fact, for each prime p we obtain ringmorphisms of the two domains into $\mathbb{Z}_p$ by sending X, $X\sqrt{2}$, Y, $Y\sqrt{3}$ to 0 and $\sqrt{3X^2+1}$, $\sqrt{2Y^2+1}$ to 1. That both domains are discretely orderable follows similarly.

For the second part of the above statement, assume that both domains are subrings of the same discretely ordered ring, and suppose that $Y \le X$. Then, with $B = \sqrt{3}\cdot Y$, $C = \sqrt{3X^2+1}$, we have $(YC-BX)(YC+BX) = Y^2$, leading to the contradiction $1 \le YC - BX = Y/(C+\sqrt{3}\cdot X) < 1$. The case $X \le Y$ is handled similarly.

§4. <u>The case of curves: Runge's Theorem</u>

(4.1) <u>How to decide for an irreducible polynomial</u> $p = p(X,Y) \in \mathbb{Z}[X,Y]$, <u>whether or not the domain</u> $\mathbb{Z}[X,Y]/(p)$ <u>is discretely orderable, or equivalently, whether or not</u> p <u>has a nonstandard zero in a discretely ordered domain?</u> (*)

The solution of this decision problem (*) is the key to deciding whether or not a given diophantine equation $p(X,Y) = 0$, $p \in \mathbb{Z}[X,Y]$, has a solution in a (valued) model of open induction. (This follows by some easy reductions from (3.7).)

A solution of (*) was given in [v. d. D. 2], and the (decidable) criterion obtained was the following (under some harmless restrictions on p):

$\mathbb{Z}[X,Y]/(p)$ <u>is discretely orderable if and only if</u> $p(X,Y)$ <u>is irreducible as a polynomial in</u> Y <u>over the field of descending power series</u> $\mathbb{Q}((X^{-1}))$. (**)

The proof in [v. d. D. 2] depends on (Riemann's part of) Riemann-Roch. Profiting from a suggestion by A. Wilkie we give here a selfcontained proof of the most interesting half of the equivalence (**) in (4.2).

Recently, I found that criterion (**) was essentially known almost a century ago to C. Runge, in the form of a <u>necessary</u> condition for p to have infinitely many integral zeros, cf. [Ru, p. 433]. Of course, a much deeper result obtained by C. Siegel in 1929 gives even a decidable necessary and sufficient condition for p to have infinitely many integral zeros, but the known proofs do not give an effective upperbound on the size of these zeros in case there are only finitely many, and even the problem of deciding whether p has an integral zero at all remains open although positive results for special types of equations have been obtained by A. Baker c.s.

We shall show here that Runge's theorem (and its converse) can be made effective in the following sense: <u>there is an algorithm which on any input</u> $p \in \mathbb{Z}[X,Y]$ <u>computes a bound</u> $B(p) \in \mathbb{N}$ <u>such that if</u> p <u>has an integral zero</u> (x,y) <u>with</u> $|x| + |y| > B(p)$, <u>then</u> p <u>has a nonstandard zero in some model of open induction</u>. See (4.4).

(4.2) <u>Theorem</u> <u>Let</u> $p = p(X,Y) \in \mathbb{Z}[X,Y]$ <u>be of positive degree</u> n <u>in</u> Y <u>and irreducible over the field</u> $Q(X)$, Q <u>the fraction field of</u> Z. <u>If</u> p <u>has a zero</u> (x,y) <u>in a discretely ordered extension</u> Z' <u>of</u> Z <u>such that</u> $|x| > a$ <u>for all</u> $a \in Z$, <u>then</u> p <u>is even irreducible over the field</u> $Q((X^{-1}))$.

<u>Proof</u> Without loss of generality we may consider the case that x is positive and we may then identify x with the element X of the real closed extension $R((X^{-\frac{1}{\infty}}))$ of Q (using that X is also positive and infinite w.r.t. Z), and y with one of the roots of p in $R((X^{-\frac{1}{\infty}}))$, where p is considered as a polynomial in Y. This identification of y with a descending Puiseux series in X is possible because the ordering on $Z[x,y] \subset Z'$ is induced by one of its $Z[x]$-embeddings in the real closed field $R((X^{-\frac{1}{\infty}}))$.

Consider the complete valued field $Q((X^{-1}))$, with valuation ring $Q[X^{-1}]_{(X-1)}$, and its valued field extension $Q((X^{-1}))(y)$. Let t, e, F and $\alpha_1, \ldots, \alpha_f$ denote respectively a local parameter, the ramification degree, the residueclass field and a basis of the residueclass field (over Q) of this valued field extension. We may take $Q \subset F \subset Q((X^{-1}))(y)$, and then the ef elements $\alpha_i t^j$, $1 \le i \le f$, $1 \le j \le e$, form a basis of $Q((X^{-1}))(y) = F((t))$ over $Q((X^{-1}))$.

Suppose now that p were reducible over $Q((X^{-1}))$. This can now be expressed expressed by: $ef < n = \deg_Y P$. We shall derive a contradiction from this inequality. Let $K \in N$ and consider

$$a_o(X) + a_1(X)y + \ldots + a_{n-1}(X)y^{n-1} \in Z[x,y] \subset Z'$$

where $a_o(X),\ldots,a_{n-1}(X) \in Z[X]$ are of degree $\leq K$ with as yet undetermined coefficients. So $n(K+1)$ coefficients are available. Choose $L \in \mathbb{N}$ such that each of $1, y,\ldots,y^{n-1}$ is of the form:

$$c_{-L}t^{-L} + c_{-L+1}t^{-L+1} + \ldots \in F((t)), \text{ all } c\text{'s in } F.$$

Then for each $0 \leq i$, $0 \leq j \leq n-1$ $x^i y^j$ is of the form: $d_{-ei-L}\, t^{-ei-L} + d_{-ei-L+1}\, t^{-ei-L+1} + \ldots \in F((t))$, all d's in F. Substituting these expressions we obtain:

$$a_o(X) + a_1(X)y + \ldots + a_{n-1}(X)y^{n-1} = \ell_{-eK-L}\cdot t^{-eK-L} + \ldots + \ell_o + \ell_1 t + \ldots ,$$

where $\ell_k = \ell_{k1}\alpha_1 + \ldots + \ell_{kf}\alpha_f$, each ℓ_{ki}, $1 \leq i \leq f$, being a linear form over Q in the $n(K+1)$ coefficients of $a_o(X),\ldots, a_{n-1}(X)$. So the system of equations $\ell_{-eK-L} = \ldots = \ell_o = 0$, is equivalent with a system of $(+eK+L+1)\cdot f$ homogeneous linear equations over Q in the $n(K+1)$ coefficients of the a_i's as unknowns. If K is sufficiently large, it follows from $ef < n$ that there are more unknowns than equations, so we can take $a_o(X),\ldots, a_{n-1}(X)y^{n-1}$ in Z[X], not all zero, such that $a_o(X) + a_1(X)y + \ldots + a_{n-1}(X)y^{n-1}$ is in the maximal ideal $t\,F[[t]]$ of $F[[t]]$. Then $a_o(X) + \ldots + a_{n-1}(X)y^{n-1}$ is non-zero infinitesimal element of the discretely ordered ring $Z[x,y]$, which is a contradiction. ∎

(4.3) <u>Corollary</u> <u>If</u> $p = p(X,Y) \in \mathbb{Z}[X,Y]$ <u>is irreducible, of positive degree in</u> X <u>as well as in</u> Y, <u>and has a non-standard zero in a discretely ordered ring, then the monomials in</u> p <u>satisfy the following relations</u>:

(i) <u>the term</u> $c(Y)X^m$ <u>of maximum degree</u> m <u>in</u> X <u>is of the form</u> cX^m, $0 \neq c \in \mathbb{Z}$;

(ii) <u>the term</u> $d(X)Y^n$ <u>of maximum degree</u> n <u>in</u> Y <u>is of the form</u> dY^n, $0 \neq d \in \mathbb{Z}$;

(iii) <u>if</u> eX^kY^ℓ, $0 \neq e \in \mathbb{Z}$, is any monomial occurring in p, <u>then</u> $kn + \ell m \leq mn$, <u>where</u> m,n <u>are as in</u> (i), (ii).

<u>Proof</u> By (4.2) the polynomial $p(X,Y)$ is irreducible over $\mathbb{Q}((X^{-1}))$, so $p(X,Y) = d(X).(Y-(a_1X^r+\ldots))\ldots(Y-(a_nX^r+\ldots))$, where $d(X) \in \mathbb{Z}[X]$ is as in (ii), $a_1,\ldots, a_n$ are non-zero complex numbers, and r is a rational. The crucial point here is that the n roots of $p(X,Y)$ over $\mathbb{Q}((X^{-1}))$ have the same order $-r$ in the valued field $\mathbb{C}((X^{-\frac{1}{\infty}}))$, because they are conjugate over $\mathbb{Q}((X^{-1}))$. From this we immediately obtain (i), and by interchanging the roles of X and Y we get (ii). this implies in particular that $d(X) = d \in \mathbb{Z}$ and $r = m/n$.

Considering now the monomials of degree ℓ in Y, we see that the one with highest degree in X is of the form $e\,X^{(n-\ell)r}Y^{\ell} = e\,X^{(n-\ell)\,m/n}Y^{\ell}$, $e \in \mathbb{Z}$. So if X^kY^{ℓ} occurs non-trivially in p, then $kn + \ell m \le (n-\ell)\,(m/n)\cdot n + \ell m = mn$. ∎

Remarks

If one replaces the hypothesis that $p(X,Y)$ has a non-standard zero in a discretely ordered ring by the much stronger one that $p(X,Y)$ has infinitely many integral zeros, then this corollary reduces to a statement due to C. Runge [Ru, p. 434].

Of course, a polynomial $p(X,Y) \in \mathbb{Z}[X,Y]$ which is irreducible in $\mathbb{Q}((X^{-1}))[Y]$ satisfies in general more conditions than those in (4.3) On the other hand, for certain equations which have in fact been studied intersively, like $Y^n = f(X)$, where $f(X) \in \mathbb{Z}[X]$ is of degree m, our theorem (4.2) gives little information; e.g., if n is prime, then $Y^n - f(X)$ is irreducible in $Q((X^{-1}))[Y]$ if either $n \nmid m$ or the leading coefficient of $f(X)$ is not an nth power in Z.

(4.4) We will not show how (4.2) admits a sort of converse, a fact which can be used to decide effectively whether any given polynomial $p(X,Y) \in \mathbb{Z}[X,Y]$ has a non-standard zero in a model of open induction. (This has been carried out in detail in [v.d.D. 2].)

Instead, let us derive here from the existence of such an algorithm the result mentioned at the end of (4.1): let $p \in \mathbb{Z}[X,Y]$ be given. First decide whether p has a non-standard zero in a model of open induction. If so, put $B(p) = 1$. If not, consider the theory of open induction together with all sentences $\exists x\, \exists y (p(x,y) = 0 \wedge |x| + |y| > n)$, $n \in \mathbb{N}$. This set of sentences is then inconsistent, so for some $B \in \mathbb{N}$ we have:

Open Induction $\vdash \forall x\, \forall y (p(x,y) = 0 \longrightarrow |x| + |y| \le B)$.

Such a B, and a proof that it has this property, will be found by systematically generating proofs from the theory of open induction.

(4.5) As to normal models of open induction: to decide whether a given (irreducible) polynomial $p(X,Y) \in \mathbb{Z}[X,Y]$ has a zero in a normal model of open induction, reduces by (3.16) and arguments in §2 of [v.d.D.2] to the effective construction of the normalization of $\mathbb{Z}[X,Y]/(p)$.

Such an effective construction does indeed exist as I intend to show elsewhere.

§5. Concluding Remarks

(5.1) Shepherdson's problem, whether the set of $f(X_1,\ldots,X_n) \in \mathbb{Z}[X_1,\ldots,X_n]$ with a zero in a model of open induction is recursive, seems to be still open, even for any particular $n > 2$. Considering polynomials of degree 2 in 3 variables, and a few other types of polynomials, I found a decision method based on substitutions which transform the polynomial into one in 2 variables; but Wilkie pointed out that this substitution trick seems too restricted to work in general.

(5.2) One might consider (projective) cubic surfaces $p(X_0, X_1, X_2, X_3) = 0$ over $\mathbb{Z}$, and their points in models of some particular weak fragment of arithmetic, and hope to discover new methods for proving the nonexistence of rational points on the surface; to quote from Swinnnerton-Dyer's [S-W]: "It would seem worthwile to look for more methods (other than p-adic) by which one could prove that a given surface had no rational points". Certainly it would be nice to reproduce in a weak fragment of arithmetic the arguments in [S-W] on the two cubic surfaces considered in that paper.

A step in this direction would be to extend the theory of valued normal models of open induction by requiring for each prime p that $\mathcal{O}_p = Z_{pZ}$, and to prove an analogue of Wilkies theorem for this extended theory.

(5.3) Number theorists like to point out that the proof of Thue's Theorem — a crucial improvement of Liouville's Theorem mentioned in the introduction — does not not allow one to obtain an upper bound on the size of the integers p,q such that p/q is a "good" approximation to a given algebraic number α ("good" in a certain technical sense), although the theorem asserts the finiteness of the set of good approximations.

In fact, the proof starts by assuming that one has <u>two</u> very good approximations (i.e. p,q are very large), and derives a contradiction from this assumption.

A sharper version of this remark by number theorists might consist of finding an algebraic number α (of degree $n \geq 3$ over $\mathbb{Q}$) and a fragment of arithmetic in which Thue's argument for α can be reproduced, presumably implying that in each model at most one non-standard good approximation exists, but such that this last possibility is indeed realized in some model of that fragment.

(5.4) What is, in fact, a weak fragment of arithmetic? The fragments considered in this paper have three common features: they extend the theory of discretely ordered rings, they have, besides the "standard" model $\mathbb{Z}$, at least one other recursive model, and they admit an analogue of Wilkies theorem, which gives an algebraic characterization of the solvable diophantine equations.

For the moment these three properties seem to me to give a reasonable characterization of the notion of weak fragment of arithmetic.

From this point of view the theory of bounded induction is not a weak fragment: it follows from a recent result due to K. McAloon that this theory does not have the second mentioned property.

References

[Ax] J. Ax, The elementary theory of finite fields, Ann. of Math. 88 (1968), 239-271.

[Ba] A. Baker, Transcendental Number Theory, Cambridge University Press, Cambridge 1975.

[v.d.D.,R] L. van den Dries and P. Ribenboim, Lefschetz' principle in Galois theory, Queen's Mathematics Preprint No. 1976-5.

[v.d.D.1] L. van den Dries, Model Theory of Fields, Dissertation, Utrecht 1978.

[v.d.D.2] L. van den Dries, Which curves over $\mathbb{Z}$ have points with coordinates in a discrete ordered ring? To appear in Trans. AMS.

[La] S. Lang, Diophantine Geometry, Interscience, New York 1961.

[Ma] A. Macintyre, Nonstandard number theory, in Proceedings ICM, Helsinki 1978.

[R,R] A. Robinson and P. Roquette, On the Finiteness Theorem of Siegel and Mahler concerning Diophantine Equations, J. of Number Theory 7 (1975), 121-176.

[Ru] C. Runge, Ueber ganzzahlige Lösungen von Gleichunger zwischen zwei Veränderlichen, Crelles Journal 100 (1887), 425-435.

[Sh] J. C. Shepherdson, A Non-standard Model for a Free Variable Fragment of Number Theory, Bull. de l' Acad. Pol. des Sci. 12 (1964), 79-86.

[S-D] H. P. F. Swinnerton-Dyer, Two special cubic surfaces, Mathematika 9 (1962), 54-56.

[Wi] A. J. Wilkie, Some results and problems on weak systems of arithmetic, in Logic Colloquium 77, Ed. by A. Macintyre, L. Pacholski, J. Paris, North-Holland Publ. Co., Amsterdam, pp. 285-296, 1978.

Yale University
Department of Mathematics
New Haven, Connecticut 06520

Applications of Complexity Theory to Σ_o-definability Problems in Arithmetic

A.J. Wilkie

Mathematical Institute, Oxford, England.

1. Introduction

1.1. Let L denote the first order language with non-logical symbols $0,1,+,\cdot$. A formula of L is called Σ_o (or bounded) if all its quantifiers occur bounded, i.e. in the form $\exists x<y$ or $\forall x<y$, where $\exists x<y\ldots$, $\forall x<y\ldots$ are abbreviations for $\exists x(\exists u(x+(u+1)=y)\wedge\ldots)$, $\forall x(\exists u(x+(u+1)=y)\to\ldots)$ respectively. In this paper we shall be concerned with Σ_o-sets, that is subsets, S, of ω^k (where $k \in \omega$) which can be represented in the form

$$S = \{<n_1,\ldots,n_k> \in \omega^k : \mathbb{N} \models \phi[n_1,\ldots,n_k]\} \qquad \ldots. (1)$$

where $\mathbb{N} = <\omega\ ; 0,1,+,\cdot>$ is the standard model of arithmetic and $\phi(x_1,\ldots,x_k)$ is a bounded formula. Wherever (1) holds we say that ϕ represents S. For $n \in \omega$, the collection of sets that can be represented by a formula of the form

$$\exists x_{11}<y_{11}\ldots\exists x_{1t}<y_{1t}\forall x_{21}<y_{21}\ldots\forall x_{2t}<y_{2t}\ldots Qx_{n1}<y_{n1}\ldots Qx_{nt}<y_{nt}\psi$$

where ψ is open (i.e. quantifier free) and Q is $\exists(\forall)$ if n is odd (even), is denoted by $\Sigma_{o,n}$.

Clearly $n < m$ implies $\Sigma_{o,n} \subseteq \Sigma_{o,m}$. However, a hierarchy theorem, or indeed the question as to whether any of these inclusions are strict (for $n \geq 1$), is open.

An interesting related problem, due to J. Paris, is whether Matijasevic's theorem (every recursively enumerable set is representable by an existential formula of L) is provable in Σ_o-induction (Peano's axions with the induction scheme restricted to Σ_o formulas). Using a result from [5][1] a positive answer to this question easily implies that the collection of Σ_o-sets is identical with $\Sigma_{o,1}$. This in turn has a significant consequence in the theory of computational complexity. For $\Sigma_{o,1}$ contains an NP-complete set[2] (e.g. $\{<a,b,c> \in \omega^3 : \exists x<c\exists y<c\exists z<c(ax^2+by=c)\}$ see [2]), and hence, if $\Sigma_{o,1} = \Sigma_o$, also contains the complement of such a set.

---

1 Namely, if ϕ is bounded and Σ_o-induction $\vdash \forall x\exists y\phi(x,y)$, then there are $k,\ell \in \omega$ such that Σ_o-induction $\vdash \forall x\exists y<x^k+\ell\,\phi(x,y)$.

2 See the paper [2] (of this volume) for notation and an introduction to complexity theory relevant to the current discussion.

However, clearly $\Sigma_{o,1} \subseteq$ NP, so NP contains the complement of an NP-complete set. Thus if $\Sigma_{o,1} = \Sigma_o$ then NP = co-NP.

1.2 In this paper we prove a rather modest hierarchy theorem for Σ_o-sets. We shall show that for any $n \in \omega$, there is a Σ_o-set $S \subseteq \omega$ which cannot be represented by any Σ_o formula containing at most n (bounded)) quantifiers. Indeed, we shall construct (for each $k \in \omega$) a Σ_o formula of k + 1 variables which 'enumerates' all subsets of ω^k representable by an open formula. Our main tools will be Nepomnjascii's theorem on log space complexity and a result of Bennett concerning the rudimentary sets of Smullyan. We remark that the difficulty in proving our theorem directly (without going via machine complexity) lies in the fact that terms of L can be polynomials of arbitrary degree. If our language used + and$\cdot$as (3-place) predicates (and a constant symbol for each natural number) our hierarchy theorem could easily be proved directly but would say considerably less. (A detailed proof of such a theorem may in fact be found in [3]).

2. Log space complexity

2.1 Let us fix a finite set, or alphabet A, containing at least the symbols 0,1,#. A* denotes the set of all finite strings of symbols from A. If $\sigma \in A^*$, $|\sigma|$ denotes the length of the string σ.

2.2. The Turing machine model, M, that we shall be using operates as follows. M has two infinite tapes, T_w and T_r and two scanning heads, one for each tape. M can only read symbols written on T_r but can both read and write symbols on T_w. (The symbols are elements of A or blanks. 0 does not represent a blank.) Thus an example of an instruction that might occur in a program for M is

$$\left\{ \begin{matrix} q_k 0 1 q_t \\ q_l \# L q_s \end{matrix} \right\}$$, which has the meaning "if the T_w head is scanning a 0 and is in state q_k, and the T_r head is scanning a # and is in state q_l, then the T_w head erases the 0 and prints a 1 in its place and goes to state q_t, and the T_r head moves one square to the left and goes to state q_s".

Amongst the states of any program for M are two distinguished ones, q_a (the accepting state) and q_r (the rejecting state).

Suppose f: $\omega \to \omega$ is a (weakly) increasing function and $S \subseteq A^*$. We say that S is f-space computable if there is a program P for M such that if any string $\sigma \in A^*$ is written on T_r, T_w is initially blank and M is started with program P then (a) M eventually halts in state q_a if $\sigma \in S$ and in state q_r if $\sigma \notin S$, and (b) no more than $f(|\sigma|)$ distinct T_w tape squares are scanned by the T_w head during the computation.

If $k \in \omega$ and $f(n) = k\cdot[\log_2 n]$ we shall just say S is k-computable if S is f-space computable.

2.3. If $n \in \omega$, we write $\bar{n}$ for the binary representation of n, i.e. if n = 0, $\bar{n} = 0$ and if n > 0 and $n = \sum_{i=o}^{k} \varepsilon_i 2^i$ with $\varepsilon_i = 0$ or 1, $\varepsilon_k \neq 0$, then $\bar{n} = \varepsilon_k \varepsilon_{k-1} \cdots \varepsilon_o$. Thus

$\bar{n} \in A^*$, $|\bar{0}| = 1$ and if $n > 0$, $\log_2 n < |\bar{n}| \leq 1 + \log_2 n$. Further, if $S \subseteq \omega^m$ (for some $m \in \omega$) we write $\bar{S}$ for the set $\{\bar{n}_1\# \ldots \#\bar{n}_m\# : \langle n_1, \ldots, n_m \rangle \in S\}$.

We now state a theorem from complexity theory that we shall need later.

2.4. Theorem

Suppose $k,m \in \omega$, $S \subseteq \omega^m$ and $\bar{S}$ is k-computable. Then S is a Σ_o-set.

In obtaining this theorem we have combined two completely separate results. Firstly, in his thesis ([1]) Bennett showed that a set $S \subseteq \omega^m$ is a Σ_o-set if and only if $\bar{S}$ is rudimentary in the sense of Smullyan (see [6], but the definition of rudimentary will not be needed here). Secondly, in [4] Nepomnjascii proves (slightly more than) every k-computable set is rudimentary. The theorem above of course immediately follows from these results.

3. Some explicit computations

3.1. Lemma

Suppose $m \in \omega$, $S \subseteq \omega^m$ and S is represented by an open formula of L. Then $\bar{S}$ is 33-computable.

Proof

By a standard easy argument we may suppose there are polynomials f,g with coefficients in ω such that S is represented by the formula $f(x_1,\ldots,x_m) = g(x_1,\ldots x_m)$. (The class of 33-computable sets is clearly closed under the boolean operations). Choose $c,d \in \omega$ such that $f(\vec{n}) + g(\vec{n}) < c + N^d$ for all $\vec{n} \in \omega^m$, where $N = \max\{n_1,\ldots,n_m\}$. Also choose a rational $b > 0$ such that $\prod_{\substack{p<x \\ p \text{ prime}}} p \geq 2^{bx}$ for all $x \in \omega$, $x \geq 2$. (That we can do this is an easy application of the prime number theorem).

Suppose now that a string $\sigma \in A^*$ is given (i.e. written on T_r). We describe an algorithm that decides whether or not $\sigma \in \bar{S}$, indicating how it may be implemented on M so that no more than $33\cdot[\log_2|\sigma|]$ distinct T_w tape squares are ever used.

First, check that σ is of the form $\bar{n}_1\# \ldots \#\bar{n}_m\#$ for some $\langle n_1,\ldots,n_m \rangle \in \omega^m$. Clearly this can be done just by scanning σ - no T_w space is required at all. If σ is not of this form, reject it. Otherwise, say $\sigma = \bar{n}\# \ldots\#\bar{n}_m\#$ and let $N = \max\{n_1,\ldots,n_m\}$, $\vec{n} = \langle n_1,\ldots,n_m \rangle$. Now check whether

$$N \leq \max\{c+2, 2^{\frac{4(d+1)}{b}}\}$$

holds, if it does, accept σ if $f(\vec{n}) = g(\vec{n})$ and reject σ if $f(\vec{n}) \neq g(\vec{n})$. Again no T_w space is required here since we can build this fixed (i.e. independent of $\vec{n}$) amount of information into our program. If

$$N \geq \max\{c+2, 2^{\frac{4(d+1)}{b}}\}$$

carry out the following instruction, I :-

I: For each $t < \frac{(d+1)}{b} \cdot \log_2 N$ compute the residues of $f(\vec{n})$ and $g(\vec{n})$ (mod t). If they are equal for each such t, accept σ, otherwise reject σ.

Before checking that we can keep within the T_w space restrictions in executing

I, let us see why it works.

Clearly if the residues of $f(\vec{n})$ and $g(\vec{n})$ (mod t) are different for some t, then $f(\vec{n}) \neq g(\vec{n})$. On the other hand, if $f(\vec{n}) \equiv g(\vec{n})$ (mod t) for all $t < \frac{(d+1)}{b} \cdot \log_2 N$, then certainly $\pi = \Pi\{p: p < \frac{(d+1)}{b} \cdot \log_2 N, \text{ p prime}\}$ divides $f(\vec{n}) - g(\vec{n})$. However, by the choice of b, we have $\pi \geq 2^{(d+1)\log_2 N} = N^{d+1} \geq c + N^d$ (since $N \geq c + 2$), so $\pi > f(\vec{n}) + g(\vec{n})$ by the choice of c,d. Hence $f(\vec{n}) - g(\vec{n}) = 0$ and our algorithm works.

Let $\alpha = [\log_2|\sigma|]$. To see how I may be executed in 33α T_w tape squares, let us first observe that for any polynomial expression, $h(x_1,\ldots,x_n)$ (with coefficients in ω), we may find a sequence $\vec{P}(h)$ of polynomials, $(P_{11},P_{12},P_{13}),\ldots,(P_{k1},P_{k2},P_{k3})$ such that (a) $P_{1i} = 0$ for $i = 1,2,3$, (b) for $1 \leq i < k$, $j = 1,2,3$, either $P_{(i+1)j} \in \{x_1,\ldots,x_m\}$, or $P_{(i+1)j} = P_{ij'} \cdot P_{ij''}$ or $P_{(i+1)j} = P_{ij'} + P_{ij''}$ for some j',j'', $1 \leq j,j' \leq 3$, and (c) $P_{k3} = h$. (Of course equality here is modulo ring theory).

Now mark off 4 consecutive sections of the tape T_w, say E,B,C and D of lengths 2α, 2α, 8α, 20α respectively. Clearly we need scan at most 33α tape squares in doing this. The T_w head will never move (from now on) left of the first square of E nor right of the last square of D, so that our space restrictions will be met.

Instruction I is carried out in stages $I_o, I_1, \ldots I_t, \ldots$. Stage I_t starts with $\bar{t}$ written in E and the rest of T_w blank. We compute the residue of $f(\vec{n})$ (mod t) as follows. Suppose that, at some sub-stage (of I_t) the binary representations for a_1,a_2,a_3 appear in C, where $a_i \equiv P_{ji}(\vec{n})$ (mod t), $0 \leq a_i < t$ for $i = 1,2,3$, and (P_{j1},P_{j2},P_{j3}) is the j'th triple in the sequence $\vec{P}(f)$. Let $b_i \equiv P_{(j+1)i}(\vec{n})$, $0 \leq b_i < t$ for $i = 1,2,3$. Copy $\bar{a}_1,\bar{a}_2,\bar{a}_3$ into D and, using only D as work tape, compute $\bar{b}_1,\bar{b}_2,\bar{b}_3$. This only involves either multiplying and/or adding two of the numbers a_1,a_2,a_3 and reducing the result mod t, which can be done in the given space (notice $|\bar{a}_i| \leq 2\alpha$, and D has 20α tape squares) using well known algorithms (although usual "long multiplication" is not good enough here since it requires roughly $(2\alpha)^2$ tape squares; however, see e.g. [3] for more efficient alternatives), or computing n_ℓ (nod t) for some ℓ, $1 \leq \ell \leq m$. The latter is achieved by dividing t (written in E) into n_ℓ (written on T_r) by long division, but only keeping track of successive remainders (or "carries") in D (i.e. forget the "quotients"). In this way we never need to store numbers of binary length greater than $|\bar{t}|$ ($\leq 2\alpha$) as "carry" and $[\log_2|\sigma|]$ ($=\alpha$) as place marker. Once $\bar{b}_1,\bar{b}_2,\bar{b}_3$ have been found, transfer them to C (erasing $\bar{a}_1,\bar{a}_2,\bar{a}_3$) and erase any scratch-work left in D. If we continue these sub-stages k times (where k is the length of $\vec{P}(f)$) we will eventually arrive at the residue $f(\vec{n})$ (mod t) which we store in B.

Repeat the whole process to find $g(\vec{n})$ (mod t), and check if the two residues are equal. If they are not, halt and reject σ. If they are, add one to t if the space in E allows, erase any contents of B,C and D and perform stage I_{t+1}. If we cannot increase t without exceeding the space in E, halt and accept σ. Notice that we will be able to increase t provided $|\bar{t}| < 2\alpha$. Hence it only remains to check that

$|[\frac{(d+1)}{b} \cdot \log_2 N]| \leq 2\alpha$ to be sure that enough stages can be performed so that instruction I will have been completely executed. But clearly $|\sigma| \geq \log_2 N$ (since $\bar{N}$ is a sub-sequence of σ) so

$$|[\frac{(d+1)}{b} . \log_2 N]| \leq |[\frac{(d+1)}{b} . |\sigma|]| \leq 1 + \log_2(\frac{d+1}{b}) + \log_2|\sigma|.$$

Also $N > 2^{\frac{4(d+1)}{b}}$, so $\log_2\log_2 N > 2 + \log_2(\frac{d+1}{b})$. Hence

$$|[\frac{(d+1)}{b} . \log_2 N]| \leq -1 + \log_2\log_2 N + \log_2|\sigma| \leq -1 + 2\log_2|\sigma| \leq 2\ [\log_2|\sigma|] = 2\alpha.$$

as required, and the proof of 3.1 is complete.

3.2. Fix some sensible coding of programs (as natural numbers) such that the set $P(M) = \text{def}\ \{\bar{e} : e$ codes a program for $M\}$ is k-computable for some $k \in \omega$. (Actually we can take $k = 0$ here, so that P(M) is recognisable on a finite state machine). For $s \in \omega$ define

$U_s = \{\bar{a}\#\sigma :\ \exists \bar{e} \in P(M)$ such that $2^e = a$, $\sigma \in A^*$, and if M operates with the program coded by e with input σ, then it halts in its accepting state and uses at most $s.[\log_2|\sigma|]$ T_w tape squares during the course of this computation$\}$.

3.3. <u>Lemma</u>

For all $s \in \omega$, there is $t \in \omega$ such that U_s is t-computable.

<u>Proof</u>

Suppose $\tau \in A^*$. To decide on M whether $\tau \in U_s$ first check if τ is of the form $2^e\#\sigma$ for some $\sigma \in A^*$, $\bar{e} \in P(M)$. If it is, print $\bar{e}$ on T_w (hence the reason for 2^e rather than e in the input), mark off $s.[\log_2|\sigma|]$ consecutive T_w tape squares, B say, and mimic the operation of the program with code e using B as the T_w tape for e. This is possible using only $t.[\log_2|\tau|]$ T_w tape squares (for some fixed $t \in \omega$) since the contents of T_r never changes during the e-computation, so we only have to keep track of the position and state of the T_r head (of the e-computation) and there are clearly at most $e.|\sigma|$ such possibilities. (In a sensible coding, the program with code e will have fewer than e states). Hence we can keep track of this information with a number of binary length at most $1 + \log_2(e.|\sigma|) \leq 2\log_2|\tau|$.

If the program e tries to move its T_w head outside E, reject τ. Otherwise, continue mimicing program e

for $(1+b)^{(s+2).[\log_2|\tau|]}$ steps of its computation, where b is the cardinality of the alphabet A. (It is possible to count this far within our space limitation since

$$|(1+b)^{(s+2).[\log_2|\tau|]}| \leq (1 + (s+2)\log_2(1+b)).[\log_2|\tau|].)$$

If program e has not halted within this time it is easy to see that it must have already cycled - and hence will never halt - so we reject τ. Otherwise, we accept or reject τ according to whether program e has accepted or rejected τ.

4. The main theorems

4.1 Theorem

Let $m \in \omega$. There is a Σ_o formula $T_m(y,x_1,\ldots,x_m)$ containing just the m + 1 free variables shown, such that if $\Delta(x_1,\ldots,x_m)$ is any open formula of L, containing just the free variables shown, then there is some $a \in \omega$ such that

$$N \models \forall\vec{x}(\Delta(\vec{x}) \leftrightarrow T_m(a,\vec{x})).$$

Proof

By 3.3 the set U_{33} (see 3.2) is k-computable for some $k \in \omega$. Hence, by 2.4 we may find a Σ_o-formula $T_m(y,\vec{x})$ such that $\mathbb{N} \models T_m[a,\vec{n}]$ if and only if $\bar{a}\#\bar{n}_1 \ldots \#\bar{n}_m\#$ $\in U_{33}$. Now if $\Delta(x_1,\ldots,x_m)$ is any open formula of L, let $S \subseteq \omega^m$ be the set it represents, and e the code of a program that accepts $\sigma \in A^*$ if and only if $\sigma \in \bar{S}$, and always halts in $33.[\log_2|\sigma|]$ space - as given by 3.1. Then for all $\vec{n} \in \omega^m$, $\mathbb{N} \models T_m[2^e,\vec{n}]$ iff $\vec{n} \in S$.

4.2 Corollary

For each $k \in \omega$, there is a Σ_o-set $S \subseteq \omega$, which cannot be represented by any Σ_o-formula containing only k bounded quantifiers.

Proof

If 4.2 were false it is easy to see that there would be $m \in \omega$ such that every Σ_o set ($\subseteq \omega$) could be represented by a formula (with free variable x) of the form

$$\forall x_1 < x\, \exists x_2 < x \ldots \forall x_{m-1} < x\, \exists x_m < x\, \Delta(x_1,\ldots,x_m,x) \qquad \ldots(1)$$

for some open formula Δ.

Consider the formula $\phi(x,y)$ defined as

$$\forall x_1 < x\, \exists x_2 < x \ldots \forall x_{m-1} < x\, \exists x_m < x\; T_{m+1}(y,x_1,\ldots,x_m,x)$$

where T_{m+1} is given by 4.1. Then $\phi(x,y)$ is Σ_o and (by 4.1 and (1)) for every Σ_o set $S \subseteq \omega$, there is $a \in \omega$ such that for all $n \in \omega$

$$\mathbb{N} \models \phi[a,n] \quad \text{iff } n \in S.$$

However, no such a exists for the Σ_o-set $\{n \in \omega : \mathbb{N} \models \neg\phi[n,n]\}$.

Of course 4.2 immediately implies

4.3 Theorem

For each $k \in \omega$, there is a set $S \subseteq \omega$ which can be represented by a Σ_o formula containing $k + 1$ (bounded) quantifiers, but not by any Σ_o formula containing only k (bounded) quantifiers.

References

[1] J.H. Bennett, "On Spectra", Ph.D. dissertation, Princeton University, (1962).

[2] K.L. Manders, "Computational Complexity of Decision Problems in Elementary Number Theory", this volume.

[3] J. Meloul, "Rudimentary Predicates, Low Complexity Classes and Related Automata", D.Phil. dissertation, Oxford University, (1979).

[4] V.A. Nepomnjascii, "Rudimentary Interpretations of Two-Tape Turing Computation", Kebernetika, 6, (1970). Translated in : Cybernetics, December (1972).

[5] R. Parikh, "Existence and Feasibility in Arithmetic", J.S.L. 36, (1971).

[6] R. Smullyan, "Theory of Formal Systems", Annals of Mathematics Studies no. 47, Princeton University Press, (1961).

MINIMALLY SATURATED MODELS

George Wilmers
Department of Mathematics
University of Manchester M13 9PL.

Abstract: A model for a complete first order theory T in a language of finite type is minimally saturated if it is recursively saturated and elementarily embeddable in every recursively saturated model of T. Such a model is unique when it exists, and may be regarded as the smallest model of T with saturation properties. (Alternatively, if T^* denotes the theory obtained from T by adding all Σ_1^1-sentences consistent with T, then a minimally saturated model for T is simply a prime model for T^*). We show that the existence of such a model is implied by the existence of a countably saturated model, and in turn implies the existence of a prime model, and that both these implications are strict. We also give an easily applicable sufficient condition for T to have no minimally saturated model. §3 includes a general result about the degrees of complete types of first-order theories.

§1 Preliminaries

Throughout this paper T denotes an arbitrary first-order theory in a language $\mathcal{L}$ of finite type. We shall identify formulae with their Gödel numbers according to some fixed recursive Gödel numbering. An <u>n-type</u> of T is a set of formulae Γ, with n free variables between them, which is consistent with T. A <u>type</u> of T is an n-type for some n. An <u>n-type</u> (resp. <u>type</u>) over a structure $\mathfrak{A}$ is an n-type (res. type) of $Th(\mathfrak{A}, a_1 \ldots a_m)$ for some $a_1 \ldots a_m \in |\mathfrak{A}|$. An n-type Γ of T with free variables $x_1 \ldots x_n$ is <u>complete</u> if for every formula $\phi(x_1 \ldots x_n)$ of $\mathcal{L}$ with free variables $x_1 \ldots x_n$ either $\phi \in \Gamma$ or $\neg\phi \in \Gamma$. Similarly T is <u>complete</u> if for every sentence ϕ of $\mathcal{L}$ either $\phi \in T$ or $\neg\phi \in T$. It is important to notice that we use the term complete in this restricted sense, so that we distinguish between theories and their deductive closures. A type over $\mathfrak{A}$ is a <u>pure type</u> if it is a type of $Th(\mathfrak{A})$.

Following Barwise and Schlipf [76], a structure $\mathfrak{A}$ for $\mathcal{L}$ is <u>recursively saturated</u> if every recursive type over $\mathfrak{A}$ is realized in $\mathfrak{A}$.

We denote by $2^{<\omega}$ the set of finite sequences of 0's and 1's. $\tau \subseteq 2^{<\omega}$ is a (binary) <u>tree</u> if

(i) $s \in \tau \Rightarrow s \upharpoonright n \in \tau \qquad \forall n < lh(s)$

(ii) τ is infinite

Here $lh(s)$ denotes the length of s. A function $f \in 2^{\omega}$ is a <u>branch of τ</u> if $f \upharpoonright n \in \tau \quad \forall n \in \omega$. We assume a fixed recursive coding of $2^{<\omega}$ so that both trees and branches are identified with subsets of ω. Notice that according to our definition a tree may have "terminating" sequences. A tree τ is <u>non-terminating</u> if any $s \in \tau$ is an initial segment of some branch of τ.

A branch f of a tree τ is <u>isolated</u> if $\exists n \in \omega$ such that for any branch g of τ, $g \neq f \Rightarrow g \upharpoonright n \neq f \upharpoonright n$. τ is <u>perfect</u> if it is non-terminating and has no isolated branches. A set X of branches of τ <u>covers τ</u> if $\tau = \{f \upharpoonright n \mid n \in \omega \text{ and } f \in X\}$. A <u>subtree</u> τ' of a tree τ is a tree $\tau' \subseteq \tau$.

<u>Definition 1.1</u>: $A \subseteq 2^{\omega}$ is a <u>Scott set</u> if

(i) (Recursive closure)
$a_1, \ldots a_n \in A$ and $b \subseteq \omega$ recursive in $a_1 \ldots a_n$ $b \in A$ $b \in A$

(ii) (König condition)
If $\tau \in A$ is a tree then τ has a branch $f \in A$.

Scott sets were first introduced in Scott[62]. Using Scott's idea Friedman [71] proved that a countable set $A \subseteq 2^{\omega}$ is a Scott set if and only if it is the set of reals coded in some countable model of Peano arithmetic. Wilmers [75] gave a complete classification of countable recursively saturated structures in terms of Scott sets, which we describe briefly below.

<u>Definition 1.2</u>: If $A \subseteq 2^{\omega}$ we call a structure $\mathfrak{A}$ for $\mathcal{L}$ <u>A-saturated</u> if for any

complete type Γ over $\mathfrak{A}$,

$$\mathfrak{A} \text{ realizes } \Gamma \iff \Gamma \in A.$$

The following three results are taken from Wilmers [75]. (See also Lessan [78] and Lessan and Wilmers [79]):

Lemma 1.3: If for some Scott set A $\mathfrak{A}$ and $\mathfrak{B}$ are both countable A-saturated structures for $\mathcal{L}$ then

$$\mathfrak{A} \equiv \mathfrak{B} \Rightarrow \mathfrak{A} \cong \mathfrak{B}.$$

[Note: this is an easy back-and-forth argument and depends only on the recursive closure of A].

Result 1.4: If A is a countable Scott set and $T \in A$ then there is an A-saturated countable model $\mathfrak{A}$ of T.

Clearly the $\mathfrak{A}$ of 1.4 is unique up to isomorphism by 1.3.

Result 1.5: For any model $\mathfrak{A}$ of T the following are equivalent:

(i) $\mathfrak{A}$ is recursively saturated

(ii) $\mathfrak{A}$ is A-saturated for some Scott set A.

Notice that it is trivially true that if $\mathfrak{A}$ is an A-saturated model of T, where A is Scott, then $T \in A$, since A is recursively closed and T is recursive in any complete type of T. Thus, taken together, 1.3, 1.4 and 1.5 completely classify the countable recursively saturated models of T.

We shall make some use of the following result due to Grilliot (et alii), which may be proved by an omitting types argument using the Scott-Friedman characterisation of countable Scott sets mentioned above:

Result 1.6: (Grilliot [72])

Let $a \in 2^\omega$ and $\{b_i \mid i \in \omega\} \subseteq 2^\omega$ be such that no b_i is recursive in a. Then there is a countable Scott set A such that $a \in A$, but $A \cap \{b_i \mid i \in \omega\} = \emptyset$.

Model theoretic definitions and results which are not mentioned in this section are classical and may be found e.g. in Chang and Keisler [73].

Finally we must mention what is undoubtedly the most fundamental characteristic of Scott sets in their relation to logic and from which most of the previous results can be derived. The following fact is implicit in Scott [62]; a proof may be found e.g. in Lessan and Wilmers [79].

Result 1.7: If S is a theory in $\mathcal{L}$, A is a Scott set, and $S \in A$, then S has a complete extension $T \in A$.

§2 The Existence Problem

In this section we define the notion of a minimally saturated model, prove a uniqueness theorem, and establish conditions for existence.

Let Ω_T denote the set of complete types of T.

Lemma 2.1: A countable recursively saturated model $\mathfrak{A}$ of T is determined up to isomorphism by the subset of Ω_T of types which are realized in $\mathfrak{A}$.

Proof. This is a standard fact about recursively saturated models which follows easily e.g. from 1.5 and 1.3 above.

Now one half of the back-and-forth argument used in the proof of 1.3 shows that:

Lemma 2.2: If A and B are Scott sets, $A \cap \Omega_T \subseteq B \cap \Omega_T$, $\mathfrak{A}$ is a countable A-saturated model of T, and $\mathfrak{B}$ is a B-saturated model of T, then $\mathfrak{A}$ is elementarily embeddable in $\mathfrak{B}$.

Putting these two lemmas together with the characterisation of recursive saturation in terms of Scott sets given in §1, we see that the relation of elementary embeddability between countable recursively saturated structures is isomorphic to the relation of inclusion between members of $\{A \cap \Omega_T \mid A \text{ Scott and } T \in A\}$.

Definition 2.3: A model $\mathfrak{A}$ of T is minimally saturated if it is recursively saturated and elementarily embeddable in every recursively saturated model of T.

Obviously such models are countable.

Equivalently we have:

Lemma 2.4: Let $\mathfrak{A}$ be a model of T. The following are equivalent:

(i) $\mathfrak{A}$ is minimally saturated.

(ii) $\mathfrak{A}$ is recursively saturated and countable, and every member of Ω_T realized in $\mathfrak{A}$ is recursive in T.

Proof: (i) $\Rightarrow$ (ii). If (i) holds but (ii) does not hold, then there is some $\Gamma \in \Omega_T$ which is realized in $\mathfrak{A}$ but is not recursive in T. By 1.6 let A be a countable Scott set such that $T \in A$ but $\Gamma \notin A$. By 1.4 there is an A-saturated model $\mathfrak{B}$ of T. $\mathfrak{B}$ is recursively saturated by 1.5. But $\mathfrak{B}$ does not realize Γ and so $\mathfrak{A}$ is not elementarily embeddable in $\mathfrak{B}$, contradicting (i).

(ii) $\Rightarrow$ (i). If (ii) holds then the pure complete types realized in $\mathfrak{A}$ are realized in any recursively saturated model $\mathfrak{B}$ of T. By 1.5 suppose $\mathfrak{A}$ is A-saturated and $\mathfrak{B}$ is B-saturated where A and B are Scott sets. Then $A \cap \Omega_T \subseteq B \cap \Omega_T$ so $\mathfrak{A}$ is elementarily embeddable in $\mathfrak{B}$ by 2.2.

Theorem 2.5: (Uniqueness)

If $\mathfrak{A}$ and $\mathfrak{B}$ are minimally saturated models of T then $\mathfrak{A} \cong \mathfrak{B}$.

Proof: By 2.4 the members of Ω_T which are realized in $\mathfrak{a}$ and $\mathfrak{b}$ are exactly those recursive in T, in each case. So the result follows by 2.1.

Theorem 2.6: (Existence)

If T has a countably saturated model, then T has a minimally saturated model.

Proof: By assumption $\Omega_T \leq \omega$. So by 1.6 there is a countable Scott set A such that $T \in A$ but such that no member of Ω_T which is not recursive in T is in A. By 1.4 let $\mathfrak{a}$ be a countable A-saturated model of T. $\mathfrak{a}$ is recursively saturated by 1.6 and hence minimally saturated by 2.4.

The above theorem bears a strong relationship to a results of Millar [78] and MacIntyre [79] which essentially answer the question: when does a complete theory T with a countably saturated model have a recursively saturated model which is recursive in T? Now, when it exists, such a model is clearly minimally saturated. In the light of this we may reformulate these results as follows:

Result 2.7: (MacIntyre [79], Millar [78])

Let T have a countably saturated model. Then the following are equivalent:

(i) The minimally saturated model of T has a presentation which is recursive in T

(ii) The set of complete types of T which are recursive in T has an enumeration which is Σ^0_1 in T.

We shall show in §3 that the implication of 2.6 does not reverse, although the counterexample we construct is necessarily pathological. Thus minimally saturated models exist more often than countably saturated ones. Now a classical result of Vaught [61] tells us that if T has a countably saturated model then T has a prime model. So it is natural to ask whether the existence of a minimally saturated model also implies the existence of a prime model. We show as a corollary to the next theorem that this is indeed the case.

Definition 2.8: Let Σ denote some set of formulae of $\mathcal{L}$ with n free variables, $\vec{x}$, and let $\phi_i(\vec{x})$ denote the i'th member of Σ according to the natural ordering of Σ induced by the Gödel numbering of $\mathcal{L}$. The Σ-tree of T is the non-terminating tree

$$\tau = \{s \in 2^{<\omega} \mid \exists \vec{x} \bigwedge_{i<\ell h(s)} \phi_i^{\text{if } s(i) = 1} \text{ is in } T\}$$

where $\phi_i^{\text{if } s(i) = 1}$ denotes ϕ_i if $s(i) = 1$ and $\neg\phi_i$ otherwise. Note that if Σ is recursive in T, then so is τ, and that each branch of τ codes an n-type of T.

Theorem 2.9: Let Σ be some set of formulae of $\mathcal{L}$ with n free variables $\vec{x}$, which is recursive in T. Then if the Σ-tree of T has a perfect subtree which is recursive in T, T has no minimally saturated model.

<u>Proof</u>: Let $\mathfrak{A}$ be a recursively saturated model of T. We show that $\mathfrak{A}$ realizes a pure complete n-type which is not recursive in T. This suffices by 2.4.

Let τ be a perfect subtree of the Σ-tree of T such that τ is recursive in T. By 1.5 $\mathfrak{A}$ is A-saturated for some Scott set A. Let $f \in 2^{\omega}$ be such that $f \in A$ but f is not recursive in T. (This is possible since by a diagonal argument no Scott set can be the set of reals recursive in one given real). We can use f to define in an obvious way a branch $g \in A$ of the perfect tree τ which is not recursive in T. Then g determines an n-type Γ of T which decides every formula of Σ and such that $\Gamma \in A$ but Γ is not recursive in T. Now by 1.7 Γ can be extended to a complete n-type $\Gamma' \in A$. Since Σ is recursive in T, Γ is recursive in Γ', and hence Γ' is not recursive in T. But $\mathfrak{A}$ is A-saturated and so Γ' is realized in $\mathfrak{A}$. This completes the proof.

In conjunction with 2.6, Theorem 2.9 settles the question of the existence of a minimally saturated model in almost all cases. Thus we have:

<u>Corollary 2.10</u>: If T has any of the following properties then T has no minimally saturated model:

(i) T is not atomic

or (ii) T is a theory in which Presburger arithmetic is interpretable

or (iii) T is a theory in which $Th\langle \mathbb{Q},<\rangle$ is interpretable in such a way that to each rational $q \in \mathbb{Q}$ we can assign a formula $\theta_q(y)$ of $\mathcal{L}$ defining q in T so that the set of defining formulae $\{\theta_q \mid q \in \mathbb{Q}\}$ is recursive in T.

<u>Proofs</u>: (i) If T is not atomic then for some n there is a formula $\phi(\vec{x})$ with n free variables such that for no formula $\theta(\vec{x})$ is $\theta \wedge \phi$ complete, and such that $\exists\vec{x}\,\phi(\vec{x}) \in T$. Let Σ be the set of all formulae with free variables $\vec{x}$. Then, using the notation of 2.8, $\phi = \phi_i$ for some $i \in \omega$. Now let τ' be the subtree of the Σ-tree, τ, of T defined by

$$\tau' = \{f{\restriction}n \mid n \in \omega \text{ and } f \text{ a branch of } \tau \text{ and } f(i) = 1\}$$

Then τ' is the required perfect subtree.

(ii) Let Σ be the set of interpretations of formulae of the form $\exists y(y \neq 0 \wedge x = \underbrace{y + y + \ldots + y}_{p \text{ times}})$ where p is a prime. Then the Σ-tree of T is perfect.

(iii) Let Σ be the set of formulae of the form $\forall y(x <_T \theta_q(y))$ where $<_T$ is the interpretation of $<$ in T. Then it is clear that the Σ-tree of T is perfect.

We remark that 2.10 (iii) shows that the theory of real closed fields has no minimally saturated model, a result due to MacIntyre (unpublished).

We also have

Corollary 2.11:

If T has a minimally saturated model then T has a prime model.

Proof: If T has a minimally saturated model then T is atomic by 2.10(i), and so has a prime model.

Clearly the implication of 2.11 does not reverse since e.g. Th$\langle\omega,+\rangle$ has a prime model, but no minimally saturated model, by 2.10 (ii).

The next section is devoted to the construction of a theory T which has a minimally saturated model, but no countably saturated model. Some reflection on 2.9 and 2.10 above indicates that such a theory must be very strange.

§3 A theory with a minimally saturated model and 2^{ω} pure types.

We start this section with a general result which answers the question as to which sets of Turing degrees can occur as the set of degrees of complete 1-types of a complete first-order theory. Jokusch and Soare [72] answered a similar question concerning the degrees of the sets of complete extensions of axiomatized first-order theories. The answers to both questions turn out to be very similar.

For any $a \in 2^{\omega}$, let $\bar{a}$ denote the degree of a.

Theorem 3.1: (i) For any non-terminating tree τ there is a complete theory T, of the same degree as τ, such that the set of degrees of complete 1-types of T is $\{\bar{f} \cup \bar{T} \mid f$ a branch of $\tau\}$.

In particular,

(ii) For any recursive tree τ there is a complete decidable T such that the set of degrees of complete 1-types of T is

$$\{\bar{f} \mid f \text{ a branch of } \tau\}$$

Proof: Let $\mathcal{L}$ be the language with two binary relation symbols R and S, and let T' be the following set of axioms:

(1) "R and S are equivalence relations"

(2) $\forall x y z \ (S(x,y) \wedge R(y,z)) \rightarrow S(x,z))$

(3) For each $n, k \in \omega$ $(n \neq 0)$ the sentence "$\exists$ at least k R-equivalence classes of cardinality n."

(4) For each $s \in 2^{<\omega}$, putting $lh(s) = n$, and for each $m > 0$, the sentence

$$\exists y \ldots y_m \bigwedge_{\substack{k,k' = 1,\ldots m \\ k \neq k'}} \neg S(y_k, y_{k'}) \wedge \bigwedge_{k=1}^{m} \bigwedge_{i=0}^{n-1} \theta_{i,s(i)+1}(y_k)$$

if $s \in \tau$,

and the sentence

$$\neg\exists y \bigwedge_{i=0}^{n-1} \theta_{i,s(i)+1}(y)$$

otherwise,

where $\theta_{i,j}(y)$ denotes the formula asserting "y is S-equivalent to exactly j R-equivalence classes of cardinality $i+1$."

(5) For each n,

$$\forall y\ (\theta_{n,1}(y) \vee \theta_{n,2}(y)) \quad .$$

Let T be the deductive closure of T'. Clearly T is consistent since if $X = \{f_i \in 2^\omega \mid i \in I\}$ is a set of branches of τ which covers τ then we can form a model of T by taking ω copies of each f_i and letting an S-equivalence class correspond to each copy. (These are not the only models however, even amongst those which have no infinite equivalence classes). Now it is easy to see that if for $\mathfrak{A} \models T$ and $n > 0$ $\mathfrak{A}_n$ denotes the submodel of $\mathfrak{A}$ whose domain is the union of the R-equivalence classes with $\leq n$ elements, then for $\mathfrak{A}$, $\mathfrak{B}$ countable models of T, $\mathfrak{A}_n \cong \mathfrak{B}_n$ for any $n > 0$. An Ehrenfeucht-Fraïssé argument then easily shows that $\mathfrak{A} \equiv \mathfrak{B}$. So T is complete. Furthermore T is clearly recursive in τ (since T' is), and τ is clearly recursively recoverable from T.

Now any branch $f \in 2^\omega$ of τ determines an obvious 1-type $\Gamma_f(y)$ given by

$$\Gamma_f = \{\theta_{i,f(i)+1}(y) \mid i \in \omega\}.$$

Γ_f is not complete, but if we add to it the formula $\exists!^{(n)} x\, R(x,y)$ for some $n > 0$, or alternatively the set $\{\exists^{(n)} x\, R(x,y) \mid n > 0\}$ we obtain a type whose deductive closure Γ'_f is complete. To see this notice that any countable recursively saturated model of T has the property that each S-equivalence class has ω different R-equivalence subclasses each of power ω. It follows that if a and a' are two realizations of Γ'_f in a countable recursively saturated model $\mathfrak{A}$ of T, then there is an automorphism of $\mathfrak{A}$ mapping a to a'. Thus Γ'_f must be complete. By construction $\bar{\Gamma}'_f \leq \bar{f} \cup \bar{T}$. But clearly both f and T can be recursively recovered from Γ'_f and so $\bar{\Gamma}'_f = \bar{f} \cup \bar{T}$. Since it is clear that every complete 1-type of T must have the form of some Γ'_f the theorem follows.

Remark 3.2: Notice that the model theoretic properties of T in the construction above correspond directly to structural properties of τ; e.g.

T has a countably saturated model $\Longleftrightarrow$ τ has countably many branches

and

T is atomic $\Longleftrightarrow$ there is a set of isolated branches of τ which covers τ.

The next result which we need in our construction is an unpublished theorem of Jeff Paris whose proof we include with his permission.

Result 3.3: (Paris)

There is a non-terminating recursive tree τ with 2^{ω} branches such that each branch is either recursive or has degree $\geq \underset{\sim}{0}'$. Hence if A is any Scott set such that $\underset{\sim}{0}' \notin A$ the branches of τ which are in A are exactly the recursive branches.

Proof: Let U_e denote the e'th Σ_1^0-set in some standard enumeration of r.e. sets, so that

$$n \in U_e \leftrightarrow \exists y \theta(e, n, y)$$

where θ is some recursive relation.

Let $W = \{s \in 2^{<\omega} \mid s(0) = 1 \text{ and } s(lh(s)-1) = 1\}$. For any $s \in W$ suppose that s consists of $k+1$ non-empty blocks of 1's separated by non-empty blocks of 0's, thus:

$$\underbrace{111\ldots11}_{n_0 \text{ 1's}}\, 0\ldots0\, \underbrace{11\ldots1}_{n_1 \text{ 1's}}\, 00\ldots0\, \underbrace{11\ldots1}\, 0\ldots0\, \underbrace{111\ldots1}_{n_k \text{ 1's}}$$

We define τ' by

$$s \in \tau' \Leftrightarrow s \in W \wedge \forall i < k \left[\exists y \leq n_i\ \theta(i, i, y) \leftrightarrow \exists y \leq lh(s)\ \theta(i, i, y)\right]$$

and we let $\tau = \{s \restriction m \mid s \in \tau' \text{ and } m \in \omega\}$. We remark that we may ignore the restriction on τ that it be non-terminating since a trivial conversion may be effected by adding an infinite sequence of 1's to each terminating sequence. Now any branch f of τ either has only finitely many 0's and so is recursive, or satisfies the condition

$$\forall i \left[\exists y \leq n_i\ \theta(i, i, y) \leftrightarrow \exists y\ \theta(i, i, y)\right]$$

where n_i denotes the $(i + 1)$'th block of 1's in f. But in the latter case the set $\{i \mid \exists y\ \theta(i, i, y)\}$ can be recursively decoded from f, from which it follows that the degree of f is $\geq \underset{\sim}{0}'$. On the other hand τ clearly has 2^{ω} branches since from any non-recursive branch of τ we can generate 2^{ω} other branches by inserting extra zeros in the existing blocks of zeros in an arbitrary manner.

From 3.1 and Paris' result above we can deduce the main theorem of this section:

Theorem 3.4:

There is a complete theory T in a language with two binary relation symbols, such that T has a minimally saturated model, but no countably saturated model.

Proof: Let τ be the tree of 3.3 and let T be the complete theory constructed from it as in 3.1(ii). Then T has 2^{ω} complete 1-types and so has no countably saturated model. By 1.6 let A be a Scott set such that $\underset{\sim}{0}' \notin A$. Since T is recursive $T \in A$ and so by 1.4. there is a countable A-saturated model $\mathfrak{A}$ of T. Then the complete 1-types of T which are realized in $\mathfrak{A}$ are exactly the recursive ones. Since it is easily seen that any complete n-type Γ of T is recursive relative to some set of n complete 1-types of T which are realized in a given model if Γ is, it follows that any complete type of T which is realized in $\mathfrak{A}$ is recursive and so by 2.4 $\mathfrak{A}$ is minimally saturated.

We conclude this section with a simple application of 3.11:

Theorem 3.5: There is a complete theory T with a countably saturated model and, a fortiori, a minimally saturated model and a prime model, such that these three models are non-isomorphic.

Proof: Let τ be a recursive tree with $\leqslant \omega$ branches and at least one non-recursive branch. (Such trees exist by results of Yates and others). Then from τ we can easily construct a tree τ' with the same properties and with the additional property that τ' has at least one recursive branch which is not isolated. Now let T be the theory constructed from τ' as in 3.1. Since τ' has $\leqslant \omega$ branches, by 3.2 T has a countably saturated model. But the minimally saturated model realizes no non-recursive complete types, and the prime model realizes only the 1-types corresponding to isolated branches of the tree τ'. Hence all three models are mutually non-isomorphic

The author wishes to thank Ken McAloon, Angus MacIntyre and Jeff Paris for several useful conversations and in particular the latter from his proof of 3.3.

Added in proof:

Anand Pillay has pointed out to the author an example of a complete theory with a countably saturated model, which is not ω-categorical, and such that the minimally saturated and prime models coincide. The obvious question which remains therefore is whether minimally saturated and prime can coincide for a complete theory having no countably saturated model. However since such a theory would have to be even more pathological than that of 3.4 above, this question seems to be of purely technical interest.

References:

J. Barwise and J. Schlipf [76], An Introduction to Recursively Saturated and Resplendent Models, J.S.L. 41, 1976.

C.C. Chang and H.J. Keisler [73], Model Theory, North Holland, 1973.

H. Friedman [71], Countable Models of Set Theories, Cambridge Summer School in Mathematical Logic, 1971. Springer Lecture Notes 337 pp.539-573.

T.J. Grilliot [72], Omitting Types: Application to Recursive Theory, JSL 37 No. 1. pp.81-89.

C.G. Jockusch, and R.I. Soare 72 , Π_1^0-classes and degrees of theories, Trans. AMS 173 Nov. 1972 pp. 33-56.

H. Lessan [78], Ph.D. Thesis, Manchester 1978,

H. Lessan and G.M. Wilmers [79], Scott sets and non-standard model theory (to appear).

A. MacIntyre [79], Decidable Recursively Saturated Models (to appear).

T.S. Millar [78], Foundations of Recursive Model Theory, Annals of Math. Logic 13 (1978) pp.45-72.

D. Scott [62], Algebras of sets binumerable in complete extensions of arithmetic, Proc. Sympos. Pure Math, Vol. 5, AMS, 1962.

R.L. Vaught [61], Denumerable models of complete theories, Infinistic Methods, Warsaw, 1961.

G.M. Wilmers [75], Ph.D. Thesis, Oxford 1975.

TOTALLY CATEGORICAL THEORIES: STRUCTURAL PROPERTIES AND THE NON-FINITE AXIOMATIZABILITY

by

B.I. Zilber

One of the main results of the paper is the proof of the non-existence of a theory which is finitely axiomatizable, complete and totally categorical (i.e. categorical in all infinite powers). The result was obtained by through investigations of structural properties of models of $\aleph_1$-categorical, and especially totally categorical, theories. We introduce a notion of an envelope of a subset of a structure, which is a natural generalization of the notion of algebraic closure (see[2]), and at least in the case of totally categorical theories has very nice properties.

More precisely, we call an envelope of a subset X of a structure $\mathfrak{M}$ with respect to a formula δ , a maximal set E(X) such that E(X) and $\delta(\mathfrak{M})$ are independent over X (in the sense of Lascar [4])

<u>Theorem A.</u> If $\mathfrak{M}$ is a model of a totally categorical theory, δ is a strongly minimal formula, E(X) is an envelope of X with respect to δ, then :

(a) $E(X) \supseteq cl(X)$ (cl denotes the algebraic closure);

(b) $E(E(X)) = E(X)$;

(c) E(X) is finite, provided X is finite ;

(d) for any natural m there is a natural number s(m) such that if card $(cl(X) \cap \delta(\mathfrak{M})) \geq s(m)$, then E(X) is m+1 -saturated

(e) if $cl(X) \cap \delta(\mathfrak{M})$ is infinite, then E(X) is an elementary substructure of $\mathfrak{M}$, prime over X.

Here we say that E(X) is m+1-saturated if every 1-type over any subset of E(X) of cardinality m is realized by an element of E(X).

Theorem A gives for every natural m a finite m+1-saturated subset E_m of $\mathfrak{M}$, provided $\mathfrak{M}$ is a model of a totally categorical theory. To get such an E_m it suffices to choose a finite subset X_m of $\mathfrak{M}$ such that card $(\mathrm{cl}(X_m) \cap \delta(\mathfrak{M})) \geq s(m)$

and put $E_m = E(X_m)$.

It is easy to see that if a structure $\mathfrak{M}$ is a model of a closed formula α with m+1 quantifiers then every m+1-saturated subset of $\mathfrak{M}$ defines a substructure which is also a model of α. Thus it is proved that if α is a sequence which has a unique (up to isomorphism) model in every infinite power, then α has a finite model, and thus the theory defined by α is not complete.

Theorem A is a consequence of a more technical result:

Theorem B. If $\mathfrak{M}$ is a model of a totally categorical theory, φ and δ are one-variable formulas with parameters from a finite A, $\delta(\mathfrak{M})$ is strongly minimal then there exists a finite subset D_φ of $\delta(\mathfrak{M})$ such that for any B there is an element $b \in \varphi(\mathfrak{M})$ which is independent from $\delta(\mathfrak{M})$ over $A \cup B \cup D_\varphi$.

Obviously, if $\mathrm{cl}(A \cup B \cup D_\varphi) \cap \varphi(\mathfrak{M}) \neq \emptyset$ then any element b of the intersection satisfies this independence condition. But for non-almost strongly minimal theories, even for large A one can not quarantee that the intersection will be non-empty. Thus, it is natural to study the connection of $\varphi(\mathfrak{M})$ and $\delta(\mathfrak{M})$ in this context.

Section 1 of the paper is preliminary. In Section 2 we deal with a notion of definability of structures, which, of course, in various versions was used in model theory earlier. We show how in a natural way one can extend a structure $\mathfrak{M}$ by adjoining to it another structure which is definable in $\mathfrak{M}$. Particularily, for ε a definable equivalence relation on $\varphi(\mathfrak{M})$ we can adjoin $\varphi(\mathfrak{M})/_\varepsilon$ to $\mathfrak{M}$ without any essential changes of model-theoretical properties of $\mathfrak{M}$.

In Section 3 it is introduced a definable equivalence relation on $\varphi(\mathfrak{M})$ characterizing the dependence between $\varphi(\mathfrak{M})$ and $\delta(\mathfrak{M})$.

Section 4 uses the previous results to construct a group, which we call the binding group of $\varphi(\mathfrak{M})$ and $\delta(\mathfrak{M})$. The binding group is definable in $\mathfrak{M}$ and its structure reflects some important properties concerning the connection of $\varphi(\mathfrak{M})$ with $\delta(\mathfrak{M})$, this connection is simplier when the binding group is abelian. Note, that by a result of Baur, Cherlin, Macintyre [3] in the case of a totally categorical theory our groups must be almost abelian.

Section 5 deals with the case of totally categorical theories only. There we use the preceding technique and some finite combinatories - like computations to prove Theorem B. The last section is devoted to the notion of envelopes in models of totally categorical theories.

Let us note that the results of Sections 2-4 are stated for $\aleph_1$-categorical theories but can be in a natutal way generalized to superstable unidimensional theories (for definition see [6]). Also the notion of an envelope can be used in a very general situation and of course, there arise many questions about it. For example, we do not know, even in the totally categorical case, under what conditions all envelopes of X with respect to a fixed δ are isomorphic over X . Does the notion depend on δ essentially ?

The paper was prepared during the author's stay at Wroclaw University. The author is very grateful to Wroclaw logicians for their hospitality and help.

1. Notation and preliminaries.

In this paper we shall always deal with structures of countable languages with equality and without functional symbols.

If $\mathfrak{M}$ is a structure, then $\mathfrak{M} = \langle M, \Omega_{\mathfrak{M}} \rangle$, where M is its universum and $\Omega_{\mathfrak{M}}$ is its signature. For $A \subseteq M$ by $\Omega_{\mathfrak{M}}(A)$ we denote the expansion of $\Omega_{\mathfrak{M}}$ by symbols for all elements from A, $F_n(\Omega_{\mathfrak{M}}(A))$ is the set of formulas over $\Omega_{\mathfrak{M}}(A)$ with n free variables. If no confusion will arise we simply write $F_n(A)$ or F_n, if $A = \emptyset$. Admitting some inaccuracy we shall otfen consider $F_n(A)$ as a Boolean algebra with $\vee$, & and $\neg$ as Boolean operations. This Boolean algebra is isomorphic to the Boolean algebra of all subsets of $\mathfrak{M}$, definable using parameters from A. $S_n(A)$ will denote the Stone space over $F_n(A)$. The Morley rank and degree of a type p from $S_n(A)$ and of a formula φ from $F_n(A)$ will be denoted $R(p)$, $Dg(p)$, $R(\varphi)$, $Dg(\varphi)$, respectively.

For a structure $\mathfrak{M}$ and $\varphi \in F_n(A)$ we define

$$\varphi(\mathfrak{M}) = \{\bar{a} \in M^n : \mathfrak{M} \models \varphi(\bar{a})\}$$

The rank and the degree are defined for subsets $\varphi(\mathfrak{M})$ of M^n accordance with our agreement to identify φ and $\varphi(\mathfrak{M})$.

If $\bar{a} \in M^n$, $A \subseteq M$ then $t(\bar{a},A)$ denotes the type from $S_n(A)$ which is realized by $\bar{a}$;

$$R(\bar{a},A) \text{ is } R(t(\bar{a},A)), \quad Dg(\bar{a},A) = Dg(t(\bar{a},A)).$$

If $a = \langle \bar{a}_1,\ldots,\bar{a}_n \rangle$ then $|a| = \{\bar{a}_1,\ldots,\bar{a}_n\}$

Now we recall the results which we use as main technical tools of the paper.

In what follows $\mathfrak{M}$ is a model of an $\aleph_1$-categorical theory.

Fact 1.1. (Baldwin [1], Zilber [7]). $R(\varphi)$ is finite for every φ from $F_n(M)$.

This theorem was first obtained by Baldwin, [7] contains an independent proof by a method of stratifications.

Definition. We say that a formula Ψ from $F_2(M)$ is a stratification of a set $\varphi(\mathfrak{M})$ over a set $\delta(\mathfrak{M})$ (φ, $\delta \in F_1(M)$) if the following holds:

$$\mathfrak{M} \models (\Psi(v_o,v_1) \to \delta(v_o)) \,\&\, ((\exists v_o)\Psi(v_o,v_1) \leftrightarrow \varphi(v_1)).$$

Subsets of $\varphi(\mathfrak{M})$ of the form $\Psi(a,\mathfrak{M})$, where $a \in \delta(\mathfrak{M})$, are called strata. The natural number

$$R_{str}(\Psi) = \max\{R(\Psi(a,\mathfrak{M})) : a \in \delta(\mathfrak{M})\}$$

is called the rank of the stratification Ψ.

If the rank of each stratum is equal to $R_{str}(\Psi)$ and

$$\mathfrak{M} \models \Psi(v_o,v_1) \,\&\, \Psi(v_o', v_1) \leftrightarrow v_o = v_o'$$

then we say that Ψ is an exact stratification.

Fact 1.2. (Zilber [7]). If $R(\varphi) > 0$, $R(\delta) > 0$, then there exists a stratification Ψ of $\varphi(\mathfrak{M})$ on $\delta(\mathfrak{M})$ with $R_{str}(\Psi) \leq R(\varphi) - 1$.

It is not difficult to see that 1.2 is another version of the following result by Shelah.

Fact 1.2" (Shelah [6], Ch.V. 6.1) If $R(\delta) > 0$ then there does not exist any infinite set in $\mathfrak{M}$ which is indiscernible over $\delta(\mathfrak{M})$.

Fact 1.3 (Zilber [7]). If Ψ is a stratification of $\varphi(\mathfrak{M})$ on $\delta(\mathfrak{M})$ then

(a) $R(\varphi) \leq R(\delta) + R_{str}(\Psi)$

if Ψ is an exact stratification then

(b) $R(\varphi) = R(\delta) + R_{str}(\Psi)$

Fact 1.4. (Baldwin [1], Shelah [6], Zilber [7]). Let $\Psi(v_o,v) \in F_{n+1}$. For every natural number m there exists a formula $\vartheta_{\Psi,m}(v) \in F_n$ such that for any $\bar{a} \in M^n$

$$R(\Psi(v_o,a)) \leq m \text{ iff } \mathfrak{M} \models \vartheta_{\Psi,m}(\bar{a}).$$

Definition. Let $\varphi, \Psi \in F_1(M)$, we say that $\varphi(\mathfrak{M})$ almost includes $\Psi(\mathfrak{M})$ if $R(\Psi(\mathfrak{M}) - \varphi(\mathfrak{M})) < R(\Psi(\mathfrak{M})$, and denote the fact by

$\varphi(\mathfrak{M}) \sqsupset \Psi(\mathfrak{M})$

$\varphi(\mathfrak{M}) \sqsupset\!\sqsubset \Psi(\mathfrak{M})$ denotes that $\varphi(\mathfrak{M}) \sqsupset \Psi(\mathfrak{M})$ & $\Psi(\mathfrak{M}) \sqsupset \varphi(\mathfrak{M})$.

It follows from Fact 1.1 and Fact 1.4 that for any two formulas $\varphi(v,v_o)$, $\Psi(v,v_o) \varepsilon F_{n+1}$ there exists a formula $\vartheta \varepsilon F_n$ such that for any $\bar{a} \varepsilon M^{n^o}$

$$\varphi(\bar{a},\mathfrak{M}) \sqsupset \Psi(\bar{a},\mathfrak{M}) \text{ iff } \mathfrak{M} \models \vartheta(\bar{a}) .$$

Proposition 1.5. Let $\varphi \varepsilon F_1(M)$, $\Psi \varepsilon F_{n+1}(M)$, $\gamma \varepsilon F_n(M)$. If for every finite subset A of $\varphi(\mathfrak{M})$ there exists $\bar{a} \varepsilon \gamma(\mathfrak{M})$ such that

$$\Psi(\bar{a},\mathfrak{M}) \supseteq A$$

then there exists a $\bar{b}$ from $\gamma(\mathfrak{M})$ such that

$$\Psi(\bar{b},\mathfrak{M}) \sqsupset \varphi(\mathfrak{M}).$$

Proof. By induction on $R(\varphi)$. If $R(\varphi) = 0$ then $\varphi(\mathfrak{M})$ is finite, and the conclusion is obvious.

Let φ be strongly minimal. By the assumption of the proposition and by trivial compactness arguments in some elementary extension $\mathfrak{M}'$ of $\mathfrak{M}$ we can find $\bar{b} \varepsilon \gamma(\mathfrak{M}')$ such that

$$\text{card}(\Psi(\bar{b},\mathfrak{M}') \quad \varphi(\mathfrak{M}')) \geq \aleph_0$$

Using Fact 1.4, we can choose such $\bar{b}$ from $\gamma(\mathfrak{M})$. Since φ is strongly minimal

$$\text{card}(\varphi(\mathfrak{M}) - \Psi(\bar{b},\mathfrak{M})) < \aleph_0$$

this implies that

$$\Psi(\bar{b},\mathfrak{M}) \sqsupset \varphi(\mathfrak{M}) .$$

Now let $R(\varphi) = m+1$. According to Fact 1.2 there exists a stratification χ of $\varphi(\mathfrak{M})$ over a strongly minimal set $\delta(\mathfrak{M})$ ($\delta \varepsilon F_1(M)$) with $R_{str}(\chi) \leq m$. From the induction hypothesis for every finite subset A of $\delta(\mathfrak{M})$ we have an element $\bar{b}_A \varepsilon \gamma(\mathfrak{M})$ such that

$$\Psi(\bar{b}_A,\mathfrak{M}) \sqsupset \bigcup_{a \varepsilon A} \chi(a,\mathfrak{M}).$$

Let us define $\Psi^o \varepsilon F_{n+1}(M)$, using Fact 1.4:

$$\mathfrak{M} \models \Psi^o(\bar{b},a) \text{ iff } R(\chi(a,\mathfrak{M}) - \Psi(\bar{b},\mathfrak{M})) \leq m \dot{-} 1 .$$

It follows from the definition that for every finite $A \subseteq \delta(\mathfrak{M})$

$$\Psi^o(\bar{b},\mathfrak{M}) \sqsupset \delta(\mathfrak{M})$$

thus $\delta(\mathfrak{M}) - \Psi^o(b,\mathfrak{M})$ is finite. Now using Fact 1.3(a), we can compute, that

$$R(\varphi(\mathfrak{M}) - \Psi(b,\mathfrak{M})) \leq m$$

and the proof is finished.

Let us say that a set X is invariant under a mapping f if $f(X) \subseteq X$.

The following proposition holds without the assumption of $\aleph_1$-categoricity. Let us only note that any uncountable model of $\aleph_1$-categorical theory is saturated (and thus homogeneous.)

Proposition 1.6. If $\mathfrak{M}$ is a λ-saturated and λ-homogeneous structure, A, B $\subseteq$ M, card(B) < λ, $\varphi \in F_k(A)$ and $\varphi(\mathfrak{M})$ is invariant under all automorphisms of $\mathfrak{M}$ which are elementary over B then there exists a formula φ^* from $F_k(B)$ such that $\varphi^*(\mathfrak{M}) = \varphi(\mathfrak{M})$.

Proof of the proposition is a standard application of compactness arguments.

Fact 1.7 (Baur, Cherlin, Macintyre, [3]). If the theory of a group $\mathcal{G}$ is totally transcendental and $\aleph_0$-categorical theh there exists a normal abelian subgroup $\mathcal{H}$ of $\mathcal{G}$ such that $\mathcal{H}$ is definable in $\mathcal{G}$ without parameters and the factor group $\mathcal{G}/\mathcal{H}$ is finite.

Except the results given above we use some techniques of Lascar [4], though for the case of $\aleph_1$-categorical theory it may be deduced from Facts 1.1 - 1.4, as was done in [8]. We will also often use the fact that any model of an $\aleph_1$-categorical theory is prime and atomic over every infinite subsets, of the model which is definable using parameters (see [2]).

2. Definability of structures.

Definition. Let $\mathfrak{M}$ and $\mathfrak{N}$ be structures, A $\subseteq$ M, and σ a mapping from a subset of M^k onto N (for some natural number k).

We say that σ is an A-interpretation of the structure $\mathfrak{N}$ in the structure $\mathfrak{M}$ if the following holds:

The domain D_σ of σ is a subset of M^k definable using parameters from A ;

the preimages of the equality relation and all predicates $P(\mathfrak{N})$ for $P \in \Omega_{\mathfrak{N}}$ are definable in $\mathfrak{M}$ using parameters from A.

We say that $\mathfrak{N}$ is definable in $\mathfrak{M}$ using parameters from A if there exists an A-interpretation of $\mathfrak{N}$ in $\mathfrak{M}$.

We will omit A, if A = M.

Lemma 2.1. Let σ be an A-interpretation of a structure $\mathfrak{N}$ in a structure $\mathfrak{M}$ then:

(a) σ-preimages of all predicates definable in $\mathfrak{N}$ without parameters are definable in $\mathfrak{M}$ using parameters from A;

(b) if $\mathfrak{M}$ is λ-saturated and $\lambda > \mathrm{card}(A)$ then $\mathfrak{N}$ is λ-saturated;

(c) if a structure $\mathfrak{L}$ is definable using parameters in $\mathfrak{N}$ then $\mathfrak{L}$ is definable using parameters in $\mathfrak{M}$.

Proof.(a) We show by indusction on the complexity of a formula $\varphi(v_1,\ldots v_m) \in F_m(\Omega_{\mathfrak{N}})$ that there exists a formula $\tilde{\varphi}(\overline{w}_1,\ldots,\overline{w}_m) \in F_{mk}(\Omega\,(A))$ such that for any $\overline{a}_1,\ldots,\overline{a}_m \in M^k$

$$\mathfrak{M} \models \tilde{\varphi}(\overline{a}_1,\ldots,\overline{a}_m) \quad \text{iff } \overline{a}_1,\ldots,\overline{a}_m \in D_\sigma \quad \text{and } \mathfrak{N} \models \varphi(\sigma(\overline{\sigma}_1),,,.\ (\overline{a}_m))$$

For atomic formulas it follows from the definition. Assume φ and Ψ exist and $D_\sigma = \delta(\mathfrak{M})$ for some $\delta \in F_k(\Omega_{\mathfrak{M}}(A))$. Then it is easy to see that we can put

$$\widetilde{\neg\varphi} = \delta(\overline{w}_i)\,\&\ldots\&\,\delta(\overline{w}_n)\,\&\,\neg\tilde{\varphi}; \quad \widetilde{\varphi\,\&\,\Psi} = \tilde{\varphi}\,\&\,\tilde{\Psi};$$

$$\widetilde{(\exists v_i)\varphi} = (\exists \overline{w}_i)\ \delta(\overline{w}_i)\,\&\,\tilde{\varphi}$$

(b) Follows immediately from (a)

(c) If ρ is an interpretation of $\mathfrak{L}$ in $\mathfrak{N}$ then the mapping τ defined as

$$\tau(\overline{a}_1,\ldots,\overline{a}_m) = \rho(\sigma(\overline{a}_1),\ldots,\sigma(\overline{a}_m))$$

is, by (a), an interpretation of $\mathfrak{L}$ in $\mathfrak{M}$.

Definition. Let σ be an interpretation of a structure $\mathfrak{N}$ in a structure $\mathfrak{M}$ and $\Omega_{\mathfrak{M}} \cap \Omega_{\mathfrak{N}} = \emptyset$, $M \cap N = \emptyset$. A join of $\mathfrak{M}$ with $\mathfrak{N}$ by σ is a structure denoted by $\mathfrak{M}^\sigma\mathfrak{N}$ such that:

the signature of $\mathfrak{M}^\sigma\mathfrak{N}$ is $\Omega_{\mathfrak{M}} \cup \Omega_{\mathfrak{N}} \cup \{P_M, P_N, P_\sigma\}$ where P_M, P_N are symbols of unary predicates and P_σ is a symbol of k+1-ary predicate;

the universum of $\mathfrak{M}^\sigma\mathfrak{N}$ is $M \cup N$; the symbols are interpreted as follows:

$$P_M(\mathfrak{M}^\sigma\mathfrak{N}) = M; \qquad P_N(\mathfrak{M}^\sigma\mathfrak{N}) = N$$

$$P(\mathfrak{M}^\sigma\mathfrak{N}) = P(\mathfrak{M}) \quad \text{for} \quad P \in \Omega_{\mathfrak{M}};$$

$$P(\mathfrak{M}^\sigma\mathfrak{N}) = P(\mathfrak{N}) \quad \text{for} \quad P \in \Omega_{\mathfrak{N}}$$

$$P_\sigma(\mathfrak{M}^\sigma\mathfrak{N}) = \mathrm{graph}(\sigma).$$

If $\Omega_{\mathfrak{N}}$ is empty, i.e. $\mathfrak{N}$ is trivial then instead of $\mathfrak{M}^\sigma\mathfrak{N}$ we write $\mathfrak{M}^\sigma$. In this case the construction of $\mathfrak{M}^\sigma$ is determined by the equivalence relation ε such that

$$\mathfrak{M} \models \varepsilon(\bar{a},\bar{b}) \text{ iff } \sigma(\bar{a}) = \sigma(\bar{b})$$

for any $a, b \in D_\sigma$. Then we can put $N = D\sigma/\varepsilon$ and $\sigma : D\sigma \to D_\sigma/\varepsilon$.

Lemma 2.2. (a) If $\mathfrak{M}^\sigma\mathfrak{N}$ is elementary equivalent to a structure $\mathfrak{L}$ then $\mathfrak{L} \cong \mathfrak{M}' \,^{\sigma'}\mathfrak{N}'$ for some structures $\mathfrak{M}'$ and $\mathfrak{N}'$ which are elementary equivalent to $\mathfrak{M}$ and $\mathfrak{N}$, respectively, and for some interpretation σ' of $\mathfrak{N}'$ in $\mathfrak{M}'$.

(b) $\mathfrak{M}^\sigma\mathfrak{N}$ is definable in $\mathfrak{M}$ using parameters from A, provided M contains two distinct elements and σ is an A-interpretation. Any subset Φ of M^m, which is definable in $\mathfrak{M}^\sigma\mathfrak{N}$, using parameters from $C \subseteq M$, is definable in $\mathfrak{M}$ using parameters from $A \cup C$.

If Φ is definable in $\mathfrak{M}^\sigma\mathfrak{N}$ using parameters then $\Phi \in F_m(\Omega_{\mathfrak{M}}(M))$.

(c) If $\mathfrak{M}$ is λ-categorical and $\lambda > \mathrm{card}(\Omega_{\mathfrak{N}})$ the so is $\mathfrak{M}^\sigma\mathfrak{N}$.

Proof. (a) Follows from the definition.

(b) Define a mapping ρ on a set

$$D_\rho = M^3 \times D_\sigma$$

in the following way

$$\rho(x_1,x_2,y,\bar{z}) = \begin{cases} y, & \text{if } x_1 = x_2 ; \\ \sigma(\bar{z}), & \text{if } x_1 \neq x_2 . \end{cases}$$

$$(x_1,x_2,y \in M, \quad \bar{z} \in D_\sigma)$$

It is easy to check, that ρ is an A-interpretation of $\mathfrak{M}^\sigma\mathfrak{N}$ in $\mathfrak{M}$.

Now let $\Psi \subseteq M^n$ be a set definable in $\mathfrak{M}^\sigma\mathfrak{N}$ without parameters. By the definition of ρ the ρ-preimage of Ψ is equal to

$$(*) \; \{ \langle\langle x_1,x_1,y_1,\bar{z}_1 \rangle ,\dots, \langle x_n,x_n,y_n,\bar{z}_n \rangle\rangle : x_1,\dots,x_n \in M;$$
$$\langle y_1,\dots,y_n \rangle \in \Psi ; \; \bar{z}_1,\dots,\bar{z}_n \in D_\sigma \}$$

By 2.1 (a) this predicate is definable in $\mathfrak{M}$ using parameters from A. Clearly, the projection of this predicate on $y_1,\dots,y_n$ is equal to Ψ. Thus we have proved that for every formula $\Psi \in F_n(\Omega_{\mathfrak{M}^\sigma\mathfrak{N}})$ such that $\Psi(\mathfrak{M}^\sigma\mathfrak{N}) \subseteq M^n$ there exists $\tilde{\Psi} \in F_n(\Omega_{\mathfrak{M}}(A))$ such that for any $c_1,\dots,c_n \in M$

$$\mathfrak{M} \models \Psi(c_1,\dots,c_n) \text{ iff } \mathfrak{M}^\sigma\mathfrak{N} \models \Psi(c_1,\dots,c_n)$$

Now take $n \geq m$, $c_1,\dots,c_{n-m} \in C$, $\Psi \in F_n(\Omega_{\mathfrak{M}^\sigma\mathfrak{N}})$ such that

$$\Psi(c_1,\dots,c_{n-m}, \mathfrak{M}^\sigma\mathfrak{N}) = \Phi \quad ,$$

it follows $\Phi \in F_m(\Omega_{\mathfrak{M}}((C))$.

Finally, if Φ is definable in $\mathfrak{M}^\sigma\mathfrak{N}$ using some parameters from $M \cup N$, then the ρ-preimage of Φ(see (*) with Φ instead of Ψ)

is definable in $\mathfrak{M}$ using some parameters from M, its projection on $y_1,\dots,y_m$ is equal to Φ.

(c) The condition of λ-categoricity of a complete theory T, as is well-known [6], is equivalent to the condition of saturation of every model of T of cardinality λ. If $\Omega_{\mathfrak{N}}$ is finite then we can choose A so, that card(A) $< \mathrm{card}(\Omega_{\mathfrak{N}}) + \aleph_0$, if $\Omega_{\mathfrak{N}}$ is infinite, then card (A) = card($\Omega_{\mathfrak{N}}$). It follows from 2.1(b) and from (a),(b) of the present lemma, that every model of Th($\mathfrak{M}^{\sigma}\mathfrak{N}$) of cardinality λ is λ- saturated.

If a structure $\mathfrak{M}^{\sigma}\mathfrak{N}$ is considered then subsets of $\mathfrak{M}$ and $\mathfrak{N}$ definable using parameters are regarded as subsets of $\mathfrak{M}^{\sigma}\mathfrak{N}$. By Morley rank of these subsets we mean the Morley rank $R_{\mathfrak{M}^{\sigma}\mathfrak{N}}$ in $\mathfrak{M}^{\sigma}\mathfrak{N}$ It is obvious that

$$R_{\mathfrak{M}^{\sigma}\mathfrak{N}} \geqslant R_{\mathfrak{N}}$$

But for subsets of $\mathfrak{M}$ we have:

<u>Lemma 2.3</u> If Φ is a subset of M definable using parameters then

$$R_{\mathfrak{M}}(\Phi) = R_{\mathfrak{M}^{\sigma}\mathfrak{N}}(\Phi)$$

<u>Proof.</u> It is well-known (see, e.g. [7]) that Morley rank of Φ for $\aleph_0$-saturated structures can be defined as a Boolean invariant of the Boolean algebra of subsets of Φ which are definable using parameters. Taking an elementary extension of $\mathfrak{M}^{\sigma}\mathfrak{N}$ we may by 2.2(a) assume that we have an $\aleph_0$-saturated structure and then by 2.1(a) and 2.2(b) the mentioned Boolean algebras in $\mathfrak{M}$ and $\mathfrak{M}^{\sigma}\mathfrak{N}$ are equal, so the ranks coincide.

3. Definability of an equivalence relation.

In what follows we assume that $\mathfrak{M}$ is a saturated model of an $\aleph_1$-categorical theory.

<u>Definition</u>. (Lascar [4]). Let A,B,C $\subseteq$ M. Subsets A and B are said to be independent over C if for every finite sequence $\bar{a}$ of elements of A

$$R(\bar{a}, B \cup C) = R(\bar{a}, C)$$

Sometimes we say also that A is independent with B over C.

It is proved by Lascar that independence is a symmetric relation

<u>Lemma 3.1</u> Let p be an n-type over C. If A and B_i are independent over C for all i ($1 \leqslant i \leqslant k$) then there exists $\bar{b}$ realizing p such that A and $|\bar{b}| \cup B_i$ are independent over C for all i ($1 \leqslant i \leqslant k$).

Proof. Choose $p' \in S(A \cup B_1 \cup \ldots \cup B_k \cup C)$ extending p, such that $R(p') = R(p)$. Let $\overline{b}$ be a sequence realizing p', then

$$R(\overline{b}, A \cup B_i \cup C) = R(\overline{b}, \quad B_i \cup C)$$

By the reciprocity principle of Lascar [4] we have

$$R(\overline{a}, |\overline{b}| \cup B_i \cup C) = R(\overline{a}, B_i \cup C)$$

for every finite sequence $\overline{a}$ of elements of A. Since

$$R(\overline{a}, B_i \cup C) = R(\overline{a}, C)$$

the following holds:

$$R(\overline{a}, |\overline{b}| \cup B_i \cup C) = R(\overline{a}, C)$$

this finishes the proof.

Definition. Let $C \subsetneq M$, $\varphi, \delta \in F_1(C)$. The function $\rho(\varphi,\delta,C)$ whose values are pairs of natural numbers is defined by:

$$\rho(\varphi,\delta,C) = \min \{ < R(x, Y \cup C \cup \delta(\mathfrak{M})), D_g(x, Y \cup C \cup \delta(\mathfrak{M})) > :$$

$\{x\} \cup Y \subseteq \varphi(\mathfrak{M})$; x and Y are independent over C $\}$ where the minimum is taken according to the lexicographic order of pairs of natural numbers.

It is easy to see that we can assume that Y ranges over finite subsets of $\varphi(\mathfrak{M})$ in the definition.
By $\rho^R(\varphi,\delta,C)$ we denote the first coordinate of $p(\varphi,\delta,C)$.

Lemma 3.2. If $\varphi(\mathfrak{M})$ and $\delta(\mathfrak{M})$ are infinite then

$$\rho^R(\varphi,\delta,C) < R(\varphi) .$$

Proof. Othervise we can construct an infinite sequence $\{y_i / i < \omega\}$ of elements of $\varphi(\mathfrak{M})$ such that

$$R(y_{i+1}, \{y_j \mid j \leq i\} \cup C \cup \delta(\mathfrak{M})) = R(\varphi)$$

Hence there is an infinite subset of M which is indiscernible over $\delta(\mathfrak{M})$ ([6]), contradicting Fact 1.2′.

Lemma 3.3. If φ is an atom of $F_1(C)$ then there exists a finite subset A of $\varphi(\mathfrak{M})$ such that for any element a of which is independent with A over C the following holds:

$$(*) < R(a, A \cup \delta(\mathfrak{M}) \cup C), \quad D_g(a, A \cup \delta(\mathfrak{M}) \cup C) > \neq \rho(\varphi,\delta,C).$$

Proof. Let $p_1, \ldots, p_m$ be all the types from $S_1(\varphi(\mathfrak{M}) \cup C)$ which contain φ and have rank equal to $R(\varphi)$. There exists a finite subset A_0 of $\varphi(\mathfrak{M})$ such that for every $i \leq m$ the retriction

$p_i \mid_{A_o \cup C}$ has p_i as the only extension in $S_1(\varphi(\mathfrak{M}) \cup C)$ of rank $R(\varphi)$. Let a_i realize the types $p_i|_{A_o \cup C}$ ($1 \leq i \leq m$).

Since φ is an atom over C, for every a_i there exists a finite subset A_i of $\varphi(\mathfrak{M})$ such that

$$< R(a_i, A_i \cup \delta(\mathfrak{M}) \cup C), \ D_g(a_i, A_i \cup \delta(\mathfrak{M}) \cup C)> \ = \rho(\varphi, \delta, C) .$$

Observe that by definition of ρ the equality still holds if we substitute A instead of A_i, provided $A_i \subseteq A \subseteq \varphi(\mathfrak{M})$ and A is independent with $\{a_i\}$ over C.

Now put $A = A_o \cup \ldots \cup A_m$. Let $\{a\}$ be independent with A over C, $a \in \varphi(\mathfrak{M})$. Then

$$R(a, A \cup C) = R(a, C) = R(\varphi)$$

and therefore $t(a, A \cup C)$ is an extension of $p_i/_{A_o \cup C}$ for some $i \leq m$. Hence the equality (*) holds for this a.

In the sequel we fix $\varphi, \delta, \in F_1(C)$, C is finite and φ is an atom of $F_1(C)$. Let $A = \{a_1, \ldots, a_n\}$ be a minimal subset of $\varphi(\mathfrak{M})$ satisfying the assertion of Lemma 3.3.

Let us fix the notation t_o for $t(< a_1, \ldots, a_n >, C)$. Note that if $\bar{a}$ is any sequence realizing t_o in $\mathfrak{M}$ then the set $A = |\bar{a}|$ satisfies the assertion of 3.3 since $\delta(\mathfrak{M})$ is definable over C.

<u>Definition.</u> Let $A \subseteq M$, $a \in \varphi(\mathfrak{M})$. Denote $[a, A] = \{x \in M : t(x, A \cup C \cup \delta(\mathfrak{M})) = t(a, A \cup C \cup \delta(\mathfrak{M}))\}$.

<u>Lemma 3.4.</u>

(a) $[a, A]$ is definable using $A \cup C \cup \delta(\mathfrak{M})$;

(b) $[a, A]$ is definable using $A \cup C \cup \{a\}$;

(c) $< R([a,A]), D_g([a,A]) > = \rho(\varphi, \delta, C)$, provided $\{a\}$ and A are independent over C and $A = |\bar{a}|$, $\bar{a}$ realizes t_o.

<u>Proof.</u> (a) follows from the fact that $\mathfrak{M}$ is atomic over $\delta(\mathfrak{M}) \cup A \cup C$;

(b) follows from (a) and Proposition 1.6;

(c) is a consequence of (a) and the definition of t_o.

<u>Lemma 3.5.</u> Let $\bar{a}$ and $\bar{a}'$ realize t_o, $b, b' \in \varphi(\mathfrak{M})$ and let $\{b\}$ and $|\bar{a}|$ as well as, $\{b'\}$ and $|\bar{a}'|$ be independent over C. If $[b, |\bar{a}|] \cap [b', |\bar{a}'|] \neq \emptyset$ then $[b, |\bar{a}|] \sqsupset [b' |\bar{a}'|]$.

<u>Proof.</u> Choose an element b" in $[b, |\bar{a}|] \cap [b', |\bar{a}'|]$. $\{b''\}$ and $|\bar{a}|$ as well as $\{b''\}$ and $|\bar{a}'|$ are independent over C since

$$t(b^\frown\bar{a}, C) = t(b''^\frown\bar{a}, C), \quad t(b'^\frown\bar{a}', C) = t(b''^\frown\bar{a}', C).$$

Choose $\bar{a}''$ using Lemma 3.1, such that $\{b''\}$ and $|\bar{a}| \cup |\bar{a}''|$ as well as, $\{b''\}$ and $|\bar{a}'| \cup |\bar{a}''|$ are independent over C and $t(a'', C) = t_0$. By 3.4 (a) $[b'', |\bar{a}|] \cap [b'', |\bar{a}''|]$ is definable using $|\bar{a}| \cup |\bar{a}''| \cup C \cup \delta(\mathfrak{M})$ hence, by the definition of $\rho(\varphi, \delta, C)$.

$$< R([b'', |\bar{a}|] \cap [b'', |\bar{a}''|]), Dg([b'', |\bar{a}|] \cap [b'', |\bar{a}''|]) \geq$$
$$\geq \rho(\varphi, \delta, C).$$

This, combined with 3.4 (c), gives

$$[b'', |\bar{a}|] \sqsubset [b'', |\bar{a}''|].$$

Similarly

$$[b'', |\bar{a}'|] \sqsubset [b'', |\bar{a}''|].$$

By transitivity $[b'', |\bar{a}|] \sqsubset [b'', |\bar{a}'|]$. Since

$$[b'', |\bar{a}|] = [b, |\bar{a}|], \quad [b'', |\bar{a}'|] = [b', |\bar{a}'|]$$

the proof is finished.

<u>Definition</u>. A binary relation $\varepsilon_{\varphi,\delta,C}$ on $\varphi(\mathfrak{M})$ is defined as follows:

for any two elements b, b′ from $\varphi(\mathfrak{M})$ $\varepsilon_{\varphi,\delta,C}(b, b')$ holds iff there exists $\bar{a}$ realizing t_0 in $\mathfrak{M}$ such that $\{b\}$ and $|\bar{a}|$ as well as $\{b'\} \cup |\bar{a}|$ are independent over C and

$$[b, |a|] = [b', |a|].$$

<u>Lemma 3.6.</u> $\varepsilon_{\varphi,\delta,C}$ is an equivalence relation on $\varphi(\mathfrak{M})$. $\varepsilon_{\varphi,\delta,C}$ is definable using parameters from C.

<u>Proof.</u> First, let us in the proof abbreviate the notation $\varepsilon_{\varphi,\delta,C}$ to ε.

It is obvious that ε is reflective and symmetric. Let us show transitivity.

Let $\varepsilon(b, b')$ and $\varepsilon(b', b'')$ hold. This means that for some $\bar{a}$ and $\bar{a}'$ realizing t_0

$$[b, |\bar{a}|] = [b', |\bar{a}|], \quad [b', |\bar{a}'|] = [b'', |\bar{a}'|]$$

and all the pairs of sets

$$\{b\} \text{ and } |\bar{a}|, \ \{b'\} \text{ and } |\bar{a}|; \ \{b'\} \text{ and } |\bar{a}'|; \ \{b''\} \text{ and } |\bar{a}'|$$

are independent over C. Choose an a'', realizing t_0, such that $\{b\}$ and $|\bar{a}''|$, $\{b'\}$ and $|\bar{a}''|$, $\{b''\}$ and $|\bar{a}''|$ are independent over C. Then by Lemma 3.5

$$[b, |\bar{a}''|] \sqsubset [b, |\bar{a}|], \quad [b'', |\bar{a}''|] \sqsubset [b'', |\bar{a}'|] .$$

Since $b \in [b, |\bar{a}|] \cap [b'', |\bar{a}'|]$, we have by 3.5

$$[b, |\bar{a}|] \sqsubset [b'', |\bar{a}'|] .$$

Hence $[b, |\bar{a}''|] \cap [b'', |\bar{a}''|] \neq \emptyset$, this is possible only if $[b, |\bar{a}''|] = [b'', |\bar{a}''|]$, thus $\varepsilon(b, b'')$ holds and transitivity is proved.

Now let us prove that ε is definable using C. Assume $\{b\}$ and $|\bar{a}|$ are independent over C, $b \in \varphi(\mathfrak{M})$, $\bar{a}$ realizes t_0. Since $\mathfrak{M}$ is atomic over $\delta(\mathfrak{M}) \cup C$ there exists an atom $\Psi(\bar{v}, u)$ of $F_{k+1}(\delta(\mathfrak{M}) \cup C)$ such that

$$\mathfrak{M} \models \Psi(\bar{a}, b) .$$

Let

$$\chi(v_0, v_1) = (\exists u)(\exists \bar{w})(\exists \bar{v})(\Psi(\bar{w}, v_0) \& \Psi(\bar{v}, v_1) \& \Psi(\bar{w}, u) \& \Psi(\bar{v}, u))$$

This formula χ defines ε . For, if $\mathfrak{M} \models \chi(b', b'')$ then for some u $\varepsilon(b', u)$ and $\varepsilon(b'', u)$ hold, thus $\varepsilon(b', b'')$ is true. Conversely, let $\varepsilon(b', b'')$ hold. Since φ is an atom over C and $\mathfrak{M} \models (\exists \bar{w})\Psi(\bar{w}, b)$ there exists $\bar{a}'$ and $\bar{a}''$ such that

$$\mathfrak{M} \models \Psi(\bar{a}', b') \& \Psi(\bar{a}'', b'') .$$

From $\varepsilon(b', b'')$ and 3.5 we have

$$[b', |\bar{a}'|] \sqsubset [b', |\bar{a}|] = [b'', |\bar{a}|] \sqsubset [b'', |\bar{a}''|]$$

for some $\bar{a}$. Hence

$$[b', |\bar{a}'|] \quad [b'', \bar{a}''] \neq \emptyset \quad \text{i.e.} \ \mathfrak{M} \models (\exists u)\ \Psi(\bar{a}', u) \& \Psi(\bar{a}'', u)$$

and thus $\mathfrak{M} \models \chi(b', b'')$.

We have proved that ε is defined by the formula χ from $F_2(\delta(\mathfrak{M}) \cup C)$. To prove that ε is definable using parameters from C only it suffices, by Proposition 1.6, to observe that ε is invariant under all automorphisms of $\mathfrak{M}$ elementary over C. This is immediate from the definition of ε .

<u>Lemma 3.7.</u> For every b from $\varphi(\mathfrak{M})$

(*) $\langle R(\varepsilon(b, \mathfrak{M})), Dg(\varepsilon(b, \mathfrak{M})) \rangle \ \rho(\varphi, \delta, C)$, i.e. the rank - degree pair of each of the equivalence classes of $\varepsilon_{\varphi,\delta,C}$ is equal to $\rho(\varphi, \delta, C)$.

<u>Proof.</u> Since $[b, |a|] \subseteq \varepsilon(b, \mathfrak{M})$ for some $|\bar{a}|$, satisfying the assumption of 3.4(c), the left hand side of (*) is not less than the right one.

To prove the converse inequality consider an arbitrary finite subset $\{b_1,\dots,b_k\}$ of $\varepsilon(b,\mathfrak{M})$. By the definition of ε there exist $\bar{a}_1,\dots,\bar{a}_k$ realizing t_0 such that for every $i \leqslant k$ $\{b_i\}$ and $|\bar{a}_i|$ are independent over C and $[b_i,|\bar{a}_i|] = [b,|\bar{a}_i|]$. Choose $\bar{a}$ realizing to such that $\{b, b_1,\dots,b_k\}$ and $|\bar{a}|$ are independent over C. By 3.5

$$[b_i,|\bar{a}|]\sqsubset[b_i,|\bar{a}_i|] = [b,|\bar{a}_i|]\sqsubset[b,|\bar{a}|].$$

Hence

$$[b_i,|\bar{a}|]\cap[b,|\bar{a}|] \neq \emptyset, \text{ i.e. } [b_i,|\bar{a}|] = [b,|\bar{a}|], \text{ thus}$$

$$\{b_i,\dots,b_k\}\subseteq[b,|a|] .$$

Let t_1 be an extension of t_o in $S_n(\{b\}\cup C)$, $R(t_1) = R(t_o)$. Without loss of generality we can assume that $\bar{a}$ reslizes t_1 since the only requirement we have for $\bar{a}$ is

$$R(a, \{b,b_1,\dots,b_k\}\cup C) = R(t_o) \text{ and } t(\bar{a},C) = t_o.$$

Now using 1.5 and taking into account that $\mathfrak{M}$ is saturated we can find an $\bar{a}$ realizing t_1 such that

$$[b,|\bar{a}|]\sqsupset\varepsilon(b,\mathfrak{M})$$

This $\bar{a}$ satisfies 3.4(c), therefore the right hand side of (*) is not less than the left one.

<u>Lemma 3.8.</u> Let ε be an arbitrary equivalence relation on $\varphi(\mathfrak{M})$ which is definable using C. Let σ be the natural mapping

$$\sigma : \varphi(\mathfrak{M}) \dashrightarrow \varphi(\mathfrak{M})/\varepsilon$$

Then in the structure $\mathfrak{M}^\sigma$ the following hold :

(a) $\varphi(\mathfrak{M})/\varepsilon$ is definable without parameters and it is an atom of $F_1(\Omega_{\mathfrak{M}^\sigma}(C))$;

(b) for every $\hat{a}\ \varepsilon\ \varphi(\mathfrak{M})/\varepsilon$ the subset $\sigma^{-1}(\hat{a})$ of $\varphi(\mathfrak{M})$ is definable using $\hat{a}$ and it is an atom of $F_1(\Omega_{\mathfrak{M}^\sigma}(C\cup\{\hat{a}\}))$.

<u>Proof.</u> (a) Definability of $\varphi(\mathfrak{M})/\varepsilon$ is given by the definition. To prove that this set is an atom over C assume the contrary and use 2.2(b).

(b) Suppose $\sigma^{-1}(\hat{a})$ is not an atom over $C\cup\{\hat{a}\}$. Then there is a formula $\chi(v,u)\ \varepsilon\ F_2(\Omega_{\mathfrak{M}^\sigma}(C\cup\{\hat{a}\}))$ such that $\chi(\hat{a},u)$ and $\neg\chi(\hat{a},u)$ are consistent with $\sigma(u) = \hat{a}$

Thus the formula

$$\Psi(u) = (\exists v)(\chi(v,u)\,\&\,\sigma(u) = v)$$

is consistent with $\varphi(u)$ in $\mathfrak{M}^{\sigma}$, but if a realizes $\neg\chi(\hat{a},u) \& \sigma(u)=a$ then a can not realite $\Psi(u)$.
Hence

$$\Psi(\mathfrak{M}^{\sigma}) \cap \varphi(\mathfrak{M}) \text{ and } \neg\Psi(\mathfrak{M}^{\sigma}) \cap \varphi(\mathfrak{M})$$

are both non-empty, this contradicts, by 2.2(b), the fact that φ is an atom of $F_1(\Omega_{\mathfrak{M}}(C))$.

4. The binding group of definable subsets.

In this section we assume that $R(\delta) > 0$, $\rho(\varphi,\delta,C) = < 0,1 >$ for fixed φ,δ from $F_1(C)$ where φ is an atom of $F_1(\delta(\mathfrak{M}) \cup C)$, The type t_0, fixed in the previous section, has a principal complete extension in $S_n(\delta(\mathfrak{M}) \cup C)$ since $\mathfrak{M}$ is atomic over $\delta(\mathfrak{M}) \cup C$, let ϑ be an atom of $F_n(\delta(\mathfrak{M}) \cup C)$, which defines the principal complete extension of t_0. In other words, if $\bar{a}$ realizes ϑ then $A = |\bar{a}|$ satisfies the requirement of 3.3. We can extend C, if necessary, and assume that $\vartheta \in F_n(C)$.

Refining the definition of algebraic closure [2], we define for any subset X of M $cl_1(X) = \{a \in M :$ there is a $\gamma \in F_1(X)$ such that $\mathfrak{M} \models \gamma(a) \& (\exists!\, v)\gamma(v)\}$
Observe that

$$\mathfrak{M} \models (\exists!\, v)\gamma(v) \text{ iff } < R(\gamma), Dg(\gamma) > = < 0,1 >$$

and

$$cl_1(cl_1(X)) = cl_1(X) .$$

Lemma 4.1. For every $\bar{a} \in \vartheta(\mathfrak{M})$

$$\varphi(\mathfrak{M}) \subseteq cl\ (|\bar{a}| \cup C \cup \delta(\mathfrak{M})).$$

Proof. Since $\rho(\varphi,\delta,C) = < 0,1 >$ we have

$$b \in cl_1\ (|\bar{a}| \cup C\ \ \delta(\mathfrak{M}))$$

for every $b \in \varphi(\mathfrak{M})$ such that $\{b\}$ and $|\bar{a}|$ are idenpendent over C. Now let c be an arbitrary element of $\varphi(\mathfrak{M})$. Choose $\bar{b}$ from $\vartheta(\mathfrak{M})$ such that $|\bar{b}|$ and $\{c\} \cup |\bar{a}|$ are independent over C. As was noted already

$$c \in cl_1(|\bar{b}| \cup C \cup \delta(\mathfrak{M})) \text{ and } b \in cl_1(|\bar{a}| \cup C \cup \delta(\mathfrak{M}))$$

for every $b \in |\bar{b}|$. Hence $c \in cl_1(|a| \cup C \cup \delta(\ \ ))$.

Lemma 4.2. There exists a formula $\chi(\bar{v},\bar{w},\bar{u}) \in F_{2n+k}(C)$ for some k such that for any $a_1,a_2 \in \vartheta(\mathfrak{M})$ there is a sequence $\bar{b}$ of length k for which

$$(*) \quad \mathfrak{M} \models \chi(\bar{a}_1,\bar{b},\bar{a}_2) \& (\exists!\ \bar{u})\ \chi(\bar{a}_1,\bar{b},\bar{u}) .$$

Proof. Since $\vartheta(\mathfrak{M}) \subseteq (\varphi(\mathfrak{M}))^n$ it follows from the previous lemma that for fixed $\bar{a}_1$, and for every $\bar{a}_2 \in \vartheta(\mathfrak{M})$ we can find χ and $\bar{b}$ (both dependent on $\bar{a}_2$) such that (*) is satisfied. By Compactness Theorem we can find a finite set $\chi_1, \dots, \chi_m$ of such formulas, which are pairwise inconsistent and such that for every $\bar{a}_2 \in \vartheta(\mathfrak{M})$ there exist $i \leq m$ and $\bar{b}$ for which

$$\mathfrak{M} \models \chi_i(\bar{a}_1, \bar{b}, \bar{a}_2) \,\&\, (\exists! \bar{u})(\bar{a}_1, \bar{b}, \bar{u}).$$

Put $\chi = \chi_1 \vee \dots \vee \chi_m$. It is easy to see that the formula χ satisfies (*) for all $\bar{a}_1, \bar{a}_2$.

Definition. Let $X, Y \subseteq M$. We call the group of all monomorphisms of X onto itself which are elementary over Y the binding group of X over Y and denote it $g(X/Y)$.

We shall examine the group $g(\varphi(\mathfrak{M})/\delta(\mathfrak{M})\,(C)$, which we denote by $g(\varphi/\delta \cup C)$ or simply by g when ambiguity can occur.

Remark. Since $\vartheta(\mathfrak{M})$ consists of sequences of elements of $\varphi(\mathfrak{M})$ the group $g(\varphi/\delta \cup C)$ acts on $\vartheta(\mathfrak{M})$ as well as on $\varphi(\mathfrak{M})$. Since φ and ϑ are atoms of $F(\delta(\mathfrak{M}) \cup C)$ and $\mathfrak{M}$ is prime over $\delta(\mathfrak{M}) \cup C \cup \{x\}$ for any x from one of the sets, the group $g(\varphi/\delta \cup C)$ acts transitively on $\varphi(\mathfrak{M})$ and $\vartheta(\mathfrak{M})$ i.e. for any two elements x_1, x_2 of one of the sets there exists a $\mu \in g(\varphi/\delta \cup C)$ such that $\mu x_1 = x_2$, About $\vartheta(\mathfrak{M})$ more can be said. It follows from Lemma 4.1 that if $\mu_1 \bar{a} = \mu_2 \bar{a}$ for some $\mu_1, \mu_2 \in g(\varphi/\delta \cup C)$ and $\bar{a} \in \vartheta(\mathfrak{M})$ then $\mu_1 = \mu_2$. I.e. for any two $\bar{a}_1, \bar{a}_2$ from $\vartheta(\mathfrak{M})$ there exists a unique $\mu \in g(\varphi/\delta \cup C)$ such that $\bar{a}_1 = \bar{a}_2$.

Proposition 4.3. For every $\bar{a}$ from $\vartheta(\mathfrak{M})$ there exists an interpretation $\pi_{\bar{a}}$ of the binding group $g(\varphi/\delta \cup C)$ with the following properties

(a) $\pi_{\bar{a}}$ is defined using parameters from C, the domain D_π of $\pi_{\bar{a}}$ is a subset of $(\delta(\mathfrak{M}))^K$.

(b) In the structure $\mathfrak{M}^{\pi_{\bar{a}}}\, g(\varphi/\delta \cup C)$ the ternary relation $\mu x = y$ ($\mu \in g(\varphi/\delta \cup C)$; $x, y \in \varphi(\mathfrak{M})$) and the 2n+1-ary relation $\mu \bar{x} = \bar{y}$ ($\bar{x}, \bar{y} \in \vartheta(\mathfrak{M})$) are definable using parameters from $C \cup |\bar{a}|$. In other words the action of $g(\varphi/\delta \cup C)$ on $\varphi(\mathfrak{M})$ and $\vartheta(\mathfrak{M})$ is definable using $C \cup |\bar{a}|$.

(c) If $\bar{b} = \lambda \bar{a}$ for some $\lambda \in g(\varphi/\delta \cup C)$ then $\pi_{\bar{b}} = \lambda \cdot \pi_{\bar{a}} \cdot \lambda^{-1}$

Proof. Let χ be the formula taken from Lemma 4.2.

Define

$$D_\pi = \{\bar{b}\ \varepsilon (\delta(\mathfrak{M}))^K : \mathfrak{M} \models (\forall\bar{v})\vartheta(\bar{v}) \longrightarrow (\exists!\bar{w})\ \chi(\bar{v},\bar{b},\bar{w})\}$$

We have noted already that for any $\mu\ \varepsilon\ g$ there is $\bar{d}\ \varepsilon\ D_\pi$ such that

$$\mathfrak{M} \models \chi(\bar{a}, \bar{d}, \mu\,\bar{a})$$

and if $\bar{a}\ \varepsilon\ \vartheta(\mathfrak{M})$, $\bar{d}\ \varepsilon\ D_\pi$ are given then μ is determined in a unique way. Given such a μ let

$$\pi_{\bar{a}}(\bar{d}) = \mu .$$

It is obvious that

$$\pi_{\bar{a}}(\bar{d}_1) = \pi_{\bar{a}}(\bar{d}_2) \quad \text{iff } \mathfrak{M} \models (\forall\bar{w})(\chi(\bar{a},\bar{d}_1,\bar{w}) \longleftrightarrow \chi(\bar{a},\bar{d}_2,\bar{w}))$$

Taking into account that $\bar{a}\ \varepsilon\ \vartheta(\mathfrak{M})$, ϑ is an atom of $F_n(\delta(\mathfrak{M}) \cup C)$ and $\bar{d}_1$, $\bar{d}_2$ are sequences of elements of $\delta(\mathfrak{M})$ we have

$$\pi_{\bar{a}}(\bar{d}_1)=\pi_{\bar{a}}(\bar{d}_2) \text{ iff } \mathfrak{M} \models (\exists\bar{v})(\vartheta(v) \,\&\, (\forall\bar{w})(\chi(\bar{v},\bar{d}_1,\bar{w}) \longleftrightarrow \chi(\bar{v},\bar{d}_2,\bar{w}))) .$$

Hence it is proved that the $\pi_{\bar{a}}$ - preimage of the equality in g is definable using C. We shall show that the $\pi_{\bar{a}}$ - preimage of the ternary relation $\mu_1\,\mu_2 = \mu_3$ is also definable using C and then the proof of (a) will be finished.

Let $\bar{d}_1, \bar{d}_2, \bar{d}_3\ \varepsilon\ D_\pi$, $\pi_{\bar{a}}(\bar{d}_i) = \bar{\mu}_i$ (i = 1,2,3). Thus

$$\mathfrak{M} \models \chi(\bar{a}, \bar{d}_2, \mu_2\,\bar{a})$$

Since μ_i is elementary over $\delta(\mathfrak{M}) \cup C$,

$$\mathfrak{M} \models \chi(\mu_1\,\bar{a}, \bar{d}_2, \mu_1\,\mu_2\,\bar{a})$$

hence

$$\mu_1\ \mu_2 = \mu_3 \text{ iff } \mathfrak{M} \models (\exists\bar{w})(\exists\bar{v})(\chi(\bar{a},\bar{d}_1,\bar{w}) \,\&\, \chi(\bar{w},\bar{d}_2,\bar{v}) \,\&\, \chi(\bar{a},\bar{d}_3,\bar{v}))$$

Using once again the fact that $\bar{a}\ \varepsilon\ \vartheta(\mathfrak{M})$ and ϑ is an atom over $\delta(\mathfrak{M}) \cup C$, we finally have $\pi_{\bar{a}}(\bar{d}_1) \cdot \pi_{\bar{a}}(\bar{d}_2) = \pi_{\bar{a}}(\bar{d}_3)$ iff

$$\mathfrak{M} \models (\exists\bar{u})(\vartheta(\bar{u}) \,\&\, (\exists\bar{v})(\exists\bar{w})(\chi(\bar{u},\bar{d}_1,\bar{w}) \,\&\, \chi(\bar{w},\bar{d}_2,\bar{v}) \,\&\, \chi(\bar{u},\bar{d}_3,\bar{v})))$$

(b) By the definition of $\pi_{\bar{a}}$ the formula

$$(\exists\bar{w}\ \varepsilon\ D_\pi)(\pi_{\bar{a}}(\bar{w}) = \mu \,\&\, \chi(\bar{a}, \bar{w}, \bar{x}))$$

defines the relation

$$\mu\,\bar{a} = \bar{x}$$

for x $\varepsilon\ \vartheta(\mathfrak{M})$, $\mu\ \varepsilon$ g. It is obvious that for $\vartheta\ \varepsilon$ g

It is obvious that for $\vartheta \varepsilon g$

$\vartheta \overline{x} = \overline{y}$ iff $(\exists \mu_1 \varepsilon g)(\exists \mu_2 \varepsilon g)(\mu_1 \overline{a} = \overline{x} \,\&\, \mu_2 \overline{a} = \overline{y} \,\&\, \vartheta\mu_1 = \mu_2)$

Thus we have proved that the action of g on $\vartheta(\mathfrak{M})$ is definable using $C \cup |\overline{a}|$. Now it can be easily seen, that action of g on $\varphi(\mathfrak{M})$ is definable using $C \cup |\overline{a}|$, too

(c) Consider interpretations $\pi_{\overline{a}}$ and $\pi_{\overline{b}}$, $\overline{a}, \overline{b} \varepsilon \vartheta(\mathfrak{M})$, $\overline{b} = \lambda\overline{a}$ for some $\lambda \varepsilon g$, Let $\pi_{\overline{a}}(\overline{d}) = \mu$, i.e.

$\mathfrak{M} \models \chi(\overline{a}, \overline{d}, \mu\overline{a})$, hence $\mathfrak{M} \models \chi(\lambda\overline{a}, \overline{d}, \lambda\mu\overline{a})$.

The latter can be rewritten as

$\mathfrak{M} \models \chi(\overline{b}, \overline{d}, \lambda\mu\lambda^{-1}\overline{b})$

thus, by the definition, $\pi_{\overline{b}}(\overline{d}) = \lambda \cdot \pi_{\overline{a}}(\overline{d}) \cdot \lambda^{-1}$.

Note, that it follows from Proposition 4.3 that the structures

$\mathfrak{M}^{\pi_{\overline{a}}} g$ and $\mathfrak{M}^{\pi_{\overline{b}}} g$

are isomorphic, therefore we shall denote them by $\mathfrak{M}^{\pi} g$.

<u>Definition</u>. Let $\mathcal{H}$ be a subgroup of $g(\varphi/\delta \cup C)$. On $\varphi(\mathfrak{M})$ and $\vartheta(\mathfrak{M})$. we define equivalence relations

$$(\exists \mu \varepsilon \mathcal{H})(\mu x = y)$$

If x,y are from $\varphi(\mathfrak{M})$ then this relation is denoted by $\varepsilon_{\mathcal{H}}(x,y)$, if $x,y \varepsilon \vartheta(\mathfrak{M})$ we denote it by $\varepsilon^{n}_{\mathcal{H}}(x,y)$.

The following lemma is true for every group g which is definable in $\mathfrak{M}$ by an interpretation π.

<u>Lemma 4.4.</u> If $\mathcal{H}$ is a subgroup of g which is definable in $\mathfrak{M}^{\pi} g$ using some parameters then $\mathcal{H}$ is definable in $\mathfrak{M}^{\pi} g$ using parameters from G.

<u>Proof.</u> Let

$$\mathcal{H} = \alpha(\overline{c}_0, \mathfrak{M}^{\pi} g), \qquad \alpha(\overline{w}, v) \varepsilon F_{m+1}(\Omega_{\mathfrak{M}^{\pi} g}).$$

Clearly, there exists a formula $\beta(\overline{w})$ without parameters such that

$$\mathfrak{M}^{\pi} g \models \beta(\overline{c}) \quad \text{iff} \quad \alpha(\overline{c}, \mathfrak{M}^{\pi} g) \text{ is a subgroup of } g.$$

Let us prove that there is a finite subset A of $\mathcal{H}$ such that for every $\overline{c}$ satisfying β

(*) if $\alpha(\overline{c}, \mathfrak{M}^{\pi} g) \supseteq A$, then $\alpha(\overline{c}, \mathfrak{M}^{\pi} g) \supseteq \mathcal{H}$.

Otherwise, by 1.5 we can find $\overline{c}$ such that

$$\mathfrak{M}^{\pi}g \models \beta(\bar{c}) \& (\exists v \in \mathcal{H})\ \neg\alpha(\bar{c},\bar{v}) \text{ and } \alpha(\bar{c}, \mathfrak{M}^{\pi}g) \sqsupset \mathcal{H}.$$

Since $\alpha(\bar{c}, \mathfrak{M}^{\pi}g)$ is a group, the intersection

$$\alpha(\bar{c}, \mathfrak{M}^{\pi}g) \cap \mathcal{H}$$

is a subgroup of $\mathcal{H}$ of the same rank and degree as $\mathcal{H}$ and, this is possible only if it coincides with $\mathcal{H}$, i.e.

$$\alpha(\bar{c}, \mathfrak{M}^{\pi}g) \supseteq \mathcal{H}$$

which contradicts the way $\bar{c}$ was chosen.

Thus there exists an A satisfying (*) hence

$$\mathcal{H} = \{\alpha(\bar{c}, \mathfrak{M}^{\pi}g) : \alpha(\bar{c}, \mathfrak{M}^{\pi}g) \supseteq A \quad \mathfrak{M}^{\pi}g = \beta(\bar{c})\}$$

and therefore $\mathcal{H}$ is definable using $A \subseteq G$.

Proposition 4.5. Let $\mathcal{H}$ be a normal subgroup of $g(\varphi/\delta \cup C)$

(a) if $\mathcal{H}$ is definable in $\mathfrak{M}^{\pi}g$ using parameters from a subset B of $M \cup G$ then $\varepsilon_{\mathcal{H}}$ and $\varepsilon^{n}_{\mathcal{H}}$ are definable using B C.

(b) if the assumption of (a) holds then $\varepsilon_{\mathcal{H}}$ and $\varepsilon^{n}_{\mathcal{H}}$ are definable using $\delta(\mathfrak{M}) \cup C$, too.

Proof. It follows from 4.2 that $\varepsilon_{\mathcal{H}}$ and $\varepsilon^{n}_{\mathcal{H}}$ are definable using $|\bar{a}| \cup B \cup C$ for some $\bar{a} \in \vartheta(\mathfrak{M})$.

Let $\gamma(\bar{w}, v_0, v_1)$ be the formula over $\Omega_{\mathfrak{M}^{\pi}g}(B \cup C)$ such that

$$\mathfrak{M} \models \gamma(\bar{a}, x, y) \text{ iff } y \in \mathcal{H}x.$$

If $\bar{b}$ is another element of $\vartheta(\mathfrak{M})$, then $\bar{b} = \lambda\bar{a}$ for some $\lambda \in g$. Clearly

$$\mathfrak{M} \models \gamma(\bar{b}, x, y) \text{ iff } \mathfrak{M} \models \gamma(\bar{a}, \lambda^{-1}x, \lambda^{-1}y) \text{ iff } \lambda^{-1}y \in \mathcal{H}\lambda^{-1}x$$
$$\text{iff } y \in \lambda\mathcal{H}\lambda^{-1}x \text{ iff } y \in \mathcal{H}x.$$

Thus the relation $y \in \mathcal{H}x$ is defined by $\gamma(\bar{b}, x, y)$ too, hence it is defined by

$$(\exists\bar{w})\vartheta(\bar{w}) \& \gamma(\bar{w}, x, y)$$

(b) follows from Lemma 4.4(a) and the fact that $G \subseteq cl_1(\delta(\mathfrak{M}))$ in the sense of $\mathfrak{M}^{\pi}g$.

Proposition 4.6. Let $g(\varphi/\delta \cup C)$ be an abelian group, then

(a) for any $\bar{a}, \bar{b} \in \vartheta(\mathfrak{M}))$ $\pi_{\bar{a}} = \pi_{\bar{b}}$;

(b) for any $a, b \in \varphi(\mathfrak{M})$ the element $\mu \in (\varphi/\delta \cup C)$ such that $\mu a = b$ is unique, in other words for every $a \in \varphi(\mathfrak{M})$

$\varphi(\mathfrak{m}) \subseteq cl_1(\{a\} \cup \delta(\mathfrak{m}) \cup C)$;

(c) the action of $g(\varphi/\delta \cup C)$ on $\varphi(\mathfrak{m})$ is definable in $\mathfrak{m}^{\pi}g$ using parameters from C only.

Proof. (a) follows from 4.3(c)

(b) indeed this is a group - theoretical fact: clearly, the uniqueness of μ such that $\mu a = b$ is equivalent to the fact that the subgroup

$$St(a) = \{\vartheta \in g: \vartheta a = a\}$$

contains no elements exept the identity. Now suppose $\vartheta \in St(a)$, $b \in \varphi(\mathfrak{m})$. As was already remarked, there is an element μ of g such that $\mu a = b$. Clearly, $\vartheta b = \vartheta \mu a = \mu \vartheta a = b$, i.e. ϑ acts trivially on every element of $\varphi(\mathfrak{m})$. From the definition of $g(\varphi/\delta \quad C)$ ϑ is the identity of the group. Thus $St(a)$ is trivial.

Since $\mu \in cl_1(\delta(\mathfrak{m}) \cup C)$ then $b = \mu a$ implies $b \in cl_1(\{a\} \cup \cup \delta(\mathfrak{m}) \cup C)$ in $\mathfrak{m}^{\pi}g$ and hence in $\mathfrak{m}$.

(c) As it was observed in the proof of 4.3 (b), the formula

$$(\exists \overline{w} \in D_{\pi}) (\pi(\overline{w}) = \mu \,\&\, \chi(\overline{y},\overline{w},\overline{x}))$$

defines the relation $\mu \overline{y} = \overline{z}$ and in our case $\overline{y}$ need not be fixed because π does not depend on it.

Proposition 4.7. Let $\mathcal{H}$ be a normal subgroup of $g(\varphi/\delta \cup C)$, definable using parameters, let σ be the factorization mapping

$$\sigma : \vartheta(\mathfrak{m}) \dashrightarrow \vartheta(\mathfrak{m})_{\varepsilon^n} = \Theta$$

Then in $\mathfrak{m}^{\sigma}$

(1) $g(\varphi(\mathfrak{m})/\delta(\mathfrak{m}) \cup \Theta \cup C) = \mathcal{H}$

(2) $g(\Theta/\delta(\mathfrak{m}) \cup C) \cong g/_{\mathcal{H}}$

Proof. By the definition every monomorphism from $\mathcal{H}$ leaves $\varepsilon^n_{\mathcal{H}}$-classes fixed, i.e. its extension to an automorphism of $\mathfrak{m}^{\sigma}$ is an identity on Θ, therefore $\mathcal{H}$ is included in the left hand side of (1). On the other hand, if μ belongs to the left hand side of (1) then for every $\overline{a}$ from $\vartheta(\mathfrak{m})$ $\sigma(\mu\overline{a}) = \sigma(\overline{a})$, hence $\mu \overline{a} \in \mathcal{H}\overline{a}$, i.e. $\mu \in \mathcal{H}$. So, (1) is proved.

Every monomorphism μ from g induces a monomorphism $\hat{\mu}$ from $g(\Theta/\delta(\mathfrak{m}) \cup C)$. Clearly $\hat{}: \mu \dashrightarrow \hat{\mu}$ is a group homomorphism. Every monomorphism λ from the group $g(\Theta/\delta(\mathfrak{m}) \cup C)$ can be extended to an automorphism of $\mathfrak{m}^{\sigma}$, thus $\lambda = \hat{\mu}$ for some $\mu \in g$.

So $\wedge$ is an epimorphism. Its kernel consists of those μ which leave the $\varepsilon^n_{\mathcal{H}}$ - classes fixed, i.e. the kernel is equal to $\mathcal{H}$.

Definition. An equivalence relation is called strict if it is proper and every class of it contains infinitely many elements.

Proposition 4.8. Let $g(\varphi/\delta \cup C)$ be abelian.
If $B \supseteq C$ and φ is not an atom of $F_1(B \cup \delta(\mathfrak{m}))$ then

$$\varphi(\mathfrak{m}) \subseteq \mathrm{cl}(B \cup \delta(\mathfrak{m}))$$

or

there is a strict equivalence relation on $\varphi(\mathfrak{m})$ which is definable using parameters from B as well as from $\delta(\mathfrak{m}) \cup C$.

Proof. Let $\alpha(\overline{a},v)$ be an atom of $F_1(B \cup \delta(\mathfrak{m}))$ of minimal rank and degree, $\overline{a} \in (\delta(\mathfrak{m}))^m$, $\alpha(\overline{w},v) \in F_{m+1}(B)$. Clearly

$$\mathcal{H} = \{\mu \in g \ : \ \mu \cdot \alpha(\overline{a},\mathfrak{m}) = \alpha(\overline{a},\mathfrak{m})\}$$

is a subgroup of g, which does not coincide with g by the assumption of the Proposition. It follows from 4.6(c) that $\mathcal{H}$ is definable using $|\overline{a}| \cup B$. Take an arbitrary $d \in \alpha(\overline{a},\mathfrak{m})$. If $\mu\, d \in \alpha(\overline{a},\mathfrak{m})$ for some $\mu \in g$ then

$$\mu \cdot \alpha(a,\mathfrak{m}) \cap \alpha(a,\mathfrak{m}) \neq \emptyset$$

and the intersection is definable using parameters from $B \cup \delta(\mathfrak{m})$ in $\mathfrak{m}^{\pi}g$, because $\mu \in \mathrm{cl}_1(\delta(\mathfrak{m}) \cup C)$. This intersection is also definable in $\mathfrak{m}$ using $B \cup \delta(\mathfrak{m})$, by 2.2(b). By the minimality of $\alpha(\overline{a},\mathfrak{m})$

$$\mu \cdot \alpha(\overline{a},\mathfrak{m}) = \alpha(\overline{a},\mathfrak{m})$$

and thus $\mu \in \mathcal{H}$. Hence

$$\alpha(\overline{a},\mathfrak{m}) = \mathcal{H} d$$

for any $d \in \alpha(\overline{a},\mathfrak{m})$. We noted in the proof of 4.6(a) that no element of g exept the identity acts trivially on an element of $\varphi(\mathfrak{m})$. Thus $\alpha(\overline{a},\mathfrak{m})$ contains the same number of elements as $\mathcal{H}$.

Now, if $\mathcal{H}$ is finite then $\alpha(a,\mathfrak{m})$ is finite and $\alpha(a,\mathfrak{m}) \subseteq \mathrm{cl}(B \cup \delta(\mathfrak{m}))$ and therefore $\varphi(\mathfrak{m}) \subseteq \mathrm{cl}(B \cup \delta(\mathfrak{m}))$ by 4.6(b).

If $\mathcal{H}$ is infinite then $\varepsilon_{\mathcal{H}}$ is a strict equivalence relation on $\varphi(\mathfrak{m})$, which is by 4.5, definable using $\delta(\mathfrak{m}) \cup C$. Now using 1.6, we shall show that $\varepsilon_{\mathcal{H}}$ is definable using parameters from B.

Under any automorphism of $\mathfrak{m}^{\pi}g$, which is an identity on B, the set $\alpha(\overline{a},\mathfrak{m})$ is mapped onto a set $\alpha(\overline{a}',\mathfrak{m})$ where $\overline{a}'$ is image of $\overline{a}$. Take $d' \in \alpha(\overline{a}',\mathfrak{m})$ and $\mu \in g$, such that $\mu d' \in \alpha(\overline{a},\mathfrak{m})$. Then

$$\alpha(\overline{a},\mathfrak{m}) \cap \mu \cdot \alpha(\overline{a}',\mathfrak{m}) \neq \emptyset .$$

This intersection is definable using parameters from $B \cup \delta(\mathfrak{M})$, therefore

$$\alpha(\bar{a},\mathfrak{M}) = \mu\, \alpha(\bar{a}',\mathfrak{M})$$

but

$$\alpha(\bar{a}',\mathfrak{M}) = \mathcal{H}' d', \quad \mu d' = d \in \alpha(\bar{a},\mathfrak{M})$$

where $\mathcal{H}'$ is the image of $\mathcal{H}$ under the automorphism. Hence

$$\mathcal{H} = \mathcal{H}',$$

therefore $\mathcal{H}$ and $\varepsilon_{\mathcal{H}}$ are definable using B.

5. The closure of finite subsets of models of totally categorical theories.

In this section we investigate the case when $\mathfrak{M}$ is a model of a theory categorical in all infinite powers.

We assume $\delta(\mathfrak{M})$ is a strongly minimal set.
We use in the section two, different in some cases, notions of independence. To avoid the ambiquity we say " cl - independent" when the Marsh's notion ([2]) is used.

For any $X \subseteq M$ and $\gamma \in F_m(M)$ let us define

$$\gamma(X) = cl(X) \cap \gamma(\mathfrak{M}) = \{\bar{d} \in \gamma(\mathfrak{M}) : |\bar{d}| \subseteq cl(X)\}$$

$$x = \text{card}\, \delta(X).$$

<u>Proposition 5.1.</u> Let B a finite subset of elements of $\mathfrak{M}$, $\gamma \in F_m(B \cup C)$. If $\gamma(\mathfrak{M}) \subseteq cl(\delta(\mathfrak{M}) \cup B \cup C)$ then there exists a one variable polynomial f_γ over rationals such that for any $X \supseteq B \cup C$:

(a) $\text{card}\ \gamma(X) = f_\gamma(x)$

(b) if $\gamma = \alpha(\bar{b},\bar{v})$, $|\bar{b}| = B$, $\alpha(\bar{w},v) \in F_{k+m}(C)$, $|\bar{b}'| \subseteq X$, $\gamma' = \alpha(\bar{b}', v)$ and $t(\bar{b}',C) = t(\bar{b},C)$ then $f_{\gamma'}(x) = f_\gamma(x)$.

<u>Proof.</u> First let $\gamma(\mathfrak{M})$ be equal to $\delta_B^{(m)}(\mathfrak{M}) = \{< d_1,\dots,d_m > :$ $\{d_1,\dots,d_m\}$ is a cl-independent over $B \cup C$ subset of $\delta(\mathfrak{M})\}$.

By induction on m we shall show that

$$\text{card}(\delta_B^{(m)}(X)) = f_{\delta^{(m)}}(x) = (x-s_0)\dots(x-s_{m-1}), \text{ where}$$

for $0 \leq i \leq m-1$

$$s_i = \text{card}(cl(B \cup C \cup \{d_1,\dots,d_i\}) \cap \delta(\mathfrak{M}))$$

and $\{d_1,\dots,d_i\}$ is some (any) subset of $\delta(\mathfrak{M})$ which is cl-independent over $B \cup C$. Obviously, (b) is satisfied by this polynomial.

If $m = 1$, then

$$\delta_B^{(m)}(X) = cl(X) \cap \delta(\mathfrak{M}) - cl(B \cup C) \cap \delta(\quad),$$

thus card $(\delta_B^{(1)}(X)) = (x-s_o)$.

For $\delta_B^{(k+1)}$ define the mapping

$$\delta_B^{(k+1)}(X) \dashrightarrow \delta_B^{(k)}(X)$$

as projection

$$< d_1,\dots,d_{k+1} > \dashrightarrow < d_1,\dots,d_k >$$

Let $<d_1,\dots,d_k> \varepsilon\ \delta^{(k)}(X)$. It was already proved that card$\{d \ \varepsilon\ \delta(X)$: d is cl-independent over $C \cup B \cup \{d_1,\dots,d_k\}\} = (x-s_k)$.
Thus each point of $\delta_B^{(k)}(X)$ has $x-s_k$ points in the preimage and then

$$f_{\delta^{(k+1)}}(x) = f_{\delta^{(k)}}(x) \cdot (x-s_k)$$

and the case $\gamma = \delta_B^{(m)}$ is proved.

Let now $\gamma(\mathfrak{M})$ be a subset of $\delta_B^{(m)}$, $B \supsetneq B_o$ and let γ be an atom over $B \cup C$. For a $<d_1,\dots,d_k,\dots,d_m> \varepsilon\ \gamma(\mathfrak{M})$ we can, by renumbering, assume that $\{d_1,\dots,d_k\}$ is cl-independent over $B \cup C$ and $cl(d_1,\dots,d_k \cup B \cup C) \supsetneq \{d_{k+1},\dots,d_m\}$.
Map $\gamma(X)$ onto $\delta_B^{(k)}(X)$ by the projection

$$p: < d_1,\dots,d_k,\dots,d_m > \dashrightarrow < d_1,\dots,d_k> \text{ for every}$$

$< d_1,\dots,d_m >$ from $\gamma(X)$. Observe, that p is a mapping of $\gamma(\mathfrak{M})$ onto $\delta_B^{(k)}(\mathfrak{M})$. Since γ is an atom, for every $\bar{d}\ \varepsilon\ \delta_B^{(k)}(\mathfrak{M})$ the set $p^{-1}(\bar{d})$ is included in $cl(|\bar{d}| \cup B \cup C)$ and since $\gamma(X)=\gamma(\mathfrak{M}) \cap cl(C)$, we have $p^{-1}(\bar{d}) \cap \gamma(\mathfrak{M}) = p^{-1}(\bar{d}) \cap \gamma(X)$ for every $\bar{d}\ \varepsilon\ \delta_B^{(k)}(X)$.

Denote by r the number of elements of any of these preimages and now it is obvious that

$$f_\gamma(x) = r \cdot f_{\delta_B^{(k)}}(x) .$$

The condition required for f_γ by (b) follows from the same condition for $f_{\delta^{(k)}}$.

Now let γ be an arbitrary atom from $F_m(B \cup C)$.
Since $\gamma(\mathfrak{M}) \subseteq cl(B \cup C \cup \delta(\mathfrak{M}))$ for some $\bar{a}\ \varepsilon\ \gamma(\mathfrak{M})$ there exists an element $\bar{d}\ \varepsilon\ \delta_B^{(k)}$ and a formula $\Psi(\bar{w},\bar{v})\ \varepsilon F_{k+m}(B \cup C)$ such that

$$\mathfrak{M} \models \Psi(\bar{a},\bar{d}) \ \&\ (\exists^{=r}_{\bar{w}})(\Psi(\bar{w},\bar{d}))$$

and Ψ is an atom over $B \cup C$.

We have already proved that

$$\mathrm{card}(\Psi(\bar{a},X)) = f_{\Psi(\bar{a},\bar{v})}(x)$$

for every $\bar{a} \in \gamma(X)$ and the polynomial does not depend on $\bar{a}$. To compute the cardinality of

$$\Psi(X,X) = \{\bar{a}\frown\bar{d}:\ |\bar{a}|\cup|\bar{d}|\ \mathrm{cl}(X),\ \mathfrak{M} \models \Psi(\bar{a},\bar{d})\}$$

take the projection

$$p_1 : \Psi(X,X) \to \delta_B^{(k)}(X), \quad p_1(\bar{a}\frown\bar{d}) = \bar{d}.$$

As was noted earlier

$$\mathrm{card}(\Psi(X,X)) = r \cdot f_{\delta_B^{(k)}}(x)$$

On the other hand, taking the projection

$$p_2: \Psi(X,X) \to \gamma(X)$$

we have

$$\mathrm{card}(\Psi(X,X)) = \mathrm{card}(\gamma(X)) \cdot f_{\Psi(\bar{a},v)}(x)$$

for some (any) $\bar{a} \in \gamma(X)$. Hence

$$\mathrm{card}\,\gamma(X) = g(x) = r \cdot f_{\delta_B^{(k)}}(x) \,/\, f_{\Psi(\bar{a},v)}(x).$$

To finish the proof let us state the following lemma which can be easily proved by elementary methods.

Lemma. If $g(x)$ is a rational function of one variable (over integers) such that the set of those integers z, for which $g(z)$ is an integer, is infinite then $g(x)$ is equal to a polynomial with rational coefficients.

So, $g(x) = f_\gamma(x)$ for a polynomial f_γ. Note that every set of $F_m(B \cup C)$ can be represented as a disjoint union of atoms of $F_m(B \cup C)$, thus the proof is finished.

Lemma 5.2. Let $\varphi \in F_1(C)$, $\rho(\varphi,\delta,C) = \langle 0,1 \rangle$. Also suppose $g(\varphi,\delta,C)$ is an abelian group and there does not exists any strict equivalence relation on $\varphi(\mathfrak{M})$ which is definable using parameters from $\delta(\mathfrak{M}) \cup C$.

If φ is not an atom of $F_1(B \cup C \cup \delta(\mathfrak{M}))$ then $\mathrm{cl}(B \cup C) \cap \varphi(\mathfrak{M}) \neq \emptyset$.

Proof. By 4.6 (b), $\vartheta(\mathfrak{M}) = \varphi(\mathfrak{M})$ and thus all the assumptions of section 4 are satisfied. Let X be an arbitrary subset of M such that $X \supseteq B \cup C$. First let us note that in $\mathfrak{M}^{\mathfrak{g}}$ if $\varphi(\mathfrak{M})\ \mathrm{cl}(X) \neq \emptyset$, then card $(\varphi(\mathfrak{M}) \cap \mathrm{cl}(X)) = \mathrm{card}(G \cap \mathrm{cl}(X))$. For this fix an element $a_0 \in \varphi(\mathfrak{M}) \cap \mathrm{cl}(X)$ and denote by μ_a the unique element from g for which $\mu_a \cdot a_o = a$ (4.6(b)). It follows easily from 4.6.(c) that $a \to \mu_a$ is a one - to - one correspondence between $\varphi(\mathfrak{M}) \cap cl(X)$ and $G \cap cl(X)$.

Now consider the polynomials $f_\varphi(x) = card(\varphi(\mathfrak{M}) \cap cl(X))$ and $f_G(x) = card(G \cap cl(X))$. The existence of these polynomials is given by Proposition 5.1 (in the case of φ also Lemma 4.8 and assumptions of the present Lemma are needed).

As we have seen $f_G(x) = f_\varphi(x)$ if X is large enough, i.e. values of the polynomials coincide for an infinite number of points, hence the polynomials coincide.

Now take $X_o = B \cup C$ and we have

$card(\varphi(\mathfrak{M}) \cap cl(X_o)) = f_\varphi(x_o) = f_G(x_o) = card(G \cap cl(X_o)) \geq 1$.

The last inequality follows from the fact that the identity of the group g belongs to $cl(\emptyset)$ in $\mathfrak{M}^\pi g$. Thus $\varphi(\mathfrak{M}) \cap cl(B \cup C) \neq \emptyset$ in $\mathfrak{M}^\pi g$ and by 2.2(b) the same holds in $\mathfrak{M}$.

Lemma 5.3. If $\delta, \varphi \in F_1(A)$, φ is an atom of $F_1(A \cup \delta(\mathfrak{M}))$ and no strict equivalence relation is definable on $\varphi(\mathfrak{M})$ using parameters from $A \cup \delta(\mathfrak{M})$ then there exists a finite subset D_φ of $\delta(\mathfrak{M})$ such that for any $B \subseteq M$ if φ is not an atom of $F_1(A \cup B \cup \delta(\mathfrak{M}))$ then $cl(A \cup B \cup D_\varphi) \cap \varphi(\mathfrak{M}) \neq \emptyset$.

Proof. Let $\varepsilon(x,y)$ be the relation

$$x,y \in \varphi(\mathfrak{M}) \quad y \in cl(\{x\} \cup A)$$

on $\varphi(\mathfrak{M})$. Let us prove that ε is an equivalence relation.

$\varepsilon(a,\mathfrak{M})$ is finite for every $a \in \varphi(\mathfrak{M})$ and the cardinality of the set does not depend on a, because φ is an atom of $F_1(A)$. Since $\varepsilon(a,\mathfrak{M}) \supseteq (b,\mathfrak{M})$ if $\varepsilon(a,b)$, we have $\varepsilon(a,\mathfrak{M}) = \varepsilon(b,\mathfrak{M})$, provided $\varepsilon(a,b)$ holds. Hence ε is symmetric, transitive and, obviously, reflexive.

Let now σ be the natural mapping

$$\sigma: \varphi(\mathfrak{M}) \to \varphi(\mathfrak{M})/\varepsilon .$$

σ is an interpretation using parameters from A. In the structure $\mathfrak{M}^\sigma$ denote $\hat{a} = \sigma(a)$, $\hat{\varphi}(\mathfrak{M}^\sigma) = \varphi(\mathfrak{M})/\varepsilon$.

It is easily seen that

$$\hat{a} \in cl(\{b\} \cup A) \quad \text{iff} \quad a \in cl(\{b\} \cup A)$$

therefore

$$\hat{a} \in cl(\{\hat{b}\} \cup A) \quad \text{iff} \quad \hat{a} = \hat{b}.$$

Note that if the statement of the lemma holds for $\hat{\varphi}(\mathfrak{M}^\sigma)$ then it holds for $\varphi(\mathfrak{M})$, therefore without loss of generality we can assume that

$$a \in cl(\{b\} \cup A) \text{ iff } a = b$$

for any $a,b \in \varphi(\mathfrak{M})$.

Now consider the equivalence relation $\varepsilon_{\varphi,\delta,A}$ defined in Section 3. Since by our assumptions $\varepsilon_{\varphi,\delta,A}$ has only a finite number of elements in each class, $\varepsilon_{\varphi,\delta,A}(a,b)$ implies that $a \in cl(\{b\} \cup A)$ thus $\varepsilon_{\varphi,\delta,A}$ is trivial and by 3.7 $\rho(\varphi,\delta,A) = < 0,1 >$.

Now choose C, $A \subseteq C \subseteq A \cup \delta(\mathfrak{M})$, as in Section 4, thus $\vartheta \in F_1(C)$ and ϑ is an atom of $F_n(C \cup \delta(\mathfrak{M})$. Take an abelian normal subgroup $\mathcal{H}$ of $g(\varphi,\delta,C)$ such that $g/\mathcal{H}$ is finite and is definable in g and thus in $\mathfrak{M}^{\pi}g$ without parameters. The existence of such subgroup is given by 1.7. By our assumptions $\varepsilon_{\mathcal{H}}$ must be trivial on $\varphi(\mathfrak{M})$ (Proposition 4.5) , evidently, only the case that any two elements of $\varphi(\mathfrak{M})$ are $\varepsilon_{\mathcal{H}}$ -equivalent is possible. Using 4.7 we have in some $\mathfrak{M}^{\sigma}$,

$$g(\varphi(\mathfrak{M})/\delta(\mathfrak{M}) \cup C \cup \Theta) = \mathcal{H}$$

where $\Theta = \vartheta(\mathfrak{M})/_{\varepsilon^n_{\mathcal{H}}}$ has the same number of elements as $g(\varphi/\delta \cup C)/\mathcal{H}$ i.e. is finite. Let $C' = C \cup \Theta$, then $g(\varphi/\delta \cup C') = \mathcal{H}$ in $\mathfrak{M}^{\sigma}$

Now take any $B \subseteq M$ such that φ is not an atom $F_1(B \cup C \cup \delta(\mathfrak{M}))$, then by 5.2

$$cl(B \cup C') \cap \varphi(\mathfrak{M}) \neq \emptyset \text{ in } \mathfrak{M}^{\sigma}$$

Note that $C' \subseteq cl(C)$ in $\mathfrak{M}^{\sigma}$ and also that by 2.2(b) $cl(B \cup C) \cap \varphi(\mathfrak{M})$ in $\mathfrak{M}^{\sigma}$ and in $\mathfrak{M}$ coincide.
Put $D_{\varphi} = C - A$ and the proof is finished.

<u>Theorem 1.</u> For any $\varphi \in F_1(A)$ there exists a finite subset $D_{\varphi} \subseteq \delta(\mathfrak{M})$, depending on A,δ,φ only, such that for any subset B of M there is an element $b \in \varphi(\mathfrak{M})$ such that $\{b\}$ and $\delta(\mathfrak{M})$ are independent over $A \cup B \cup D_{\varphi}$.

<u>Proof</u>. Assume the Theorem is not true. Choose $\mathfrak{M},\delta,\varphi$ such that the required D_{φ} does not exist, and $< R(\varphi), Dg(\varphi) >$ is minimal.

By the minimality, φ is an atom of $F_1(\delta(\mathfrak{M}) \cup A)$.

Suppose there is a strict equivalence relation ε on $\varphi(\mathfrak{M})$ which is definable using parameters from $A \cup \delta(\mathfrak{M})$. Let σ be the natural mapping :

$$\varphi(\mathfrak{M}) \dashrightarrow \varphi(\mathfrak{M})/\varepsilon \quad .$$

Then σ is an interpretation in $\mathfrak{M}$ using parameters from $A \cup D_0$, for a finite subset D_o of $\delta(\mathfrak{m})$.

Consider the structure $\mathfrak{m}^\sigma$, and put $\varphi(\mathfrak{m})/\varepsilon = \hat{\varphi}(\mathfrak{m}^\sigma)$, $\hat{\varphi}(\mathfrak{m}^\sigma)$ is a subset of $\mathfrak{m}^\sigma$ definable without parameters. By 1.3(b) and 2.3 $R(\hat{\varphi}) < R(\varphi)$ and for any $\hat{a} \varepsilon \hat{\varphi}(\mathfrak{m}^\sigma)$

$$< R(\sigma^{-1}(\hat{a})), Dg(\sigma^{-1}(\hat{a})) > << R(\varphi), Dg(\varphi) >$$

Thus by the minimality of φ there exists $D_{\hat{\varphi}}$ and $D_{\sigma^{-1}(a)}$. Note that $D_{\sigma^{-1}(\hat{a})}$ fits for $D_{\sigma^{-1}(\hat{b})}$ for any $\hat{b} \varepsilon \hat{\varphi}(\mathfrak{m}^\sigma)$ because $D_{\sigma^{-1}(\hat{a})} \subseteq \delta(\mathfrak{m})$ and $\hat{\varphi}$ is an atom over $A \cup \delta(\mathfrak{m})$.

Take $D_\varphi = D_o \cup D_{\sigma^{-1}(a)} \cup D_{\hat{\varphi}}$ and let B be an arbitrary subset of M. By the definition of $D_{\hat{\varphi}}$ and $D_{\sigma^{-1}(\hat{a})}$ there exist $\hat{b} \varepsilon \varphi(\mathfrak{m}^\sigma)$ such that $\{\hat{b}\}$ and $\delta(\mathfrak{m})$ are independent over $A \cup B \cup D_{\hat{\varphi}}$, and there exists $b \varepsilon \sigma^{-1}(\hat{b})$, such that $\{b\}$ and $\delta(\mathfrak{m})$ are independent over $A \cup B \cup \{\hat{b}\} \cup D_{\sigma^{-1}(\hat{a})}$. Since

$$\delta(\mathfrak{m}) \supseteq D_\varphi \supseteq D_{\hat{\varphi}} \cup D_{\sigma^{-1}(\hat{a})} ,$$

we have : $\delta(\mathfrak{m})$ and $\{\hat{b}\}$ are independent over $A \cup B \cup D_\varphi$ and $\delta(\mathfrak{m})$ and $\{b\}$ are independent over $A \cup B \cup D_\varphi \cup \{\hat{b}\}$.
It follows easily that $\delta(\mathfrak{m})$ and $\{b\}$ are independent over $A \cup B \cup D_\varphi$. This contradicts our assumption.

Thus the assumption of 5.3 are satisfied. Take D_φ as in 5.3, Since we have supposed that the negation of the statement of Theorem holds, there exists a subset B such that φ is an atom of $F_1(A \cup B \cup \delta(\mathfrak{m}))$ but for every $b \varepsilon \varphi(\mathfrak{m})$

$$R(b, A \cup B \cup \delta(\mathfrak{m})) < R(b, A \cup B \cup D_\varphi).$$

This means that there are at least two different extension of the type $\{\varphi\}$ over $A \cup B \cup \delta(\mathfrak{m})$. So we get a Contradiction, which finishes the proof.

6. Envelopes

Definition. A subset N of $\mathfrak{m}$ is called m-saturated if for any subset A of N of cardinality $<$ m each type of $S_1(A)$ is realized by an element of N.

Remark. If N is r+k - saturated, then for any subset A of N of cardinality $<$ r each type of $S_k(A)$ is realized by an element of N.

Proposition 6.1. Let $\mathfrak{n}$ be a substructure of $\mathfrak{m}$ such that N is r+k-saturated. Then for every formula $\alpha \varepsilon F_{r-1}$ with k quantifiers for any $\bar{a} \varepsilon N^{r-1}$

$$\mathfrak{n} \models \alpha(\bar{a}) \text{ iff } \mathfrak{m} \models \alpha(\bar{a})$$

Proof. By an easy induction on k.

Corollary. Let T be a complete theory having infinite models, such that for every natural m there exists a finite m-saturated subset of a model of T. Then T is not finitely axiomatizable.

Definition. Let $A_0 \subseteq X \subseteq M$, $\delta \in F_1(A_0)$.
An envelope of X with respect to δ is a maximal set $E(X) \subseteq M$ such that $\delta(\mathfrak{m})$ and $E(X)$ are independent over X.

Theorem 2. Let $\mathfrak{m}$ be a model of a theory categorical in all infinite powers, $E(X)$ an envelope of a subset $X \subseteq M$ with respect to a strongly minimal formula $\delta \in F_1(A_0)$. Then

(a) $E(X) \supseteq X$;

(b) $E(E(X)) = E(X)$;

(c) $E(X) \supseteq cl(X)$;

(d) $E(X)$ is finite, provided X is finite;

(e) for any natural number m there exists a natural number s(m) such that
if $card(cl(X) \cap \delta(\mathfrak{m}) \geq s(m)$, then $E(X)$ is m+1-saturated.

(f) if $cl(X) \cap \delta(\mathfrak{m})$ is infinite then $E(X)$ is an elementary submodel of $\mathfrak{m}$, which is prime over X.

Proof. (a) cl(X) follows immediately from the definition.

(b) By the definition of an evenlope the following equalities hold for every finite sequence $\bar{d}$ of elements of $\delta(\mathfrak{m})$:

$$R(\bar{d}, E(E(X)) \cup E(X)) = R(\bar{d}, E(X))$$

$$R(\bar{d}, E(X) \cup X) = R(\bar{d}, X)$$

Thus, using (a), we have

$$R(\bar{d}, E(E(X)) \cup X) = R(\bar{d}, X)$$

i.e. $\delta(\mathfrak{m})$ and $E(E(X))$ are independent over X. Now the equality (b) follows from the maximality of $E(X)$

(c) Note, that if $b \in cl(X)$, then $\delta(\mathfrak{m})$ and $\{b\} \cup E(X)$ are independent over X.

(d) Suppose not. Then, using the finiteness of

$F_1(X \cup \{y_1,\ldots,y_k\})$ for any finite set $\{y_1,\ldots,y_k\}$, we can construct an infinite subset $\{\bar{y}_i \mid i < \omega\}$ of $E(X)$ which is indiscernible over X. Since $E(X)$ and $\delta(\mathfrak{M})$ are independent over X the following holds for every natural number k :

$$R(y_{k+1}, \{y_1,\ldots,y_k\} \cup \delta(\mathfrak{M}) \cup X) = R(y_{k+1}, \{y_1,\ldots,y_k\} \cup X).$$

Hence $\{y_i \mid i < \omega\}$ is indiscernible over $\delta(\mathfrak{M}) \cup X$, which contradicts Fact 1.2′.

(e) Let

$$\bar{a} \in M^m, \quad \varphi = \varphi(\bar{a},v), \ \varphi(\bar{w},v) \in F_{m+1}(A_o).$$

Choose a finite subset D_φ of $\delta(\mathfrak{M})$ as defined in Theorem 1. Let $D_\varphi = |\vec{d}_\varphi|$ and γ be the formula over $|\bar{a}| \cup A_o$ which defines the type $t(d_\varphi, |\bar{a}| \cup A_o)$. It follows from 5.1 that for some polynomial f_γ if $x = \mathrm{card}(\mathrm{cl}(X) \cap \delta(\mathfrak{M}))$ and $f_\gamma(x) > 0$ then $\mathrm{cl}(X) \cap \gamma(\mathfrak{M}) \neq \emptyset$. Let s_φ be a natural number such that $x \geq s_\varphi$ implies $f_\gamma(x) > 0$. Thus

$$\text{if } \mathrm{card}(\mathrm{cl}(X) \cap \delta(\mathfrak{M})) > s_\varphi \text{ then } D_\varphi \text{ can be found in } \mathrm{cl}(X).$$

Since f_γ depends only on the type of $|\bar{a}|$ over A_o and on $\varphi(\bar{w},v)$ so does s_φ. Since there is only a finite number of non-equivalent formulas in $F_{m+1}(A_o)$ and since there exists only a finite number of m-types over A_o, there exists a number s(m) which is greater than s_φ for any $\varphi(\bar{a},v)$, $\bar{a} \in M^m$.

Now let $\mathrm{cl}(X) \cap \delta(\mathfrak{M}) \geq s(m)$, $A \subseteq E(X)$, $\mathrm{card}(A) \leq m$.

Since any type over a finite subset of $\mathfrak{M}$ is principal, to get m+1-saturatedness of $E(X)$ it suffices to prove that for any consistent formula $\varphi \in F_1(A \cup A_o)$ there exists an element $b \in \varphi(\mathfrak{M}) \cap E(X)$.

Take a subset D_φ of $\mathrm{cl}(X) \cap \delta(\mathfrak{M})$, note that by (C). $D_\varphi \subseteq E(X)$. By Theorem 1 there exists $b \in \varphi(\mathfrak{M})$ such that $\delta(\mathfrak{M})$ and and $\{b\}$ are independent of over $A \cup A_o \cup E(X) \cup D_\varphi = E(X) \cup A_o$. Hence, it is easily seen, that $\delta(\mathfrak{M})$ and $\{b\} \cup E(X)$ are independent over A_o. By the maximality of $E(X)$ $b \in E(X)$.

(f) It follows from 6.1 and (e) that $\varepsilon = \langle E(X), \Omega_{\mathfrak{M}} \rangle$ is an elementary substructure of $\mathfrak{M}$, provided $\mathrm{cl}(X) \cap \delta(\mathfrak{M})$ is infinite. If $X \subseteq N \subseteq E(X)$ and $\mathfrak{N}$ is an elementary substructure of $\mathfrak{M}$ then $\delta(\mathfrak{N}) \supseteq \mathrm{cl}(X) \cap \delta(\mathfrak{M})$, i.e. $\delta(\mathfrak{N}) = \delta(\varepsilon)$, since, obviously $E(X) \cap \delta(\mathfrak{M}) = \mathrm{cl}(X) \cap \delta(\mathfrak{M})$. Thus $\mathfrak{N} = \varepsilon$, so ε is prime over X.

References

[1] Baldwin J.T., α_T is finite for $\aleph_1$-categorical T, Trans.Amer.Math.Soc., 181 (1973), 37-52.

[2] Baldwin J.T., Lachlan A.H., On strongly minimal sets. J.Symb. Log., 36(1971), 79-96.

[3] Baur W., Cherlin G., Macintyre A., Totally categorical groups and rings. J.Algebra, 57(1979), 407-440.

[4] Lascar D., Ranks and definability in superstable theories, Isr. J.Math., 23 (1976), 53-87.

[5] Makowsky J.A. On some conjectures connected with complete sentences, Fund. Math., 81(1974), 193-202.

[6] Shelah S., Classification Theory and the number of Non-Isomorphic Models, North-Holland Publ. Comp., 1978.

[7] Zilber B.I., The transcendentence rank of the formulae of an $\aleph_1$-categorical theory (Russian), Math.Zametki 15 (1974) 321-329.

[8] Zilber B.I., The structure of models of categorical theories and the finite-axiomatizability problem. Preprint, mineographed by VINITI, Dep. N 2800-77, Kemerovo, 1977.

Institute of Mathematics
University of Wroclaw
Wroclaw, Poland

Kemerowo State University
Kemerowo, 650043, USSR

Vol. 729: Ergodic Theory. Proceedings, 1978. Edited by M. Denker and K. Jacobs. XII, 209 pages. 1979.

Vol. 730: Functional Differential Equations and Approximation of Fixed Points. Proceedings, 1978. Edited by H.-O. Peitgen and H.-O. Walther. XV, 503 pages. 1979.

Vol. 731: Y. Nakagami and M. Takesaki, Duality for Crossed Products of von Neumann Algebras. IX, 139 pages. 1979.

Vol. 732: Algebraic Geometry. Proceedings, 1978. Edited by K. Lønsted. IV, 658 pages. 1979.

Vol. 733: F. Bloom, Modern Differential Geometric Techniques in the Theory of Continuous Distributions of Dislocations. XII, 206 pages. 1979.

Vol. 734: Ring Theory, Waterloo, 1978. Proceedings, 1978. Edited by D. Handelman and J. Lawrence. XI, 352 pages. 1979.

Vol. 735: B. Aupetit, Propriétés Spectrales des Algèbres de Banach. XII, 192 pages. 1979.

Vol. 736: E. Behrends, M-Structure and the Banach-Stone Theorem. X, 217 pages. 1979.

Vol. 737: Volterra Equations. Proceedings 1978. Edited by S.-O. Londen and O. J. Staffans. VIII, 314 pages. 1979.

Vol. 738: P. E. Conner, Differentiable Periodic Maps. 2nd edition, IV, 181 pages. 1979.

Vol. 739: Analyse Harmonique sur les Groupes de Lie II. Proceedings, 1976–78. Edited by P. Eymard et al. VI, 646 pages. 1979.

Vol. 740: Séminaire d'Algèbre Paul Dubreil. Proceedings, 1977–78. Edited by M.-P. Malliavin. V, 456 pages. 1979.

Vol. 741: Algebraic Topology, Waterloo 1978. Proceedings. Edited by P. Hoffman and V. Snaith. XI, 655 pages. 1979.

Vol. 742: K. Clancey, Seminormal Operators. VII, 125 pages. 1979.

Vol. 743: Romanian-Finnish Seminar on Complex Analysis. Proceedings, 1976. Edited by C. Andreian Cazacu et al. XVI, 713 pages. 1979.

Vol. 744: I. Reiner and K. W. Roggenkamp, Integral Representations. VIII, 275 pages. 1979.

Vol. 745: D. K. Haley, Equational Compactness in Rings. III, 167 pages. 1979.

Vol. 746: P. Hoffman, τ-Rings and Wreath Product Representations. V, 148 pages. 1979.

Vol. 747: Complex Analysis, Joensuu 1978. Proceedings, 1978. Edited by I. Laine, O. Lehto and T. Sorvali. XV, 450 pages. 1979.

Vol. 748: Combinatorial Mathematics VI. Proceedings, 1978. Edited by A. F. Horadam and W. D. Wallis. IX, 206 pages. 1979.

Vol. 749: V. Girault and P.-A. Raviart, Finite Element Approximation of the Navier-Stokes Equations. VII, 200 pages. 1979.

Vol. 750: J. C. Jantzen, Moduln mit einem höchsten Gewicht. III, 195 Seiten. 1979.

Vol. 751: Number Theory, Carbondale 1979. Proceedings. Edited by M. B. Nathanson. V, 342 pages. 1979.

Vol. 752: M. Barr, *-Autonomous Categories. VI, 140 pages. 1979.

Vol. 753: Applications of Sheaves. Proceedings, 1977. Edited by M. Fourman, C. Mulvey and D. Scott. XIV, 779 pages. 1979.

Vol. 754: O. A. Laudal, Formal Moduli of Algebraic Structures. III, 161 pages. 1979.

Vol. 755: Global Analysis. Proceedings, 1978. Edited by M. Grmela and J. E. Marsden. VII, 377 pages. 1979.

Vol. 756: H. O. Cordes, Elliptic Pseudo-Differential Operators – An Abstract Theory. IX, 331 pages. 1979.

Vol. 757: Smoothing Techniques for Curve Estimation. Proceedings, 1979. Edited by Th. Gasser and M. Rosenblatt. V, 245 pages. 1979.

Vol. 758: C. Năstăsescu and F. Van Oystaeyen; Graded and Filtered Rings and Modules. X, 148 pages. 1979.

Vol. 759: R. L. Epstein, Degrees of Unsolvability: Structure and Theory. XIV, 216 pages. 1979.

Vol. 760: H.-O. Georgii, Canonical Gibbs Measures. VIII, 190 pages. 1979.

Vol. 761: K. Johannson, Homotopy Equivalences of 3-Manifolds with Boundaries. 2, 303 pages. 1979.

Vol. 762: D. H. Sattinger, Group Theoretic Methods in Bifurcation Theory. V, 241 pages. 1979.

Vol. 763: Algebraic Topology, Aarhus 1978. Proceedings, 1978. Edited by J. L. Dupont and H. Madsen. VI, 695 pages. 1979.

Vol. 764: B. Srinivasan, Representations of Finite Chevalley Groups. XI, 177 pages. 1979.

Vol. 765: Padé Approximation and its Applications. Proceedings, 1979. Edited by L. Wuytack. VI, 392 pages. 1979.

Vol. 766: T. tom Dieck, Transformation Groups and Representation Theory. VIII, 309 pages. 1979.

Vol. 767: M. Namba, Families of Meromorphic Functions on Compact Riemann Surfaces. XII, 284 pages. 1979.

Vol. 768: R. S. Doran and J. Wichmann, Approximate Identities and Factorization in Banach Modules. X, 305 pages. 1979.

Vol. 769: J. Flum, M. Ziegler, Topological Model Theory. X, 151 pages. 1980.

Vol. 770: Séminaire Bourbaki vol. 1978/79 Exposés 525–542. IV, 341 pages. 1980.

Vol. 771: Approximation Methods for Navier-Stokes Problems. Proceedings, 1979. Edited by R. Rautmann. XVI, 581 pages. 1980.

Vol. 772: J. P. Levine, Algebraic Structure of Knot Modules. XI, 104 pages. 1980.

Vol. 773: Numerical Analysis. Proceedings, 1979. Edited by G. A. Watson. X, 184 pages. 1980.

Vol. 774: R. Azencott, Y. Guivarc'h, R. F. Gundy, Ecole d'Eté de Probabilités de Saint-Flour VIII-1978. Edited by P. L. Hennequin. XIII, 334 pages. 1980.

Vol. 775: Geometric Methods in Mathematical Physics. Proceedings, 1979. Edited by G. Kaiser and J. E. Marsden. VII, 257 pages. 1980.

Vol. 776: B. Gross, Arithmetic on Elliptic Curves with Complex Multiplication. V, 95 pages. 1980.

Vol. 777: Séminaire sur les Singularités des Surfaces. Proceedings, 1976-1977. Edited by M. Demazure, H. Pinkham and B. Teissier. IX, 339 pages. 1980.

Vol. 778: SK_1 von Schiefkörpern. Proceedings, 1976. Edited by P. Draxl and M. Kneser. II, 124 pages. 1980.

Vol. 779: Euclidean Harmonic Analysis. Proceedings, 1979. Edited by J. J. Benedetto. III, 177 pages. 1980.

Vol. 780: L. Schwartz, Semi-Martingales sur des Variétés, et Martingales Conformes sur des Variétés Analytiques Complexes. XV, 132 pages. 1980.

Vol. 781: Harmonic Analysis Iraklion 1978. Proceedings 1978. Edited by N. Petridis, S. K. Pichorides and N. Varopoulos. V, 213 pages. 1980.

Vol. 782: Bifurcation and Nonlinear Eigenvalue Problems. Proceedings, 1978. Edited by C. Bardos, J. M. Lasry and M. Schatzman. VIII, 296 pages. 1980.

Vol. 783: A. Dinghas, Wertverteilung meromorpher Funktionen in ein- und mehrfach zusammenhängenden Gebieten. Edited by R. Nevanlinna and C. Andreian Cazacu. XIII, 145 pages. 1980.

Vol. 784: Séminaire de Probabilités XIV. Proceedings, 1978/79. Edited by J. Azéma and M. Yor. VIII, 546 pages. 1980.

Vol. 785: W. M. Schmidt, Diophantine Approximation. X, 299 pages. 1980.

Vol. 786: I. J. Maddox, Infinite Matrices of Operators. V, 122 pages. 1980.